William M. (William Miller) Barr

Boilers and furnaces considered in their relations to steam engineering

William M. (William Miller) Barr

Boilers and furnaces considered in their relations to steam engineering

ISBN/EAN: 9783743467026

Manufactured in Europe, USA, Canada, Australia, Japa

Cover: Foto ©berggeist007 / pixelio.de

Manufactured and distributed by brebook publishing software (www.brebook.com)

William M. (William Miller) Barr

Boilers and furnaces considered in their relations to steam engineering

STEAM ENGINEERING SERIES

VOLUME I.

BOILERS AND FURNACES

BY THE SAME AUTHOR

PUMPING MACHINERY

A PRACTICAL HAND-BOOK
RELATING TO THE CONSTRUCTION
AND MANAGEMENT OF

STEAM AND POWER PUMPING MACHINES

BY

WILLIAM M. BARR

Member American Society Mechanical Engineers

WITH UPWARDS OF TWO HUNDRED AND SEVENTY ENGRAVINGS,
COVERING EVERY ESSENTIAL DETAIL IN PUMP CONSTRUCTION

One Volume. Octavo. 450 pages
Price, $5.00

Boilers and Furnaces

CONSIDERED IN THEIR RELATIONS TO

STEAM ENGINEERING

BY

WILLIAM M. BARR

MEMBER AMERICAN SOCIETY MECHANICAL ENGINEERS

WITH UPWARDS OF FOUR HUNDRED AND FIFTY ENGRAVINGS OF BOILER AND FURNACE DETAILS FROM DRAWINGS EXECUTED EXPRESSLY FOR THIS WORK

PHILADELPHIA
THE FLORENCE COMPANY
1898

PREFACE.

A WORD of explanation may not be out of place in regard to the bearing which this volume may have upon a former treatise of mine on "High-Pressure Steam Boilers" (1880), and which is now out of print. Many requests have come to me from time to time to revise that book and bring it up to date. But twenty years is a long interval. Boiler pressures were much lower then than now. Wrought iron was the material then employed in boiler construction, now it is rarely met with. The conditions having wholly changed, the problem of revision is impracticable. The first book must be set aside as a thing of the past and the subject taken up anew, with special reference to the exacting conditions which obtain at this time in both design and performance. This book differs somewhat from the former one in being essentially one of constructive detail, in the preparation of which I have endeavored to present, by means of well-chosen illustrations, the latest and best practice in steam-boiler design.

Mild steel as a material for steam boilers has practically displaced wrought iron. In its physical qualities it leaves scarcely anything to be desired. Tensile tests of mild steel plates, possessing those chemical properties which have given the best physical results, are allotted a considerable space in this book. They are made to cover as wide a range as seems necessary for ordinary stationary boiler work, or sufficiently so for the preparation of any specifications requiring plates not less than one-fourth inch, and not more than three-fourths inch thick.

Riveted joints have always been of first importance in boiler construction. The numerous records of tests of purely experimental joints, or those not used in boiler-making, as well as those made to the actual working dimensions usually employed when riveting together plates of the various thicknesses, given in Chapter III., are mainly upon specimens prepared by direction of the Bureau of Steam Engineering of the Navy Department. These experimental tests were made at the Watertown Arsenal, and extend over more than ten years. They are of the greatest practical value to persons engaged in the designing and construction of steam boilers for high pressures, and it is much to be regretted that the records of these tests are not more generally accessible. In the selection of examples for illustration and record in this volume great care has been exercised to include all, and only, such as shall be useful in stationary steam boiler design and construction. Probably no such similar amount of accurate and reliable data on this important detail in mechanical engineering has ever before been submitted in a single volume.

The limitations imposed as to the size of the completed volume prevented the introduction of subject matter or illustrations relating to the early history and development of steam boilers, as well as illustrated reference to some of the more recent examples of design and construction. This increase in subject matter could only have been accomplished by adding to the number of pages or by reduction in the size of the engravings. No doubt many of the latter could have been slightly reduced without detriment, but it was thought that a fewer number of clearly executed fundamental details would be of more value to a designer or student than a larger number wanting in proper mechanical execution or clearness of detail. After all, it will be seen that the omission is not so much in types of boilers, to which considerable space has been given, as it has been in the variations of these types.

Boilers for steamships will be described in the volume on "Marine Engines," to be included in this series, they requiring more or less special treatment in connection with other subjects beyond the scope of this volume; so also the design and construction of locomotive boilers, which will receive special consideration in the volume on "Locomotive Engines," also to be included in this series.

The consideration of chimneys in this volume includes only the necessary dimensions of diameter and height for boiler plants from 20 to 1000 horse-power, no space having been given to their consideration as an isolated structure, for the reason that an illustrated volume on the design and construction of "Brick and Metal Chimneys" is in preparation for immediate publication.

WILLIAM M. BARR.

PHILADELPHIA, March, 1898.

CONTENTS.

BOILERS AND FURNACES.

CHAPTER I.

FURNACE COMBUSTION.

Combustion in steam engineering means the controlled chemical combination of the elements carbon and hydrogen in the fuel with the oxygen of the atmosphere, by which an evolution of heat is secured and maintained in a suitably constructed furnace for the purpose of generating steam.

The Unit of Work in steam engineering is a gravitation unit known as the foot-pound, or the amount of work required to raise one pound one foot high against the force of gravity, and is entirely independent of the time it takes to do it.

Horse-Power is a unit of work employed in steam engineering to denote the power-rating of a steam boiler and engine, or the power transmitted by a belt, shaft, etc. In computing work done it is always independent of the time taken to do it, but in computing horse-power time is an essential element. The unit of horse-power is a gravity measurement, and represents 33,000 pounds raised one foot high in one minute against the force of gravity.

Chemical and Physical Changes.—If a combustible like wood is thrown upon a fire it disappears, and nothing visible remains but ashes. Changes of this kind, in which a substance disappears and something else is formed in its place, are known as chemical changes.

Changes which do not affect the composition of substances are called physical changes. The freezing of water into ice or the evaporation of water into steam is a physical change, because the composition of the substance in these three states is the same.

Chemical Changes due to combustion always involve the conversion of the substances burnt into new substances, and these final products are compounds; the constituent elements which enter into or make the compound always combine according to certain definite proportions, either by weight or measure. In the combustion of carbon, for example, we may have either of two possible combinations: One atom of carbon uniting with one atom of oxygen produces carbonic oxide, CO; the atomic weights would be, Carbon, 12 + Oxygen, 16 = Carbonic Oxide, 28. By percentages instead of the atomic weights we would have in 100 parts of carbonic oxide, Carbon, 42.86 + Oxygen, 57.14 = 100.00.

In the case of carbonic acid gas, CO_2, one atom of carbon unites with two atoms of oxygen, thus: Carbon, 12 + 2 Oxygen, 16 = 44 = the atomic weight of CO_2. By percentages instead of atomic weights we have in 100 parts of carbonic acid gas 27.27 parts of carbon and 72.73 parts of oxygen. These proportions constitute the only two direct inorganic compounds of carbon and oxygen.

Hydrogen gas burnt in oxygen combines in the proportion of 11.1 parts of hydrogen with 88.9 parts of oxygen to form 100 parts of water. The ratio of the weights of equal volumes of hydrogen and oxygen is as 1 : 16. If, then, two volumes of hydrogen combined with one volume of oxygen, the ratio between the weights is 2 : 16, or 1 : 8, and these gases will combine in no other proportions to form water; any excess of either gas will remain unchanged.

Furnace Combustion requires a combustible such as carbon or hydrogen, which is supplied by the fuel, and a supporter of combustion, —oxygen,—supplied by the atmosphere; such combustion is accompanied by flame and incandescence, the former a result of the combustion of the hydro-carbon gases of the fuel, the latter, glowing carbon from which the gases have been expelled. The color and intensity of incandescence is dependent upon the temperature, red indicating the lowest and a dazzling white light the highest. Most solids emit light or become a dull red at about 750° Fahr. The temperature of coke burned under favorable conditions approximates 3600° Fahr., and at this temperature emits an intensely brilliant white light.

Hydrogen, H. At. Wt., 1.—When pure, hydrogen is colorless, tasteless, and inodorous. It occurs in nature in combination with carbon in varying proportions. The compound which contains it in greatest abundance is marsh-gas, of which hydrogen forms four parts, the formula being CH_4. Under ordinary temperatures hydrogen has no tendency to enter into combination with other substances. It combines with eight parts of oxygen to form water, but this combination does not take place spontaneously. Pure hydrogen burns in the atmosphere with a pale blue light scarcely perceptible in full daylight, giving off an intense heat. The heat units of one pound of hydrogen burned in oxygen were ascertained by Favre & Silbermann to be 62,032. This is not equalled by any other known substance.

Carbon, C. At. Wt., 12.—This is one of the most widely diffused and abundant of the elements, being the central element in organic nature. It exists in three different forms, as diamond, graphite, and charcoal, each having its own physical properties, the chemical properties being the same. Carbon is an inactive element at ordinary temperatures, but when raised to its kindling temperature, about red heat, it unites with oxygen and combustion ensues.

Sulphur, S. At. Wt., 32.—Nearly all coals contain sulphur in combination with iron in the form of golden-yellow crystals, commonly known as iron-pyrites. It burns with a clear blue, feebly luminous flame, being converted into sulphurous oxide, SO_2. The heat developed by the combustion of any sulphur present in coal is not taken into account in steam engineering. The amount of sulphur in coal ranges from 0.3 to 5 per cent.

Oxygen, O. At. Wt., 16.—When free or uncombined, oxygen is known only in the gaseous state ; when pure, it is colorless, tasteless, and inodorous. It is the sustaining element in all the ordinary phenomena of combustion.

Ignition is simply the incandescence of a body unattended by chemical change, and must not be confused with combustion : the ignition of solids is a source of light, the combustion of solids is a source of heat. Every combustible must be heated to a certain definite temperature before it will combine with oxygen. This temperature is usually called the point of ignition, or its kindling temperature.

Flame is simply gas burning on its exterior surface ; its color and brightness depend not only on its degree of temperature, but upon the presence of solid incandescent particles in the flame ; purely gaseous substances do not become highly luminous when burning. The structure of flame from burning wood or coal consists of three parts : first, a central core of unburned hydro-carbon gas ; second, an envelope of burning hydrogen and carbon, the oxidizing portion of the flame ; third, still another envelope consisting of aqueous vapor and other products corresponding to the combustibles contained in the issuing gas.

The Air.—The oxygen needed for furnace combustion is supplied by the air, which consists essentially of two gases, nitrogen and oxygen, in the proportion of 79 volumes of nitrogen to 21 volumes of oxygen, or, by weight, 77 per cent. of nitrogen to 23 per cent. of oxygen. Air is a mechanical mixture of these two gases and not a chemical compound ; their union in the proportion given above is distinguished by no properties which may not be attributed individually to these gases. One of the most important properties of the atmosphere is that of weight, without which furnace combustion would be a much more complex operation than at present, because a different mechanical agency would be required for producing draught in the furnace. The mean pressure of the atmosphere at the mean level of the sea is equal to 14.7 pounds per square inch, or 2116.8 pounds per square foot.

Nitrogen, N. At. Wt., 14.—By volume and by weight nitrogen is the principal constituent of air ; it is a colorless, tasteless, inodorous gas ; its specific gravity is .971, air = 1. The specific heat of nitrogen is 0.244 at constant pressure. It is an inert gas in the furnace, not being itself combustible, nor will it support combustion, but the presence of nitrogen is useful, inasmuch as it greatly modifies the intensity of combustion, and at the same time does not alter the chemical relations of the oxygen to the combustible substances. If it were not for the presence of nitrogen in the air, or suppose the atmosphere were wholly oxygen, combustion would be very hazardous,—in fact, a fatal occurrence, because the coals would burn so violently as to have a combustion wholly uncontrollable; nor would the combustion stop with the fuel, but the oxygen would attack the grate bars, the furnace front, the boiler, and everything else made of iron, for the latter substance burns even more violently in oxygen gas than coal. Nitrogen is useful, then, in the good offices which it performs in lessening the avidity of oxygen during combustion, making fire a moderate, useful, and easily controlled servant in the service of man; its negative qualities make nitrogen a safe substance, and, while it plays no active part in combustion, it is the means by which the oxygen is delivered in the body of incandescent fuel, where it parts company with the oxygen and passes on through the fire, its less specific gravity aiding materially in producing the draught necessary for furnace combustion.

Fuel is a term used in steam engineering to include combustibles of every sort that can be utilized to generate heat. The combustibles in common use consist almost wholly of carbon and hydrogen.

Wood.—Commercially, woods are distinguished as hard or soft, the former including the heavy, compact woods, as oak, hickory, etc.; the latter, pine, spruce, etc. The average composition of five dry woods, including beech, oak, birch, poplar, willow, by Chevandier, was: Carbon, 50.00; hydrogen, 6.00; oxygen, 41.00; nitrogen, 1.00; ash, 2.00 = 100.00.

The hydrogen present in wood is not available as fuel, owing to the presence of oxygen, these two gases uniting to form water. Carbon averaging 50 per cent. is present in all woods, and is the only combustible available for generating heat. The American Society of Mechanical Engineers, in their rules for boiler tests, assume one pound of wood to equal 0.4 pound of coal.

Tan is the spent bark from which the tannic acid has been extracted in the process of tanning leather; the barks commonly used are oak and hemlock. The principal drawback to tan as a fuel is its contained moisture, and for this reason special furnaces are made for burning it. Tan bark, as commonly used for fuel, will yield about 3600 heat units per pound, which is one-half the value of ordinary dry wood and about one-fourth the value of good bituminous coal.

Bagasse is the woody fibre of sugar-cane after the saccharine juices have been expelled for sugar-making. Special furnaces have been contrived for burning it, and with fair results; the contained water is about 50 per cent. of the gross weight; the remaining fibre is not unlike wood in its heat-giving power. On an average, six pounds of bagasse are equivalent to one pound of good bituminous coal.

Peat is organic matter undergoing a gradual carbonization, the oxygen of the plants being liberated under special conditions of moisture and heat, leaving a spongy carbonaceous mass. When dried, peat consists of about 58 per cent. of carbon. Scarcely any peat is used for fuel in this country because of the abundance and superior heating power of coal.

Lignite occupies a position historically between peat and bituminous coal; it is of later origin than bituminous coal and in a less advanced state of decomposition. It varies considerably in appearance and structure: the fracture is uneven, presenting a brown to a very dark brown-black color, with a dull and frequently fatty lustre; it crumbles easily in handling and will not bear rough transportation, nor will it bear long-continued exposure to the weather without crumbling. It is non-caking in the fire and yields but a moderate heat, below the average of bituminous coals. In its natural state, lignite contains from 10 to 30 per cent. of water.

Brown Coal is a term sometimes used, though not correctly, as interchangeable with lignite; it is in a more advanced stage of decomposition,—that is, it is nearer the bituminous coal series than the lignites. Brown coals contain less fixed carbon than the coals of the carboniferous epoch, and usually a much larger percentage of moisture when freshly mined, both of which tell against the coal commercially.

Bituminous Coal.—In physical properties bituminous coals vary so widely that a single description cannot include them all. The color averages from brown to pitch black; the lustre is vitreous, resinous, sometimes silky; the structure may be compact, slaty, columnar, and even fibrous; the fracture, irrespective of structural joints and cleavage, is conchoidal, often flat and rectangular, and sometimes fibrous. It is distinctive of these coals to burn with a more or less yellow flame and smoke.

In composition, bituminous coals range in volatile matter from 15 to 50 per cent., including 2 to 12 per cent. of contained moisture. The fixed carbon varies from 40 to 75 per cent., the earthy matter from 2 to 20 per cent. Sulphur is almost always present in bituminous coals as pyrite.

Bituminous coals are broadly classed as caking and non-caking coals.

Caking Coal.—The characteristic circumstance that lumps of coal, either large or small, are rendered pasty by the action of the heat, and will cohere in the fire to form a spongy-looking mass which may cover the whole surface of the grate, is the property called caking; the fixed carbon remaining after the expulsion of the gases is called coke. It is more difficult to completely and economically burn caking coals in a boiler furnace than those of non-caking variety because of this tendency to run together and prevent the free flow of air through the fuel; such masses of burning coal must be broken up with a slice-bar at frequent intervals to facilitate its combustion.

Non-caking coals burn free in the fire, not unlike charcoal or soft coke. The action of heat does not cause the coal to fuse or run together into large masses which require afterwards to be broken up to allow free access of air through the body of the burning coal to secure quick combustion. Free-burning coal is the same as non-caking coal.

Cannel Coal is a variety of bituminous coal very rich in hydrogen. This coal kindles readily and burns without melting, emitting a bright flame. When thrown upon an active fire the piece splits into fragments, producing a crackling noise. In appearance this coal differs from all bituminous coals : its structure is a compact mass and more nearly homogeneous than others ; it varies from brown to black in color, and has a dull resinous lustre. When broken it does not preserve any distinct order of fracture, and is liable to split in any direction. It is highly esteemed as a gas coal, but is not much used as a steam coal except locally near the mines.

Block Coal is a representative non-caking bituminous coal. It occurs in several of the Western States, but is found at its best in Indiana. It has a laminated structure, composed of alternate thin layers of vitreous dull black coal and mineral charcoal. Chemically it does not appear to differ from caking coal, but in burning it behaves quite differently : it does not swell, shoot out jets of gas, nor form a cake by running together, but retains its shape until entirely consumed to a white ash which contains no trace of clinker, its behavior in the furnace being quite like that of hickory wood, burning with a uniform flame that spreads evenly over the exposed surface.

Coke is the fixed carbon and earthy matter remaining after the distillation of the gases from bituminous coal. The only coke of any commercial value is made from caking coals. The quality of the coke is affected by the temperature at which it is made : the higher the temperature and the longer it is exposed to that temperature, the harder, more dense, and less easily combustible will be the coke.

Semi-Bituminous Coal partakes somewhat of the nature of anthracite coal. Modified by volatile matter, it forms an excellent fuel for steam-boiler furnaces. In appearance it resembles anthracite coal rather than bituminous ; its fracture, however, as compared with anthracite, is less conchoidal ; it is not so hard, and is of less specific gravity. When thrown upon the fire it kindles more readily and burns faster than anthracite, but without the smoke and soot characteristic of bituminous coal.

Semi-Anthracite Coals are restricted to such as average from 6 to 8 per cent. of volatile combustible matter, but otherwise have the physical characteristics of true anthracites. These coals are in high estimation for steaming purposes, as they kindle easily and burn much more freely than do the harder anthracite coals.

Anthracite Coal is slow to ignite ; it does not soften or swell in the fire ; the flame is quite short and nearly transparent, having a yellowish tinge at first, changing to a soft blue, with occasionally a red tinge, and gives off no smoke. When broken it presents a conchoidal appearance and is quite homogeneous in structure. In general it is compact, slaty, grayish-black, splendent, varying somewhat according to the locality at which it is mined. It is not found in large quantities outside of Eastern Pennsylvania.

Culm is fine anthracite coal. Formerly this was waste product and had no commercial value. Culm heaps are now being carefully screened and assorted into sizes for use in steam-boiler furnaces. The sizing must be uniform or the smaller pieces will drop into and clog the passages between the larger ones and obstruct the free passage of air through the fire. The percentage of ash in sizes recovered from culm is much greater than in the larger sizes, such as chestnut or stove coal. A very strong draught is required for burning fine anthracites, and a forced draught is often resorted to or required for their rapid combustion.

Petroleum is a natural hydro-carbon oil found in large quantities, especially in Pennsylvania and Ohio, the weight per gallon varying from 6 to 7 pounds. The specific gravity of petroleum averages about 0.8, with variations on either side. The composition of crude Pennsylvania oil averages: Carbon, 84.00; hydrogen, 13.75; water, 2.25 = 100. Heat units, 20,746 per pound. Evaporative power of one pound of oil from and at 212° Fahr. = 21.47 pounds of water.

One of the interesting exhibits at the World's Fair, Chicago, was the boiler plant which was furnished with crude oil from the Lima, Ohio, district for fuel. Never before was there such an opportunity in this country for testing liquid fuel on so large a scale. The quantity of petroleum used for firing the main boiler plant amounted to upwards of 31,000 tons, and the work done is stated to have been 32,316,000 horse-power hours, or about 2.1 pounds of oil per horse-power per hour.

Natural Gas is found locally in Western Pennsylvania, Northern Ohio, and Central Indiana in paying quantities; in lesser quantities it is found in many other localities. Natural gas is an ideal fuel if used near the source of supply, as no labor is required in its use except to regulate the supply in the furnace; it is not difficult to regulate the supply of air to insure perfect combustion; there is no soot, ashes, or other débris. The composition of natural gas at Findlay, Ohio, is,—

	By weight.	By volume.
Hydrogen	0.27	2.18
Marsh gas	90.38	92.60
Carbonic oxide	0.86	0.50
Olefiant gas	0.53	0.31
Carbonic acid	0.70	0.26
Nitrogen	6.18	3.61
Oxygen	0.66	0.34
Sulphydric acid	0.42	0.20
	100.00	100.00

The heat units in one pound of Findlay, Ohio, natural gas = 21,520.

Evaporative power of one pound of the above gas from and at 212° F. = 22.27 pounds of water.

Natural-gas tests under boiler for steam-making at Pittsburg, Pennsylvania, show that one pound of good bituminous coal equals from 7½ to 12¼ cubic feet of natural gas; other experiments show that 1000 cubic feet of natural gas equal from 80 to 133 pounds of bituminous coal, a variation of more than 60 per cent. between the two extremes; quality of coal and manipulation of furnace accounts for much of this difference.

PRODUCTS OF COMBUSTION.

The combustible elements of wood and coal are carbon, hydrogen, and sulphur. The supporter of combustion is the oxygen of the air.

Carbon when burnt in oxygen yields two products, depending upon the supply of oxygen in the furnace, viz.:

	Formula.	Combustion.	Product.
Carbonic acid gas	CO_2	Complete.	Incombustible.
Carbonic oxide gas . . .	CO	Incomplete.	Combustible.

Carbonic Acid Gas, CO_2, is the first product of carbon combustion formed in the furnace. It is a colorless gas, with a slightly acid taste and smell, and is incombustible, because it holds in combination all the oxygen it has the power to combine with. Its specific gravity is 1.529. This gas in passing through or over a bed of red-hot carbon will take up additional carbon, changing the original product, CO_2, into a lower oxide, CO, which is a combustible gas. This change, if it occurs in the furnace, is very wasteful of fuel in case the lower oxide escapes unburned; for example, carbon burned to CO_2 = 14,500 heat units per pound, but if burned to CO, = 4452 heat units per pound, equivalent to a loss of two-thirds of the heating power of the carbon.

Carbonic Oxide Gas, CO, is a product of incomplete combustion; the gas is colorless, tasteless, and inodorous. Its specific gravity is 0.967. It burns in the air with a blue flame, forming carbonic acid gas, CO_2. At high temperatures, such as obtain in steam-boiler furnaces, it has a very strong tendency to combine with oxygen, but at ordinary temperatures this gas does not combine readily with it. Carbon burnt to carbonic oxide gas = 4452 heat units per pound of carbon, which is approximately 10,000 units less than if burned to carbonic acid gas. To remedy this it has long been recommended that air be admitted over the fire or at the bridge wall in quantity sufficient to burn the CO and convert it to CO_2. As this combination can only occur at a high temperature, no less than that of red-hot coals, efforts of this kind have not always been successful; the best practice now is to carry a moderately thick fire with a strong draught and less air over the fire than formerly, compelling the air to pass through the fire rather than over it.

Aqueous Vapor.—Hydrogen unites with oxygen to form water, H_2O, in which the combustion is complete and the product incombustible. The water formed in the furnace passes off as aqueous vapor, condensing in the atmosphere above the chimney. Any excess of hydrogen in the furnace over that combining with oxygen as above passes off uncombined. The specific heat of gaseous steam is 0.622.

Sulphurous Oxide.—Sulphur combines with oxygen to form sulphurous oxide, SO_2, a colorless gas with a suffocating odor. It is a non-supporter of combustion, instantly extinguishing flame when brought within its influence. Sulphurous oxide in absorbing the vapor of water changes to sulphurous acid, $SO_2 + H_2O$, which may, and often does, become a direct cause of the external corrosion of boilers, mud-drums, feed-pipes, etc.

Nitrogen is a neutral element in the furnace. It is incombustible and has no affinity for any of the products of combustion. It acts simply as a dilutant of the gases in the furnace. Its specific gravity, 0.9736, being less than that of air, it performs the useful office of assisting the draught.

Free Air.—There is always an excess of air passing through the fire above that required for combustion. Chemically, about 12 pounds of air are required for the combustion of 1 pound of coal; practically, from 18 to 24 pounds actually pass through the furnace. This excess is, of course, waste product and must be regarded as a dilutant of the furnace gases.

Smoke, when taken collectively, includes all the gaseous products of combustion escaping from the furnace; specifically it means the colored gases accompanying combustion discharging into the atmosphere. In the combustion of anthracite coal and coke very little smoke appears at the chimney top; but in the combustion of bituminous coal the products often become a veritable nuisance in the neighborhood where the chimney happens to be located. The coloring matter is carbon in a finely divided state, small particles of soot, so small and of so little weight that they are carried off mechanically out of the furnace, up the chimney, and into the atmosphere. The number of these sooty particles determines the color of the smoke, and may vary in density from light gray to black. An excess of sooty particles indicates generally a low temperature in the furnace.

Ashes.—Whatever incombustible substances originally in the fuel remain after complete combustion are called ashes, irrespective of composition.

An average analysis of ash from a number of anthracite and bituminous coals, the percentage of ash averaging approximately 5 per cent., gave the following:

	Bituminous.	Anthracite.
Silica	56.22	49.68
Alumina	36.17	39.83
Iron, Oxide	2.74	7.51
Lime	2.24	2.17
Magnesia	.92	.72
Potash and Soda	1.13	
Sulphur	.58	.09
	100.00	100.00

The specific heat of ashes may be assumed to be 0.215 without sensible error in engineering calculations.

The colors of ashes are designated as red, brown, yellow, or white, as they appear to the observer. Red or reddish-brown ashes indicate the presence of iron in the coal, probably in the form of pyrites.

Clinker is formed by fusing together the impurities in the coal, such as iron, silica, lime, potash, etc. Each of these substances being differently fusible and affecting differently the fusion of each other, their final form will depend somewhat on the intensity of combustion, or, in other words, the temperature of the fire in which they are formed. There are few colored ashes that will not soften under the action of intense heat and form clinker; and this fact in itself should affect the commercial value of coals. Those which burn to a nearly pure white are the best, because they contain little or no alkali, lime, or oxide of iron.

Heat Developed by Combustion.—Knowing the elementary constituents of coal, the heat developed may be calculated, as shown in the following example of the analysis of one pound of semi-anthracite coal, containing: Carbon, .83; hydrogen, .05; oxygen, .04; sulphur, .02; ashes, .06 = 1.00.

Carbon, .83 × 14,500 heat units = 12,035 heat units.

Hydrogen and oxygen unite to form water; therefore all the oxygen must be deducted, together with one-eighth of the hydrogen, thus: $\frac{1}{8}$ of $\frac{4}{100} = \frac{4}{800} = .005$ pound of hydrogen neutralized by the presence of oxygen in the coal, leaving .05 — .005 = .045 pound of available hydrogen; then, proceeding as before:

Hydrogen, .045 × 62,032 = 2,791 heat units.
Sulphur, .02 × 4,000 = 80 heat units.

We have then:

Carbon, = 12,035 heat units.
Hydrogen, = 2,791 heat units.
Sulphur, = 80 heat units.

The theoretic calorific value of the coal = 14,906 heat units.

The ash, being inert, is not taken into account.

This calorific value is had on the supposition that the carbon has been burnt to carbonic acid gas, CO_2; if, however, the carbon has been incompletely burnt, the product being carbonic oxide gas, CO, instead, a less number of heat units would be had, thus:

Carbon burnt to CO, .83 × 4452	=	3695.15 heat units.
Hydrogen as above,	=	2791.00 heat units.
Sulphur as above,	=	80.00 heat units.
Calorific value of the coal burnt to CO	=	6566.15 heat units.

This result is most likely to occur, at least in part, when carrying thick fires with an insufficient air supply.

The available hydrogen in coal after deducting the combining portion of oxygen also present may conveniently be expressed thus:

$$\frac{\text{Hydrogen, } 62{,}032 \text{ heat units}}{\text{Carbon, } 14{,}500 \text{ heat units}} = 4.28.$$

Hydrogen may be taken, then, as 4.28 times as valuable as carbon in thermal calculations. Practically three times the value of carbon is as high as should be taken, and even this is greatly in excess of what is realized.

E. T. Cox, formerly State Geologist, Indiana, informed the writer that the net result of his investigations gave 20,115 heat units as the average thermal value of one pound of the volatile matter liberated from Indiana bituminous coals by heat during the process of combustion, or a little less than olefiant gas (21,300 heat units). Mr. Cox's figures are used in the following example.

A sample of bituminous coal by proximate analysis can be calculated with tolerable accuracy, thus:

	Per cent.		Per cent.
Volatile matter	41	Water	3
		Gas	38
Coke	59	Fixed carbon	50
		Ash	9
	100		100

The volatile matter requires the expenditure of heat for its liberation, which in a large number of experiments approximated 3600 heat units per pound of volatile matter.

The volatile matter is not all combustible gas; more or less aqueous vapor passes off with it. Experimentally, the calorific value of the volatile matter of bituminous coals was found to approximate 20,115 heat units per pound.

The theoretical calorific value of one pound of such coal may be determined thus:

Gas	.38 × 20,115 =	7,643.70 heat units.
Less	.38 × 3,600 =	1,368.00 heat units.
Net value of gas . . .		6,275.70 heat units.
Fixed carbon	.50 × 14,500 =	7,250.00 heat units.
Total calorific value .		13,525.70 heat units.

Dynamical Value of Combustion.—A British thermal unit is that quantity of heat necessary to raise the temperature of one pound of pure water from 39° to 40° Fahr., the former being the temperature of its greatest density.

The mechanical equivalent of heat is equal to raising 772 pounds one foot high against the action of gravity; 33,000 pounds raised one foot high per minute = 1 horse-power.

The combustion of one pound of carbon yields 14,500 heat units; then 14,500 × 772 = 11,194,000 foot-pounds.

Hydrogen yields 62,032 heat units for each pound burnt to water. The dynamic value of hydrogen is: 62,032 × 772 = 47,882,704 foot-pounds.

Coal fed to steam-boiler furnaces is usually reckoned in pounds per hour. If it be required to know the dynamic value of one pound of coal per hour expressed in horse-power, the coal containing say 13,680 thermal units, we have

$$\frac{13{,}680 \times 772}{33{,}000 \times 60} = 5.303 \text{ horse-power.}$$

Of this amount, however, only about 10 per cent. is available for doing useful work.

Temperature of Fire.—Carbon and hydrogen are the principal heat-giving constituents of coal, and of these carbon is the most effective in steam-boiler practice, because an incandescent bed of it can be maintained at all times, the direct effect of which is to prevent violent fluctuations of temperature in the furnace. To illustrate the method of calculating the temperature of fire, we will take the analysis of coal given on page 18, the carbon of which was 83 per cent. Carbon requires for its complete combustion 2.67 times its own weight of oxygen; then 1 pound carbon + 2.67 pounds oxygen = 3.67 pounds carbonic acid gas. In order to get this oxygen from the air there would remain in the furnace 8.94 pounds of nitrogen, which must be taken into account thus:

Products.	Pounds.		Specific heat.		Heat units.
Carbonic acid gas	3.67	×	.2164	=	.794
Nitrogen	8.94	×	.244	=	2.181
	12.61		Total . . .		2.975

heat units absorbed in raising the temperature of one pound of carbon 1° Fahr. The combined weight of the two gases = 12.61 pounds; then,

$$\frac{\text{Heat units, } 2.975}{\text{Pounds, } 12.61} = .236, \text{ the mean specific heat.}$$

Carbon yields in its perfect combustion 14,500 heat units per pound; this divided by the 2.975 heat units absorbed as above gives 14,500 ÷ 2.975 = 4874° Fahr. as the highest temperature attainable by the combustion of one pound of carbon, and with the exact amount of air (11.61 pounds) needed to furnish the necessary oxygen, a much smaller allowance than is possible in the actual generation of steam by ordinary furnaces.

Eighteen pounds of air is, on an average, as little as passes through the furnace for each pound of carbon burnt. A reduction in temperature follows as here shown:

Products.	Pounds.		Specific heat.		Heat units.
Carbonic acid gas	3.67	×	.2164	=	.794
Nitrogen	8.94	×	.244	=	2.181
Air in excess, uncombined .	6.39	×	.2377	=	1.519
Totals	19.00				4.494

heat units absorbed in the furnace, being 1519 more per pound of carbon than in the previous example. We have, then, 14,500 ÷ 4.494 = 3226° Fahr., a reduction of 1648° Fahr. from that obtained in the preceding example; but this accords more nearly with the best practice and is as high a temperature as can ordinarily be expected.

Rate of Combustion.—This is commonly expressed in pounds of coal burnt per square foot of grate surface per hour. The weight of coal burnt will depend, other things being equal, upon the quantity of air passing through the fire; the rate of combustion varies between wide limits: horizontal tubular-boiler furnaces burn from 8 to 12 pounds of anthracite coal per square foot of grate surface per hour with natural draught; bituminous coals range from 12 to 20 pounds, and occasionally more. Internally fired boilers by reason of their relatively smaller proportion of grate to heating surface have a rate of combustion varying from 12 to 40 pounds of coal per square foot of grate per hour, depending upon the design of the boiler and the ratio of the grate to heating surfaces, the smaller ratio of grate requiring a higher rate of combustion. Boilers of this type are usually supplied with a higher chimney, requiring a stronger draft than boilers of the former type. The rate of combustion will vary with the quality of the fuel, draft, ability, and watchful care of the fireman.

Efficiency.—The efficiency of a steam-boiler is a percentage indicating how nearly the actual performance attains to the theoretical possibilities; if the latter be expressed by 100, the efficiency will always be a less number. For example: a coal used for generating steam by calorimeter test yields 12,500 heat units per pound; the equivalent evaporation from and at 212° Fahr. would be $12,500 \div 966 = 12.95$ pounds of water per pound of coal, the theoretical possibility; but by actual test only 8.75 pounds of water were evaporated, then:

$$\text{Efficiency} = \frac{8.75 \times 100}{12.95} = 68.62 \text{ per cent.}$$

The loss of heat in this case is 31.38 per cent. of the total, accounted for by heat escaping by the chimney, by radiation, the contact of the hot surfaces with the air, as well as imperfect combustion.

Cylinder and flue boilers, externally fired, set in brick-work, have an efficiency ranging from 45 to 60 per cent.; tubular boilers from 50 to 70 per cent.; internally fired boilers from 60 to 70 per cent.; water-tube boilers from 65 to 75 per cent. Good boilers properly set and well managed will average nearly the same efficiency, approximating 65 per cent.

Calorific Value of Coal.—Table I. gives proximate analyses and calorific values of selected American coals. Only a few of the numerous coals of the United States can be mentioned, but enough are given to show the average analysis for localities named. In quality, the range of fuels for steaming purposes covers everything from the softest lignites, which sometimes contain as much as 15 per cent. of water, to Lehigh anthracite, which is nearly pure carbon. Bituminous coals are most abundant, and while these vary much in calorific value, they are, in general, good steaming coals.

The sulphur contained in some of the softer coals is a mischievous element one would gladly be rid of, inasmuch as from 1½ to 4 per cent. is not uncommon, and in some localities as much as 10 per cent. is recorded, notably specimens analyzed from mines in Summit County, Utah; but this high percentage is quite unusual. No account is made of the contained sulphur in the coal in any calculations connected with this table.

The calorific values in this table are based upon the experimental determination that each pound of carbon will yield 14,500 heat units when burned in oxygen to carbonic acid gas. The volatile portions of the coal are calculated upon the experimentally ascertained fact, by E. T. Cox, that the total average calorific value of the gases obtained by the destructive distillation of bituminous coal is approximately 20,000 heat units per pound, and that 3600 heat units are absorbed in the process of disassociation of the gases from the coal; this leaves 16,400 heat units per pound as the net calorific value for the volatile portion of bituminous coals. The calorific value given in the next to the last column in the table is the sum of the carbon and volatile gases as thus ascertained, expressed in British thermal units.

TABLE I.

PROXIMATE ANALYSES AND CALORIFIC VALUE OF SELECTED AMERICAN COALS.

Coals and Locality.	Volatile Matter, per cent.		Coke, per cent.		Heat Units per pound.	Evaporative Power per pound, from and at 212° F.
	Water.	Gas.	Fixed Carbon.	Ash.		
ALABAMA.						
Bibb Co., Bit., Helena Vein	1.74	35.48	58.96	3.82	14,368	14.87
Jefferson Co., Bit., Birmingham	3.01	42.76	48.30	5.93	14,017	14.51
Tuscaloosa Co., Bit.	1.59	38.33	54.64	5.44	14,209	14.71
ARKANSAS.						
Franklin Co., Bit., Falker Slope	1.13	13.21	81.28	4.38	13,952	14.44
Johnson Co., Bit., Coal Hill	1.52	14.73	74.49	9.26	13,217	13.68
Sebastian Co., Bit., Huntington Slope	.93	15.55	77.54	5.98	13,793	14.28
CALIFORNIA.						
Alameda Co., Bit., Livermore	18.08	39.30	35.61	7.01	11,608	12.01
COLORADO.						
Boulder Co., Bit.	12.01	35.19	46.24	6.56	12,476	12.92
Tremont Co., Bit.	3.93	42.43	47.16	6.48	13,797	14.28
Las Animas, Bit.	1.26	36.40	53.10	9.24	13,670	14.15
GEORGIA.						
Dade Co., Bit.	1.20	23.05	60.50	15.25	12,553	12.99
ILLINOIS.						
Macoupin Co., Bit., Mount Olive	10.38	36.38	46.10	7.14	12,651	13.10
McLean Co., Bit., Bloomington	7.90	34.02	53.12	4.96	13,281	13.75
Mercer Co., Bit.	8.40	31.20	54.80	5.60	13,063	13.52
Peoria Co., Bit., Elmwood	1.36	27.69	35.41	35.54	9,675	10.02
Stark Co., Bit., Lombardville	9.42	31.38	51.74	7.46	12,648	13.09
Trenton Co., Bit., Clinton	9.95	31.04	51.96	7.05	12,625	13.07
Vermilion Co., Bit., Danville	5.78	43.70	45.37	5.15	13,746	14.23
INDIANA.						
Block Coal, Lafayette	13.05	32.34	48.78	5.83	12,377	12.81
Clay Co., Bit., McClellan & Zeller	8.50	32.50	56.50	2.50	13,523	14.00
Davies Co., Buckeye Cannel Coal Co.	3.50	48.00	42.00	6.50	13,962	14.45
Greene Co., Bit.	7.00	29.50	63.00	.50	13,973	14.46
Owen Co., Bit.	2.00	38.50	57.50	2.00	8,969	9.28
Vermillion Co., Bit.	5.50	44.00	46.00	4.50	13,886	14.37
INDIAN TERRITORY.						
Choctaw Nation, Bit.	1.59	23.31	66.85	8.25	13,517	13.99
Choctaw Nation, Bit., Atoka	6.66	35.42	51.32	6.60	13,248	13.71
IOWA.						
Marion Co., Bit., Oscaloosa	5.73	46.54	45.60	2.13	14,245	14.75
Monroe Co., Bit., Albia	5.16	40.21	45.88	8.75	13,247	13.71
Wapello Co., Bit., Ottumwa	6.50	41.35	48.25	3.90	13,777	14.26
KANSAS.						
Cherokee Co., Bit.	1.94	36.77	52.45	8.84	13,585	14.06
KENTUCKY.						
Fulton Co., Bit.	2.00	47.85	47.73	2.42	14,768	15.29
Hancock Co., Bit., Hawesville	3.30	39.00	50.50	7.20	13,719	14.20
Lawrence Co., Bit., Peach Orchard	3.24	36.56	49.24	10.96	13,136	13.60
Muhlenberg Co., Bit.	3.60	30.60	58.80	7.00	13,544	14.02
MARYLAND.						
Cumberland, Bit.	1.23	15.47	73.57	9.79	13,205	13.67
Garrett Co., Semi-Bit., George's Creek	.59	18.52	74.31	6.58	13,812	14.30
MISSOURI.						
Bates Co., Bit.	2.54	42.62	41.14	13.70	12,955	13.41
Caldwell Co., Bit., Hamilton	5.06	34.24	47.69	13.04	12,530	12.97
Putnam Co., Bit., Mendota	9.03	37.48	46.24	7.25	12,852	13.30
MONTANA.						
Cascade Co., Bit., Sandcoulee	3.01	30.23	59.71	7.05	13,616	14.10
NEBRASKA.						
Adams Co., Bit., Hastings	0.21	27.82	60.88	11.09	13,390	13.86
NEW MEXICO.						
Colfax Co., Bit., Ranton	3.10	35.00	51.50	10.40	13,208	13.67
NORTH CAROLINA.						
Guilford Co., Bit., Deep River	1.79	29.56	58.30	10.35	13,302	13.77

TABLE I.—*Continued.*

Coals and Locality.	Volatile Matter, per cent.		Coke, per cent.		Heat Units per pound.	Evaporative Power per pound, from and at 212° F.
	Water.	Gas.	Fixed Carbon.	Ash.		
OHIO.						
Columbiana Co., Bit., Salineville	2.32	39.08	52.78	5.82	14,062	14.56
Hocking Valley, Bit.	8.25	35.88	53.15	2.72	13,591	14.07
Holmes Co., Bit., Walnut Creek	4.49	42.50	47.27	5.74	13,824	14.31
Jefferson Co., Bit., Brilliant	1.85	37.82	55.62	4.71	14,267	14.77
Mahoning Co., Bit., Brier Hill	2.47	31.83	64.25	1.45	14,537	15.05
Perry Co., Bit., New Straitsville	7.09	36.61	52.00	4.30	13,544	14.02
Trumbull Co., Bit., Liberty	5.91	35.01	55.70	3.38	13,819	14.31
OREGON.						
Grant Co., Bit., John Day River	4.55	40.00	48.19	7.26	13,547	14.02
Tillamook Co., Bit., Nehalem	8.00	37.83	45.17	9.00	12,754	13.20
Benton Co., Bit., Yaquina Bay	13.04	46.70	32.60	7.66	8,386	8.68
PENNSYLVANIA.						
Anthracite, Upper and Lower Measures	1.35	3.45	89.06	6.14	13,480	13.95
Anthracite, Carbon Co., Beaver Meadow	1.50	2.38	88.94	7.18	13,286	13.75
Anthracite, Carbon Co., Buck Mountain	3.04	3.95	82.66	10.35	12,634	13.08
Anthracite, Lackawanna Co., Scranton, 40′ Shaft	1.12	4.99	83.98	9.91	12,995	13.45
Anthracite, Lackawanna Co., Scranton, Mt. Pleasant	1.27	7.54	80.54	10.65	12,915	13.37
Anthracite, Lehigh Co.	1.01	5.28	88.15	5.56	13,648	14.13
Anthracite, Luzerne Co., Drifton	2.97	2.30	87.96	6.77	13,131	13.59
Anthracite, Luzerne Co., Jeanesville	4.04	2.99	88.20	4.77	13,279	13.75
Anthracite, Luzerne Co., Wilkes-Barre, Lehigh Valley Buckwheat	1.34	6.42	76.94	15.30	12,209	12.64
Bituminous, Allegheny Co., Pittsburg, Average	1.80	35.34	54.94	7.92	13,762	14.25
Bituminous, Armstrong Co.	0.96	38.20	52.03	8.81	13,809	14.30
Bituminous, Fayette Co., Connellsville	1.93	28.71	63.26	6.10	13,881	14.37
Bituminous, Greene Co., Main Bench	1.04	37.23	56.61	5.12	14,314	14.82
Bituminous, Indiana Co., Lower Bench	1.46	32.00	53.79	12.75	13,048	13.51
Bituminous, Jefferson Co., Freeport, Average	1.23	32.66	58.18	7.93	13,792	14.28
Bituminous, Westmoreland Co., Loyal Hanna, Average	0.73	24.11	69.11	6.05	13,975	14.48
Bituminous, Westmoreland Co., Youghiogheny	1.00	35.00	58.40	5.60	14,208	14.71
TENNESSEE.						
Campbell Co., Bit., Newcomb	2.00	33.77	60.64	3.59	14,331	14.84
Franklin Co., Bit.	1.77	25.41	62.00	10.82	13,157	13.62
Hamilton Co., Bit., Melville	2.74	26.50	67.08	3.68	14,073	14.57
Marion Co., Bit.	3.16	31.94	54.81	10.09	13,185	13.65
TEXAS.						
Maverick Co., Bit., Eagle Pass	3.67	35.51	41.70	19.12	11,971	12.39
Palo Pinto Co., Bit., Strawn	6.67	40.20	43.54	9.59	12,906	13.36
Rusk Co., Bit., Stevens	10.40	35.94	49.46	4.20	13,066	13.53
Tarrant Co., Bit., Fort Worth	4.60	34.72	49.27	11.41	12,838	13.29
UTAH.						
Emery Co., Bit., Castle Dale	3.42	42.81	47.81	5.95	13,953	14.44
Iron Co., Bit., Cedar City	3.50	43.66	43.11	9.73	13,411	13.88
Summit Co., Bit., Coalville	0.43	38.90	56.37	4.30	14,554	15.07
VIRGINIA.						
Halifax Co., Bit., Elmo	1.05	23.62	72.67	2.66	14,411	14.92
Rockingham Co., Bit., Clover Hill	1.34	30.98	56.83	10.85	13,321	13.79
Scott Co., Bit., Clinch Valley	0.91	34.33	59.89	4.87	14,314	14.82
Wise Co., Bit., Big Stone Gap	1.80	33.90	59.25	5.05	14,151	14.65
WEST VIRGINIA.						
Logan Co., Bit., Dingess	1.22	41.50	54.58	2.70	14,720	15.24
Mineral Co., Bit., Elk Garden	0.76	19.39	72.99	6.86	13,764	14.25
Pocahontas Co., Semi-Bit.	0.50	19.83	75.63	4.04	14,218	14.72
WASHINGTON.						
Kittitas Co., Bit., Ellensburgh	2.00	39.10	54.40	4.50	14,300	14.80
Pierce Co., Bit., Wilkeson	1.10	35.10	54.50	9.30	13,659	14.14
Stevens Co., Bit., Calispell	2.39	41.18	42.92	3.51	10,977	11.36
Whatcom Co., Bit., Bellingham Bay	3.98	29.54	59.90	6.58	13,531	14.01
WYOMING.						
Carbon Co., Bit., Dana	11.30	42.01	39.69	7.00	12,645	13.09
Sheridan Co., Bit., Sheridan	6.04	42.37	35.57	16.02	12,107	12.53
Weston Co., Bit., Cambria	4.20	40.60	41.50	13.70	12,676	13.12

CHAPTER II.

MATERIALS OF CONSTRUCTION.

PART I.—CAST IRON.

THE principal materials entering into the construction of steam boilers are limited by commercial and practical considerations to cast iron, wrought iron, and steel. Copper was formerly used in boiler construction, especially for fire-boxes in locomotive boilers and the internal heating surfaces in marine boilers. Its use is practically abandoned at this time because of its want of hardness and tensile strength as compared with either wrought iron or steel, while its cost is much greater.

Cast Iron.—This product is had by a remelting together of two or more kinds of pig-iron in order to secure castings having certain qualities determined approximately in advance. This presupposes a knowledge of the constituents of the pig-iron to be used, which is, taken altogether, a very complex material, because pig-iron is always combined with extraneous substances, which, taken collectively, are called impurities; and of these the principal ones are carbon, silicon, sulphur, phosphorus, and manganese. These may be combined with the iron both chemically and mechanically, and this is especially true of carbon. There is no good reason for regarding carbon and silicon as impurities as we ordinarily use that word, for the presence of both are beneficial, if not altogether necessary, in the manufacture of iron castings. The amounts of sulphur, phosphorus, and manganese ordinarily present are small in amount and do no particular harm.

Carbon in Cast Iron.—Carbon, by reason of its chemical and physical effects, is the most important element in cast iron. It is always present in pig-iron, and repeated meltings does not sensibly diminish its quantity, which varies from at least 1.5 per cent. up to 4.5 per cent. As carbon may be present in a combined form or may be present as graphite, it is best when referring to carbon in iron to have it understood that such reference means total carbon, the latter determining also the melting-point of the iron.

Combined Carbon.—Cast iron in a fluid state will take up at least 3.5 per cent. of carbon and hold it in solution, part of which is expelled during the process of cooling. For small and medium castings there should be very little combined carbon, because it makes the iron hard and brittle and influences adversely the amount of shrinkage.

Graphitic Carbon.—The total carbon in cast iron must be either combined or in the graphitic form: the latter is characteristic of gray irons. Carbon in fluid iron is combined with the iron and has no tendency to separate from it; but during the process of cooling marked changes occur, depending somewhat on whether the cooling be rapid or slow; if the latter, the combined carbon will separate from the iron, but remain in it mechanically in the form of graphite between the crystals. This precipitation of carbon renders iron soft, changing the color of white iron into a darker hue,—hence the name of gray iron.

Silicon in Cast Iron.—Next after carbon the substance most commonly met with in pig-iron is silicon, the quantity ranging from 0.5 to more than 4 per cent. The presence of silicon in cast iron is important, because of its effect upon the contained carbon, especially in the conversion of the combined carbon into graphite, by which white iron is changed to gray, the hardness and brittleness of the former being thus greatly modified, showing marked increase in both the transverse and tensile strength of iron castings. Silicon tends also towards the elimination of blow-holes, and thus contributes materially towards the production of sounder castings than can be had in its absence.

Sulphur in Cast Iron.—Sulphur is almost always present in pig-iron. During the process of remelting an additional quantity is absorbed by the fluid metal from the coke fuel in the cupola. The unexplainable peculiarities of cast iron have long been attributed to sulphur; but recent investigations, combining physical and chemical tests, made for the purpose of ascertaining what the precise action of sulphur is upon cast iron, show that sulphur in the quantities usually found in gray iron does not injuriously affect cast iron; and, further, that if an excess of sulphur should be found to exert a pernicious influence upon the iron, a slight increase in the quantity of silicon pig used would counteract any such effect.

Castings composed of Southern gray irons containing 0.088 to 0.100 per cent. of sulphur show them to be of good quality, with no chill, no blow-holes, very low shrinkage, and high strength; but this latter quality is not to be attributed to any action of the sulphur present, but simply shows that it does not within the above permissible limits of percentage detract from the strength of cast iron.

Phosphorus in Cast Iron.—Phosphorus enters into chemical union with iron, the effect of which is to render iron close and compact, with a tendency to become cold-short at low temperatures. Phosphorus is not known to influence the change of carbon in pig-iron one way or the other. In producing hardness in casting, such change is to be ascribed to the influence of phosphorus alone. The permissible allowance of phosphorus in cast iron is confined to narrow limits; its presence may be beneficial in some mixtures of pig-iron, but from 0.5 to 1 per cent. is the beneficial limit. In moderate quantities phosphorus lessens the tendency to form blow-holes in castings; it also lessens shrinkage and prolongs the period of fluidity.

Manganese.—The physical properties of cast iron are not greatly altered by the addition of manganese if the latter does not much exceed 1 per cent., but as much as 1.5 per cent. makes it very hard. The presence of say 1 per cent. of manganese is beneficial in foundry practice, increasing the fluidity of the iron when melted; but when the proportion of manganese is much greater than that, it renders cast iron less plastic, more hard and brittle when cold, and increases the shrinkage. The effect of manganese when used alone is, as stated above, to harden cast iron, but it does not turn gray iron white, nor does it increase the combined carbon, nor does it increase the tendency to chill; its hardening effect is to be ascribed to its one influence, which is to harden iron.

Ferro-Manganese, when added in small quantities to molten metal in a foundry ladle, softens and improves the iron. The probable explanation is, that the manganese counteracts the effect of sulphur and silicon, tending to eliminate the former and neutralize the latter, and so, when common iron with a tendency to hardness is employed, it sometimes happens that ferro-manganese may be used as a softener. The hardness, however, generally returns when the iron is remelted, because the manganese is oxidized and more sulphur absorbed. The good effects of manganese appear to be twofold: by its action it leads directly to a measure of hardness and closeness of grain which is beneficial, while indirectly it is useful in preventing the absorption of sulphur during remelting.

Slow Cooling.—Gray iron machinery castings should be slowly cooled, and particularly when they are small and thin. Rapid cooling tends to brittleness by preventing the separation of the combined carbon into graphite; slow cooling tends to make the grain coarser, and such castings can have any necessary machine-work done much more rapidly and better than if the metal was hard.

Cooling Strains.—Fractures in cooling are likely to occur where two portions of a casting join each other at right angles and with square corners. Thick castings joining each other at right angles or nearly so are almost certain to have cavities occur at their points of intersection by reason of an irregular grouping of crystals; the direction of the shrinkage not being parallel will also produce distortion. The importance of slow cooling after pouring a casting of considerable size, and especially when of varying thickness, is known to every foundryman: gray irons, having a natural tendency to hardness, are made harder by rapid cooling; on the other hand, slow cooling tends to soften such castings.

Blow-Holes.—These are serious defects in a casting, because they are generally below the surface; there is seldom any outward indication as to their location or to what extent they exist. A blow-hole not only lessens the area of cross-section of the casting in which it occurs, but its presence in the casting may be additionally harmful in setting up internal strains within the casting which might not otherwise occur.

The surface of a casting should be smooth, free from bits of slag, scabs, and unusual roughness; small, sharp indentations in the surface of a casting, however caused, may, as far as they go, be considered incipient fractures, and if undue stress be applied to such a casting, a fracture may begin at any such point at a stress much lower than would otherwise break the casting.

Shrinkage of Cast Iron.—Shrinkage represents the difference in size between a mould and the casting made in it. When molten iron is poured into a mould it expands at the moment of solidification, and if properly vented the metal will take a sharp impression of the mould. The cooling of a casting begins at and along its outer surface; so, also, crystallization begins at the surface and proceeds towards the centre. It sometimes happens in the case of thick castings that the interior portion may be in a semi-molten state, while the whole exterior surface has been solidified to a considerable depth. As cooling progresses and the casting parts with its heat, it diminishes in bulk; but this contraction is not uniform throughout; thin portions cool first and take their permanent form, the thicker portions of the casting cooling later. It is to be expected, and it often actually occurs, that the cooling of the thick portions will show lines of incipient, if not actual, fracture along the lines of intersection where a thicker portion joins a thinner one. To prevent this, it is customary to place quarter-round concave fillets at all such intersections. Shrinkage is at best an uncertain thing to deal with, depending not only on the size and shape of the pattern, but upon the temperature at which the iron is poured, the quality of the iron, and especially upon the quality of hardness; some observers state that the amount of shrinkage corresponds closely to that of the total quantity of carbon present.

The ordinary shrinkage of gray iron castings is $\frac{1}{8}$ inch per foot; this is the graduation on standard shrinkage rules. Some irons shrink in the proportion of $\frac{1}{16}$ inch per foot, others still less; the standard rule is, however, sufficiently accurate for all ordinary purposes.

Strength of Castings.—Combined carbon was long thought to be the medium by which strength was imparted to castings, and that its conversion to graphite was an occasion of weakness; but recent chemical and physical tests show that the reverse is true, and that, if anything, combined carbon weakens castings and never strengthens them. This is especially true of small castings, where it has been observed that as the percentage of combined carbon was decreased by the addition of silicon, that the hardness and brittleness of the casting was also decreased, and that the strength of the casting had been increased at the same time.

The strength of cast iron increases when uninfluenced by any other element than carbon by an increase of silicon up to as much as 3.5 per cent. of the total weight, and this is as high as will be found in any ordinary silicon pig; with this increase of strength there is also an absence of brittleness and an increase in the size of the grain. This latter must not be carried too far, however, or weakness may result from this cause alone.

Crushing Strength.—That of cast iron varies with the chemical and physical properties of any given sample, the range being as great as from 50,000 to 75,000 pounds per square inch, averaging so high as to require no special calculation,—that is to say, when patterns are properly dimensioned for tensile strains, any probable amount of compression can be safely carried by cast iron, as the carrying capacity of the latter is four or five times as great as the former. From experimental data it appears that the maximum crushing strength of cast iron would be obtained with about 0.75 per cent. of silicon and 2 per cent. of combined carbon. In ordinary calculations, 60,000 pounds per square inch may be assumed as the ultimate crushing strength of medium hard gray iron machinery castings in short lengths.

Transverse Strength of Cast Iron.—The usual method of testing cast iron is to break test-bars one inch square by one foot in length; or, if other sizes are used, the results are reduced for comparison to the one inch square section. Transverse tests when applied to the centre of a test-bar supported at each end combine both a crushing and a tensile test, the transverse test being an intermediate of the two.

The breaking strength of gray iron castings of good quality, the bars one inch square by twelve inches long, loaded at the centre, will vary from 1600 to 3600 pounds.

Tensile Strength of Cast Iron.—Tests for tensile strength are not nearly so common as are the transverse tests, and for the reason that cast iron is seldom employed where its tensile strength alone is brought into use. Tensile tests do not as a rule exhibit physical qualities which transverse tests do not bring out equally well, and such tests vary, of course, with the physical and chemical properties of the iron, and may, for good gray iron castings, vary anywhere from 14,000 to 22,000 pounds per square inch of section when the test-bars approximate one square inch of fractured area. As the quality of the casting can only be approximately arrived at by an inspection of the grain at the points where the gates are knocked off, it seems unwise to accord to cast iron as high an average tensile strength as would be assigned wrought iron or mild steel covering the same percentage of variations in tensile strength. The writer does not, therefore, recommend more than 16,000 pounds per square inch of section for the best quality gray iron castings when not less than ½ inch nor more than 1½ inches thick.

Elastic Limit of Cast Iron.—This quality in cast iron is seldom taken into account for ordinary machine-work, consequently but few tests have been made concerning it. The comparatively few tests do show, however, that the elastic limit of cast iron is about ⅓ of its tensile strength. If we regard cast iron as good for 16,000 pounds per square inch tensile strength, the elastic limit would then be: 16,000 ÷ 3 = 5333 pounds per square inch.

Cast Iron as a Material for Steam Boilers.—The advantages of cast iron as a material for steam boilers in preference to steel or wrought iron are thus set forth by the principal manufacturer of such boilers in this country.

1. The spherical or globular form of parts can be manufactured at reasonable prices, which would be impossible with any wrought metal.

2. Any number of such parts can be made uniform and duplicate; likewise impossible with wrought metal, unless at enormous expense.

3. Cast metal resists much better the corrosive action of acids in the water or in the products of combustion.

4. Cast metal is much less affected by oxidation.

5. Cast metal cannot blister, furrow, or pocket.

6. Cast metal also transmits heat more readily than wrought iron, and while it may have to be a trifle thicker than ordinary boiler tubes, it is in the case of this boiler ($\frac{5}{16}$ to $\frac{3}{8}$ inch) much thinner than ordinary crown sheets.*

Objections to Cast Iron.—This material has been objected to as a material for steam boilers for the following reasons:

1. Because it is a crude product to begin with, it is brittle and of low tensile strength, and there is no certainty that castings can be made uniform in strength or in other qualities.

2. The cooling strains in castings often produce flaws or other defects which are hidden to the eye and thus escape detection in the workshop; further, that such defects do not become apparent even when under test by hydraulic pressure, but which may when under the influence of heat and unequal expansion and without a moment's warning end in sudden and disastrous failure.

3. Its unyielding nature is thought to especially unfit it for parts of a boiler subject to unequal expansion arising from differences of temperature.

4. Cast iron when cold seldom or never exhibits indications of weakness or fracture in advance of actual breakage. It has been further objected to because it loses coherence and will crumble under moderate loads, and not infrequently by its own weight, when subjected to high temperatures or those approximating red heat.

Cast Iron in the Fire.—The effects of ordinary boiler-furnace heat on unprotected cast iron is to change the characteristic granular structure common to all gray iron castings of good quality to coarse, uneven grains having scarcely any metallic lustre and little or no granular

* This was long believed to be true, but Isherwood's experiments show the relative conductivity of cast iron to be less than that of wrought iron.—Ed.

coherence. Burnt cast iron shows a change in the color of fracture from gray to brown and gray mixed, the metal is usually twisted out of shape, and is so extremely brittle and lifeless that it is utterly unfit for further use in the foundry in the production of castings requiring strength. The continued heating and reheating of any metal would in time destroy it, but cast iron, by reason of its relatively coarser grain, seems to be less able to withstand the effects of reheating and cooling than either wrought iron or mild steel; but no metal of which iron is the basis should be subjected to the continued action of intense heat,—that is to say, temperatures approximating red heat. Cast iron yields to the fire sooner than wrought iron; it loses strength at temperatures much below those which similarly affect wrought iron. When red hot, cast iron will scarcely sustain its own weight, and when so heated is liable to crumble to pieces if under compression.

Use of Cast Iron.—The fact in regard to the use of cast iron as a material for steam boilers is that for steam- and hot-water-heating purposes it is more largely employed than any other material. It is not more injuriously affected by heat than are other materials when the parts are properly proportioned and no thick flanges are exposed to the heat of the furnace. Boilers for heating purposes work under moderate pressures, seldom more than 20 pounds per square inch for water and still less for steam. The Harrison boiler has been continuously and successfully in the market for more than thirty years, and at this time is in good repute as a safety boiler, as a steam-maker, and has through all these years suffered no deterioration which is not common to all boilers. This is the only high-pressure cast-iron boiler known to the writer which has been continuously on the market for ten years.

PART II.—WROUGHT IRON.

Wrought Iron.—This product is commercially pure iron prepared from selected pig-iron by a succession of processes, such as puddling, squeezing, hammering, rolling, etc., to rid the iron of its original impurities. The operation of puddling consists in stirring a mass of spongy iron in the midst of a bath of cinder, which prevents the intimate approximation of its particles. This cinder, adhering to the iron, opposes a thorough welding of the mass and favors the production of fibrous texture, since during subsequent working the molecules of the iron slide over each other, giving the iron a characteristic fibrous appearance in fracture. If a bar of wrought iron is nicked on one side and then bent over double upon itself, it will expose the longitudinal grain of the metal; and if the iron is of good quality, the fracture thus obtained will be a dense mass of fibre, slightly interspersed with fine grains of iron.

Physical Properties of Wrought Iron.—Boiler-plates should possess the properties of tenacity, ductility, and welding. Each of these is influenced in some measure by the impurities of the iron. That quality of boiler-plate is judged to be the best which has the greatest tensile strength combined with ductility and freedom from brittleness; such an iron will be at once strong, tough, and fibrous. When the above properties are well combined, wrought iron will resist strains due to unequal expansion remarkably well.

Tensile Strength.—Boiler-plates may possess high tensile strength at the expense of other qualities, such as homogeneousness and toughness. Wrought-iron plates possessing all the necessary qualities suitable for steam boilers will not have a tensile strength much, if anything, above 55,000 pounds per square inch;* a higher tensile strength is liable to have associated with it the undesirable qualities of hardness and brittleness, either of which will more than offset any gain in tensile strength above the figures given. Wrought iron having a tensile strength of less than 45,000 pounds per square inch should not be employed in boiler construction.

Tensile tests made in the direction of the grain of fibrous iron show greater strength than those made across the grain. The following data was obtained from United States government tests of wrought-iron plates, which it will be observed are of very high quality. These were short specimens:

Thickness.	Tensile Strength.	Reduction of Area.
$\frac{1}{4}$ inch with the grain	58,373 pounds.	38 per cent.
$\frac{1}{4}$ inch across the grain	53,333 pounds.	9 per cent.
$\frac{5}{16}$ inch with the grain	62,195 pounds.	43 per cent.
$\frac{5}{16}$ inch across the grain	60,202 pounds.	10 per cent.
$\frac{3}{8}$ inch with the grain	56,270 pounds.	25 per cent.
$\frac{3}{8}$ inch across the grain	56,461 pounds.	17 per cent.

Test-Pieces.—Wrought-iron test-pieces have usually been short specimens, the tests being made with reference to tensile strength only. Fig. 1 represents the size and form prescribed by the United States

FIG. 1.

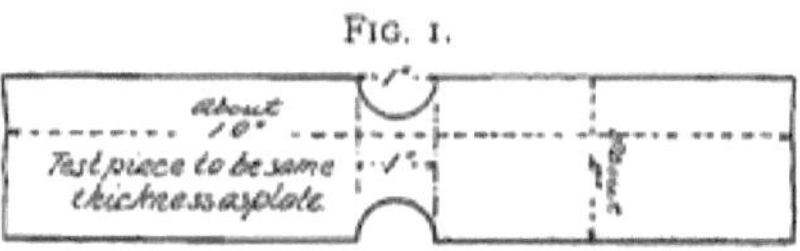

Board of Supervising Inspectors of Steam-Vessels,—viz., 10 inches long, 2 inches wide, cut out in the centre as indicated.

All sample pieces of iron plate $\frac{5}{16}$ inch thick and under shall be one inch wide at reduced section. Plate over $\frac{5}{16}$ inch thick shall be reduced

* It will be understood that wrought iron made by the usual processes is meant, and not homogeneous iron or ingot iron, which have physical properties differing from ordinary wrought iron.

in width at centre to an aggregate area approximating 0.4 of one square inch ; but such reduced area shall in no case exceed 0.45 nor be less than 0.35 of an inch ; and the force at which the piece can be parted in the direction of the fibre or grain (when of iron), represented in pounds avoirdupois in proportion to the ratio of its area, shall be deemed the tensile strength per square inch of the plate from which the sample was taken. When more than one sample shall be tested from one sheet, the sample showing the lowest tensile strength shall be allowed as the tensile strength of the plate.

Ductility is the property which enables a material to be drawn out without breaking. It is also called elongation or extension in reports on the mechanical tests to which plates or bars are subjected. Elongation occurs when a ductile material is subjected to a tensile stress higher than its elastic limit, after which a permanent change of form takes place. It may be measured in a tensile testing-machine in two ways,—by the actual amount of elongation in inches and parts of an inch, and by reducing the amount so found to percentage extension of its original length.

Wrought-iron boiler-plates under 45,000 pounds tensile strength should show a reduction of area of not less than 12 per cent. ; 45,000 to 50,000 pounds, 15 per cent. ; 50,000 to 55,000, 25 per cent. ; 55,000 pounds and over should show 35 per cent. reduction of area.

Elastic Limit.—This limit may be determined in the same manner and at the same time as when making a tensile test by simply applying progressive loads and then removing them. After the removal of each increment of load the specimen is measured to determine whether or not it has returned to its original length. The weight required to give the specimen its permanent set divided by the area of the specimen will approximate its elastic limit. From a large range of experiments made upon plates and bars it appears that the elastic limit of wrought iron is approximately one-half its tensile strength.

Welding.—Wrought iron possesses the property of welding when the two parts to be joined are brought up to a white heat. Welded joints are, when well made, scarcely inferior to the original bar ; but stays, braces, etc., for boilers should be made from whole stock if possible, because there is always more or less uncertainty about welded joints, particularly when the parts to be joined are of considerable diameter or thickness.

Fibre.—The best wrought iron has a characteristic fibre resulting from its method of manufacture, the fineness and uniformity of which is taken as a practical indication of the quality of the iron.

A fibrous fracture in wrought iron is always taken to indicate a high grade of iron, and this is sometimes associated with the act of pulling in a tensile testing-machine. But stresses of this kind do not cause an extension of the grain to produce fibre,—this is already present in the iron ; the act of pulling simply draws these fibres out in clusters ; and as they do not all break in the same time nor in the same plane, a fibrous fracture is thus secured.

The manner in which test-pieces are broken is of importance when fractures are broadly classed as fibrous or crystalline. To illustrate : two samples may be cut from the same sheet, one of which can be so broken as to present a fibrous fracture, if the fibre be originally in the plate, by simply applying the load gradually,—such a test will give maximum elongation and reduction of area at the point of fracture ; on the other hand, by applying the load suddenly, or so rapidly that the flow of the iron will not follow the effect of the stress, the piece is liable to break with a snap, the fracture in this case having little or no fibre and presenting a more or less crystalline appearance, indicating a lower grade of iron than was actually under test. It will be seen that the element of time is important when making tensile tests, because it gives the metal a chance to adjust itself to the conditions imposed by the stress beyond the elastic limit.

Bending Test.—This is one of the severest tests to which wrought-iron boiler-plates can be subjected. It is made by shearing off a strip of any convenient width from the plate, say one or two inches, and bending it down cold upon itself without fracture on the outer curve, as shown in Fig. 9. Very few irons except the better grades of flange iron will stand a test of this kind ; but the latter grade of iron $\frac{3}{8}$ inch thick and less should bend double without fracture ; plates $\frac{7}{16}$ to $\frac{5}{8}$ inch thick of a similar grade should bend down cold upon a plate of its own thickness.

Hot bending tests are in no respect different from cold bending tests, except that the iron is heated to at least a cherry-red heat before the operation of bending is begun. Any iron which when heated to redness will not bend over double upon itself, either with or against the grain, and without fracture on the outer curve or along the edges of the plate, is not fit to enter into steam-boiler construction.

Hammer Test.—This test is sometimes resorted to for ascertaining the internal defects of a plate when caused by lamination or imperfect welding during manufacture, and consists in lightly tapping the entire surface of the plate with a light hand-hammer. The plate is usually placed on edge, its weight resting on any convenient bearing at each end to keep it off the floor ; the upper edge is supported in any manner that will not interfere with the vibration of the plate. If a plate thus arranged be struck with a hand-hammer, it will give a clear, ringing sound, which will be the tone of that particular plate ; if now the hammering be proceeded with on say 4-inch centres over the entire surface and on both sides of the plate, and this same characteristic tone due to the size and weight of the plate be observed, the plate is judged to be solid ; but if a dull sound is emitted, it is reasonably certain that a defect exists in the plate at that point.

Chemical Properties of Wrought Iron.—The operation of puddling does not eliminate all the impurities in cast iron during its conversion into wrought iron, but to show how nearly such elimination occurs in practice, reference is directed to the following analysis of a sample of 55,000 tensile strength wrought-iron plate:

Iron	99.20
Carbon	.04
Manganese	.17
Silicon	.15
Sulphur	.03
Phosphorus	.21
Oxygen	.20
	100.00

Of the above substances, those which act most injuriously upon iron are sulphur and phosphorus, the former making the iron red-short, and the latter cold-short.

Red-short or hot-short iron is a defect generally attributed to the sulphur present in the finished plate or bar. This sulphur may have been present in the pig-metal from which the wrought iron was made, but much of it comes from the coal used as fuel in puddling. Red-short irons are often tenacious and otherwise good when cold, but become brittle and are easily broken when hot; such irons weld with great difficulty.

Cold-short iron is very brittle when cold, cracking badly, or breaking if bent at a sharp angle or doubled; but such irons may be forged and welded at a high heat. This defect is a much more serious one than red-shortness. Irons which have an excess of phosphorus are cold-short.

Commercial Qualities of Boiler-Plate.—The quality of boiler-plate is dependent upon the impurities in the crude metal from which it is made, and upon the amount of working which it receives to prepare it for the market. Wrought iron is improved in quality by judicious heating and working, especially hammering, but it soon reaches its greatest strength, after which reheating and rolling is found to reduce its strength and ductility. Unless portions of plate have been actually tested, or the plates are known to have been made from blooms of the very best quality, it is not safe to assume a greater tensile strength than 45,000 pounds per square inch of section. This applies to such irons only as are stamped by reputable makers C. H. No. 1 and higher grades. These latter are usually designated by some private brand or trade-mark.

C. Iron, or charcoal iron, is the common boiled or puddled iron rolled into bars or plates. This grade of iron is porous, and will become very brittle with repeated heating and cooling. It will not stretch much before breaking, and will break suddenly. Its tensile strength ranges usually from 30,000 to 40,000 pounds per square inch. It is only suited for tank-work, and ought never to enter into any portion of boiler construction.

C. No. 1 Iron, or C. H. iron (charcoal hammered, as it is oftener known), is the same iron as the above, except that it is subjected to more careful working and is hammered into suitable blooms before rolling. This iron very much resembles the common iron in its general qualities, having but little elasticity and breaking suddenly. Like the above, it becomes very brittle by repeated heating and cooling, though somewhat stronger than C. iron, its tensile strength ranging from 35,000 to 45,000 pounds per square inch. It is not a suitable iron for boiler construction.

C. H. No. 1 Shell Iron is made from C. H. blooms, with the addition of selected scrap, the whole being thoroughly welded under a heavy steam-hammer and afterwards rolled into plates. This iron, like the other two just described, is injuriously affected by repeated heating and cooling, which has the effect to render it brittle. This is the quality of plate generally used in the construction of land boilers using pressures of steam below 100 pounds per square inch. It rarely enters into the construction of boilers for river or ocean service, its principal defect being a lack of homogeneity and imperfect welding. Its tensile strength is from 45,000 to 50,000 pounds per square inch.

C. H. No. 1 Flange Iron is similar to the above, the difference being that only the very best scrap iron and charcoal hammered blooms are used. The greatest care is exercised in the selection of materials, and the working is such as to insure thorough welding. In texture it is less fibrous and more granular than any of the irons preceding it. On account of its nearer approach to a homogeneous structure it is less liable to blister or crack in the fire. It will stand repeated heating and cooling and should have good flanging qualities. The tensile strength should never fall below 50,000 pounds per square inch and rarely exceeds 60,000 pounds. The elastic limit will vary from 18,000 to 25,000 pounds per square inch, and will stretch from 25 to 30 per cent. in ordinary 2-inch specimens. This is the highest grade of iron regularly offered in the market, and was extensively used in the construction of marine boilers and for the heads and other flange-plates of land boilers.

Brand.—The brand of plate iron is a commercial index to its quality. Plates of a grade corresponding to C. H. No. 1 iron should always be stamped with the maker's name and the guaranteed tensile strength, thus :

SMITH, JONES & CO.,
C. H. No. 1 Flange,
55,000 T. S.

This method of stamping was introduced in order to meet the requirements of the United States Government regulations with reference to the quality of plates entering into steam boilers intended for use on steam vessels in the United States.

Defects in Iron Boiler-Plates.—These are principally imperfect welding, brittleness, and low ductility, all of which may be largely overcome by a proper selection of materials in the first stage of manufacture and by a careful manipulation during the successive operations of reheating, welding, and especially by a thorough working under a heavy steam-hammer. As boiler-plates are always cut to dimensions at the mill, one defect of plate iron is not known to the purchaser,—that is, rough edges, which is a sign of red-shortness, the plate containing an excess of sulphur, which may have been absorbed from the coal during the process of puddling. Such plates do not weld readily.

Surface Defects in Plates caused by imperfections in the rolls should be carefully examined to ascertain whether the quality of the plate is likely to be affected thereby; so, also, cracks in the scale should be carefully examined, though usually they do not amount to much, rarely entering into the iron. Scabs, occasioned by the imbedding of hard pieces of cinder while the metal is at a high heat, may extend some distance below the surface and prove a serious defect.

Of interior defects lamination is perhaps the most serious. This defect occurs when welding is imperfect, which may have been occasioned by the presence of cinder, sand, or other impurity in the pile, preventing welding contact. This is the defect which later on, when the plate is exposed to the fire, produces a blister. Lamination is difficult of detection, because both sides of a sheet may be faultless and give no surface indication of its occurrence. The object of the hammer test is to lead to the detection and location of such defects.

Blisters.—This defect seldom manifests itself until the plate has been put into the boiler and subjected to the repeated action of heating and cooling: it is directly traceable to lamination, already referred to. Blisters are much more common in wrought-iron plates than in those of mild steel. The remedy for a blister is either a patch or a new plate.

PART III.—MILD STEEL.

Steel occupies a position intermediate between cast iron, which contains so much carbon that it is very brittle, and wrought iron, which contains so little carbon that its influence is imperceptible. The difference between mild steel and wrought iron does not wholly depend upon the larger quantity of carbon contained in the former over the latter, but rather that in the preparation of steel it has acquired that property which permits casting into a malleable ingot.

Steel is characterized by fine granular texture, and when the contained carbon amounts to 0.40 to 0.50 per cent. it has the property of hardening, which unfits it for use in steam-boiler construction: steel for that purpose should not contain more than from 0.12 to 0.24 per cent. of carbon. Mild steel is recommended as a material for steam boilers because of its homogeneity, high tensile strength, malleability, ductility, and freedom from lamination and blisters.

Open-Hearth Steel.—This process of steel-making was introduced in this country in 1868, and by 1871 its manufacture had attained a high degree of perfection. The physical qualities of open-hearth steel were so admirably adapted for boiler-plates that it received the prompt recognition to which its merits fully entitled it, rapidly displacing both crucible and Bessemer boiler-plates in open market.

Carbon.—Mild steel plates for boilers vary as to the total quantity of carbon from 0.12 to 0.24 per cent., depending upon the thickness of the plate. In preparing specifications for mild steel plates less than ½ inch thick the carbon limit should not exceed 0.15 per cent., as additional carbon is likely to reduce the ductility of the plate. The ultimate tensile strength of open-hearth flange or boiler steel should not vary outside the limits of 52,000 to 62,000 pounds per square inch, with an elongation of not less than 25 per cent. A steel having 0.15 per cent. of carbon is quite as high as should be used for boiler-plates from ¼ to ½ inch thick; but for plates say 1¼ inches thick the carbon limit may be as high as 0.25 per cent., with intermediate percentages of carbon for intermediate thicknesses of plate.

TABLE II.

SHOWING THE CARBON PROPERTIES OF OPEN-HEARTH STEEL.

Carbon.	Manganese.	Phosphorus.	Average Original Sectional Area.	Average Ultimate Tensile Strength per Square Inch.	Average Final Elongation.	Average Final Area.
Per cent.	Per cent.	Per cent.	Square Inch.	Pounds.	Per cent.	Per cent.
.12	.351	.0505	.6180	58,226	27.18	45.29
.13	.340	.0491	.5491	58,352	26.72	46.73
.14	.375	.0486	.5942	60,569	26.90	46.18
.15	.383	.0518	.5719	61,618	26.75	48.16
.16	.393	.0528	.5641	62,517	25.65	49.46
.17	.404	.0461	.5492	63,333	25.89	50.45
.18	.416	.0487	.5719	65,169	24.81	48.71

Three percentages above and below 0.15 per cent. are given to show how the physical properties of the steel change with the varying carbon.

The above steel developed through a series of more than 100 tests an average increase of tensile strength per 0.01 per cent. of carbon of 1387.5 pounds.

Average decrease of final elongation per 0.01 per cent. of carbon, 0.425 per cent.

Average increase of final area per 0.01 per cent. of carbon, 0.600 per cent.

Manganese.—In the above series of tests it was found that with increasing carbon there was also an increase of manganese, which may probably be accounted for in diminished oxidation in the furnace. The exact function of manganese in steel is not clearly understood; the belief is, however, that it deoxidizes the bath as well as removes the sulphur. This is inferred from the disappearance of most of the sulphur from the iron in the bath and partly from the circumstance that only about one-half the metallic manganese added at the last of the charge is found in the analysis of the resultant steel.

Phosphorus.—This element in steel has the property of rendering it cold-short, and as boiler-plates are usually worked cold, the less there is of it in the plates the better. As a hardener of steel, phosphorus is generally considered to be more effective than carbon, but its secondary effects are very different. Other things remaining the same, increase of phosphorus, besides raising the tensile strength, notably raises the elastic ratio, diminishes the elongation, and more especially diminishes the reduction of area. Its effect in diminishing elongation, and probably also reduction of area, appears to be largely dependent on the amount of other elements present, especially of silicon. For convenience, the latter element must be supposed to vary little and be present in quantity not above 0.04 per cent., as is common in open-hearth steel. The effect of phosphorus is then identical in nature to that of cold-rolling or finishing, and is to be taken account of in much the same manner.

Sulphur.—The presence of this element renders steel hot-short, and thus affects the working in the steel-works rather than in the boiler-shop, except in flange-plates. Sulphur should not exceed 0.04 per cent. in steel boiler-plates, and even at this percentage the plates should contain at least 0.25 per cent. of manganese in order to counteract the hot-short effects of the sulphur.

Silicon.—This element, even in small quantities, renders mild steel hard and decreases its ductility. It ought not to exceed 0.05 per cent. in any steel intended for steam boilers.

PHYSICAL PROPERTIES OF MILD STEEL.

The qualities to be determined and for which physical tests are undertaken upon samples of steel boiler-plates are mainly for elastic limit, elongation, ultimate strength, reduction of area at point of fracture, cold-bending, the drifting test, and the quenching test. The first four are determined consecutively and are classed usually under tensile tests; the others are separate tests.

Tensile Tests.—The forms and dimensions of test-pieces for boiler-plate in this country do not conform to any standard. The short test-piece so long included in the Rules and Regulations prescribed by the Board of Supervising Inspectors of Steam Vessels, shown in Fig. 1, is still in use. For tensile strength only this test-piece is perhaps good enough, the groove simply indicating where the break shall occur. Such a break will also give the fractured area of the specimen ; but it is not the best form of test-piece for obtaining the ductility of the material, though this may be measured by making prick-punch marks on either side of the centre at any distance which will be unaffected by the stretch of the piece, and comparing the final distance with the original before the test began. This form of test-piece has the advantage of being easily taken from small pieces, and for pieces of small ductility the value of the tensile strength so obtained is practically correct.

The short filleted form, shown in Fig. 2, was quite extensively used by steel-makers in this country prior to the adoption of the standard

FIG. 2.

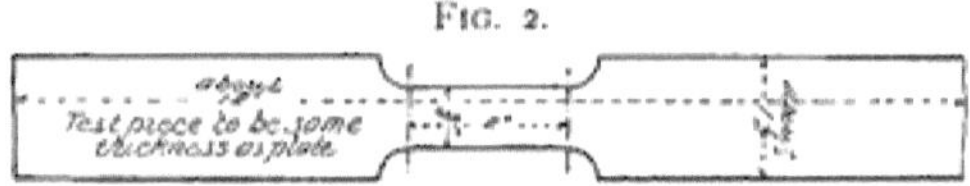

specifications, August 9, 1895, and, unless otherwise stated, may be regarded as the size used. The tensile strength and fractured area for a ductile material like mild steel are much nearer the results obtained from long pieces than in the groove form and more uniform, but are still too high. The ductility is somewhat arbitrary in that the measured length includes the short end fillets, which are in a different condition of stress from the straight portion. Being very largely dependent on the proportions of the test-piece, the ductility results from these specimens always read very high, especially for the softer materials. The

FIG. 3.

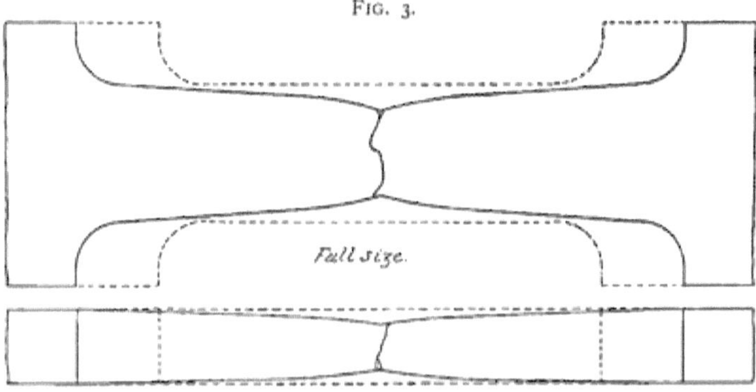

drawing, Fig. 3, is a full-size reproduction of a test-specimen of basic open-hearth boiler-steel by the Lukens Iron and Steel Company. The dotted lines indicate the original size and the full lines the final outlines after the test. Elongation and reduction of area are both clearly exhibited.

The long filleted form, shown in Fig. 4, is one recommended by the American Society of Civil Engineers as being well adapted to show all the

FIG. 4.

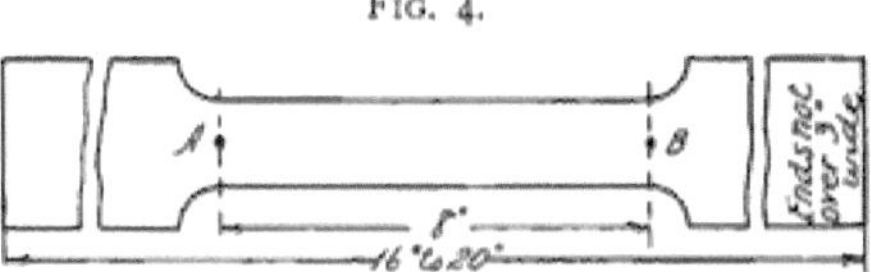

physical properties likely to be developed by a tensile test. The length between the fillets is commonly 8 inches, though there are variations on either side, such as 6 and 10 inches. The 8-inch test-pieces are used in the French navy, by the English Admiralty, and by the United States Navy Department. This uniformity in the dimensions of test-pieces makes the comparison of American and foreign results a simple matter.

TABLE III.

TESTS OF ANNEALED MILD STEEL BOILER-PLATES MADE TO SHOW THE EFFECT OF LENGTH OF TEST-PIECE, UNITED STATES NAVY DEPARTMENT.

FIG. 5.

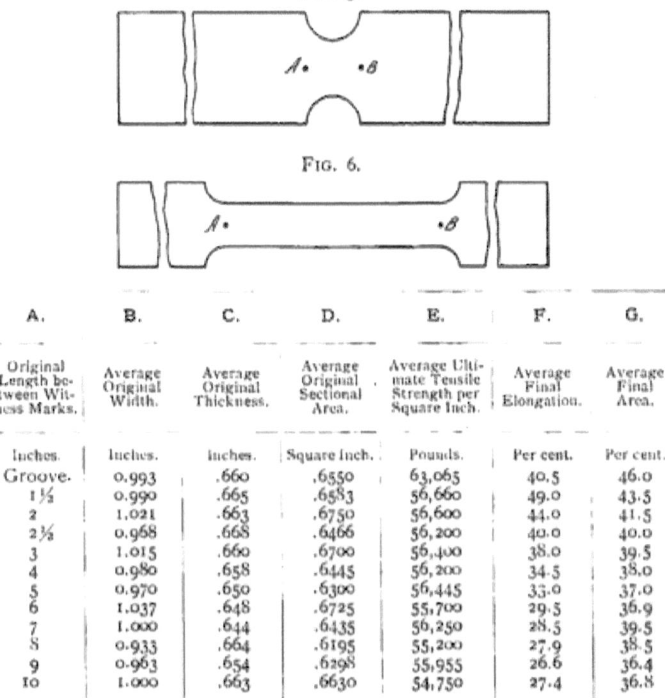

FIG. 6.

A.	B.	C.	D.	E.	F.	G.
Original Length between Witness Marks.	Average Original Width.	Average Original Thickness.	Average Original Sectional Area.	Average Ultimate Tensile Strength per Square Inch.	Average Final Elongation.	Average Final Area.
Inches.	Inches.	Inches.	Square Inch.	Pounds.	Per cent.	Per cent.
Groove.	0.993	.660	.6550	63,065	40.5	46.0
1½	0.990	.665	.6583	56,660	49.0	43.5
2	1.021	.663	.6750	56,600	44.0	41.5
2½	0.968	.668	.6466	56,200	40.0	40.0
3	1.015	.660	.6700	56,400	38.0	39.5
4	0.980	.658	.6445	56,200	34.5	38.0
5	0.970	.650	.6300	56,445	33.0	37.0
6	1.037	.648	.6725	55,700	29.5	36.9
7	1.000	.644	.6435	56,250	28.5	39.5
8	0.933	.664	.6195	55,200	27.9	38.5
9	0.963	.654	.6298	55,955	26.6	36.4
10	1.000	.663	.6630	54,750	27.4	36.8

The standard test-piece adopted by the Association of American Steel Manufacturers, August 9, 1895, as shown in Fig. 7, is for sheared plates of special open-hearth steel for boilers. The tests and inspections are required to be made at the place of manufacture prior to shipment;

FIG. 7.

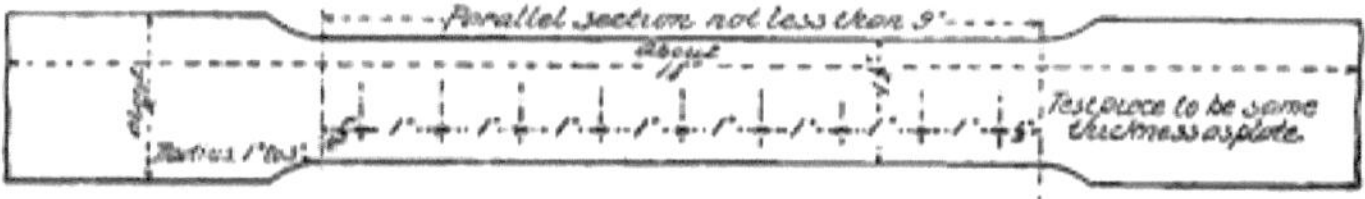

the tensile strength, limit of elasticity, and ductility are to be determined from a standard test-piece of the above dimensions cut from the finished material.

Test-pieces for plates for the new United States war-ships were employed as in the form shown in Fig. 4. These pieces were as nearly as possible in the same condition as finished at the rolls. The length A B was 8 inches, and of uniform cross-section, in which the sectional area was not less than ½ or more than 0.8 of one square inch.

The test-piece was not annealed unless the finished material was to be annealed, and no steel for boilers which had to be worked at a heat or which had to be annealed in the boiler-shop was annealed at the steel-works.

Each test-piece was submitted to a direct tensile stress until it broke. The initial stress applied was 25,000 pounds per square inch, and this first load was kept in continuous action for one minute. An observation was then made of the corresponding elongation measured upon the original length of 8 inches.

The stress was then slowly increased until the principal elastic limit was determined; then additional loads sufficient to produce an increase of stress of 5000 pounds per square inch were added at intervals of half a minute until the stress reached 50,000 pounds per square inch of the original section, after which the increments of stress were reduced to 1000 pounds per square inch. Upon close approach to the probable ultimate strength, the load was increased gradually and its maximum value carefully noted. The final elongation was that obtained after rupture.

The acceptance of any mild steel boiler-plates under the contract was conditioned on an ultimate tensile strength of not less than 57,000 pounds, nor more than 63,000 pounds per square inch of original section, and a final elongation in 8 inches of not less than 25 per cent.; but the acceptance of plates under these tests did not relieve the contractor from the necessity of making good any material which afterwards failed in working.

Direction of Grain.—In testing wrought iron it is quite common to prepare test-pieces from plates in the direction of rolling and at right angles to this direction; these are commonly spoken of as being with or across the grain. Steel, being a more homogeneous material, shows little difference between the two, as indicated in the following memoranda of mild-steel tests having 0.16 per cent. carbon and 0.40 per cent. manganese:

Stress applied lengthwise of the plate,—

Average tensile strength in pounds per square inch	59,400
Average final elongation, per cent.	27.00
Average final area, per cent.	47.00

Stress applied crosswise of the plate,—

Average tensile strength in pounds per square inch	58,995
Average final elongation, per cent.	23.15
Average final area, per cent.	47.00

Elastic Limit.—For mild steels intended for steam boilers, the elastic limit is of much importance, because it is practically its measure of strength; for boiler-plates the elastic limit will vary from 50 to 60 per cent. of its ultimate tensile strength. The standard specifications provide that all special open-hearth plate and rivet steels for boilers shall have an elastic limit not less than one-half the ultimate tensile strength.

Elastic Ratio.—This is obtained by dividing the elastic limit per square inch by the ultimate strength per square inch. For example, if a plate have an ultimate tensile strength per square inch of 57,000 pounds and an elastic limit of 26,500 pounds, the elastic ratio would be $26,500 \times 100 \div 57,000 = 46.49$ per cent. Examples of this kind are interesting as showing any departure from the average as to physical condition other than lack of homogeneity—the condition as to internal strain or comparative annealing, *i.e.*—by the ratio which the elastic limit bears to the tensile strength.

Elongation.—The average value of the extension of mild steel boiler-plates at the tensile limit varies with the quality of the steel, the length of the specimen, and the thickness of the plate: a final elongation of 44 per cent. on a 2-inch specimen would show probably 28 per cent.; on an 8-inch specimen the elongation of a ¼-inch plate would exceed that of a 1-inch plate by at least 5 per cent., both specimens from the same ingot. The standard specifications for elongation call for 28 per cent. in extra soft steel, 26 per cent. for fire-box steel, and 25 per cent. for flange or boiler steel.

Reduction of Area.—There are two methods of recording this quantity: one by giving the actual area of the fracture, irrespective of the original area,—this is the ordinary method; the other records the fractured area in percentage of the original area. In order that no confusion of terms shall arise, the latter is called in the United States navy reports on mild steel the "final area."

To measure the final area, the piece should be fitted, the least width measured, and thickness at each edge and in the centre in the plane of least width. On account of the hollow in the section, as shown in Fig. 8, it is necessary to introduce a formula for the thickness. This is best done by applying Simpson's rule for three ordinates. Thus, if t_1, t_2, t_3 be the thickness at A, C, and B respectively, and b the least width, we shall have,—

Fig. 8.

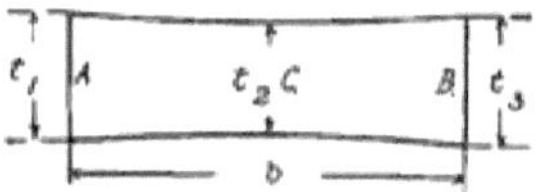

$$\text{Final area} = \frac{b}{6}(t_1 + 4t_2 + t_3).$$

Obtaining data from results given below, the above formula may be worked out arithmetically thus:

$$\begin{aligned} t_1 & & &= .300, \\ t_2 = .272 & & 4t_2 &= 1.088, \\ t_3 & & &= .308 = 1.696. \end{aligned}$$

$1.696 \div 6 = .283 =$ mean thickness.

Fractured width $b = .936 \times .283 = .2649 =$ fractured area, then $100 \times .2649 \div .5229$ original area $= 50.66 =$ final area in per cent. of the original area.

TENSILE TESTS OF MILD STEEL, UNITED STATES NAVY DEPARTMENT.

CARBON, 0.14 per cent.

Original section:	
Width	1.254 inches.
Thickness	.417 inch.
Area	.5229 square inch.
Fractured section:	
Width	.936 inch.
Thickness at edges and centre, .300, .272, .308	inch.
Area	.2646 square inch.
Elastic limit, pounds per square inch	40,160
Ultimate tensile strength, pounds per square inch	62,700
Elastic ratio	64.06 per cent.
Final elongation	25.5 per cent.
Final area in per cent. of original area	50.60 per cent.
Modulus of elasticity	30,200,000
Elongation at tensile limit	19.0 per cent.
Stress at rupture:	
Pounds per square inch of original area	54,120
Pounds per square inch of fractured area	106,950
Time of test	17 minutes.

Cold-Bending Test.—Any plate which will stand the foregoing tests will stand any cold bending to which it is ever likely to be subjected. An open-hearth steel of 60,000 pounds tensile strength and 25 per cent. elongation in 8 inches will bend back upon itself, as shown in Fig. 9, which is drawn from a specimen of open-hearth basic boiler-plate, 1⅜ inches thick, made by the Lukens Iron and Steel Company. An example of double-bending of ⅜-inch open-hearth basic boiler-plate shown in Fig. 10 is also from the same company. Where T-bars are used for stiffeners or for joining braces or stays in a boiler, the flanges should stand cold bending to the dotted lines shown in Fig. 11. In the cold bending of any specimen prepared by shearing, care should be taken that the edges are smooth and free from sharp indentations, any one of which under high stress may develop a fracture. The hammering should be gradual and the blows delivered square to the surface. Failure in cold-bending tests may occur through too high carbon percentages or by the chemical action of hardening impurities in the steel.

FIG. 9.

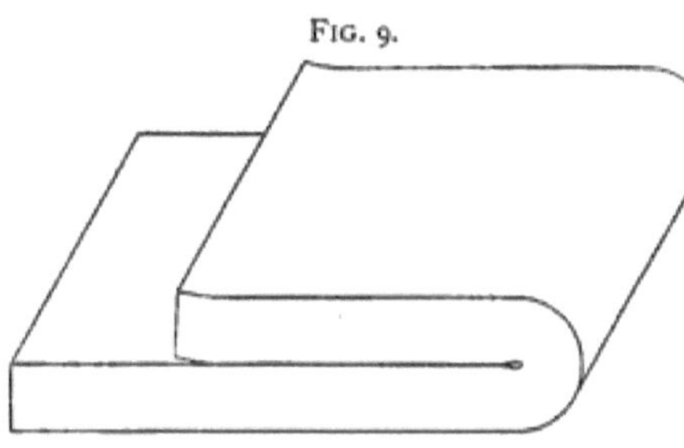

FIG. 10.

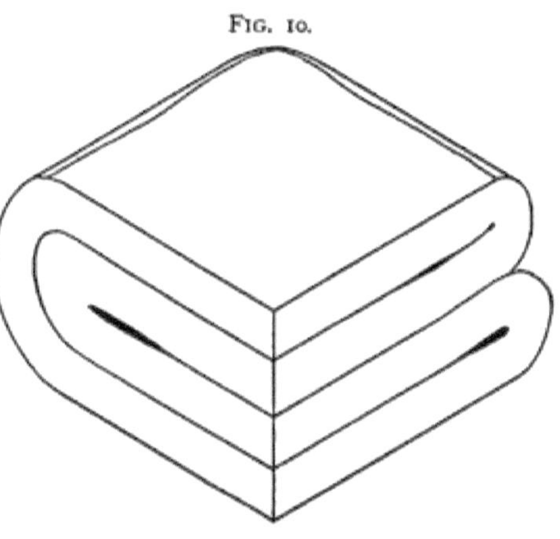

Drifting Test.—This test consists in taking a strip of steel of any convenient width (2 or 3 inches), drilling a hole ½ or ⅝ inch diameter for the insertion of a drift-pin, and by a succession of drifts through the cold plate enlarging the hole to at least twice its original diameter. Fig. 12 shows the effect of drifting a piece of open-hearth steel by the Lukens Iron and Steel Company. The original opening was ½ inch diameter, the drifted hole 1$\frac{1}{16}$ inch diameter. The effect upon the material is shown in the cross-section.

Quenching Test.—The object of this test is to determine whether or not the specimen tested possesses any hardening qualities. The standard quenching test consists in heating the specimen to a cherry-red heat and plunging it in water at a temperature of 82° Fahr., after which the quenched specimen must be capable of being bent back upon itself, as shown in Fig. 9, without signs of fracture anywhere in the bend. The United States navy requirements are that a specimen thus prepared shall bend under a press or hammer so that it shall be doubled round a curve of which the diameter is not more than one and a half times the thickness of the piece tested without presenting any trace of cracking; further, these test-pieces must not have their sheared edges rounded off; the only treatment permitted is taking off the sharpness of the edge with a fine file.

Fig. 11.

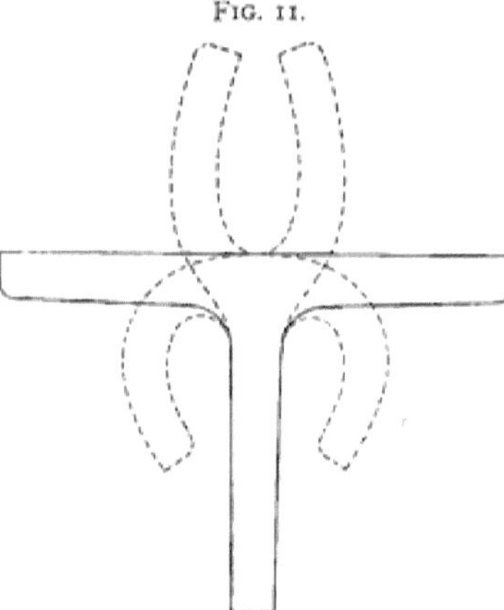

Fig. 12.

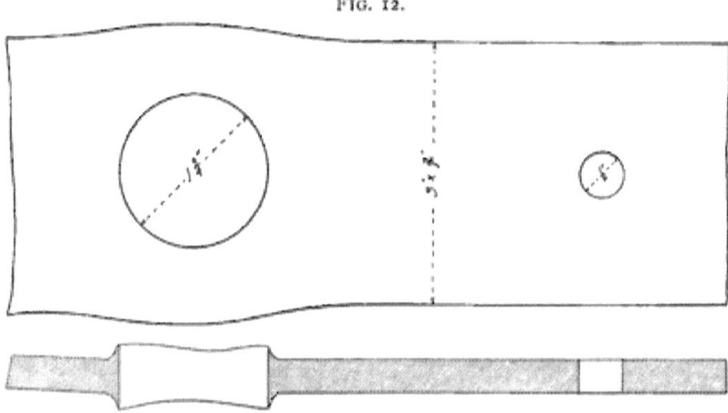

Bulging Test.—This test is seldom or never required in specifications for steam boilers. Experiments have been made, however, to determine to what extent open-hearth steel plates may be subjected to stresses of this kind. Fig. 13 shows a method employed under the direction of the United States Navy Department upon open-hearth steel plates made by the Nashua Iron and Steel Company. The steel thus tested analyzed as follows:

Carbon, combined	.130
Manganese	.100
Sulphur	.028
Phosphorus	.011
Silica	.071
Iron, difference	99.660
Total	100.000

A $\frac{9}{16}$-inch plate was placed over a hole 9 inches square in a piece of wrought iron; a 6-inch cast-iron shot was then driven down by a 2000-pound steam-hammer having a mean stroke of 24 inches until the calotte, or cup, thus formed had a depth of 4⅜ inches. The lower surface was then thoroughly examined and no signs of fracture could be detected. A second trial was then similarly made upon a plate $\frac{11}{16}$ inch thick, with a view to ascertain to what extent this test could be carried without fracture.

FIG. 13.

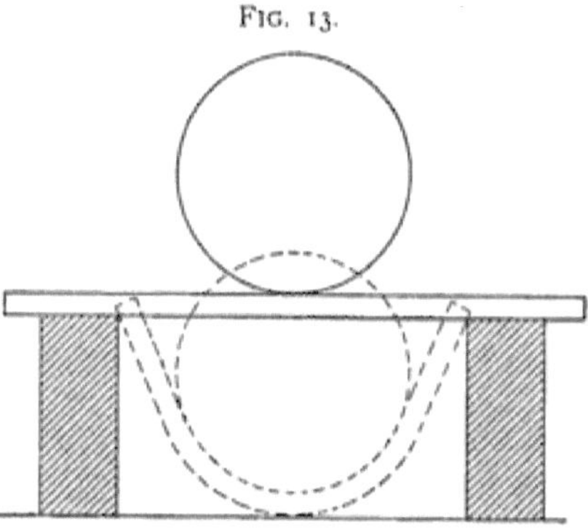

After breaking three shot, a wrought-iron cylinder with a spherical end was substituted, and at about the sixtieth blow disintegration took place along one side of the cup at a distance of 2 inches from its bottom. The thickness of the plate at the point of fracture was reduced to ¼ inch, the depth of the cup being 4⅞ inches.

When heated it was found that the plate could be folded over until the surfaces met and then bent in the opposite direction to a similar position without fracture, and after repeating this operation four times only a slight fracture took place.

Annealing.—This detail in manufacture consists in slowly heating the material to a cherry-red heat and allowing it to cool slowly, commonly under a bed of ashes. The object in annealing plates is to neutralize the interior strains which may have arisen from any cause, such as unequal heating or cooling in manufacture, and especially strains set up in a boiler-plate when allowed to cool on a cast-iron straightening-plate. Plates which undergo partial and local heating for flanging by hand, or when successive portions require to be heated for machine-flanging, or plates rather high in carbon which have undergone considerable cold-working and punching ought always to be annealed. A reverberatory furnace especially constructed for the purpose is the best method of annealing because it insures uniform heating over the entire surface, and if properly attended to there need be little or no danger of local overheating.

Manufacturers sometimes resort to annealing to prevent rejection of plates when undergoing rigid inspection. For example, a plate may be wanting in ductility partly through chemical and partly through physical causes: annealing will in all probability increase the ductility sufficiently to allow it to pass; on the other hand, a plate may be slightly above a specified tensile limit: by proper precautions in annealing the plate may, within narrow limits, be reduced in tensile strength and thus come within the limits of acceptance.

The effect of annealing is shown in the annexed memoranda relative to certain plates intended for the United States cruiser "Chicago." Eight test-pieces from four plates, representing four separate heats, were made from the original sheets; these plates contained 0.11, 0.13, 0.13, and 0.14 per cent. of carbon respectively.

Average ultimate tensile strength per square inch for the four heats	60,588 pounds.
Average final elongation	29.54 per cent.
Average final area	42.44 per cent.

The annealed test-pieces from the same plates were twenty in number, in which the following differences were observed:

Average decrease of tensile strength per square inch	6680 pounds.
Average increase in final elongation in 8 inches	.06 per cent.
Average decrease of final area in per cent. of original area	4.75 per cent.

The experiences of officers in charge of inspection and tests was, in the matter of annealing, of somewhat contradictory nature: the average effect of annealing open-hearth steel plates having a tensile strength of 55,000 to 65,000 pounds per square inch was a reduction varying from 5 to 19 per cent. of the original values for tensile strength and final area and an increase of 12.5 per cent. in the original value for ductility in 8 inches, the average original area being practically identical for both conditions. This shows a diminution of tensile strength and increase of capacity for local distortion. It was observed that changes were greatest in those plates which originally appeared the softest.

Certain precautions are necessary in annealing steel plates: the fuel should be comparatively free from ordinary impurities, the flame should be kept neutral, and the products of combustion should be kept as much as possible out of contact with the material. The heating up should be rather slow, the metal must not be soaked, and, indeed, should remain in the furnace the shortest time required to effect the desired results. The temperature ordinarily applied is too high, and especially apt to be so locally; it need not be very much above the temperature of finishing at the rolls, and should never be above a medium cherry. It cannot be too carefully borne in mind that the temperature at which the original physical structure is destroyed and replaced by a coarser and weaker structure is considerably below a bright yellow, and the metal should never be heated so high in annealing.

The conclusion arrived at from a consideration of results of annealing plates which have not been punched or otherwise worked so as to necessitate the removal of purely local strains is that the effect is as apt to be deleterious as beneficial as the process is ordinarily carried out. Good metal shows little improvement in any case; and while inferior metal may be doctored up to show somewhat better test, the improvement in intrinsic quality is uncertain. But the fact that, as commonly done, annealing may continually and with reasonable certainty lower the working quality, and sometimes excessively so, should prevent any general resort to the practice for boiler or other plate at the mills.

CHAPTER III.

RIVETED JOINTS.

PLATES for the shells of steam boilers are commonly joined by rivets, and occasionally by welding; but the latter method of joining the edges of plates is practically confined to the manufacture of tubes, flues, and cylindrical fire-boxes for internally fired boilers. Riveting is at best an unsatisfactory way of making a joint; but, notwithstanding its disadvantages in the matter of cost, and inherent weaknesses by reduction of sectional area and bad workmanship, it has become one of the most important details in mechanical engineering.

Strength of Iron and Steel Plates.—C. H. No. 1 shell iron of good quality has a tensile strength ranging from 40,000 to 50,000 pounds per square inch; in the absence of a specific test it may be assumed to be 45,000 pounds in all calculations relating to riveted joints. Table IV. gives the physical properties of wrought-iron plates referred to in this chapter in certain tests made for the Navy Department.

TABLE IV.

TENSILE STRENGTH OF WROUGHT IRON USED IN RIVETED-JOINT TESTS, REFERRED TO LATER IN THIS CHAPTER, WATERTOWN ARSENAL.

FIG. 14.

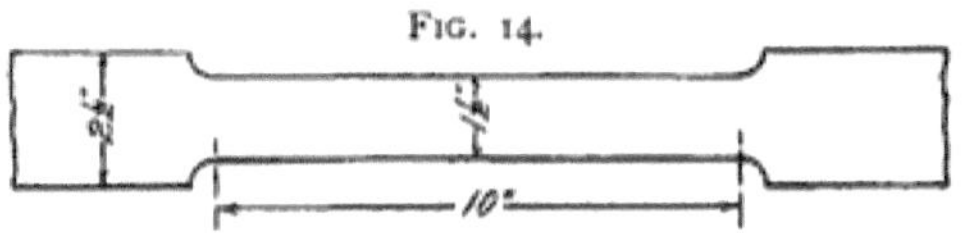

Nominal.	Actual Sectional Area.	Elastic Limit per Square Inch.	Ultimate Strength per Square Inch.	Elongation in Ten Inches.	Area of Fracture.	Contraction of Area.
Inch.	Square Inch.	Pounds.	Pounds.	Per cent.	Square Inch.	Per cent.
¼ L	.389	36,370	48,350	17.0	.300	22.9
¼ C	.397	25,700	38,160	4.5	.365	8.1
⅜ L	.560	32,679	47,590	13.6	.454	18.9
½ L	.768	25,789	45,830	13.8	.635	17.3
⅝ L	.935	25,450	45,240	12.4	.787	15.8
¾ L	1.047	27,890	46,750	12.8	.866	17.3

L—Lengthwise. C—Crosswise.

Mild steel for boilers, whether fire-box or flange, should have an ultimate strength of 52,000 to 62,000 pounds per square inch. In all riveted-joint calculations the tensile strength may be taken as 55,000 pounds per square inch.

As closely agreeing with the above, reference is had to Table V. of physical tests of steel from ¼ to ¾ inch thick, in which the average tensile strength is 54,885 pounds per square inch. The steel plates were supplied from one heat cast in ingots of same size, the thin plates differing from the thicker plates only in the amount of reduction given the metal by the rolls.

TABLE V.

TENSILE STRENGTH OF MILD STEEL USED IN RIVETED-JOINT TESTS, REFERRED TO LATER IN THIS CHAPTER, WATERTOWN ARSENAL.

FIG. 15.

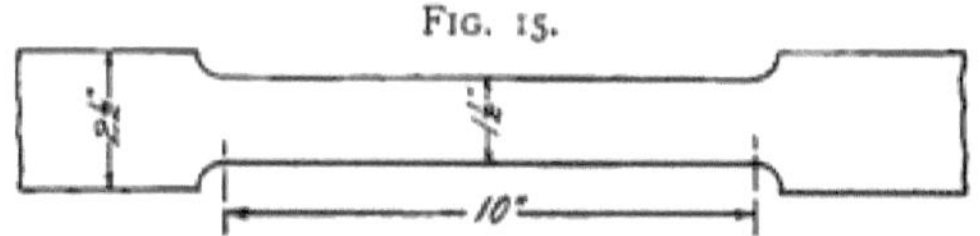

Nominal Thickness.	Actual Sectional Area.	Elastic Limit per Square Inch.	Ultimate Strength per Square Inch.	Elongation in Ten Inches.	Area of Fracture.	Contraction of Area.
Inch.	Square Inch.	Pounds.	Pounds.	Per cent.	Square Inch.	Per cent.
¼	.378	39,680	58,360	19.2	.141	62.7
⅜	.559	31,810	54,025	27.1	.208	62.8
½	.751	31,290	57,790	26.1	.360	52:1
⅝	.934	32,100	52,570	29.0	.350	62.5
¾	1.102	30,440	51,680	28.7	.394	64.3

Loss by Reduction of Area.—The total strength of a plate in a riveted joint will be lessened by that area of metal taken from it to supply places for the rivets. In good quality of wrought iron, punching, if properly done, has little or no effect upon the plates, and no other loss occurs than that due to reduction of area along the line of rivet-holes, so that the net strength of what remains of an iron plate, whether punched or drilled, is equal to that of the original plate for a corresponding sectional area; but in the case of steel plates the value of the net sectional area remaining between the rivet-holes is somewhat uncertain. In some cases there has been observed a marked falling off in tensile strength, the amount depending upon the hardness and thickness of the plate. Inasmuch as this loss of strength is partially or wholly restored by reaming, or by the annealing effect produced by the hot rivets, it will be near enough to assume in ordinary calculations that the net section between the holes is equal to that of the original plate for the same area of cross-section.

Rivet-Holes.—Whether the rivet-holes shall be punched or drilled is, so far as the manufacturer is concerned, largely a matter of convenience. Punching is more quickly accomplished than drilling, consequently punched holes are cheaper than drilled; but the former are more likely to contain errors in centring than is the case with drilled holes. A punch once started cannot usually be recalled, and if the plate should be out of centre at the start, the punched hole will be out of centre at the finish, and it is precisely such errors that the drift-pin is called into requisition to correct.

In large establishments where multiple drilling-machines are employed, drilled holes are supplied at a low cost, quite as low as that of punching, especially for thick plates, which are first punched and afterwards reamed. There can be no question as to the superiority of drilled over punched work, especially when two or more plates are fastened together after rolling and drilled at a single operation.

Punch and Die.—Two forms of punch are in common use,—the plain punch, with projecting teat or centre, and the spiral punch, as shown in Fig. 16.

When plates are laid off in the boiler-shop the centres of the hole are commonly marked with a prick-punch. To facilitate centring in the punching-machine the projecting centre above referred to is of great assistance to the workman, insuring greater accuracy in the spacing of punched holes than would occur if this detail was omitted.

FIG. 16.

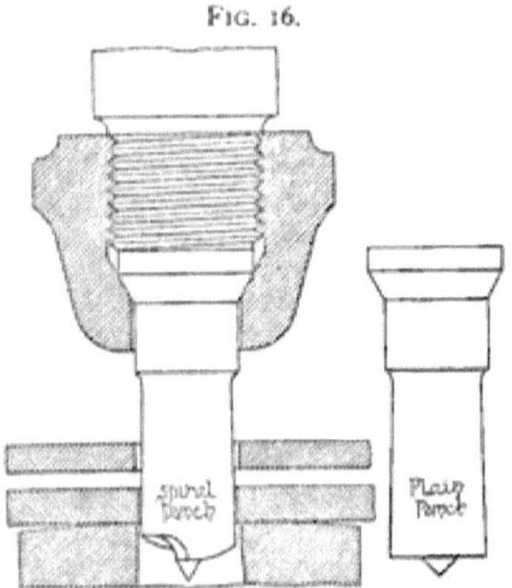

It is a common practice to make the die of a larger diameter than the punch, say $\frac{1}{16}$ of an inch, and for all sizes of rivets and all thicknesses of plates above $\frac{1}{4}$ inch and less than $\frac{5}{8}$ inch a tapering hole is the result, but if care be taken in riveting, the rivet will conform to the shape of and completely fill the hole. It is claimed that the enlargement of the die lessens the power required to punch the hole, and that the punching is performed with less strain upon the plate.

The power required to punch a hole in a plate may be estimated by multiplying the circumference of the punch in inches by the thickness of the plate in fractions of an inch, and this by the tensile strength of the material to be punched. For example, to punch a $\frac{15}{16}$-inch hole in a $\frac{1}{2}$-inch plate of steel having a tensile strength of 55,000 pounds per square inch would require $2.945 \times .5 \times 55{,}000, = 80{,}987$ pounds.

Spiral Punch.—In shearing metal plates it is the universal practice to set the top blade of the shear at an angle to the bottom one, the object being to bring about a gradual separation of the plate; in other words, to take a little more time to do the same amount of work than if the shears were parallel. Spiral punches are constructed on this principle, and for plates less than one-half the diameter of the punch the operation is very satisfactory in practice. For thicker plates the advantages are not so marked, and when the plate is three-fourths the diameter of the punch any advantage wholly disappears.

Comparison of Plain and Spiral Punches.—Professor Benjamin's experiments upon Otis steel boiler-plates ¼ inch thick, punched with ordinary flat punches, showed a tensile strength of 7.5 per cent. less, elastic limit 5 per cent. higher, and contraction 30 per cent. less than a drilled plate of the same material, showing that the effect of the punching is to render the metal around the hole more brittle and less ductile than before. When a spiral punch was used, the ultimate strength was only 3 per cent. less than in the drilled plate.

Effects of Punching Steel Plates.—The observed changes in the material in the line of punched holes are increased hardness, alteration of physical structure, and loss of ductility. In tests made several years ago from steel containing more carbon than is now allowed for boiler-plates, from specimens which had been cut from different portions of the same plate and in the same line of punched holes it appeared that the disturbance of the material was confined to within a very short distance around the hole, extending from $\frac{1}{16}$ to $\frac{3}{16}$ of an inch, and that by drilling and reaming out such punched holes and then testing the plate, making proper allowance for the reduced area, no perceptible decrease of strength was noted. Some more recent experiments on mild steel plates containing 0.12 carbon, 0.37 manganese, testing 64,200 pounds tensile strength, with a ductility of 25.15 per cent. in 8 inches, showed the following effect in punching, with or without countersinking:

Punched $\frac{11}{16}$-inch hole in 2-inch × .510-inch plate, as in Fig. 17.

FIG. 17.

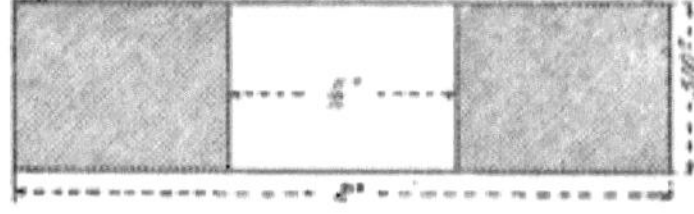

Effective dimensions of test-piece	1.313 × .510 inch.
Effective sectional area	.6696 square inch.
Ultimate tensile strength	55,000 pounds.
Reduction of strength	14.33 per cent.

Punched $\frac{11}{16}$-inch hole and countersunk, as shown in Fig. 18.

Fig. 18.

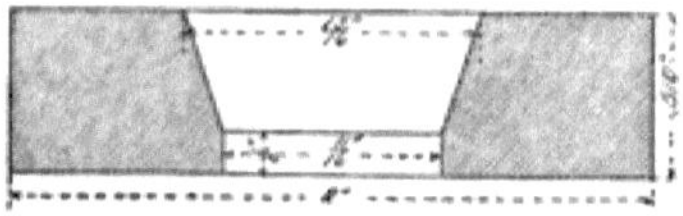

Effective sectional area	.6215 square inch.
Ultimate tensile strength	61,400 pounds.
Reduction of strength	4.36 per cent.

Commenced to crack at part of hole not countersunk, at 58,200 pounds per square inch.

Punched $\frac{9}{16}$-inch hole and countersunk, as shown in Fig. 19.

Fig. 19.

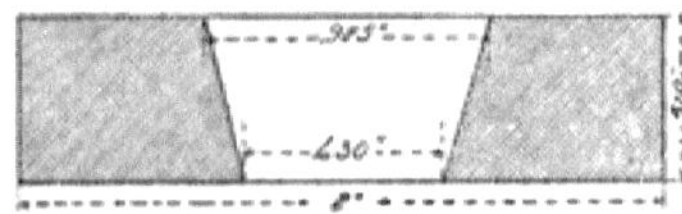

Effective sectional area	.6163 square inch.
Ultimate tensile strength	65,560 pounds.
Increase of strength	2.12 per cent.

Commenced to crack at small edge of hole, at 61,980 pounds per square inch.

These results show, first, a reduction of strength of 14.33 per cent. due to ordinary punching of a $\frac{11}{16}$-inch hole in a ½-inch steel plate of a quality suited for steam boilers; second, a recovery of strength, 4.36 per cent. of the original, by partial countersinking; third, a gain of 2.12 per cent. when a punched hole is countersunk all the way through. The loss in strength in the first example may be attributed to the formation of a thin ring of highly strained metal forming the walls of the hole; in general, this thickness will depend upon the carbon properties or hardness of the plate, upon its thickness, and the relative size of punch and die. The harder the steel, the thicker the plate, and the larger the die relatively to the punch, the thicker is the overstrained ring of metal, and the greater the amount of subsequent reaming and drilling necessary to remove it.

The increase of strength shown in the last of the three examples accords with a fact now generally recognized that the effect of a hole produced by a cutting tool in a steel plate is to increase the ultimate strength of the net section; and while the effect of punching is always to overstrain the adjacent material, yet if the damage to the material is equal to or less than the gain due to the difference of distribution of the resisting area owing to the presence of the hole, no apparent loss of strength will ensue.

Countersunk Rivets.—Experiments prove that the lap in riveted joints is an element of weakness irrespective of the loss of strength by rivet-holes. The thicker the plate the greater is the distorting leverage shown in Figs. 25 and 25 A. Clark reasons that because the absolute strength of a ½-inch lap-welded joint under test was not greater than that of a ⅜-inch joint under similar test, the principle here noticed may account for the practically equal strength of joints made with countersunk rivets compared with those having external rivet-heads, notwithstanding the greater reduction of solid section by countersinking; the leverage is shortened, and it may be measured from the centre of the cylindrical part of the rivet in the line A B, in Fig. 20, or thereabouts, towards the inner side of the plate.

FIG. 20.

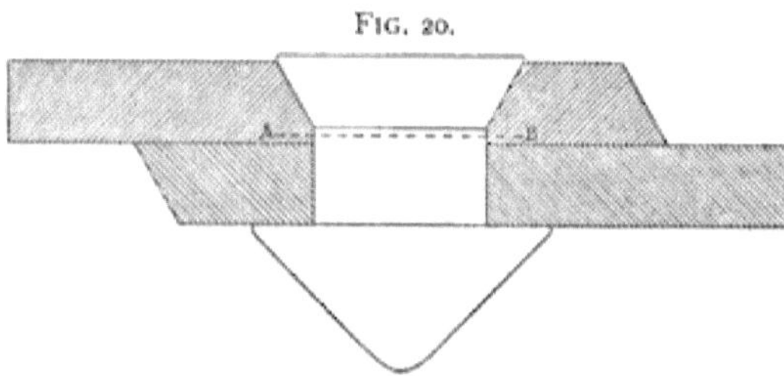

Size of Rivet.—The thinnest plate used in ordinary boiler-work is ¼ inch, and for such plates ⅝-inch rivets are recommended, which, with the customary allowance of $\frac{1}{16}$-inch clearance, makes the rivet-hole $\frac{11}{16}$ inch in diameter, the proportion of diameter of hole to thickness of plate being 2.75 to 1. Nearly all English writers give ½ inch as the diameter of rivet, or a $\frac{9}{16}$-inch hole for ¼-inch plates,—a ratio of 2.25 to 1; but these proportions have not been in favor and are not commonly used in this country.

For thick plates, say ¾-inch and thicker, the size of the hole is governed somewhat by the ability to properly upset the rivet to fill the hole and form a proper head, especially in such places as require the work to be done by hand. For ¾-inch plates the diameter of rivet is commonly 1⅛ inches, or $1\frac{3}{16}$ hole. This gives a ratio of diameter of hole to thickness of plate of 1.58 to 1. For intermediate thicknesses the diameter of rivet may be found by interpolation, as in Table VI.

When rivets pass through three or four plates overlapping each other at a joint, it is in accordance with good practice to increase the diameter of such rivets $\frac{1}{16}$ or ⅛ inch, at the discretion of the designer. Incidentally it may be remarked that if the plates are to be lap-jointed on a continuous plate, as shown in Fig. 21, one of the sheets must be forged wedge-shaped and the top sheet bent to fit the taper of the wedge. If, however, all the four corners meet at a common point, as in Fig. 22, the two inner sheets are forged wedge-shaped, right and left, the two equalling a single thickness, or approximately so. The upper and lower plates require no forging. Joints of this kind should be drilled after all the plates are assembled in place.

TABLE VI.

RIVET-HOLES.

Thickness of Plate.	Diameter.		Area of Hole.	Ratio of Diameter of Hole to Thickness of Plate.
	Rivet.	Hole.		
Inch.	Inch.	Inch.	Square Inch.	
1/4	5/8	11/16	.371	2.75 to 1
5/16	11/16	3/4	.442	2.40 to 1
3/8	3/4	13/16	.518	2.17 to 1
7/16	13/16	7/8	.601	2.00 to 1
1/2	7/8	15/16	.690	1.87 to 1
9/16	15/16	1	.785	1.78 to 1
5/8	1	1 1/16	.887	1.70 to 1
11/16	1 1/16	1 1/8	.994	1.64 to 1
3/4	1 1/8	1 3/16	1.108	1.58 to 1

Pitch of Rivets.—The distance apart from the centre of one rivet to the centre of the next rivet is called its pitch. As the gross strength of any plate is directly affected by the removal of a portion of its sectional area necessary for the rivet-holes, it follows that the fewer holes there are for a given width of plate the greater will be the strength of the

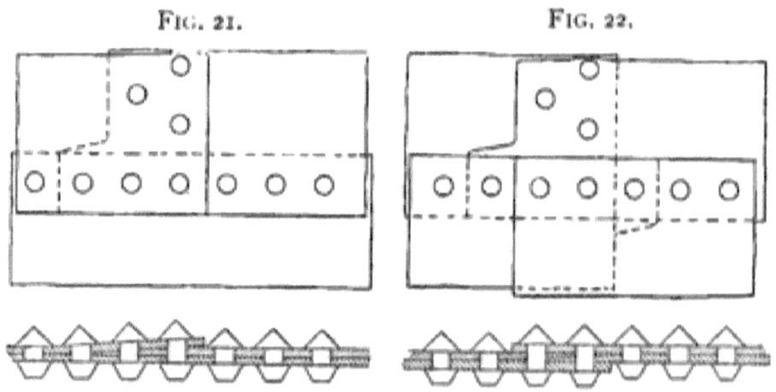
Fig. 21. Fig. 22.

plate remaining. If the pitch of rivets be too close, the strength of the plate is unnecessarily diminished ; if too wide, there is danger of shearing the rivets, unless they are increased in diameter corresponding to the increased pitch, which is not always practicable.

When designing a joint, the diameter and pitch of rivets should approximate in strength that of the plate remaining between the rivet-holes, taking into account the tensile strength per square inch of the latter and the shearing strength of the former; but such a joint can only be secured by the use of large rivets and wide spacing. Such joints are difficult to keep tight under high pressures, and tightness of joint is a detail quite as important as that of tensile strength.

Iron Plates and Iron Rivets.—Safe working dimensions for single-riveted joints have been ascertained by actual tests. For example, tests were made upon a ¼-inch plate having a tensile strength of 47,925 pounds per square inch. Specimen was 10 inches in width, with 5 ⅝-inch rivets on 2-inch centres, punched holes, the punch 0.695 inch diameter, the die 0.752 inch diameter, making a hole slightly tapered. The rivets were of iron of the best quality. This joint failed by shearing four of the rivets and breaking a corner out of the plate at the fifth rivet, as shown in Fig. 23. The elongation of holes ranged from 0.04 to 0.08 inch. The efficiency of this joint was 64.1 per cent. The weakness was in the rivets. The maximum shearing stress on the rivets was 38,640 pounds per square inch, which exhibits the usual proportion between tensile strength of plate and shearing strength of rivets when both are made of first-quality material.

FIG. 23.

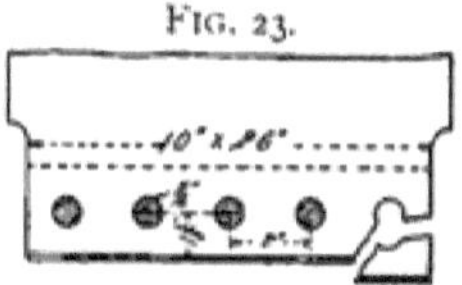

A second experiment made from the same ¼-inch plate, 9¾ inches wide, with 6 ⅝-inch iron rivets from the same lot, holes on 1⅝-inch centres, punched as above, failed by fracturing one of the plates directly across the line of rivet-holes. The efficiency of this joint was 64 per cent. The maximum shearing stress on the rivets was 35,200 pounds. In this case the pitch could have been widened to advantage. It appears, therefore, that the correct spacing for ⅝-inch iron rivets in ¼-inch iron plates should be within these two limits, say 1¾ or 1⅞ inches. The efficiency of such a joint, if the workmanship is good, may be taken at 60 per cent. of the original plate.

Steel Plates and Steel Rivets.—An open-hearth steel plate ¼ inch thick, tensile strength 55,765 pounds per square inch; punched holes, punch 0.69 inch diameter, die 0.75 inch diameter, making a hole slightly conical; pitch of rivets 2 inches. In a plate 10 inches in width, containing 5 rivets, failed in a single-riveted joint by tearing one plate through the metal, as shown in Fig. 24. The efficiency of this joint was 69.2 per cent. The first fracture which appeared in sight was at the end section; the holes in the under plate elongated about $\frac{3}{16}$ inch each. The lap end of the plate bent about 12 degrees, similar to that indicated in Fig. 25 A.

FIG. 24.

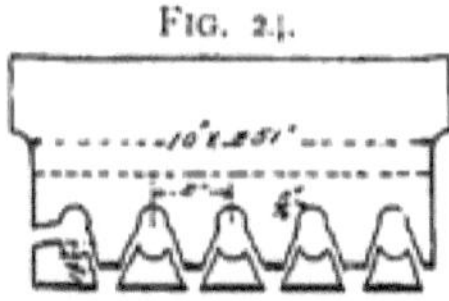

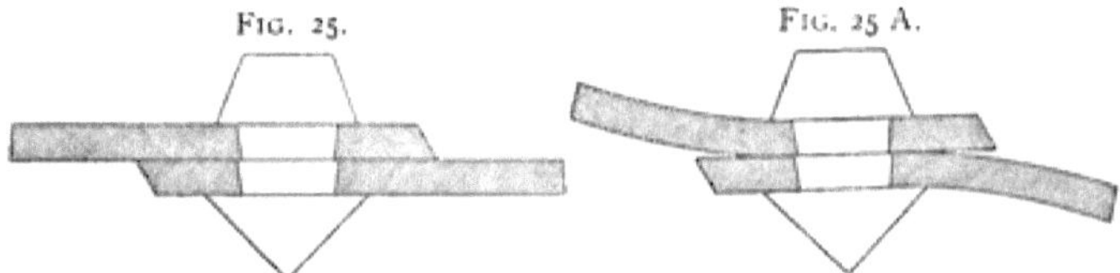
FIG. 25. FIG. 25 A.

Another joint was made from the same plate; drilled holes 0.69 inch diameter, 1⅝ inch pitch; the specimen 9.75 inches wide, containing 6 rivet-holes; failed by fracturing one plate through the rivet-holes, as indicated in Fig. 26. The efficiency of this joint was 68.8 per cent. Scales started on rivet-heads, otherwise they appear undisturbed. After the test the rivets were loose in the plate, not fractured. Fracture began at one edge and in first section between rivets.

FIG. 26.
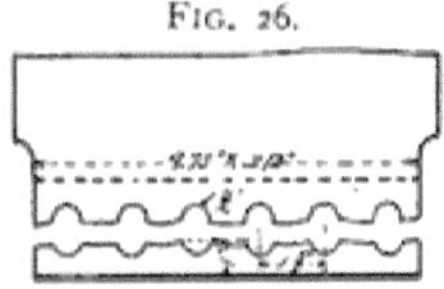

It will be observed that in both of the above tests failure occurred in the plate where, by preference, it ought to fail. It appears that the limit of pitch may be placed at 2 inches for ¼-inch steel plates, using ⅝-inch steel rivets.

Strength of Rivet Iron.—Tests made to determine the physical properties of wrought iron commonly supplied for rivets, as well as tests made of rivets after manufacture, make it appear that rivet iron varies in tensile strength from 45,000 to 52,000 pounds per square inch, excluding a few exceptionally high specimens about which there was some doubt as to the fact of their not being steel. Taking everything into consideration, it appears that 47,500 pounds per square inch of section is as high a tensile strength as ought to be ordinarily ascribed to wrought-iron rivets.

The elastic limit of rivet iron varies from 50 to 70 per cent. of the tensile strength, averaging closely to 58 per cent. There is no definite line upon which elastic limit can be established for wrought iron, but the percentage named is near enough for all practical purposes.

The shearing strength of rivet iron approximates closely to 72 per cent. of its tensile strength, with variations on either side, diminishing somewhat with increasing tensile strength.

In Table VII., 47,500 pounds tensile strength per square inch is used; so also the percentages for elastic limit and shearing strength as given above, calculated for each rivet from ⅝ to 1⅛ inches diameter. These figures are approximately correct for commercial rivets.

Strength of Rivet Steel.—The tensile strength of open-hearth steel rivets for steam boilers under the standard specifications indicate that an average of 50,000 pounds per square inch would be an acceptable tensile strength; as a matter of fact, ordinary steel rivets more nearly approach 55,000 pounds. In Table X. the tensile strength is taken at 52,500, which, it is believed, accords very closely with the average tensile strength of steel rivets.

TABLE VII.

WROUGHT-IRON RIVETS.

Thickness of Plate.	Rivets.			Tensile Strength of each Rivet at 47,500 Pounds per Square Inch.	Elastic Limit of each Rivet at 27,550 Pounds per Square Inch.	Shearing Strength of each Rivet at 34,200 Pounds per Square Inch.	
	Diameter of Rivet.	Diameter of Hole.	Area of Hole.			Single Shear.	Double Shear.
Inch.	Inch.	Inch.	Sq. Inch.	Pounds.	Pounds.	Pounds.	Pounds.
1/4	5/8	11/16	.3712	17,632	10,227	12,695	24,120
5/16	11/16	3/4	.4418	20,986	12,150	15,110	28,709
3/8	3/4	13/16	.5185	24,643	14,285	17,733	33,693
7/16	13/16	7/8	.6013	28,562	16,566	20,564	39,072
1/2	7/8	15/16	.6903	32,749	19,018	23,608	44,855
9/16	15/16	1	.7854	37,307	21,638	26,861	51,036
5/8	1	1 1/16	.8866	42,114	24,426	30,322	57,612
11/16	1 1/16	1 1/8	.9940	47,215	27,385	33,995	64,591
3/4	1 1/8	1 3/16	1.1075	52,606	30,511	37,877	71,966

The elastic limit must not be less than one-half the ultimate tensile strength according to the standard specifications; but an examination of a large number of tests show that the elastic limit approximates 65 per cent., or 34,125 pounds per square inch. When testing steel rivet-bars the same observation as in wrought-iron rivet-bars is had,—that there is no clearly defined line of elastic limit in mild steel.

The shearing strength of steel rivets as compared with the tensile strength varies considerably, from 70 to 90 per cent., but from a large number of carefully conducted experiments in riveted joints and shearing tests the shearing strength of open-hearth steel rivets closely approximated 85 per cent. of the tensile strength.

TABLE VIII.

SHOWING CHEMICAL ANALYSIS AND PHYSICAL QUALITIES OF "VICTOR" STEEL RIVETS.

Phosphorus . . . . average 20 samples, .015 per cent.
Manganese . . . . average 20 samples, .46 per cent.
Sulphur . . . . average 20 samples, .033 per cent.
Silicon . . . . average 20 samples, .005 per cent.
Carbon . . . . average 20 samples, .11 per cent.
Elastic limit . . . . average 10 samples, 36,252 pounds per sq. in.
Tensile strength . . . . . average 10 samples, 51,565 pounds per sq. in.
Elongation in 8 inches . . . average 10 samples, 31.9 per cent.
Reduction of area . . . . average 10 samples, 79.7 per cent.
Shearing strength, single . . average 8 samples, 48,277 pounds per sq. in.
Shearing strength, double . average 7 samples, 45,720 pounds per sq. in.

The preceding may be taken as representative of good quality steel rivets now offered by rivet manufacturers to the trade. Table IX. gives results of tests of rivets made for the Navy Department under a higher carbon percentage than is commonly employed for rivets.

TABLE IX.

TENSILE TESTS OF OPEN-HEARTH STEEL-RIVET METAL USED IN THE CONSTRUCTION OF RIVETED JOINTS, NAVY DEPARTMENT.

Diameter.	Elastic Limit per Square Inch.	Tensile Strength per Square Inch.	Elongation in Ten Inches.	Contraction of Area.
Inches.	Pounds.	Pounds.	Per cent.	Per cent.
$\frac{9}{16}$	39,040	53,230	21.5	39.3
$\frac{11}{16}$	39,330	55,120	21.3	40.4
$\frac{13}{16}$	41,240	56,470	25.1	38.4
$\frac{15}{16}$	36,060	52,450	26.2	38.8
$1\frac{1}{16}$	34,810	52,850	28.1	42.9
$1\frac{3}{16}$	33,860	53,600	27.4	38.1

TABLE X.*

STEEL RIVETS.

Thickness of Plate.	Rivets.			Tensile Strength of each Rivet at 52,500 Pounds per Square Inch.	Elastic Limit of each Rivet at 34,125 Pounds per Square Inch.	Shearing Strength of each Rivet at 44,625 Pounds per Square Inch.	
	Diameter of Rivet.	Diameter of Hole.	Area of Hole.			Single Shear.	Double Shear.
Inch.	Inch.	Inch.	Sq. Inch.	Pounds.	Pounds.	Pounds.	Pounds.
$\frac{1}{4}$	$\frac{5}{8}$	$\frac{11}{16}$	.3712	19,488	12,667	16,565	31,474
$\frac{5}{16}$	$\frac{11}{16}$	$\frac{3}{4}$	.4418	23,195	15,076	19,715	37,459
$\frac{3}{8}$	$\frac{3}{4}$	$\frac{13}{16}$	.5185	27,221	17,694	23,138	43,962
$\frac{7}{16}$	$\frac{13}{16}$	$\frac{7}{8}$	.6013	31,568	20,519	26,833	50,983
$\frac{1}{2}$	$\frac{7}{8}$	$\frac{15}{16}$	.6903	36,241	23,556	30,805	58,530
$\frac{9}{16}$	$\frac{15}{16}$	1	.7854	41,234	26,802	35,048	66,591
$\frac{5}{8}$	1	$1\frac{1}{16}$	.8866	46,648	30,255	39,299	74,668
$\frac{11}{16}$	$1\frac{1}{16}$	$1\frac{1}{8}$	.9940	52,185	33,920	44,357	84,278
$\frac{3}{4}$	$1\frac{1}{8}$	$1\frac{3}{16}$	1.1075	58,144	37,793	49,422	93,940

Tests and Inspection of Rivets.—In large and important contracts, sample bars of rivet-metal 18 inches long are usually required and furnished for tensile test, elastic limit, elongation and contraction of

* In the computation of this table, special steels and special tests have been avoided. The values per square inch as given in the four last columns are believed to be those which commonly obtain in mild steel rivets.—W. M. B.

area ; but in ordinary business routine this is quite impracticable. The following tests for steel rivets may be carried out in any boiler-shop :

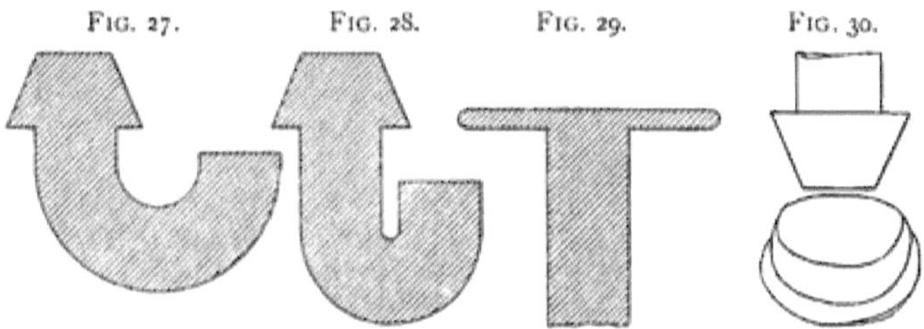

Two rivets to be taken at random from each keg or box of 200 pounds, as a portion of a lot of 10 rivets out of 1000 pounds, which may be subjected to the following tests, to be made in pairs :

TABLE XI.

BOILER RIVETS.

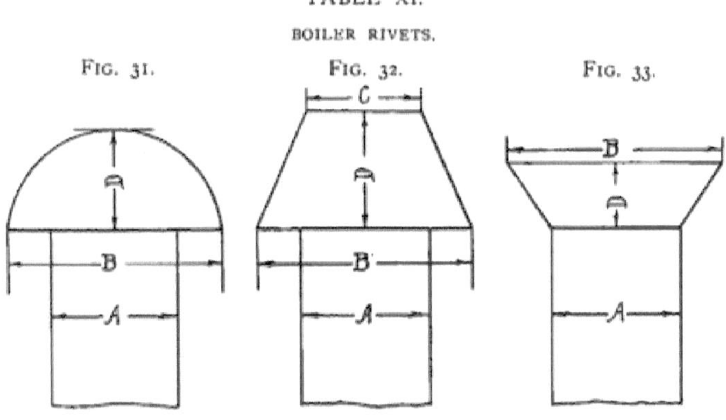

DIAMETER OF RIVET.	BUTTON-HEAD, FIG. 31.		CONE-HEAD, FIG. 32.			COUNTERSUNK HEAD, FIG. 33.	
	Diameter.	Thickness.	Diameter.	Diameter.	Thickness.	Diameter.	Thickness.
A.	B.	D.	B.	C.	D.	B.	D.
5/8	1 1/16	7/16	1 1/16	19/32	9/16	1 1/16	9/32
11/16	1 1/8	1/2	1 1/8	[illegible]	[illegible]	1 3/16	5/16
3/4	1 1/4	9/16	1 1/4	[illegible]	[illegible]	1 1/4	3/8
13/16	1 3/8	21/32	1 3/8	3/4	[illegible]	1 5/16	13/32
7/8	1 7/16	3/4	1 7/16	13/16	3/4	1 3/8	7/16
15/16	1 9/16	3/4	1 9/16	7/8	13/16	1 1/2	15/32
1	1 5/8	3/4	1 5/8	15/16	[illegible]	1 5/8	1/2
1 1/16	1 11/16	13/16	1 3/4	1	[illegible]	1 3/4	9/16
1 1/8	1 3/4	7/8	1 25/32	1 1/16	[illegible]	1 7/8	9/16

1. Two rivets to be bent cold in the form of a hook, as shown in Fig. 27, without showing cracks or flaws.

2. Two rivets to be bent hot in the form of a hook with parallel sides, as shown in Fig. 28, without showing cracks or flaws.

3. Flatten the heads of 2 rivets while hot, in the manner shown in Fig. 29, without cracking at the edges, the head to be flattened until its diameter is 2½ times the diameter of the shank.

4. Heat 2 rivets to a low cherry-red, and quench in water at 82° Fahr.; afterwards upset them cold, and forge cold to a flattened disk of 1½ diameters of shank, as shown in Fig. 30, without cracks or flaws.

5. The shearing-test to consist of riveting up 2 bars of steel by a single rivet, as in Fig. 38, and submitting it to a tensile strain, the rivet not to shear under a stress of less than 45,000 pounds per square inch.

Rivet Dimensions.—Rivets, like other manufactured articles, have their proportions fixed by their manufacturers, which proportions have been adapted to the needs of the trade. Table XI. gives the dimensions of three kinds of rivet-heads in general use. These dimensions may not exactly agree for all makes of rivets throughout the country, but they are near enough for construction purposes and for preliminary drawings.

Rivet-heads should be large enough, so that no serious distortion shall occur when riveting, and especially by hand. A large head lessens the distortion of the plates in tension when in single shear, and thus contributes to the strength of joint.

Rivet-Points.—In hand-riveting the points are usually finished conical, as in Fig. 34, which makes a good joint if the rivet is upset sufficiently to fill the hole before the spreading of the cone is begun.

FIG. 34. FIG. 35. FIG. 36.

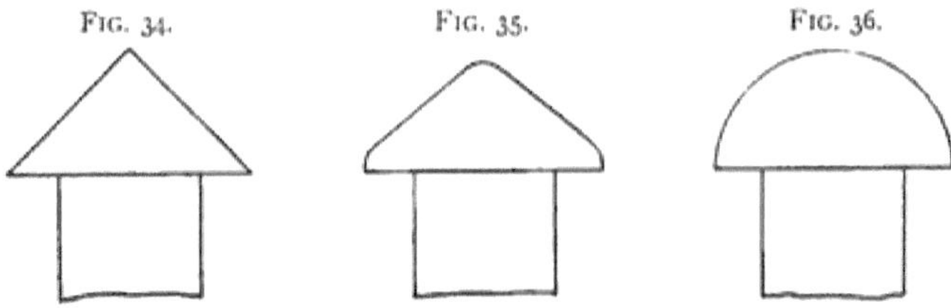

In some boiler-shops the rivet is upset to fill the hole and then finished with a snap-point, as in Fig. 36, by means of a button-head set having a concave depending for its size upon the diameter of the rivet; the proportions may follow the dimensions given in Table XI. It is not recommended that hand-riveting be finished with snap-points for rivets larger than ¾ inch diameter. In machine-riveting a conical point similar to Fig. 35 is in common use, so also the button-head in Fig. 36.

It is important in machine-riveting that the work be done not too rapidly, or the rivet will not upset sufficiently to fill the hole. It is also important that the rivet be exactly central to the machine, or bad work will be done, as in Fig. 37.

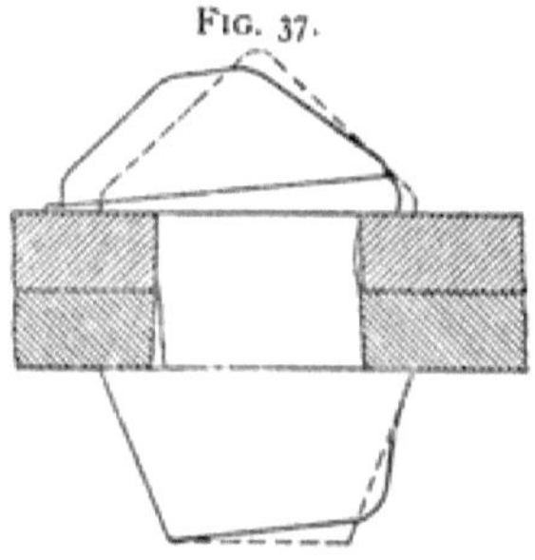
Fig. 37.

Countersunk heads or points should be avoided as far as possible when designing a boiler, because of the extra cost of making the hole and the lesser strength of the rivet. Such a joint should not be used where the stress on the rivet is in the direction of its length instead of shearing stress. There are portions of a boiler in which it is necessary to use a countersunk hole,—for example, around the strengthening ring of a manhole and other portions of a boiler where fittings must be attached which intersect a riveted joint.

Length of Rivets.—The amount of rivet-shank projecting beyond the plates to be joined may be, for—

Countersunk points for 2 sheets	1 diameter.
Countersunk points for 3 sheets	1 diameter + ⅛ inch.
Snap-points	1¼ diameters.
Conical points, small, hand-driven	1¼ diameters.
Conical points, medium, large, machine-driven .	1½ diameters.

Shearing Strength of Rivets.—The shearing resistance of a rivet is seriously interfered with by the compression between the rivet and the bearing sides of the holes, the effect being to increase the shearing resistance offered per unit of area as the size of the rivet diminishes. Difference of quality between plate and rivet, proportion of thickness and diameter, and the sharpness of the edges of the hole will also somewhat affect the shearing resistance.

Fig. 38.

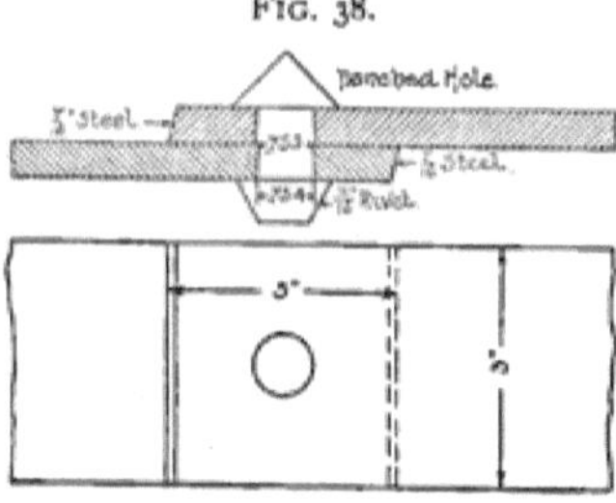

Rivets in Single Shear.—Experiments made upon iron and steel rivets in single shear is by means of 2 steel plates, as shown in Fig. 38. These plates are of such dimensions that no distortion occurs in them and all the pull of the machine is exerted in shearing the rivet.

The shearing resistance of 2 steel rivets in single shear was as follows :

First rivet.—7/16-inch steel plate; 11/16-inch steel rivets.

Punched holes.—0.753 punch; 0.754 die.

Ultimate shearing strength, 25,750 pounds = 57,220 pounds per square inch.

Second rivet.—Ultimate shearing strength, 26,490 pounds = 58,870 pounds per square inch.

TABLE XII.

IRON IN SINGLE AND DOUBLE SHEAR, WASHINGTON NAVY YARD.

Diameter.	Single Shear.			Double Shear.		
	Lowest, Pounds.	Highest, Pounds.	Mean Pounds per Sq. Inch.	Lowest, Pounds.	Highest, Pounds.	Mean Pounds per Sq. Inch.
.51 inch	8,900	9,400	44,149	16,050	17,600	82,186
.64 inch	12,650	13,300	39,253	23,600	25,650	77,348
.78 inch	18,400	19,650	39,553	36,400	39,400	79,536
.91 inch	25,500	27,600	41,503	46,200	52,000	75,789
1.03 inch	32,900	35,800	40,708	61,700	64,000	75,293
Average	....	....	41,033	....	....	78,030

Six specimens of each size were subjected to both single and double shear. The smaller diameter shows larger shearing strength per square inch than larger ones, but the decrease is not regular or uniform. The increase of average strength per square inch of sectional area for double shear over that of single shear was, for—

	Per cent.
½-inch specimen	86.2
⅝-inch specimen	97.0
¾-inch specimen	101.1
⅞-inch specimen	82.6
1-inch specimen	85.0
Average for all sizes	90.2

TESTS OF MILD STEEL RIVETS AND RIVET-BARS IN SINGLE AND DOUBLE SHEAR, NAVY DEPARTMENT.

Nominal size of rivet, ¾ inch; 4 tests.

Tensile strength of rivet-bars, 60,375 pounds per square inch.
Final elongation, 26.48 per cent. Final area, 44.2 per cent.

Shearing tests of rivets in single shear; 6 tests.

Diameter of hole, 13/16 inch. Shearing area, 0.5185 square inch.
Average shearing strength, 46,450 pounds per square inch.
Ratio of shearing to tensile strength, 76.94 per cent.

Shearing tests of rivets in double shear; 3 tests.

Diameter of hole, 13/16 inch. Shearing area, 1.037 square inches.
Average shearing strength, 44,647 pounds per square inch.
Ratio of shearing to tensile strength, 73.64 per cent.

Nominal size of rivet, ⅝ inch; 2 tests.

Tensile strength of rivet-bars, 70,120 pounds per square inch.
Final elongation, 24.10 per cent. Final area, 55.7 per cent.

Shearing tests of rivets in double shear; 2 tests.
Diameter of hole, 15/16 inch. Shearing area, 1.380 square inches.
Average shearing strength, 56,100 pounds per square inch.
Ratio of shearing to tensile strength, 80 per cent.

Rivets in Double Shear.—It may be remarked in connection with the foregoing that in testing the shearing strength of rivets under the United States specifications the test-piece consists of a double lap-joint, with a single rivet, arranged as shown in Fig. 39. The plate steel used is in no case less than half the diameter of the rivet, in order to insure shearing the rivet instead of the plate. The distance of the nearest edge of the rivet-hole to the end of the plate is not less than 1¼ diameters of rivet. Width of plate not less than three times the diameter of the rivet. The inner end of the filling-piece between the plates not to have an open space of more than 2 inches between it and the end of the single plate. The sharp edges of the rivet-hole are not to be filed down. Snap-riveted points will in no case be allowed in shearing tests, but invariably the point must be thoroughly and carefully worked down. The rivet-holes may be either punched or drilled, as desired, and the riveting may be either by hand or machine. Great care is always exercised in testing the sample to insure a fair stress on the rivet.

FIG. 39.

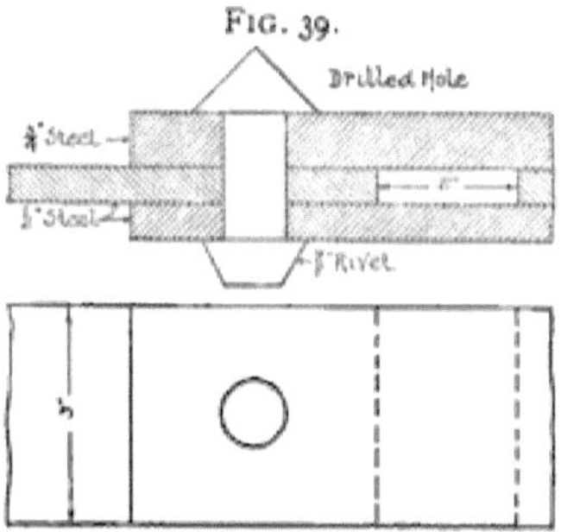

The shearing stress of 8 samples of steel rivets in private test was 48,277 pounds per square inch for single shear, and 45,270 pounds per square inch when subjected to double shear. For chemical and physical qualities of this steel see page 59.

Friction in Riveted Joints.—Friction is at best an uncertain quantity in riveted joints; it is, therefore, seldom or never made use of in calculations relating to the strength of steam boilers. But it is interesting to know how much friction there is in a joint, because to whatever extent friction exists it adds that much to the factor of safety, provided a very considerable portion of it is not eliminated by springing the plates apart in the operation of calking the edges to make a tight joint.

Four experimental joints tested at the Watertown Arsenal from designs similar to Fig. 40 show the friction of 5 ⅝-inch iron rivets in a ¼-inch iron plate 9¾ inches wide, the holes drilled 0.69 inch diameter, the corners of rivet-holes being rounded about 0.05 inch, presenting a shearing area of rivets of 1.87 square inches; this joint slipped when a load of 26,000 pounds had been applied. The friction per rivet was $\frac{26,000}{5} = 5200$ pounds. Continuing the test, the rivets sheared at 70,900 pounds. The shearing strength of the rivets, 37,914 pounds per square inch.

In a similar experiment with a steel plate 10 inches wide, ¼ inch thick, 5 ⅝-inch iron rivets, as above, the joint made a sudden slip when a load of 27,000 pounds was reached; therefore $\frac{27,000}{5} = 5400$ pounds per rivet of friction. The rivets sheared at 72,700 pounds total load; the shearing strength of the rivets, 38,880 pounds per square inch.

FIG. 40.

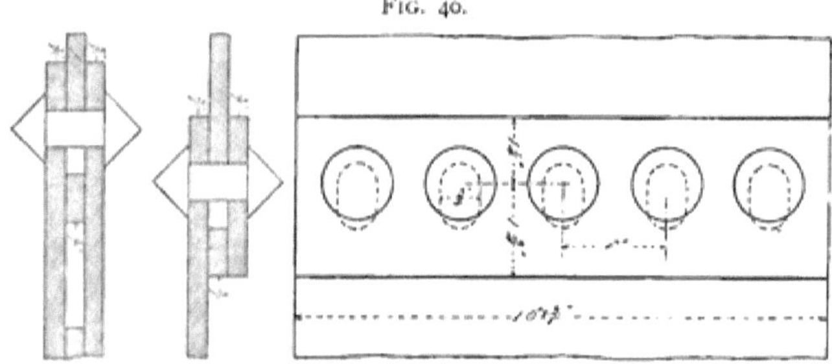

A second series of tests was made, using a ⅜-inch iron plate 10 inches wide, with 5 $\frac{11}{16}$-inch iron rivets in ¾-inch drilled holes, with clearance in front of rivets, corners of holes rounded about 0.05 inch; shearing area of the 5 ¾-inch rivets, 2.21 square inches. Rapid slipping began when the load passed 31,000 pounds. The resistance to slipping continued at about 30,000 pounds till the total slip was 0.15 inch. Friction per rivet, $\frac{31,000}{5} = 6200$ pounds. The total load applied to the joint was 92,000 pounds, when it failed by shearing the rivets. The shearing strength of the rivets, 41,630 pounds per square inch.

A ⅜-inch steel plate 10 inches wide, with 5 $\frac{11}{16}$-inch iron rivets in ¾-inch drilled holes, as above, slipped at joint at 31,000 pounds, and continued slipping until the resistance fell to 28,000 pounds, the friction per rivet being $\frac{31,000}{5} = 6200$ pounds. The ultimate strength of the joint was reached at 88,000 pounds, when the rivet sheared. Shearing strength of rivets, 39,820 pounds per square inch.

RESULTS OF TESTS OF RIVETED JOINTS.

Investigations into the strength of riveted joints have been until a few years past mathematical rather than experimental, as no machines were in use capable of pulling wide joints or large sectional areas, as now made possible by the installation of large testing-machines at the principal steel-works, and especially at the Watertown Arsenal. The tests made upon iron and steel bars, plates, and riveted joints for the United States Navy Department, which extended over several years at the Watertown Arsenal, are the most extended and comprehensive ever undertaken in this country. As these tests were undertaken for the purpose of ascertaining how riveted joints fail, a number of engravings are here introduced, copied from the records of the arsenal, which, in connection with the tabulated data, give much valuable information regarding this important detail in steam-boiler construction.

FIG. 41. FIG. 42. FIG. 43.

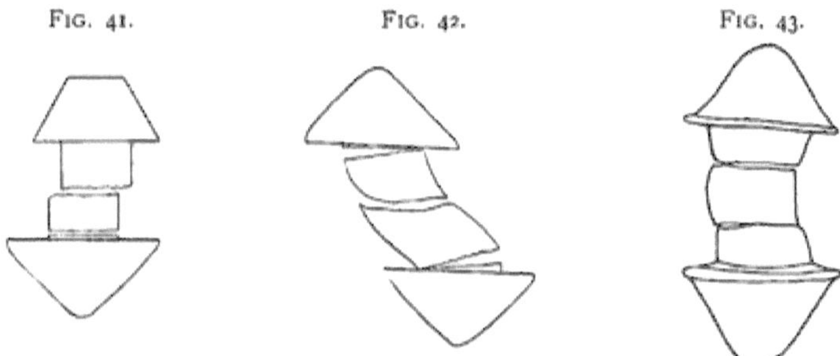

A riveted joint may fail by (1) the shearing of the rivets (Figs. 41, 42, 43), in which the rivet area and shearing resistance are too small for the thickness of plate and pitch of rivets; to remedy which a larger rivet may be introduced, or, in the case of iron rivets, the substitution of steel rivets, the latter having a shearing strength more than 20 per

FIG. 44.

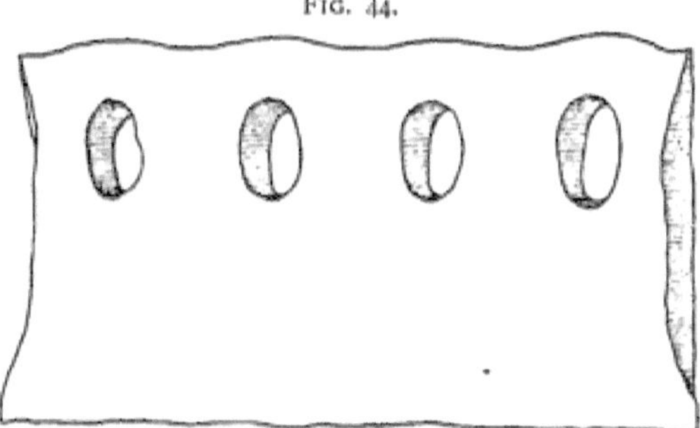

cent. greater than iron rivets. The engraving (Fig. 44) is a typical illustration of the elongation of rivet-holes where the rivets were sheared by the plate. This illustration was prepared from a tracing drawn over a photograph of a joint in a ⅝-inch steel plate with 4 1⅛ rivets on

3⅛-inch pitch. The elongations of rivet-holes were 0.64, 0.68, 0.73, and 0.83 decimal parts of an inch respectively.

(2) By crushing and tearing out the plate in front of the rivet (Figs. 45 and 46). This method of failure depends usually upon the thickness of plate relative to diameter of rivet. Such a fracture suggests that smaller rivets be used if the distance from the edge of the plate to the side of the hole be not less than the diameter of the hole. The engraving (Fig. 46) was made from a tracing drawn over a photograph of a steel plate ½ inch thick with 5 1⅛ rivets on 3⅛-inch pitch. The plate tore out in front of the rivets as shown.

Fig. 45.

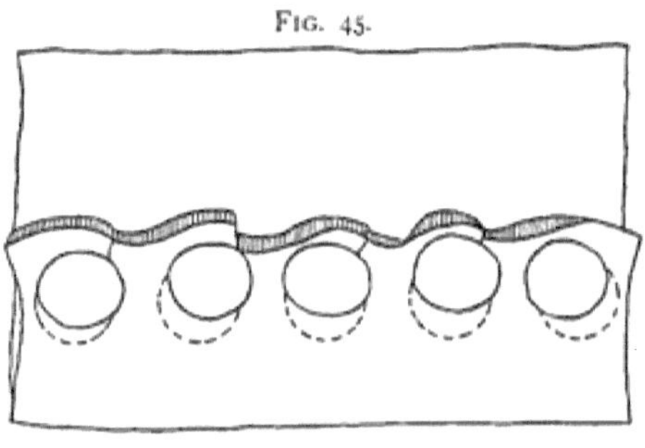

(3) By tearing the plate along the line of rivet-holes (Fig. 47). This fracture depends upon the thickness of the plate and the pitch of the rivets. The resistance to this mode of fracture equals the effective sectional area of plate between holes multiplied by its tensile strength per square inch. The engraving (Fig. 47) was made from a tracing drawn over a photograph of a steel plate $\frac{5}{16}$ inch thick with 1-inch drilled holes on 3½-inch pitch. The efficiency of this joint was 80.1 per cent.

Fig. 46.

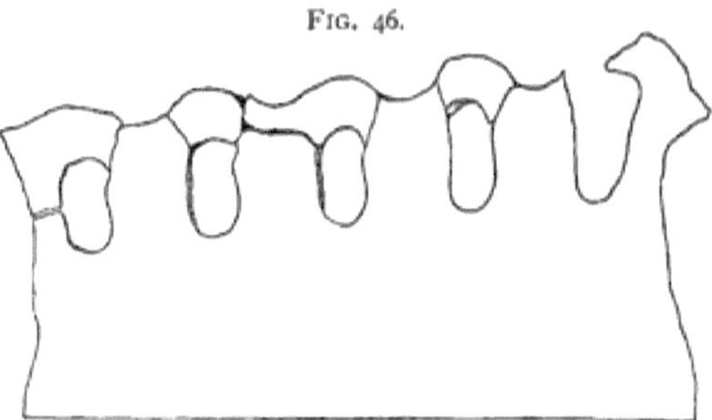

A riveted joint should fail preferably by fracturing the plate along the line of rivets from hole to hole; if failure should occur by either of the first two methods it indicates an excess of strength in the net section of the plate through the line of rivet-holes which has not been made use of; if, on the other hand, fracture occurs along the line of rivet-holes in a well-proportioned joint, it is immaterial whether there is an excess of strength in other directions or not.

FIG. 47.

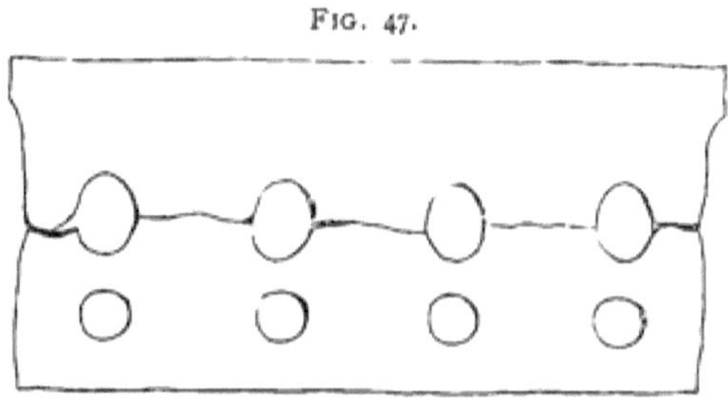

It was observed during the tests of riveted joints at the Watertown Arsenal that the failure of a joint was generally marked by three well-defined periods. In the first period the greatest rigidity was found, and it was thought that the joint was then held entirely by friction of the rivet-heads, and the movement of the joint was principally that due to the elasticity of the metal.

The second period was distinguished by a rapid increase in the stretch of the joint, attributed to the overcoming of the friction under the rivet-heads and closing up any clearance about the rivets, bringing them into bearing condition against the fronts of the rivet-holes. Rivets which are said to fill the holes can hardly do so completely, on account of the contraction of the metal of the rivet from a higher temperature than that of the plate after the rivet is driven.

After a brief interval the movement of the joint was retarded, and the third period was reached. The strength of the joint was then believed to be due to the distortion of the rivet-holes and of the rivets themselves. The movement began slowly, and so continued until the elastic limit of the metal about the rivet-holes was passed, and general flow took place over the entire cross-section, and rupture was reached.

These stages in the test of a joint were well defined except when the plates were in a warped condition initially, when abnormal micrometer readings were observed. The difference in behavior of a joint and the solid metal suggested the propriety of arranging tension-joints in boiler construction and elsewhere as nearly in line as possible.

ENGRAVINGS BELONGING TO TABLE XIII.

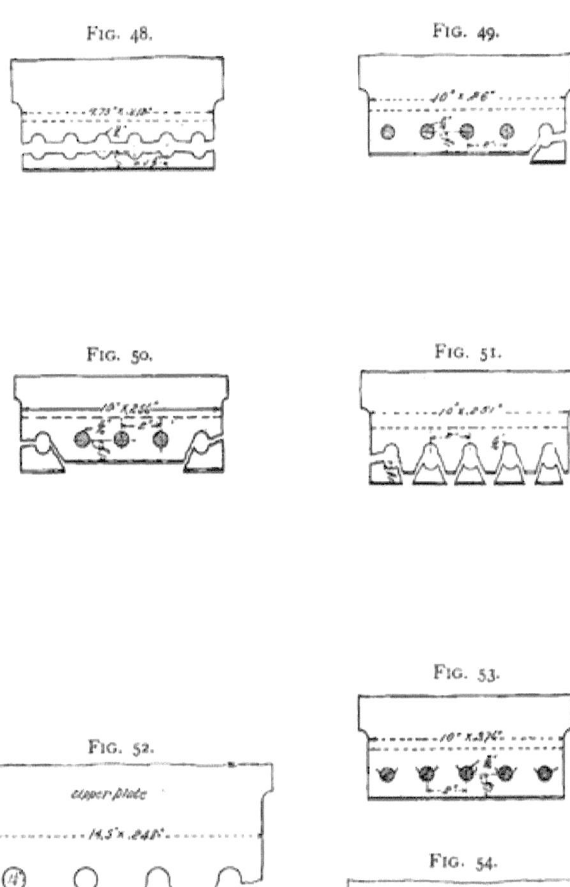

Fig. 48. Fig. 49.

Fig. 50. Fig. 51.

Fig. 52. Fig. 53.

Fig. 54.

ENGRAVINGS BELONGING TO TABLE XIII.—*Continued.*

Fig. 55.

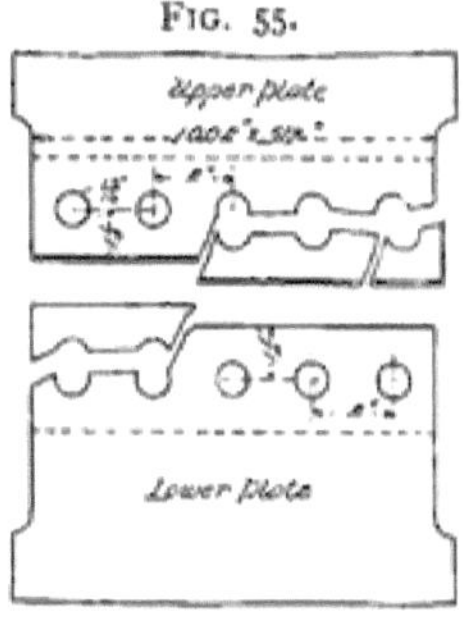

Fig. 56.

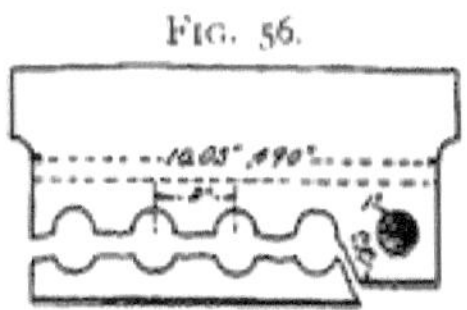

ENGRAVINGS BELONGING TO TABLE XIV.

Fig. 57.

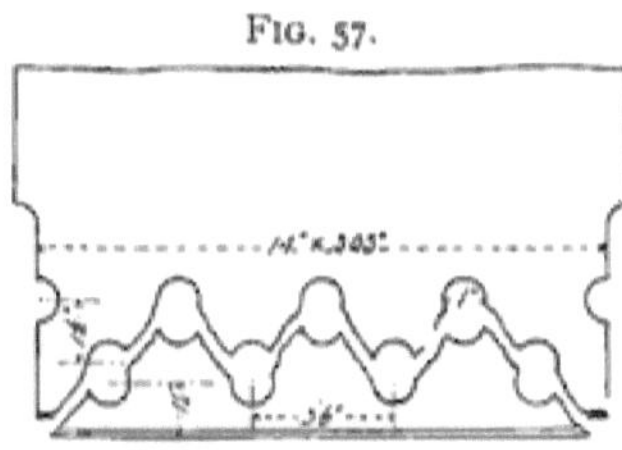

Fig. 58. Fig. 59.

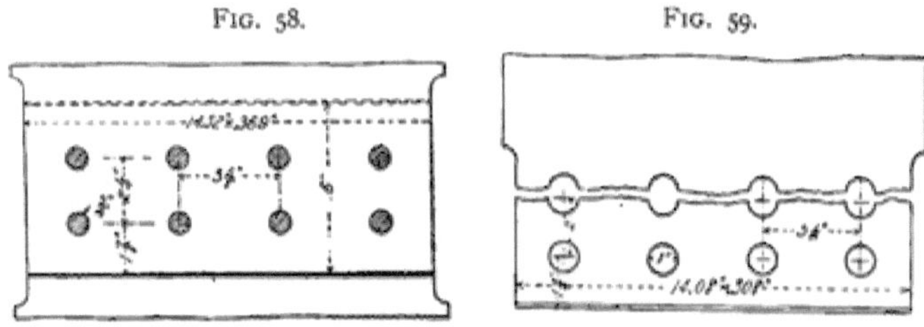

TABLE XV.

TESTS OF TRIPLE-RIVETED LAP-JOINTS, $\frac{7}{16}$-INCH IRON PLATES, WATERTOWN ARSENAL.

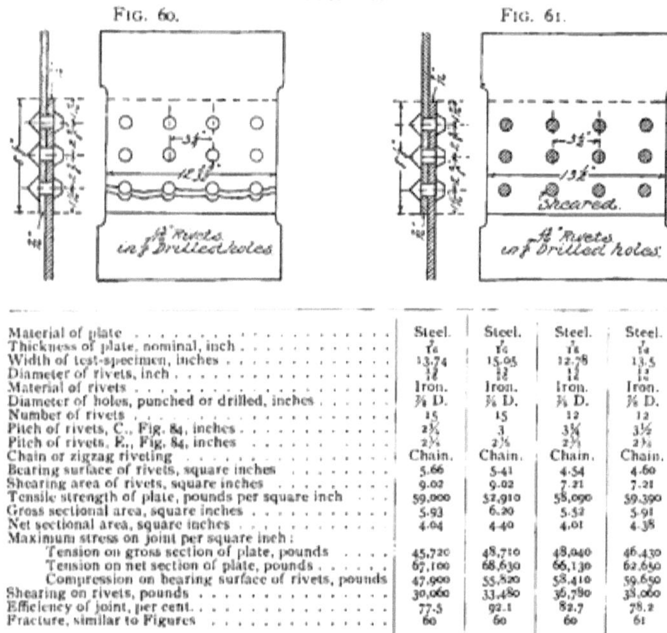

Fig. 60. Fig. 61.

Material of plate	Steel.	Steel.	Steel.	Steel.
Thickness of plate, nominal, inch	$\frac{7}{16}$	$\frac{7}{16}$	$\frac{7}{16}$	$\frac{7}{16}$
Width of test-specimen, inches	13.74	15.05	12.78	13.5
Diameter of rivets, inch	$\frac{13}{16}$	$\frac{13}{16}$	$\frac{13}{16}$	$\frac{13}{16}$
Material of rivets	Iron.	Iron.	Iron.	Iron.
Diameter of holes, punched or drilled, inches	⅞ D.	⅞ D.	⅞ D.	⅞ D.
Number of rivets	15	15	12	12
Pitch of rivets, C., Fig. 84, inches	2¾	3	3⅛	3½
Pitch of rivets, E., Fig. 84, inches	2⅛	2⅛	2⅓	2⅛
Chain or zigzag riveting	Chain.	Chain.	Chain.	Chain.
Bearing surface of rivets, square inches	5.66	5.41	4.54	4.60
Shearing area of rivets, square inches	9.02	9.02	7.21	7.21
Tensile strength of plate, pounds per square inch	59,000	52,910	58,090	59,390
Gross sectional area, square inches	5.93	6.20	5.52	5.91
Net sectional area, square inches	4.04	4.40	4.01	4.38
Maximum stress on joint per square inch:				
Tension on gross section of plate, pounds	45,720	48,710	48,040	46,430
Tension on net section of plate, pounds	67,100	68,630	66,130	62,650
Compression on bearing surface of rivets, pounds	47,900	55,820	58,410	59,650
Shearing on rivets, pounds	30,060	33,480	36,780	38,060
Efficiency of joint, per cent.	77.5	92.1	82.7	78.2
Fracture, similar to Figures	60	60	60	61

Butt-Joints.—It was experimentally shown at the Watertown Arsenal that the behavior of joints in different thicknesses of steel plates, single riveted, as in Fig. 62, was substantially the same whether ¼-inch or ¾-inch plates were used.

Fig. 62.

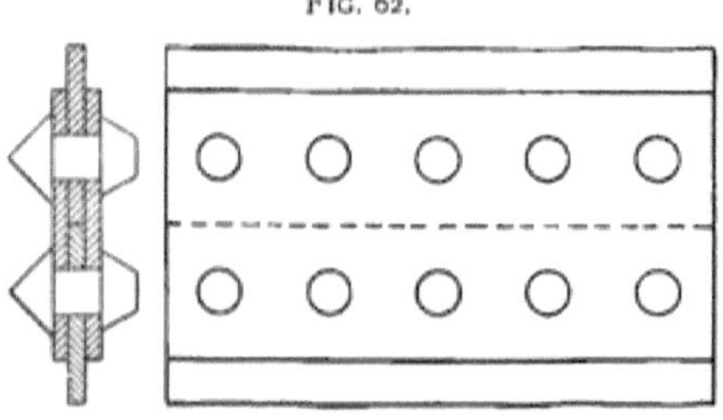

It will not be understood from this, however, that, as a consequence, the same efficiency may be obtained in different thicknesses of plate for single-riveted work, because certain essential conditions change as we approach the stronger joints in different thicknesses of plate.

TOWN ARSENAL.

Material of plate	Iron.	Iron.	Steel.	Steel.	Iron.	Steel.	Iron.	Steel.
Thickness of plate, nominal, i	½	½	½	½	⅝	⅝	¾	¾
Width of test-specimen, inches	10.03	10.02	10.08	10.03	10.5	9.5	12.02	10
Diameter of rivets, inches . .	¾	¾	¾	13/16	1	1	1⅛	1⅛
Material of rivets	Iron.	Steel.	Iron.	Iron.	Iron.	Steel.	Iron.	Steel.
Diameter of holes, punched or	13/16 P.	13/16 D.	13/16 P.	1 P.	1 1/16 P.	1 1/16 P.	1 3/16 P.	1 3/16 P.
Number of rivets	5	5	5	5	4	4	4	4
Pitch of rivets, inches	2	2	2	2	2⅝	2⅝	3	2⅝
Bearing surface of rivets, squa	2.18	2.06	2.13	2.55	2.65	2.72	3.44	3.64
Shearing area of rivets, square	2.64	2.58	2.64	3.93	3.55	3.55	4.43	4.43
Tensile strength of plate, pou	44,615	44,615	57,215	57,215	44,635	52,445	46,590	51,545
Gross sectional area, square in	5.10	5.10	4.97	4.97	6.28	5.83	8.38	7.38
Net sectional area, square inch	2.91	3.04	2.84	2.42	3.62	3.11	4.93	3.73
Maximum stress on joint per								
Tension gross section of	17,760	24,200	21,830	24,145	19,750	26,480	17,230	23,940
Tension net section of p	31,100	40,590	38,204	49,590	34,230	49,650	29,290	47,370
Compression on bearing	41,500	59,900	50,940	47,060	46,790	56,760	41,980	48,540
Shearing on rivets, pounds	34,280	47,830	41,100	30,540	34,930	43,490	32,600	39,890
Efficiency of joint, per cent.	39.8	54.2	38.2	51.2	42	50.5	37	46.4
Fracture, similar to Figures	48	55	53	56	48	53	48	48

TOWN ARSENAL.

Material of plate	Steel.	Steel.	Steel.	Steel.	Steel.	Iron.	Iron.	Steel.
Thickness of plate, nominal, i	⅜	7/16	7/16	½	⅝	⅝	¾	¾
Width of test-specimen, inche	14.36	15.01	13.77	10	10.5	10.5	12	9.98
Diameter of rivets, inches	13/16	13/16	13/16	13/16	1	1	1⅛	1⅛
Material of rivets	Iron.	Iron.	Iron.	Iron.	Steel.	Iron.	Iron.	Steel.
Diameter of holes, punched o	1 D.	⅞ D.	⅞ D.	1 P.	1 1/16 P.	1 1/16 P.	1 3/16 P.	1 3/16 P.
Number of rivets	10	12	10	10	8	8	8	8
Pitch of rivets, C., Fig. 82, in	2⅞	2½	2¾	2	2⅜	2⅝	3	2⅝
Pitch of rivets, E., Fig. 82, in	2½	2⅜	2⅜	2	2½	2½	2¾	1¾
Chain or zigzag riveting . . .	Chain.	Chain.	Chain.	Chain.	Chain.	Chain.	Chain.	Chain.
Bearing surface of rivets, squ	3.70	4.29	3.69	5.26	5.43	5.48	6.92	7.30
Shearing area of rivets, squa	7.85	7.21	6.01	7.85	7.09	7.09	8.86	8.86
Tensile strength of plate, pou	56,670	52,910	59,000	57,215	52,445	44,635	46,590	51,545
Gross sectional area, square i	5.31	6.14	5.81	5.02	5.81	6.48	8.41	7.39
Net sectional area, square inc	3.46	3.99	3.96	2.39	3.10	3.74	4.95	3.74
Maximum stress on joint per								
Tension gross section o	43,050	43,530	38,830	30,757	35,800	25,150	24,720	24,050
Tension net section of	66,070	66,990	56,990	64,602	67,100	43,580	42,000	47,510
Compression on bearin	61,780	62,310	61,170	29,354	38,300	29,740	30,040	24,340
Shearing on rivets, pounds	29,120	37,070	37,550	19,670	29,340	22,990	23,460	20,060
Efficiency of joint, per cent.	76.0	82.3	65.8	53.8	68.3	56.3	53.1	46.7
Fracture, similar to Figures	59	58	58	59	59	59	59	59

TABLE XIII.

TESTS OF SINGLE-RIVETED LAP-JOINTS, ¼-INCH TO ¾-INCH IRON AND STEEL PLATES, WATERTOWN ARSENAL.

Material of plate	Iron.	Iron.	Iron.	Steel.	Steel.	Steel.	Steel.	Iron.	Iron.	Steel.	Steel.	Steel.	Steel.	Iron.	Iron.	Steel.	Steel.	Iron.	Steel.	Iron.	Steel.
Thickness of plate, nominal, inch	[illegible]	[illegible]	[illegible]	[illegible]	[illegible]	[illegible]	[illegible]	[illegible]	[illegible]	[illegible]	[illegible]	[illegible]	[illegible]	[illegible]	[illegible]	[illegible]	[illegible]	[illegible]	[illegible]	[illegible]	[illegible]
Width of test-specimen, inches	[illegible]	[illegible]	[illegible]	[illegible]	[illegible]	[illegible]	[illegible]	[illegible]	[illegible]	[illegible]	[illegible]	[illegible]	[illegible]	[illegible]	[illegible]	[illegible]	[illegible]	[illegible]	[illegible]	[illegible]	[illegible]
Diameter of rivets, inches	[illegible]	[illegible]	[illegible]	[illegible]	[illegible]	[illegible]	[illegible]	[illegible]	[illegible]	[illegible]	[illegible]	[illegible]	[illegible]	[illegible]	[illegible]	[illegible]	[illegible]	[illegible]	[illegible]	[illegible]	[illegible]
Material of rivets	Iron.	Iron.	Iron.	Steel.	Steel.	Steel.	Iron.	Iron.	Iron.	Iron.	Steel.	Iron.	Iron.	Iron.	Steel.	Iron.	Iron.	Iron.	Steel.	Iron.	Steel.
Diameter of holes, punched or drilled, inches	[illegible]	[illegible]	[illegible]	[illegible]	[illegible]	[illegible]	[illegible]	[illegible]	[illegible]	[illegible]	[illegible]	[illegible]	[illegible]	[illegible]	[illegible]	[illegible]	[illegible]	[illegible]	[illegible]	[illegible]	[illegible]
Number of rivets	[illegible]	[illegible]	[illegible]	[illegible]	[illegible]	[illegible]	[illegible]	[illegible]	[illegible]	[illegible]	[illegible]	[illegible]	[illegible]	[illegible]	[illegible]	[illegible]	[illegible]	[illegible]	[illegible]	[illegible]	[illegible]
Pitch of rivets, inches	[illegible]	[illegible]	[illegible]	[illegible]	[illegible]	[illegible]	[illegible]	[illegible]	[illegible]	[illegible]	[illegible]	[illegible]	[illegible]	[illegible]	[illegible]	[illegible]	[illegible]	[illegible]	[illegible]	[illegible]	[illegible]
Bearing surface of rivets, square inches	[illegible]	[illegible]	[illegible]	[illegible]	[illegible]	[illegible]	[illegible]	[illegible]	[illegible]	[illegible]	[illegible]	[illegible]	[illegible]	[illegible]	[illegible]	[illegible]	[illegible]	[illegible]	[illegible]	[illegible]	[illegible]
Shearing area of rivets, square inches	[illegible]	[illegible]	[illegible]	[illegible]	[illegible]	[illegible]	[illegible]	[illegible]	[illegible]	[illegible]	[illegible]	[illegible]	[illegible]	[illegible]	[illegible]	[illegible]	[illegible]	[illegible]	[illegible]	[illegible]	[illegible]
Tensile strength of plate, pounds per square inch	[illegible]	[illegible]	[illegible]	[illegible]	[illegible]	[illegible]	[illegible]	[illegible]	[illegible]	[illegible]	[illegible]	[illegible]	[illegible]	[illegible]	[illegible]	[illegible]	[illegible]	[illegible]	[illegible]	[illegible]	[illegible]
Gross sectional area, square inches	[illegible]	[illegible]	[illegible]	[illegible]	[illegible]	[illegible]	[illegible]	[illegible]	[illegible]	[illegible]	[illegible]	[illegible]	[illegible]	[illegible]	[illegible]	[illegible]	[illegible]	[illegible]	[illegible]	[illegible]	[illegible]
Net sectional area, square inches	[illegible]	[illegible]	[illegible]	[illegible]	[illegible]	[illegible]	[illegible]	[illegible]	[illegible]	[illegible]	[illegible]	[illegible]	[illegible]	[illegible]	[illegible]	[illegible]	[illegible]	[illegible]	[illegible]	[illegible]	[illegible]
Maximum stress on joint per square inch:																					
Tension gross section of plate, pounds	[illegible]	[illegible]	[illegible]	[illegible]	[illegible]	[illegible]	[illegible]	[illegible]	[illegible]	[illegible]	[illegible]	[illegible]	[illegible]	[illegible]	[illegible]	[illegible]	[illegible]	[illegible]	[illegible]	[illegible]	[illegible]
Tension net section of plate, pounds	[illegible]	[illegible]	[illegible]	[illegible]	[illegible]	[illegible]	[illegible]	[illegible]	[illegible]	[illegible]	[illegible]	[illegible]	[illegible]	[illegible]	[illegible]	[illegible]	[illegible]	[illegible]	[illegible]	[illegible]	[illegible]
Compression on bearing surface of rivets, pounds	[illegible]	[illegible]	[illegible]	[illegible]	[illegible]	[illegible]	[illegible]	[illegible]	[illegible]	[illegible]	[illegible]	[illegible]	[illegible]	[illegible]	[illegible]	[illegible]	[illegible]	[illegible]	[illegible]	[illegible]	[illegible]
Shearing on rivets, pounds	[illegible]	[illegible]	[illegible]	[illegible]	[illegible]	[illegible]	[illegible]	[illegible]	[illegible]	[illegible]	[illegible]	[illegible]	[illegible]	[illegible]	[illegible]	[illegible]	[illegible]	[illegible]	[illegible]	[illegible]	[illegible]
Efficiency of joint, per cent	[illegible]	[illegible]	[illegible]	[illegible]	[illegible]	[illegible]	[illegible]	[illegible]	[illegible]	[illegible]	[illegible]	[illegible]	[illegible]	[illegible]	[illegible]	[illegible]	[illegible]	[illegible]	[illegible]	[illegible]	[illegible]
Fracture, similar to Figures	[illegible]	[illegible]	[illegible]	[illegible]	[illegible]	[illegible]	[illegible]	[illegible]	[illegible]	[illegible]	[illegible]	[illegible]	[illegible]	[illegible]	[illegible]	[illegible]	[illegible]	[illegible]	[illegible]	[illegible]	[illegible]

TABLE XIV.

TESTS OF DOUBLE-RIVETED LAP-JOINTS, ¼-INCH TO ¾-INCH IRON AND STEEL PLATES, WATERTOWN ARSENAL.

Material of plate	Steel.	Steel.	Steel.	Steel.	Steel.	Steel.	Iron.	Steel.	Steel.	Steel.	Steel.	Steel.	Steel.	Steel.	Steel.	Steel.	Iron.	Iron.	Steel.
Thickness of plate, nominal, inch	[illegible]	[illegible]	[illegible]	[illegible]	[illegible]	[illegible]	[illegible]	[illegible]	[illegible]	[illegible]	[illegible]	[illegible]	[illegible]	[illegible]	[illegible]	[illegible]	[illegible]	[illegible]	[illegible]
Width of test-specimen, inches	[illegible]	[illegible]	[illegible]	[illegible]	[illegible]	[illegible]	[illegible]	[illegible]	[illegible]	[illegible]	[illegible]	[illegible]	[illegible]	[illegible]	[illegible]	[illegible]	[illegible]	[illegible]	[illegible]
Diameter of rivets, inches	[illegible]	[illegible]	[illegible]	[illegible]	[illegible]	[illegible]	[illegible]	[illegible]	[illegible]	[illegible]	[illegible]	[illegible]	[illegible]	[illegible]	[illegible]	[illegible]	[illegible]	[illegible]	[illegible]
Material of rivets	Iron.	Iron.	Iron.	Iron.	Iron.	Iron.	Iron.	Iron.	Iron.	Steel.	Iron.	Iron.	Iron.	Iron.	Iron.	Steel.	Iron.	Iron.	Steel.
Diameter of holes, punched or drilled, inches	[illegible]	[illegible]	[illegible]	[illegible]	[illegible]	[illegible]	[illegible]	[illegible]	[illegible]	[illegible]	[illegible]	[illegible]	[illegible]	[illegible]	[illegible]	[illegible]	[illegible]	[illegible]	[illegible]
Number of rivets	[illegible]	[illegible]	[illegible]	[illegible]	[illegible]	[illegible]	[illegible]	[illegible]	[illegible]	[illegible]	[illegible]	[illegible]	[illegible]	[illegible]	[illegible]	[illegible]	[illegible]	[illegible]	[illegible]
Pitch of rivets, C., Fig. 8a, inches	[illegible]	[illegible]	[illegible]	[illegible]	[illegible]	[illegible]	[illegible]	[illegible]	[illegible]	[illegible]	[illegible]	[illegible]	[illegible]	[illegible]	[illegible]	[illegible]	[illegible]	[illegible]	[illegible]
Pitch of rivets, R., Fig. 8a, inches	[illegible]	[illegible]	[illegible]	[illegible]	[illegible]	[illegible]	[illegible]	[illegible]	[illegible]	[illegible]	[illegible]	[illegible]	[illegible]	[illegible]	[illegible]	[illegible]	[illegible]	[illegible]	[illegible]
Chain or zigzag riveting	Chain.	Zigzag.	Chain.	Chain.	Zigzag.	Chain.	Zigzag.	Chain.	Zigzag.	Zigzag.	Chain.	Chain.	Chain.	Chain.	Chain.	Chain.	Chain.	Chain.	Chain.
Bearing surface of rivets, square inches	[illegible]	[illegible]	[illegible]	[illegible]	[illegible]	[illegible]	[illegible]	[illegible]	[illegible]	[illegible]	[illegible]	[illegible]	[illegible]	[illegible]	[illegible]	[illegible]	[illegible]	[illegible]	[illegible]
Shearing area of rivets, square inches	[illegible]	[illegible]	[illegible]	[illegible]	[illegible]	[illegible]	[illegible]	[illegible]	[illegible]	[illegible]	[illegible]	[illegible]	[illegible]	[illegible]	[illegible]	[illegible]	[illegible]	[illegible]	[illegible]
Tensile strength of plate, pounds per square inch	[illegible]	[illegible]	[illegible]	[illegible]	[illegible]	[illegible]	[illegible]	[illegible]	[illegible]	[illegible]	[illegible]	[illegible]	[illegible]	[illegible]	[illegible]	[illegible]	[illegible]	[illegible]	[illegible]
Gross sectional area, square inches	[illegible]	[illegible]	[illegible]	[illegible]	[illegible]	[illegible]	[illegible]	[illegible]	[illegible]	[illegible]	[illegible]	[illegible]	[illegible]	[illegible]	[illegible]	[illegible]	[illegible]	[illegible]	[illegible]
Net sectional area, square inches	[illegible]	[illegible]	[illegible]	[illegible]	[illegible]	[illegible]	[illegible]	[illegible]	[illegible]	[illegible]	[illegible]	[illegible]	[illegible]	[illegible]	[illegible]	[illegible]	[illegible]	[illegible]	[illegible]
Maximum stress on joint per square inch:																			
Tension gross section of plate, pounds	[illegible]	[illegible]	[illegible]	[illegible]	[illegible]	[illegible]	[illegible]	[illegible]	[illegible]	[illegible]	[illegible]	[illegible]	[illegible]	[illegible]	[illegible]	[illegible]	[illegible]	[illegible]	[illegible]
Tension net section of plate, pounds	[illegible]	[illegible]	[illegible]	[illegible]	[illegible]	[illegible]	[illegible]	[illegible]	[illegible]	[illegible]	[illegible]	[illegible]	[illegible]	[illegible]	[illegible]	[illegible]	[illegible]	[illegible]	[illegible]
Compression on bearing surface of rivets, pounds	[illegible]	[illegible]	[illegible]	[illegible]	[illegible]	[illegible]	[illegible]	[illegible]	[illegible]	[illegible]	[illegible]	[illegible]	[illegible]	[illegible]	[illegible]	[illegible]	[illegible]	[illegible]	[illegible]
Shearing on rivets, pounds	[illegible]	[illegible]	[illegible]	[illegible]	[illegible]	[illegible]	[illegible]	[illegible]	[illegible]	[illegible]	[illegible]	[illegible]	[illegible]	[illegible]	[illegible]	[illegible]	[illegible]	[illegible]	[illegible]
Efficiency of joint, per cent	[illegible]	[illegible]	[illegible]	[illegible]	[illegible]	[illegible]	[illegible]	[illegible]	[illegible]	[illegible]	[illegible]	[illegible]	[illegible]	[illegible]	[illegible]	[illegible]	[illegible]	[illegible]	[illegible]
Fracture, similar to Figures	[illegible]	[illegible]	[illegible]	[illegible]	[illegible]	[illegible]	[illegible]	[illegible]	[illegible]	[illegible]	[illegible]	[illegible]	[illegible]	[illegible]	[illegible]	[illegible]	[illegible]	[illegible]	[illegible]

If the strength per unit of metal of the net section was constant, it would be a very simple matter to compute the efficiency of any joint, as it would merely be the ratio of the net to the gross area of the plates.

The tenacity of the net section was observed to vary, and this variation extended over wide limits, as shown in the accompanying tables.

The efficiencies shown in Table XVII. are obtained by dividing the tensile stress on the gross area of plate by the tensile strength as represented by the tensile test-strip, stating the values in per cent. of the latter.

This table is valuable as showing at once the efficiencies of different joints wherein the pitch of the rivets and their diameter vary.

It is seen there is considerable latitude allowed in the choice of rivets and pitch without materially changing the efficiency of the joint; thus in ¼-inch plate,—

> ⅝-inch rivets (driven), 1⅞-inch pitch, 72.4 per cent. efficiency,*
> ¾-inch rivets (driven), 2¼-inch pitch, 73.3 per cent. efficiency,
> ⅞-inch rivets (driven), 2⅜-inch pitch, 71.5 per cent. efficiency,
> 1-inch rivets (driven), 2½-inch pitch, 70.3 per cent. efficiency,
> 1-inch rivets (driven), 2⅞-inch pitch, 73.8 per cent. efficiency,

gave nearly the same results.

In these examples the ratios of net to gross areas of plate range from 60 to 67 per cent., while the rivet areas range from 0.3067 square inch to 0.7854 square inch. The actual areas of net sections of plate and rivets are as follows:

	⅝-inch Rivets.	¾-inch Rivets.	⅞-inch Rivets.	1-inch Rivets.
	Square Inch.	Square Inch.	Square Inch.	Square Inch.
Rivets	.3067	.4418	.6013	.7854
Plate	1.486	2.207	2.232	{ 2.259 2.319

The areas of the rivets stand to each other as the following numbers:

100	144	196	256

and the net areas of the plate to each other as

100	149	150	{ 152 156

From these illustrations it appears that to obtain the same degree of efficiency in this quality of metal, although that efficiency is probably

* The difference in per cent. shown in this summary over that of Table XVII. is explained by the fact that this summary covers a larger number of joints resulting as above.

6

not the highest attainable, a fixed ratio between rivet-metal and net section of plate is not essential.

In ½-inch plate with ⅞-inch rivets the efficiencies of the joints are nearly constant over the ranges of pitches tested.

The efficiencies and the ratio of net to gross areas of plate are as follows:

PITCH OF RIVETS.

	1⅞ inches.	2 inches.	2⅛ inches.	2¼ inches.
	Per cent.	Per cent.	Per cent.	Per cent.
Efficiency	64.5	66.3	66.3	66.4
Ratio of areas . . .	53.4	56.3	58.9	61.1

In this we have illustrated a case which, in passing from the widest pitch, having 61.1 per cent. of the solid plate left, to the narrowest pitch, which had 53.4 per cent. of the solid plate, the gain or excess in strength in the net section almost exactly compensated for the loss of metal.

Table XVIII. exhibits the differences between the efficiencies of the joints and the ratios of net to gross areas of plate. If the tenacity of net section remained constant per unit of area, the efficiencies of Table XVII. would, as above explained, be identical with the ratios of net to gross areas of plate, and the values in this table reduced to zero.

Table XIX. shows the excess in strength of the net section of the joint over the strength of the tensile test-strip in per cent. of the latter.

In this table the average of all the joints shows the highest per cent. of excess of strength in the narrowest pitch, and a tendency to lose this excess as the pitch increases.

The maximum gain in strength on the net section was 21.2 per cent., the minimum value 2.5 per cent. of the tensile test-strip. On other forms of joints and with punched holes in both iron and steel plate, illustrations are numerous in which there have been large deficiencies, the metal of the net section falling far below the strength of the plate.

It is believed to have been amply shown that increasing the net width diminishes the apparent tenacity of the plate, although other influences may tend to counteract this tendency in some joints.

In order to compare the excess of strength of one thickness of plate with another having the same net widths, we have Table XVI., rejecting those joints that failed otherwise than along the line of riveting in making the averages.

The excess in strength is generally well maintained in each of the several thicknesses, and were it possible to retain the same ratio of net to gross areas of plate, and at the same time equal net widths between rivets, it would seem from this point of view feasible to obtain the same degree of efficiency in thick as in thin plates.

TABLE XVI.

SHOWING EXCESS OF STRENGTH OF PLATE OF DIFFERENT THICKNESSES AND PITCH OF RIVETS.

Thickness of Plate.	Width of Plate between Rivet-Holes.								
	1"	1⅛"	1¼"	1⅜"	1½"	1⅝"	1¾"	1⅞"	2"
Inch.	Per cent.	Per cent.	Per cent.	Per cent.	Per cent.	Per cent.	Per cent.	Per cent.	Per cent.
¼	16.7	12.6	11.4	12.0	13.4	8.9	11.5	13.1	10.6
⅜	18.4	13.7	12.7	13.5	14.6	12.9	9.0	13.6	. . .
½	16.7	14.3	9.3	10.7	9.1	8.8	8.2	12.2	. . .
⅝	17.7	16.3	14.2	14.5	14.6	12.7	9.9	9.8	. . .
¾	11.4	15.1	13.8	14.1	7.6	11.8	10.0	10.1	3.5
Average of all thicknesses . .	16.2	14.4	12.3	12.9	11.9	11.0	9.7	11.8	7.0

The following causes, however, tend to prevent this consummation:

For equal net widths thick plates require larger rivets to avoid shearing than thin ones, the diameters of the rivets being somewhat increased for this cause; and, again, it has become necessary to increase the metal of the net section in order to retain a suitable ratio of net to gross areas of plate. There results from these considerations such an increase in net width of plate that the excess in strength displayed by narrower sections is lost, and consequently the result is a joint of lower efficiency.

Hot Tests.—Single-riveted butt-joints, steel plate, and wrought-iron rivets, tested at the Watertown Arsenal at temperatures ranging between 200° to 700 Fahr., showed an increase in tensile strength when heated over the duplicate cold joints at each temperature except 200°. From 200° there was a gain in strength up to 300°, when the resistance fell off some at 350°, increased again at 400°, and reached the maximum effect observed at 500°; from this point the strength fell very rapidly, especially at 600° and 700° Fahr.

In per cent. of the cold joint there was a loss at 200° of 3.2 per cent., the average of 3 joints; at 500° the gain was 22.6 per cent., the average of 4 joints. The maximum and minimum joints at this temperature showed gains of 27.6 per cent. and 18.3 per cent. respectively. The highest tensile strength of the net section of plate was found in the joint tested at 500° Fahr., where 81,050 pounds was reached against a strength of 58,000 pounds per square inch in the cold tensile strip.

The hot joints showed less ductility than the cold ones; those tested at 200° Fahr. not being exempt from this behavior, although there was no near approach to brittleness in any.

The shearing strength of iron rivets was also increased by an elevation of temperature. The rivets in one joint of 350° sheared at 43,060 pounds per square inch, while in the duplicate cold joint they sheared at 38,530 pounds per square inch.

TABLE XVII.

TABLE OF EFFICIENCIES OF SINGLE-RIVETED BUTT-JOINTS, STEEL PLATES, WROUGHT-IRON RIVETS, WATERTOWN ARSENAL.

Thickness of Plate.	Pitch of Rivets.															Diameter of Rivet-Holes.
	1⅝"	1¾"	1⅞"	2"	2⅛"	2¼"	2⅜"	2½"	2⅝"	2¾"	2⅞"	3"	3⅛"	3¼"	3⅜"	
Inch.	Per cent.	Per cent.	Per cent.	Per cent.	Per cent.	Per cent.	Per cent.	Per cent.	Per cent.	Per cent.	Per cent.	Per cent.	Per cent.	Per cent.	Per cent.	Inch.
¼	67.5	68.9	72.8	. .	. .	. .	. .	. .	. .	. .	. .	. .	. .	. .	. .	⅝
¼	. .	64.8	67.1	68.6	70.3	74.0	. .	. .	. .	. .	. .	. .	. .	. .	. .	¾
¼	. .	. .	62.6	64.4	67.5	70.1	70.7	76.8	69.8	. .	. .	. .	. .	. .	. .	⅞
¼	. .	. .	. .	58.7	60.6	63.4	68.4	68.5	68.4	75.0	74.1	73.4*	. .	. .	. .	1
⅜	74.6	73.3*	. .	. .	. .	. .	. .	. .	. .	. .	. .	. .	. .	. .	. .	⅝
⅜	. .	68.3	70.1	70.3	68.5*	. .	. .	. .	. .	. .	. .	. .	. .	. .	. .	¾
⅜	. .	. .	63.0	65.2	67.4	68.2	70.4	75.6	. .	. .	. .	. .	. .	. .	. .	⅞
⅜	. .	. .	. .	57.3	59.5	64.1	68.3	69.5	68.3	69.8	75.0	. .	. .	. .	. .	1
½	. .	68.1	68.6*	64.7	. .	. .	. .	. .	. .	. .	. .	. .	. .	. .	. .	¾
½	. .	. .	64.5	66.3	66.3	66.4	66.7	. .	. .	. .	. .	. .	. .	. .	. .	⅞
½	. .	. .	. .	60.1	59.2	61.5	64.5	66.9	67.2	67.6	. .	. .	. .	. .	. .	1
½	. .	. .	. .	. .	51.5	56.5	58.0	62.3	61.2	64.5	67.0	70.0	68.2*	. .	. .	1⅛
⅝	. .	. .	60.1	66.0*	66.9	68.6	. .	. .	. .	. .	. .	. .	. .	. .	. .	⅞
⅝	. .	. .	. .	58.3	60.0	63.6	65.8	66.6*	65.0*	. .	. .	. .	. .	. .	. .	1
⅝	. .	. .	. .	. .	56.5	58.0	60.7	62.1	64.7	66.5	66.6	68.6	69.9*	. .	. .	1⅛
¾	. .	. .	. .	54.2	58.3	59.3*	64.6	. .	. .	. .	. .	. .	. .	. .	. .	1
¾	. .	. .	. .	. .	54.2	57.2	60.6	62.4	61.7	62.2*	68.8	. .	. .	. .	. .	1⅛
¾	. .	. .	. .	. .	. .	49.3	55.6	57.5	59.7	57.0	62.7	65.8	63.6*	63.7	67.7*	1¼

* Denote that joint did not fracture along line of riveting.

TABLE XVIII.

TABLE OF DIFFERENCES BETWEEN THE EFFICIENCIES AND RATIOS OF NET TO GROSS AREAS. SINGLE-RIVETED BUTT-JOINTS, STEEL PLATE, WROUGHT-IRON RIVETS, WATERTOWN ARSENAL.

Thickness of Plate.	Width of Plate Between Rivet-Holes.									Diameter of Rivet-Holes.
	1"	1⅛"	1¼"	1⅜"	1½"	1⅝"	1¾"	1⅞"	2"	
Inch.	Per cent.	Per cent.	Per cent.	Per cent.	Per cent.	Per cent.	Per cent.	Per cent.	Per cent.	Inches.
¼	6.0	4.6	6.1	. .	.	. .	. .	. .	. .	⅝
¼	7.7	7.0	6.1	5.6	8.8	. .	. .	. .	. .	¾
¼	8.0	8.1	8.6	11.7	7.6	11.8	3.1	. .	. .	⅞
¼	18.8	7.6	7.8	10.5	8.5	6.6	13.3	8.8	6.7*	1
⅜	13.1	9.1*	. .	. .	. .	. .	. .	. .	. .	⅝
⅜	11.2	10.1	7.8	3.8*	. .	. .	. .	. .	. .	¾
⅜	9.7	8.9	8.6	7.1	7.2	10.6	. .	. .	. .	⅞
⅜	7.3	6.6	5.9	10.4	9.5	6.4	6.2	9.8	. .	1
½	11.1	8.6	6.4	. .	. .	. .	. .	.	. .	¾
½	11.1	10.0	17.2	5.3	12.3	. .	. .	. .	. .	⅞
½	12.8	6.2	4.7	14.9	17.2	5.2	17.8	. .	. .	1
½	4.6	6.5	5.3	21.7	4.1	5.4	6.1	7.5	4.2*	1⅛
⅝	6.4	9.7*	17.4	7.5	. .	. .	. .	. .	. .	⅞
⅝	8.3	7.1	8.0	8.1	6.6*	3.1*	. .	. .	. .	1
⅝	9.4	8.0	8.0	7.1	7.6	7.4	5.8	6.1	5.9*	1⅛
¾	4.1	5.3	3.7*	6.7	. .	. .	. .	. .	. .	1
¾	7.1	7.2	7.9	7.3	4.7	3.1*	7.9	. .	. .	1⅛
¾	4.8	8.2	7.4	6.8	2.4	6.2	7.6	3.5*	2.1	1¼

* Denote that joints did not fracture along line of riveting.

TABLE XIX.

EXCESS IN STRENGTH OF NET SECTION IN JOINT OVER STRENGTH OF TENSILE TEST-STRIP. SINGLE-RIVETED BUTT-JOINTS, STEEL PLATE, WROUGHT-IRON RIVETS, WATERTOWN ARSENAL.

Thickness of Plate.	Width of Plate Between Rivet-Holes.									Diameter of Rivet-Holes.
	1"	1⅛"	1¼"	1⅜"	1½"	1⅝"	1¾"	1⅞"	2"	
Inch.	Per cent.	Per cent.	Per cent.	Per cent.	Per cent.	Per cent.	Per cent.	Per cent.	Per cent.	Inches.
¼	9.8	7.2	9.1		. .	. .	. .	. .	. .	⅝
¼	13.4	11.7	9.7	8.5	11.0	. .	. .	. .	. .	¾
¼	17.4	14.4	14.7	14.7	11.9	16.1	4.7	. .	. .	⅞
¼	17.6	14.5	14.2	18.0	14.1	10.7	17.7	13.5	10.1*	1
⅜	21.2	14.1*	. .	. .	. .	. .	. .	.	. .	⅝
⅜	19.5	16.8	12.5	5.8*	. .	. .	. .	. .	. .	¾
⅜	18.2	15.9	13.5	11.7	11.4	16.5	. .	.	. .	⅞
⅜	14.6	12.5	10.6	17.9	15.9	10.3	9.8	15.1	. .	1
½	15.8	14.4*	10.3	. .	. .	. .	. .	. .	. .	¾
½	20.9	17.8	28.9	8.7	19.5*	. .	. .	. .	. .	⅞
½	25.5	11.8	5.5	25.8	28.7	8.6	28.1	. .	. .	1
½	9.8	13.2	10.1	39.7	7.3	9.1	10.1	12.2	6.5*	1⅛
⅝	15.1	18.8	29.6	12.2	. .	. .	. .	. .	. .	⅞
⅝	16.6	17.2	14.6	14.0	11.1*	5.2*	. .	. .	. .	1
⅝	20.0	18.4	15.3	13.0	13.5	12.7	9.7	9.8	9.2	1⅛
¾	8.3	10.1	6.8*	11.5	. .	. .	. .	. .	. .	1
¾	15.2	14.5	15.2	13.6	8.5	5.3*	11.5	. .	. .	1⅛
¾	10.8	17.4	15.0	13.2	4.7	11.0	13.0	5.7*	3.5	1¼

* Denote that joints did not fracture along line of riveting.

TABLE XX.

TESTS OF SINGLE-RIVETED BUTT-JOINTS, ¼-INCH TO ¾-INCH STEEL PLATES, WROUGHT-IRON RIVETS, WATERTOWN ARSENAL.

Material of plate	Steel.	Steel.	Steel.	Steel.	Steel.	Steel.	Steel.	Steel.	Steel.	Steel.
Thickness of plate, nominal, inch	¼	¼	⅜	⅜	½	½	⅝	⅝	¾	¾
Width of test-specimen, inches	9.75	11.25	10.50	12	11.25	14.25	12	14.25	12.75	14.375
Diameter of rivets, inches	⅝	⅝	¾	¾	⅞	⅞	1	1	1⅛	1⅛
Material of rivets	Iron.	Iron.	Iron.	Iron.	Iron.	Iron.	Iron.	Iron.	Iron.	Iron.
Diameter of holes, punched or drilled, inches	11/16	11/16	13/16	13/16	15/16	15/16	1 1/16	1 1/16	1 3/16	1 3/16
Number of rivets	6	6	6	6	6	6	6	6	6	5
Pitch of rivets, inches	1⅝	1⅞	1¾	2	1⅞	2⅜	2	2⅜	2⅛	2⅞
Bearing surface of rivets, square inches	.908	.915	1.652	1.755	2.525	2.546	3.744	3.660	5.076	4.168
Shearing area of rivets, square inches	3.68	3.68	5.30	5.30	7.22	7.22	9.43	9.43	11.93	9.94
Tensile strength of plate, pounds per square inch	61,740	61,740	54,260	59,730	57,180	60,000	55,000	57,290	59,000	59,000
Gross sectional area, square inches	2.36	2.75	3.85	4.68	5.42	6.92	7.49	8.66	9.60	10.66
Net sectional area, square inches	1.45	1.83	2.20	2.93	2.89	4.37	3.74	5.00	4.52	6.50
Maximum stress on joint per square inch:										
Tension on gross section of plate, pounds	41,690	44,920	37,050	42,000	36,890	40,050	32,080	37,720	32,000	40,630
Tension on net section of plate, pounds	67,770	67,380	64,840	67,200	69,110	63,390	64,150	65,310	67,940	65,810
Compression on bearing surface of rivets, pounds	108,370	134,750	86,430	112,000	79,130	108,800	64,150	89,260	65,340	90,600
Shearing on rivets, pounds	26,720	33,490	26,940	37,090	27,690	38,530	25,480	38,490	25,740	42,970
Efficiency of joint, per cent.	67.5	72.8	68.3	70.3	64.5	66.7	58.3	65.8	54.2	68.8
Failure occurred in	Plate.	Plate.	Plate.	Plate.	Plate.	Rivets.	Plate.	Plate.	Plate.	Plate.

ENGRAVINGS BELONGING TO TABLE XXI.

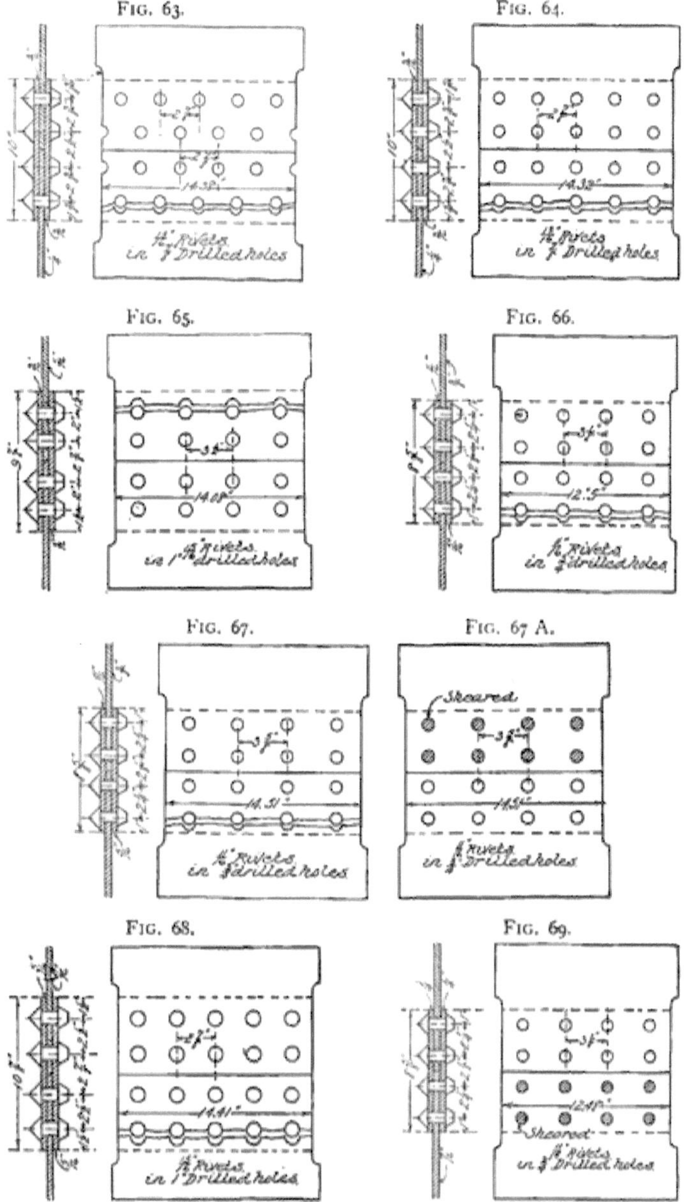

Fig. 63. Fig. 64. Fig. 65. Fig. 66. Fig. 67. Fig. 67 A. Fig. 68. Fig. 69.

ENGRAVINGS BELONGING TO TABLE XXI.—*Continued.*

FIG. 70.

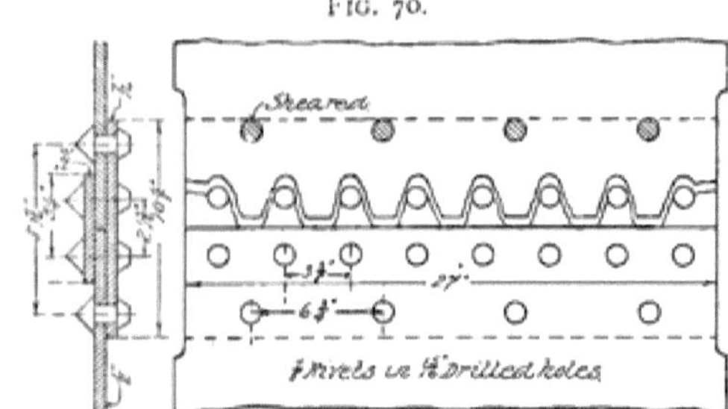

FIG. 71. FIG. 72.

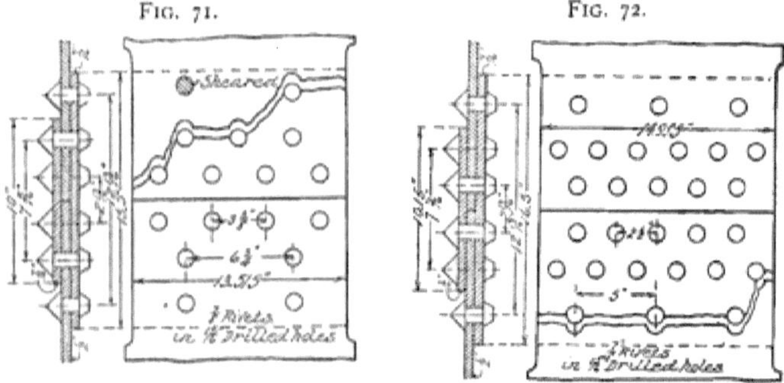

FIG. 73. FIG. 74.

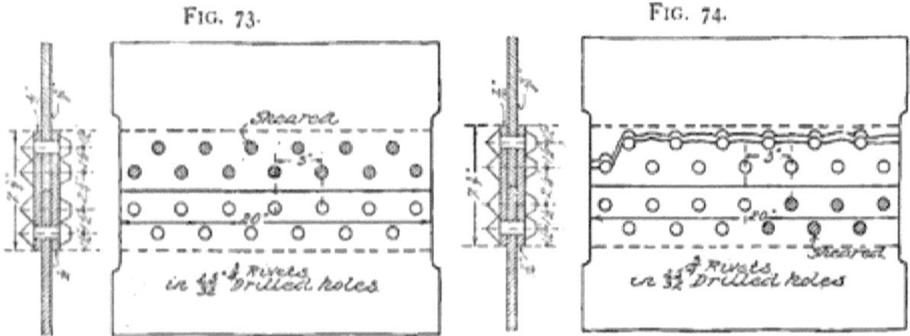

ENGRAVINGS BELONGING TO TABLE XXI.—*Continued.*

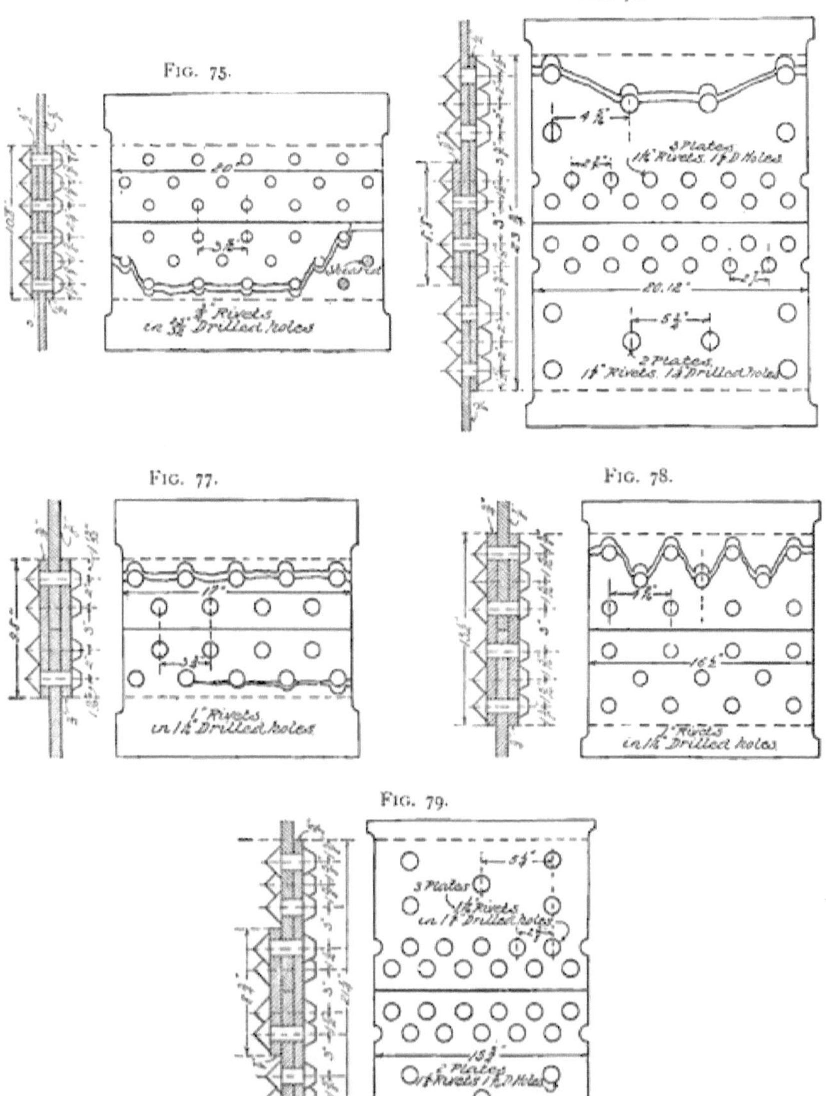

FIG. 75.

FIG. 76.

FIG. 77.

FIG. 78.

FIG. 79.

PROPORTIONING RIVETED JOINTS.

Single-Riveted Lap-Joint.—Such a joint consists of two plates overlapping each other and secured by a single row of rivets, as shown in Fig. 80. This is the simplest and weakest form of riveted joint. In estimating the value of a rivet under shearing stress the diameter of the hole should be taken, because the rivet is upset to fill the hole. The shearing strength of rivets is given in Tables VII. and X.

FIG. 80.

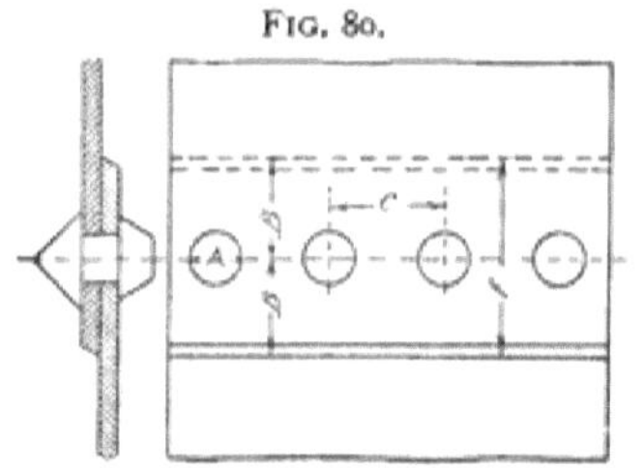

The strength of a single-riveted lap-joint using $\frac{1}{4}$-inch steel of 55,000 pounds tensile strength, steel rivets $\frac{5}{8}$ inch diameter in $\frac{11}{16}$-inch holes, 2 inches pitch, may be calculated thus:

Gross section of plate, 2 inches × .25 inch	.5000 square inch.
Section removed by hole, .6875 inch × .25 inch	.1719 square inch.
Net section of plate between holes	.3281 square inch.
Area of upset rivet, $\frac{11}{16}$ inch	.3712 square inch.
Shearing strength of rivet at 44,625 pounds per square inch	16,565 pounds.
Plate (gross), .5 square inch × 55,000 pounds	27,500 pounds.
Plate (net), .3281 square inch × 55,000 pounds	18,045 pounds.

Strength of joint in per cent. of solid plate would be

$$\frac{18{,}045 \times 100}{27{,}500} = 65.6 \text{ per cent.}$$

if it were not for the fact that the shearing strength of the rivet is less than the strength of the plate; we then have

$$\frac{16{,}565 \times 100}{27{,}500} = 60.02 \text{ per cent.}$$

An examination of the above figures shows that the net strength of the plate is 18,045 pounds and the shearing strength of the rivets 16,565 pounds. This proportion of strength of plate to that of rivet is permissible in single-riveted joints, because shearing of rivets or failure of any kind at the joint seldom occurs except by corrosion. The excess strength ought therefore to go into the plate rather than into the rivet, because the plate is subject to corrosion or other deterioration which affects the rivets to a less degree.

Single-riveted lap-joints when both plates and rivets are of iron are weaker than similar plates and rivets of steel of the same dimensions, because the latter material will resist tensile and shearing stresses to a greater degree than iron. Single riveting is little used in longitudinal seams, except for small boilers and low pressures; but single-riveted

Material of plate	Steel.	Steel.	Steel.	Steel.	Steel.	Steel.
Thickness of plate, nominal, i	⅝	⅝	⅝	⅞	⅞	⅞
Width of test-specimen, inche	20	20	20.12	17	16.5	15¾
Thickness of welt-strips, inch	½	½	Up. .6, L. .43	¾	¾	Up. ⅝, L. ⅝
Width of upper welt, inches	7.375	10.8	8.8	9.8	13½	8¾
Width of lower welt, inches .	7.375	10.8	23.8	9.8	13½	21¾
Diameter of rivets, inches .	¾	¾	3 pl. = 1 2 pl. = 1⅛	1	1	3 pl. = 1 1/16 2 pl. = 1⅛
Material of rivets	Steel.	Iron.	Iron.	Steel.	Steel.	Iron.
Diameter of holes, punched o	13/16	13/16	1 1/16 and 1¼	1 1/16 D.	1 1/16 D.	1⅛ and 1¼
Number of rivets in one-half	13	16	19	9	11	16
Pitch of rivets, middle row, in	3	3 9/16	2⅝	3¾	4 9/16	2⅝
For details of riveting, see Fi	74	75	76	77	78	79
Bearing surface of rivets, squ	6.30	7.73	13.58	8.23	10.23	16.38
Shearing area of rivets, squar	12.46	15.34	30.41	15.96	19.51	28.00
Tensile strength of plate, pour	53,710	53,710	53,710	51,190	51,190	51,190
Gross sectional area, square i	12.41	12.36	12.81	14.646	14.438	13.852
Net sectional area, square inc	9.50	9.94	10.19	10.07	10.72	10.99
Maximum stress on joint per						
Tension on gross section	42,720	43,600	46,070	36,190	38,400	41,740
Tension on net section o	55,800	54,220	57,940	52,640	51,720	52,640
Compression on bearing	84,140	69,720	43,470	64,410	54,190	35,300
Shearing on rivets, pounds .	42,540	35,130	19,410	33,210	28,420	20,650
Efficiency of joint, per cent. .	79.5	81.2	85.8	70.7	75.0	81.5
Fracture, similar to Figures	74	75	76	77	78	79

TABLE XXI.

TESTS OF DOUBLE-, TRIPLE-, AND QUADRUPLE-RIVETED BUTT-JOINTS, STEEL PLATES, IRON AND STEEL RIVETS, WATERTOWN ARSENAL.

Material of plate	Steel.	Steel.	Steel.	Steel.	Steel.	Steel.	Steel.	Steel.	Steel.	Steel.	Steel.	Steel.	Steel.	Steel.	Steel.	Steel.	Steel.	Steel.	Steel.
Thickness of plate, nominal, inch	[illegible]	[illegible]	[illegible]	[illegible]	[illegible]	[illegible]	[illegible]	[illegible]	[illegible]	[illegible]	[illegible]	[illegible]	[illegible]	[illegible]	[illegible]	[illegible]	[illegible]	[illegible]	[illegible]
Width of test-specimen, inches	[illegible]	[illegible]	[illegible]	[illegible]	[illegible]	[illegible]	[illegible]	[illegible]	[illegible]	[illegible]	[illegible]	[illegible]	[illegible]	[illegible]	[illegible]	[illegible]	[illegible]	[illegible]	[illegible]
Thickness of welt-strips, inch	[illegible]	[illegible]	[illegible]	[illegible]	[illegible]	[illegible]	[illegible]	[illegible]	[illegible]	[illegible]	[illegible]	[illegible]	[illegible]	[illegible]	[illegible]	[illegible]	[illegible]	[illegible]	[illegible]
Width of upper welt, inches	[illegible]	[illegible]	[illegible]	[illegible]	[illegible]	[illegible]	[illegible]	[illegible]	[illegible]	[illegible]	[illegible]	[illegible]	[illegible]	[illegible]	[illegible]	[illegible]	[illegible]	[illegible]	[illegible]
Width of lower welt, inches	[illegible]	[illegible]	[illegible]	[illegible]	[illegible]	[illegible]	[illegible]	[illegible]	[illegible]	[illegible]	[illegible]	[illegible]	[illegible]	[illegible]	[illegible]	[illegible]	[illegible]	[illegible]	[illegible]
Diameter of rivets, inches	[illegible]	[illegible]	[illegible]	[illegible]	[illegible]	[illegible]	[illegible]	[illegible]	[illegible]	[illegible]	[illegible]	[illegible]	[illegible]	[illegible]	[illegible]	[illegible]	[illegible]	[illegible]	[illegible]
Material of rivets	Iron.	Iron.	Iron.	Iron.	Iron.	Iron.	Iron.	Iron.	Iron.	Steel.	Steel.	Steel.	Steel.	Steel.	Iron.	Iron.	Steel.	Steel.	Iron.
Diameter of holes, punched or drilled, inches	[illegible]	[illegible]	[illegible]	[illegible]	[illegible]	[illegible]	[illegible]	[illegible]	[illegible]	[illegible]	[illegible]	[illegible]	[illegible]	[illegible]	[illegible]	[illegible]	[illegible]	[illegible]	[illegible]
Number of rivets in one-half of joint	[illegible]	[illegible]	[illegible]	[illegible]	[illegible]	[illegible]	[illegible]	[illegible]	[illegible]	[illegible]	[illegible]	[illegible]	[illegible]	[illegible]	[illegible]	[illegible]	[illegible]	[illegible]	[illegible]
Pitch of rivets, middle row, inches	[illegible]	[illegible]	[illegible]	[illegible]	[illegible]	[illegible]	[illegible]	[illegible]	[illegible]	[illegible]	[illegible]	[illegible]	[illegible]	[illegible]	[illegible]	[illegible]	[illegible]	[illegible]	[illegible]
For details of riveting, see Figures	[illegible]	[illegible]	[illegible]	[illegible]	[illegible]	[illegible]	[illegible]	[illegible]	[illegible]	[illegible]	[illegible]	[illegible]	[illegible]	[illegible]	[illegible]	[illegible]	[illegible]	[illegible]	[illegible]
Bearing surface of rivets, square inches	[illegible]	[illegible]	[illegible]	[illegible]	[illegible]	[illegible]	[illegible]	[illegible]	[illegible]	[illegible]	[illegible]	[illegible]	[illegible]	[illegible]	[illegible]	[illegible]	[illegible]	[illegible]	[illegible]
Shearing area of rivets, square inches	[illegible]	[illegible]	[illegible]	[illegible]	[illegible]	[illegible]	[illegible]	[illegible]	[illegible]	[illegible]	[illegible]	[illegible]	[illegible]	[illegible]	[illegible]	[illegible]	[illegible]	[illegible]	[illegible]
Tensile strength of plate, pounds per square inch	[illegible]	[illegible]	[illegible]	[illegible]	[illegible]	[illegible]	[illegible]	[illegible]	[illegible]	[illegible]	[illegible]	[illegible]	[illegible]	[illegible]	[illegible]	[illegible]	[illegible]	[illegible]	[illegible]
Gross sectional area, square inches	[illegible]	[illegible]	[illegible]	[illegible]	[illegible]	[illegible]	[illegible]	[illegible]	[illegible]	[illegible]	[illegible]	[illegible]	[illegible]	[illegible]	[illegible]	[illegible]	[illegible]	[illegible]	[illegible]
Net sectional area, square inches	[illegible]	[illegible]	[illegible]	[illegible]	[illegible]	[illegible]	[illegible]	[illegible]	[illegible]	[illegible]	[illegible]	[illegible]	[illegible]	[illegible]	[illegible]	[illegible]	[illegible]	[illegible]	[illegible]
Maximum stress on joint per square inch:																			
Tension on gross section of plate, pounds	[illegible]	[illegible]	[illegible]	[illegible]	[illegible]	[illegible]	[illegible]	[illegible]	[illegible]	[illegible]	[illegible]	[illegible]	[illegible]	[illegible]	[illegible]	[illegible]	[illegible]	[illegible]	[illegible]
Tension on net section of plate, pounds	[illegible]	[illegible]	[illegible]	[illegible]	[illegible]	[illegible]	[illegible]	[illegible]	[illegible]	[illegible]	[illegible]	[illegible]	[illegible]	[illegible]	[illegible]	[illegible]	[illegible]	[illegible]	[illegible]
Compression on bearing surface of rivets, pounds	[illegible]	[illegible]	[illegible]	[illegible]	[illegible]	[illegible]	[illegible]	[illegible]	[illegible]	[illegible]	[illegible]	[illegible]	[illegible]	[illegible]	[illegible]	[illegible]	[illegible]	[illegible]	[illegible]
Shearing on rivets, pounds	[illegible]	[illegible]	[illegible]	[illegible]	[illegible]	[illegible]	[illegible]	[illegible]	[illegible]	[illegible]	[illegible]	[illegible]	[illegible]	[illegible]	[illegible]	[illegible]	[illegible]	[illegible]	[illegible]
Efficiency of joint, per cent.	[illegible]	[illegible]	[illegible]	[illegible]	[illegible]	[illegible]	[illegible]	[illegible]	[illegible]	[illegible]	[illegible]	[illegible]	[illegible]	[illegible]	[illegible]	[illegible]	[illegible]	[illegible]	[illegible]
Fracture, similar to Figures	[illegible]	[illegible]	[illegible]	[illegible]	[illegible]	[illegible]	[illegible]	[illegible]	[illegible]	[illegible]	[illegible]	[illegible]	[illegible]	[illegible]	[illegible]	[illegible]	[illegible]	[illegible]	[illegible]

joints are in common use for the circumferential seams in steam boilers, of which an example is here given:

Should the circumferential seams of a 72-inch boiler be single or double riveted to withstand a pressure of 120 pounds?

A boiler 72 inches in diameter has an area of 4071.5 square inches. The steam, 120 pounds per square inch, would make total pressure on the head of 4071.5 × 120 = 488,580 pounds.

If the shell be of steel $\frac{7}{16}$ inch thick, the diameter of steel rivets may be $\frac{13}{16}$ inch in $\frac{7}{8}$-inch holes. The area of each hole is 0.601 square inch. The shearing strength of each rivet is 26,833 pounds, as per Table X. If the pitch of the rivets be 2 inches, there will be required half as many rivets as there are inches in circumference of the boiler:

$$\frac{226}{2} = 113 \text{ rivets.}$$

Then 26,833 × 113 = 3,038,129 pounds. If a factor of safety of 6 be chosen, the safe working pressure would be

$$\frac{3,038,129}{6} = 506,355 \text{ pounds,}$$

which is 17,775 pounds additional to the required strength of 488,580 pounds,—showing that the single-riveted joint has ample strength for the conditions given.

A series of experiments to determine the efficiencies of single-riveted lap-joints for pitches from 1⅝ to 3⅝ inches, with holes $\frac{11}{16}$ to 1¼ inches diameter, upon iron and steel plates ¼ inch thick, resulted as follows:

TABLE XXII.

EFFICIENCIES OF SINGLE-RIVETED LAP-JOINTS IN PER CENT. OF THE STRENGTH OF SOLID PLATE.

I, Iron; S, Steel; D, Drilled holes; P, Punched holes.

Plate.		Pitch of Rivets.	Diameter of Rivet-Holes. Rivets $\frac{1}{16}$ Inch Less.				
Thick.	Tensile Strength.		$\frac{11}{16}$	¾	1	1⅛	1¼
¼ I	47,925	1⅝	64.0		. .	. .	. .
¼ S	55,765	1⅝	68.8		. .	.	. .
¼ I	47,925	2	57.1 D 64.1 P		. .	. .	. .
¼ S	55,765	2	51.4 I 71.8 S	. .	. .	. .	. .
¼ S	61,000	2⅛	. . .	65.1	. .	. .	. .
¼ S	58,150	2⅜	. .	. .	69.7	. .	. .
¼ S	61,000	2⅝	. . .	. .	. .	70.9	. .
¼ S	58,150	2⅞	. .	. .	. .	. .	70.7
¼ S	58,150	3⅛	. .	. .	. .		69.1
¼ S	55,740	3⅜	. . .	. .	. .	. .	75.7
¼ S	55,740	3⅝	. .	. .	. .	. .	67.3

The above results show that no practical advantage is had in using rivets larger than ⅝ inch in ¼-inch plates.

TABLE XXIII.

SINGLE-RIVETED LAP-JOINTS, IRON PLATES AND IRON RIVETS.

Table of least distance between rivet-holes for iron plates of 45,000 pounds tensile strength, and iron rivets of 34,200 pounds shearing strength, that the plates and rivets shall approximate each other in strength.

Proportionate Strength of Plate at 45,000 Pounds per Square Inch.	Rivets.		Width of Plate to equal Shearing Strength of Rivet.	Pitch of Rivets.		Percentage of Joint.
	Diameter of Hole.	Shearing Strength.		Decimal.	Nearest Working Fraction.	
Inch. Pounds.	Inches.	Pounds.	Inches.	Inches.		
1/4 = 11,250	11/16 = .6875	12,695	1.111	1.799 =	1 13/16	61.76
5/16 = 14,063	3/4 = .75	15,110	1.074	1.824 =	1 13/16	58.88
3/8 = 16,875	13/16 = .8125	17,733	1.051	1.864 =	1 7/8	56.38
7/16 = 19,688	7/8 = .875	20,564	1.044	1.919 =	1 15/16	54.40
1/2 = 22,500	15/16 = .9375	23,608	1.049	1.986 =	2	52.82
9/16 = 24,313	1 = 1.000	26,861	1.105	2.105 =	2 1/8	52.49
5/8 = 28,125	1 1/16 = 1.0625	30,322	1.078	2.141 =	2 1/8	50.35
11/16 = 30,938	1 1/8 = 1.125	33,995	1.099	2.224 =	2 1/4	49.41
3/4 = 33,750	1 3/16 = 1.1875	37,877	1.122	2.310 =	2 5/16	48.57

TABLE XXIV.

SINGLE-RIVETED LAP-JOINTS, STEEL PLATES AND STEEL RIVETS.

Table of least distance between rivet-holes for steel plates of 55,000 pounds tensile strength and steel rivets of 44,625 pounds shearing strength, that the plates and rivets shall approximate each other in strength.

Proportionate Strength of Plates at 55,000 Pounds per Square Inch.	Rivets.		Width of Plate to equal Shearing Strength of Rivet.	Pitch of Rivets.		Percentage of Joint.
	Diameter of Hole.	Shearing Strength.		Decimal.	Nearest Working Fraction.	
Inch. Pounds.	Inches.	Pounds.	Inches.	Inches.		
1/4 = 13,750	11/16 = .6875	16,565	1.205	1.892 =	1 7/8	63.69 P.
5/16 = 17,188	3/4 = .75	19,715	1.147	1.897 =	1 7/8	60.46 P.
3/8 = 20,625	13/16 = .8125	23,138	1.121	1.934 =	1 15/16	57.97 P.
7/16 = 24,063	7/8 = .875	26,833	1.115	1.990 =	2	56.03 R.
1/2 = 27,500	15/16 = .9375	30,805	1.120	2.058 =	2 1/16	54.42 P.
9/16 = 30,938	1 = 1.000	35,048	1.133	2.133 =	2 1/8	53.12 P.
5/8 = 34,375	1 1/16 = 1.0625	39,299	1.143	2.205 =	2 3/16	51.84 P.
11/16 = 37,813	1 1/8 = 1.125	44,357	1.173	2.298 =	2 5/16	51.04 P.
3/4 = 41,250	1 3/16 = 1.1875	49,422	1.198	2.386 =	2 3/8	50.21 P.

P—Joint will probably fail by tearing the plate between rivet-holes.
R—Joint will probably fail by shearing the rivets.

Double-Riveted Lap-Joint.—In this form of joint the plates lap over each other far enough to admit two rows of rivets, as shown in Fig. 81. A double-riveted joint is stronger than a single-riveted joint, because of wider spacing or pitch of rivets giving a larger net section of metal, larger surfaces of frictional contact, and a larger rivet area under shearing stress. Longitudinal seams in steam boilers should be at least double riveted, no plates less than ¼ inch thick, and no rivets less than ⅝ inch diameter, however small the diameter of the boiler.

FIG. 81.

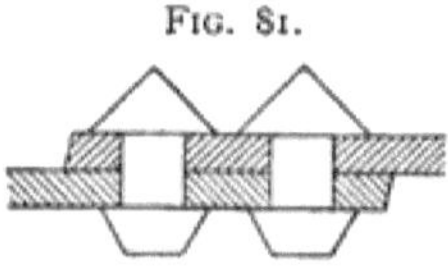

The strength of a double-riveted joint may be calculated thus: Let us assume that steel plates of 55,000 pounds tensile strength are to be joined by iron rivets of 34,200 pounds per square inch shearing strength, the riveting and spacing to be as given below:

Thickness of plate, ¼ inch	.25 inch.
Diameter of rivet-hole, 11/16 inch	.6875 inch.
Area of rivet-hole	.3712 square inch.
Pitch of rivets	2.5 inches.

The strength of the whole plate would be $2.5 \times .25 \times 55{,}000 = 34{,}375$ pounds.

The strength of net section of plate would be $(2.5 - .6875) \times .25 \times 55{,}000 = 24{,}915$ pounds.

The strength of two rivets in single shear would be $.3712 \times 2 \times 34{,}200 = 25{,}390$ pounds, showing that the plate is slightly the weaker of the two.

The percentage of strength of the joint is

$$\frac{24{,}915 \times 100}{34{,}375} = 72.48 \text{ per cent.}$$

Zigzag Riveting is that arrangement of rivets in which one row is placed over the centre of the intervening space, as shown in Fig. 82. This is the style of riveting in almost universal use, having two good qualities,—strength and tightness under pressure. Correctly made, zigzag joints are equal in shearing strength to the net section of the punched plates; that is to say, the value of a joint of this kind approximates 70 per cent. of the whole plate for ¼-inch to ⅜-inch iron plates with iron rivets or steel plates with steel rivets.

Chain Riveting is that arrangement of rivets in which one row is placed exactly above the other, as in Fig. 83. Experiments conducted with a view to ascertaining the comparative strength of chain and zigzag riveting showed that, for the same spacing of the rivets from centre to centre across the sheet, the chain riveting was the strongest. Chain riveting requires a broader lap than zig-zag riveting, and no doubt the friction of this wider joint contributes towards the observed increase in strength. The commonly accepted notion is that the second row of rivets counts for little or nothing in adding to the strength of the joint over that of single riveting for the same pitch ; but this has been proven experimentally not to be true, and the fact is that the arrangement of rivets as in a chain-riveted joint is actually stronger than a zigzag joint of the same relative proportions of rivet to plate area.

Strength of Double-Riveted Joints.—In Table XXV. are given the percentages for joints made up of iron plates with iron rivets, steel plates with iron rivets, and steel plates with steel rivets. It will be noted that most of the percentages fall below 70. As this latter figure is commonly assumed to be the strength of a double-riveted lap-joint, it may be said in explanation that percentages of strength are controlled by the pitch and diameter of rivets, the wider the spacing and the smaller the diameter of the rivet the greater will be the percentage of strength of joint when relative areas of solid and perforated plates are concerned ; but a due regard must be had for shearing strength of rivets which enter into the construction of a joint : narrow pitches and large diameter of rivets means larger resistance against shearing, but the tendency to crush thin plates before large rivets must not be disregarded. Taken altogether, the spacing and dimensioning of rivets in riveted joints is a matter upon which a large amount of care and judgment has been bestowed, resulting in the adoption of practically the dimensions as given in the table, which is fairly representative of the best boiler practice at this time.

Referring to Table XXV., in the column "iron plate and iron rivets," the riveting is shown to be stronger than the plate. The next column, "steel plate and iron rivets," shows that by reason of the higher tensile strength of the steel plates the first joint in the table is weakest through the line of rivet-holes, while the remaining figures with reference letter show that the rivets are weaker than the plates. By referring to the last column it will be seen that by using steel rivets having a higher shearing strength than iron the joints are strengthened against the shearing action of the plates throughout the whole series. An examination of these figures in connection with the horizontal distance from centre to centre of rivet would seem to indicate that a revision of the pitch of rivets would be advisable, that a higher percentage of joint might be secured by increasing the pitch of the rivets for the thinner plates and slightly decreasing the pitch for the thicker ones ; but the practical consideration of securing a tight joint is one of great importance, and it is not recommended that there be any considerable deviation from centre to centre of the horizontal rows of rivets as given.

Triple-Riveted Lap-Joints.—In any case where a double-riveted joint is deficient in rivet area, an increase in strength is had by simply extending the lap of the joint sufficiently to admit another row of rivets ;

Fig. 82.

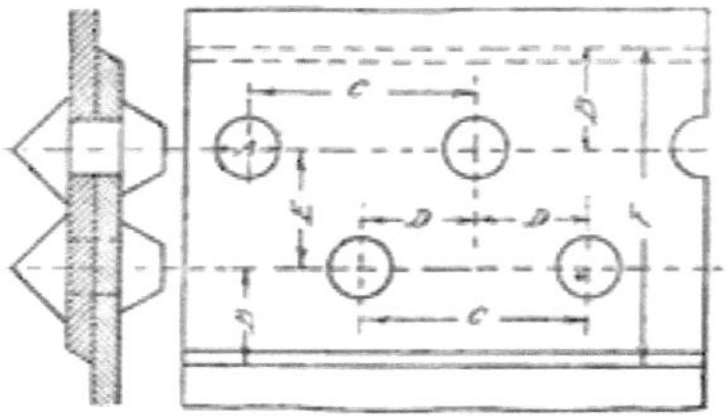

TABLE XXV.

PROPORTIONS FOR DOUBLE-RIVETED LAP-JOINTS, ZIGZAG AND CHAIN RIVETING.

Fig. 83.

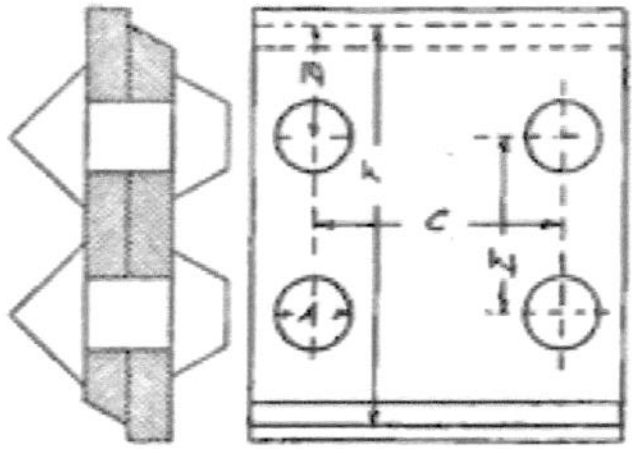

Thickness of Plate.	Diameter.		Centre of Hole to Edge of Plate.	Pitch of Rivets.				Lap of Plates.		Percentage of Joint.		
				Centre to Centre, Zigzag Riveting.		Centre to Centre, Chain Riveting.		Zigzag Riveting.	Chain Riveting	Iron Plate, Iron Rivets.	Steel Plate.	
	Rivet.	Hole.		Horizontal.	Vertical.	Horizontal.	Vertical.				Iron Rivets.	Steel Rivets.
	A.	A.	B.	C.	E.	C.	E.	F.	F.			
1/4	5/8	11/16	1	2 1/2	1 3/8	2 1/2	1 7/8	3 3/8	3 7/8	72.48 P	72.48 P	72.48 P
5/16	11/16	3/4	1 1/8	2 5/8	1 5/16	2 5/8	2	3 9/16	4 1/4	71.46 P	67.01 R	71.46 P
3/8	3/4	13/16	1 3/16	2 3/4	1 3/8	2 3/4	2 1/16	3 3/4	4 7/16	70.47 P	62.54 R	70.47 P
7/16	13/16	7/8	1 5/16	2 7/8	1 7/16	2 7/8	2 1/8	4 1/16	4 3/4	69.55 P	59.44 R	69.55 P
1/2	7/8	15/16	1 3/8	2 15/16	1 15/32	2 15/16	2 3/16	4 7/32	4 13/16	68.07 P	58.44 R	68.07 P
9/16	15/16	1	1 1/2	3	1 1/2	3	2 5/16	4 1/2	5 5/16	66.65 P	57.87 R	66.65 P
5/8	1	1 1/16	1 9/16	3 1/8	1 9/16	3 1/8	2 3/8	4 11/16	5 1/2	66.00 P	56.46 R	66.00 P
11/16	1 1/16	1 1/8	1 11/16	3 5/16	1 21/32	3 5/16	2 7/16	5 1/32	5 13/16	66.05 P	54.29 R	66.05 P
3/4	1 1/8	1 3/16	1 3/4	3 3/8	1 11/16	3 3/8	2 1/2	5 5/16	6	64.82 P	54.42 R	64.82 P

P—Joint will probably fail by tearing the plate between the rivet-holes. R—Joint will probably fail by shearing the rivets.

by so doing the pitch of the rivets in the horizontal line can usually be widened, especially in the case of zigzag riveting, without impairing the tightness of the joint.

Referring to Table XXVI., it will be seen that with but two exceptions the percentages of strength is above 75 for "iron plate and iron rivets." The riveting is in every case stronger than the net area of the plates; this also holds good for "steel plate with steel rivets." In the column of "steel plate with iron rivets," the riveting is in every instance weaker than the plate. Triple-riveted joints are not much used except for plates less than $\frac{7}{16}$ inch thick, the preference being given to butt-joints with double welts for plates $\frac{3}{8}$ inch thick and over.

The strength of a triple-riveted joint of say $\frac{1}{2}$-inch steel plates of 55,000 pounds tensile strength, rivet-holes $\frac{15}{16}$ inch diameter, spaced on 4 inches pitch, rivets of iron having a shearing resistance of 34,200 pounds per square inch, may be calculated thus:

Thickness of plate, $\frac{1}{2}$ inch	.5 inch.
Diameter of rivet-hole, $\frac{15}{16}$ inch	.9375 inch.
Area of rivet-hole	.6903 square inch.
Pitch of rivets, 4 inches	4.000 inches.

The strength of the whole plate would be $4 \times .5 \times 55{,}000 =$ 110,000 pounds.

The strength of the net section of plate would be $(4 - .9375) \times .5 \times 55{,}000 = 84{,}219$ pounds.

The strength of 3 rivets in single shear would be $.6903 \times 3 \times 34{,}200 = 70{,}824$ pounds, showing that the net section of the plate is stronger than the riveting.

The percentage of strength of joint is

$$\frac{70{,}824 \times 100}{110{,}000} = 64.39 \text{ per cent.}$$

If the rivets had been of steel the shearing resistance would have been increased to 44,625 pounds per square inch, or $.6903 \times 3 \times 44{,}625 = 92{,}414$ pounds. The shearing resistance of the rivets being greater than the strength of net section of plate, we then have as the percentage of joint

$$\frac{84{,}219 \times 100}{110{,}000} = 76.5 \text{ per cent.}$$

Lap-Joint with Reinforced Welt.—A modified form of lap-joint combining some of the features of both lap- and butt-joints has a reinforced welt or strap on one side, as shown in Fig. 86. A central row of rivets secures the three thicknesses of metal; the strap and plates are further secured by a row of rivets in each plate.

FIG. 84.

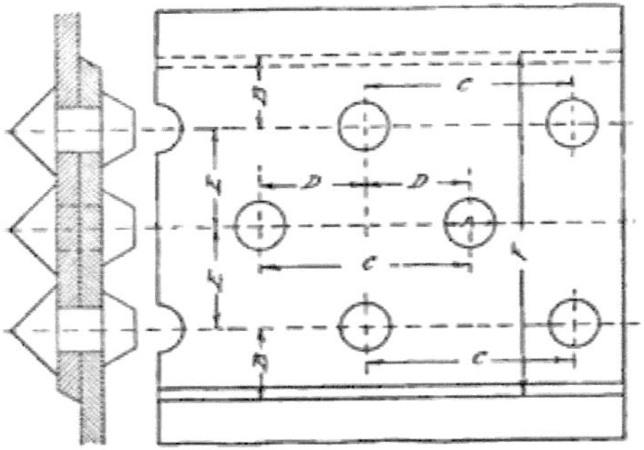

FIG. 85.

TABLE XXVI.

PROPORTIONS FOR TRIPLE-RIVETED LAP-JOINTS, ZIGZAG AND CHAIN RIVETING.

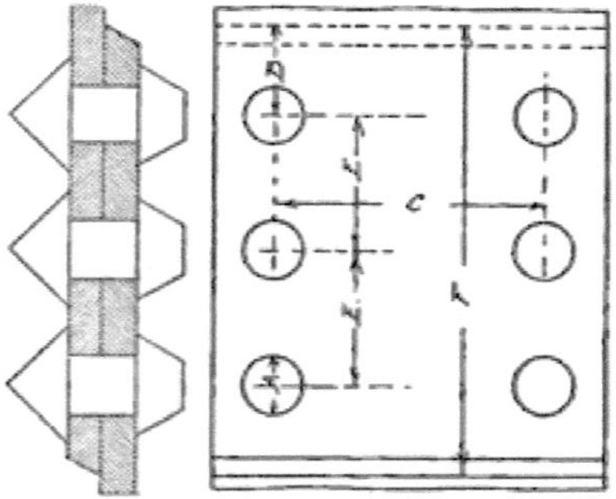

THICKNESS OF PLATE.	DIAMETER.		Centre of Hole to Edge of Plate.	PITCH OF RIVETS.				LAP OF PLATES.		PERCENTAGE OF JOINT.		
				Centre to Centre, Zigzag Riveting.		Centre to Centre, Chain Riveting.		Zigzag Riveting.	Chain Riveting.	Iron Plate, Iron Rivets.	Steel Plate.	
	Rivet.	Hole.		Horizontal.	Vertical.	Horizontal.	Vertical.				Iron Rivets.	Steel Rivets.
	A.	A.	B.	C.	E.	C.	E.	F.	F.			
1/4	5/8	11/16	1	3 1/2	1 1/2	3 1/2	1 13/16	5	5 5/8	80.34 P	79.14 R	80.34 P
5/16	11/16	3/4	1 1/8	3 5/8	1 9/16	3 5/8	1 7/8	5 3/8	6	79.25 P	72.74 R	79.25 P
3/8	3/4	13/16	1 3/16	3 3/4	1 5/8	3 3/4	1 15/16	5 5/8	6 1/4	78.37 P	68.79 R	78.37 P
7/16	13/16	7/8	1 5/16	3 7/8	1 11/16	3 7/8	2	6	6 5/8	77.46 P	66.18 R	77.46 P
1/2	7/8	15/16	1 3/8	4	1 13/16	4	2 1/16	6 3/8	6 7/8	76.56 P	64.39 R	76.56 P
9/16	15/16	1	1 1/2	4 1/8	1 7/8	4 1/8	2 1/8	6 3/4	7 1/4	75.78 P	63.15 R	75.78 P
5/8	1	1 1/16	1 9/16	4 1/4	2	4 1/4	2 3/16	7 1/8	7 1/2	75.04 P	62.27 R	75.04 P
11/16	1 1/16	1 1/8	1 11/16	4 3/8	2 1/16	4 3/8	2 5/16	7 1/2	8	74.27 P	61.64 R	74.27 P
3/4	1 1/8	1 3/16	1 3/4	4 1/2	2 1/8	4 1/2	2 3/8	7 3/4	8 1/4	73.63 P	61.22 R	73.63 P

P—Joint will probably fail by tearing the plate between the rivet-holes. R—Joint will probably fail by shearing the rivets.

The central row of rivets must be spaced for tightness under pressure as well as for strength of joint, so that a closer spacing is necessary than in the case of zigzag riveting, whether in double- or triple-riveted joints. The outer rivets are twice the pitch of those in the central row. This is a good form of joint, though expensive to make, and for the reason that butt-joints with double welts are simpler and no more expensive than this joint they are commonly preferred. The efficiency of this joint is about the same as a triple-riveted joint for the same arrangement and size of rivets.

FIG. 86.

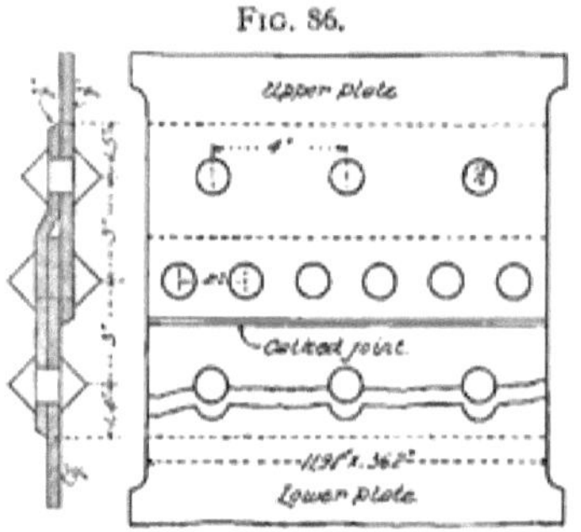

A duplicate of this joint fractured the upper plate through the outside line of rivet-holes. Fracture began at end rivet-hole. Efficiency of joint 90.1 per cent.

TABLE XXVII.

SINGLE-RIVETED LAP-JOINT WITH REINFORCED WELT, ⅜-INCH IRON AND STEEL PLATES, WATERTOWN ARSENAL.

Material of plate	Iron.	Steel.
Thickness of plate, nominal, inch	⅜	⅜
Thickness of welt-plate, nominal, inch	⅜	⅜
Width of test-specimen, inches	12	11.98
Diameter of rivets, inch	¾	⅞
Material of rivets	Iron.	Steel.
Diameter of holes, punched or drilled, inch	.82 D.	.94 D.
Number of rivets in each plate	9	9
Pitch of rivets, Fig. 86, outside rows, inches	4	4
Pitch of rivets, Fig. 86, middle row, inches	2	2
Pitch of rivets, Fig. 86, vertically, inches	3	3
Bearing surface of 9 rivets, square inches	2.89	3.06
Shearing area of 9 rivets, square inches	4.75	6.25
Tensile strength of plate, pounds per square inch	47,180	53,330
Gross sectional area of under plate, square inches	4.69	4.34
Net sectional area of under plate through line of rivet-holes, B, square inches	2.77	2.30
Net sectional area of under plate through line of rivet-holes, C, square inches	3.74	3.32
Maximum stress on joint per square inch:		
Tension on gross section of under plate, pounds	34,900	47,465
Tension on net section of under plate through line of rivets, B, pounds	59,100	89,565
Tension on net section of under plate through line of rivets, C, pounds	43,770	62,050
Compression on bearing surface of 9 rivets, pounds	56,640	32,960
Shearing strain on 9 rivets, pounds	34,660	67,320
Efficiency of joint, per cent.	74	89
Fracture, similar to Figure	86	86

Another joint similar to the preceding, except that the welt-plate was ¼-inch steel and the ⅞-inch rivets of iron in 0.93-inch drilled holes, failed by shearing 8 rivets in both rows and tearing out one hole in the corner of the ¼-inch welt-plate. Efficiency of joint 87.8 per cent.

Butt-Joints.—This form of joint is commonly used for plates ⅜ inch thick and over. It is the lap-joint repeated. The plates to be joined are placed edge to edge with outer and inner welt-strips, the whole secured together by through-going rivets, as shown in Fig. 87. Butt-joints may be single-, double-, triple-, and quadruple-riveted, according to the thickness of plates and steam pressure to be carried. This kind of a joint has an advantage over ordinary lap-joints in the fact that the pull of the joint is in the direction of the centres of the plates, and that the rivets are in double instead of single shear.

In Table XXVIII. no iron plates are given, as butt-joints are seldom used for plates thinner than ⅜ inch. Plates of this thickness and thicker are almost invariably made of mild steel. There is a choice of rivet material without loss of efficiency for the diameter and pitch of rivets, as given in the table.

The strength of a double-welt butt-joint, triple-riveted, as in Fig. 87, plates and rivets of mild steel, may be calculated thus:

Thickness of plates, ⅝ inch	.625 inch.
Diameter rivet-hole, 1$\frac{1}{16}$ inches	1.0625 inches.
Area of rivet-hole	.8866 square inch.
Pitch of rivets, 7⅜ inches	7.375 inches.

The spacing of the joint shows 1 rivet in single and 4 rivets in double shear.

The strength of the whole plate would be 7.375 × .625 × 55,000 = 253,516 pounds.

The strength of net section of plate at outer row of rivets would be (7.375 — 1.0625) × .625 × 55,000 = 216,992 pounds.

The strength of 1 rivet in single shear, Table X.	39,299
Four rivets in double shear	298,672
Total shearing resistance	337,971

It will be seen that the shearing resistance of the rivets is greater than the strength of net section of plate. We have, then,

$$\frac{216,992 \times 100}{253,516} = 85.59 \text{ per cent.}$$

"Where butt-straps are used in the construction of marine boilers, the straps for single butt-strapping shall in no case be less than the thickness of the shell-plates; and where double butt-straps are used, the thickness of each shall in no case be less than five-eighths (⅝) the thickness of the shell-plates."—*U. S. Rule.*

Method of Riveting.—Originally all riveted joints were hand-made, and hand-riveting is still largely practised, because small boiler-shops are not usually equipped with riveting-machines. In certain portions of machine-riveted boilers difficult of access, repairs, etc., are of necessity performed by hand.

FIG. 87.

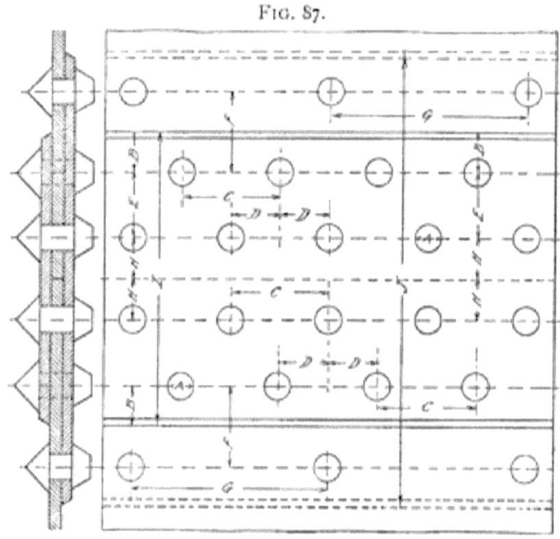

Riveting-machines are now in very general use, and include crank and cam machines, which are not much used; pneumatic machines, which are very convenient, especially in field work; steam-riveters, which are extensively used, and hydraulic riveting-machines, which are now especially in favor. Steam and hydraulic machines permit a slow and gradually controlled movement of the ram, and when the rivet-head is formed a pressure can be maintained upon it until it is fully set and

FIG. 88.

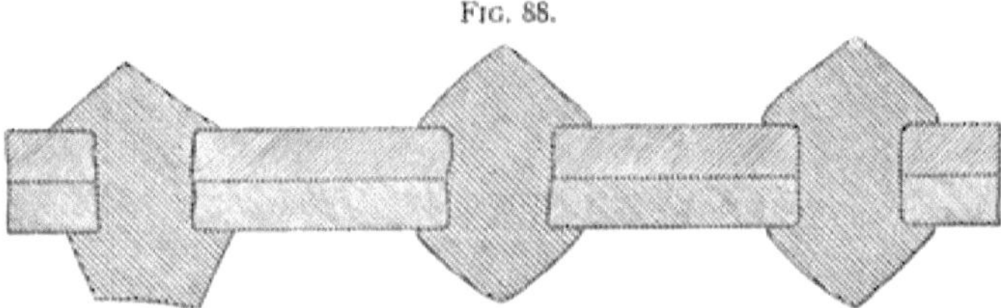

the rivet sufficiently cooled to permit the withdrawal of the ram without risk of stretching the shank of a hot rivet by the springing apart of the plates. The flow of the metal to fill the hole is usually complete if the pressure is sufficient and not too rapidly applied. The complete filling of the hole is a matter of the utmost importance. Sections from actual plates are shown in Fig. 88, in which the left-hand illustration represents

TABLE XXVIII.

PROPORTIONS FOR BUTT-JOINTS WITH DOUBLE WELTS, ZIGZAG AND CHAIN RIVETING.

Thickness of Plate.	Diameter.		Centre of Hole to Edge of Plate.	Pitch of Rivets.						Width of Welt-Plates.				Percentage of Joint.	
	Rivet.	Hole.		Centre to Centre, Zigzag Riveting.			Centre to Centre, Chain Riveting.			Zigzag Riveting.		Chain Riveting.		Steel Plate.	
				Horizontal.	Vertical.		Horizontal.	Vertical.		Narrow.	Wide.	Narrow.	Wide.	Iron Rivets.	Steel Rivets.
	A.	A.	B.	G.	E.	F.	G.	E.	F.	I.	J.	I.	J.		
1/4	5/8	11/16	1	5	1 1/4	1 7/8	5	1 13/16	1 7/8	6 1/2	10 1/4	7 5/8	11 3/8	86.47 P.	86.47 P.
5/16	11/16	3/4	1 1/8	5 3/8	1 11/32	2	5 3/8	1 7/8	2	7 1/4	11 1/4	8 1/4	12 1/4	86.31 P.	86.31 P.
3/8	3/4	13/16	1 3/16	5 3/4	1 7/16	2 3/16	5 3/4	1 15/16	2 3/16	7 5/8	12	8 5/8	13	86.05 P.	86.05 P.
7/16	13/16	7/8	1 5/16	6 1/8	1 17/32	2 3/8	6 1/8	2	2 3/8	8 1/4	13	9 1/4	14	85.82 P.	85.82 P.
1/2	7/8	15/16	1 3/8	6 1/2	1 5/8	2 1/2	6 1/2	2 1/16	2 1/2	8 3/4	13 3/4	9 5/8	14 5/8	85.54 P.	85.54 P.
9/16	15/16	1	1 1/2	6 7/8	1 23/32	2 11/16	6 7/8	2 1/8	2 11/16	9 7/16	14 7/8	10 1/4	15 5/8	85.27 P.	85.27 P.
5/8	1	1 1/16	1 9/16	7 3/8	1 27/32	2 13/16	7 3/8	2 3/16	2 13/16	10	15 5/8	10 5/8	16 1/4	85.59 P.	85.59 P.
11/16	1 1/16	1 1/8	1 11/16	7 5/8	1 29/32	3	7 5/8	2 5/16	3	10 5/8	16 5/8	11 3/8	17 3/8	85.31 P.	85.31 P.
3/4	1 1/8	1 3/16	1 3/4	8	2	3 1/8	8	2 3/8	3 1/8	11	17 1/4	11 3/4	18	85.17 P.	85.17 P.

All of these joints will probably fail by tearing the plate between rivet-holes.

For details of spacing zigzag joint, see Fig. 87. Distance H in Fig. 87 is the same as B.

Dimensions are also given for chain riveting. The zigzag is preferred because of a better distribution of rivets, contributing to a tighter joint. The strength is the same in either case.

a hand-driven rivet in a punched hole, the middle one a machine-driven rivet in a punched hole, the right-hand one a machine-driven rivet in a drilled hole. Plates should be bolted together through alternate holes, metal to metal in flanged work, to get the spring out of them; otherwise a thin film of metal is likely to be forced into the space between the plates, reducing frictional contact and lowering the efficiency of the joint.

Heating Steel Rivets.—It is important that steel rivets be uniformly heated throughout, and not the points merely, as is the ordinary method of heating iron rivets; neither should they be heated as highly as iron rivets, and should never exceed a bright cherry-red. Particular attention must be given to the thickness of the fire. Steel, of whatever kind, should never be heated in a thin fire, especially in one having a forced blast, such as an ordinary blacksmith or riveting fire. The reason for this is, that more air passes through the fire than that needed for combustion, and in consequence there is a considerable quantity of free oxygen in the fire, which will oxidize the steel, or, in other words, burn it. If excluded from this free oxygen, steel cannot be burned; if the temperature is high enough, it can be melted, and will run down through the fire; but burning is impossible in a thick fire with moderate draft. This is an important matter in using steel rivets, and should not be overlooked. The same principle applies to the heating of steel plates for flanging.

CHAPTER IV.

WELDING AND FLANGING.

It has long been the desire of both makers and users of steam boilers that a seamless shell might take the place of the aggregation of plates held together by riveted joints which now constitutes the ordinary method of constructing a boiler. It is true that cylindrical shells of steam boilers have been constructed by welding and without horizontal and circumferential riveted seams, excepting those fastening the heads and shells together, and even these are not necessary, as many digesters and pressure-tanks are made with the heads welded in place. A recent construction in this country is an internally fired boiler of large dimensions, 8½ feet diameter by 27 feet long, having welded horizontal and vertical seams, but with heads riveted in. Such examples are rare, but they serve to show what is possible by this method of construction if proper facilities are at command, coupled with knowledge of how to successfully use the facilities towards the desired end.

An advantage to be gained by making the cylindrical shells of boilers seamless is that they may be rerolled after welding, producing a perfectly cylindrical shell. This is, of course, impossible in a riveted joint. So, also, if a shell could be thus welded the objectionable two thicknesses of plate in the fire would be removed, together with the trouble incident to the accumulation of deposit which is likely to form around the joints and rivet-heads; further, if there is no jointed seam, the corrosion caused by the leakage of the lap-joints or around loose or imperfectly fitted rivets could not occur.

Reduction in thickness of plates has been advocated if welded joints be used instead of riveted joints, and such reduction in thickness has been placed as much as one-half; but this is not possible unless a welded joint is known to equal the strength of the original plate, which cannot under any conditions of workmanship be assumed to be true.

The ordinary claims made for perfectly welded joints are that welding approximates more nearly the original strength of the plates than the best form of riveted joints, relieving the plates from loss of strength due to punching and the additional loss occasioned by drifting and cold hammering. Calking could be dispensed with, and thus relieve the shell of incipient fractures occasioned by bad workmanship.

Wrought Iron and Mild Steel possess the property of welding when brought into perfect contact while the surfaces are in a state of partial fusion. To further insure a complete contact the surfaces thus joined are pressed, rolled, or hammered until they are united in a single piece at the weld. Small quantities of the impurities usually found in wrought iron, such as sulphur, phosphorus, etc., exert a marked influence upon the properties of wrought iron. These foreign substances do not wholly prevent the welding of wrought iron when parts are brought together in a state of fusion, but they do have the effect of lowering the efficiency of the welded joint, especially when the heating and welding is undertaken in an atmosphere containing free oxygen. In a non-oxidizing atmosphere these influences are less marked. What has just been said in regard to wrought iron is also true of mild steel, which does not readily weld except in a non-oxidizing atmosphere. After many trials and many failures in attempting to weld mild-steel boiler-plates Adamson found it necessary to ascertain in all cases the composition of the metal before putting any labor on it, and from a large experience he considered it desirable that the carbon should not exceed ⅛ of 1 per cent., while the sulphur and phosphorus should, if possible, be kept as low as 0.04 per cent., silicon being admissible to the extent of 0.1 of 1 per cent.

Temperature.—A high temperature is essential in welding, approximately 1600° Fahr. for wrought iron; but different irons require a temperature adapted to each varying composition, because with such variation in composition there is also a variation in point of fusion. Temperature alone is not sufficient for securing the best results. It is true that a high temperature promotes welding in a non-oxidizing atmosphere, but it is also true that in an atmosphere in which there is free oxygen, the latter, being the cause of burning the metal, not only prevents welding, but destroys the strength of the metal wherever it may occur. The temperature, then, must be regulated if welding is to be done in an oxidizing atmosphere, so as to insure the fusion of the metal surfaces to be joined and avoid covering such surfaces with a coating of iron oxide, which will either imperfectly weld or wholly prevent metallic contact.

Oxide of Iron—The presence of oxide of iron in a joint is one of the principal causes of non-welding. It is difficult to prepare iron plates for welding without the presence of this objectionable material, and it is for this reason that upset scarfed edges should have a swell in the middle of the angular face, as shown in Fig. 89, the object being to bring the metal into immediate contact at the middle of the weld, so that any subsequent hammering will force out the vitreous oxide on either side, securing a better weld than if it were allowed to lodge on either surface of the plates, for any lodgement would mean defective welding at that point.

FIG. 89.

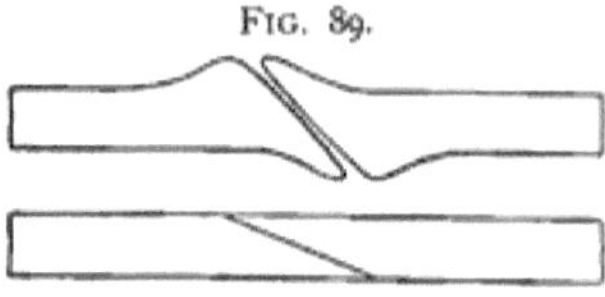

Flux.—As oxidation always occurs in a greater or less degree, the heated surfaces must be protected by means of a flux. The one generally used is sand ; this is composed of silicon and oxygen. The action of the flux is twofold,—in forming a vitreous coating over the iron and in reducing the temperature of the parts to which it is applied. This arises from the circumstance that iron is usually "scarfed" at the place where it is to be welded, as in Fig. 89. We thus have a thick and a thinner portion of the same plate exposed to the action of heat. Ordinarily the thinner portion of the plate is nearest the centre of the fire, consequently it attains welding heat before the thicker portion does. If the action of the heat was not modified in some manner this thinner edge would be burned away long before the thicker portion was brought to the welding point. The sand or other flux coming in contact with the highly heated iron is melted and absorbs so much heat from the iron that it gives the latter a vitreous coating, combining with the iron and covering that portion which is of sufficiently high temperature to melt the sand. Silicon, being very refractory in its nature, will last some time in the fire before it burns off the iron ; it thus serves to protect the thinner parts of the iron while the thicker portion is absorbing heat and arriving at a welding condition. In using sand as a flux care must be exercised that it be cleaned off the faces of the joint where two scarfed edges are to be welded, because its presence in the weld would prevent perfect contact and weaken the joint. For small work, borax is the flux generally employed in the forge for welding ; it prevents oxidation in the same manner as already described for sand.

Edges of Plates.—Scarf-welding, shown in Fig. 89, is to be preferred to lap-welding, Fig. 90, because the strain on the scarf-joint is direct, while on the lap-joint it is indirect and tends to distort the joint when under pressure, as in Fig. 91. A scarf-weld is best made by upsetting the edges to about double the thickness of the plate and bevelling the edges to about 45°, as shown in Fig. 89. In scarfing and thinning down the plate the sharp edge may be about one-sixteenth of an inch thick, perhaps less. An exact thickness of the upset portion is not a material part of making a good joint, neither is the thinning to a sharp edge of special importance. All that is necessary is the upsetting of the edge to a thickness considerably more than that of the plate itself, the object being that when the weld is made the plate may then be hammered and finished down to the regular thickness. The edges should be heated simultaneously to a white heat, and when joined the

FIG. 90.

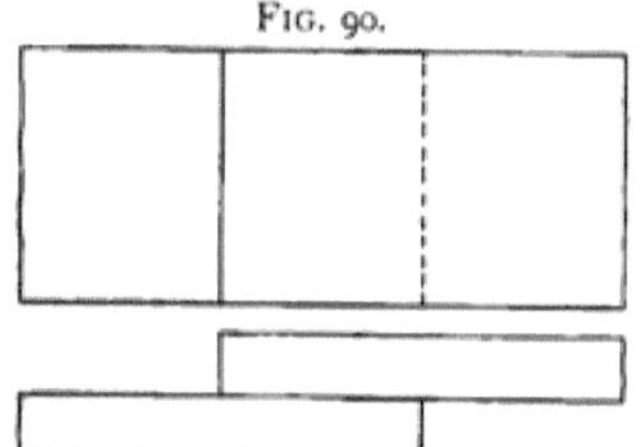

FIG. 91.

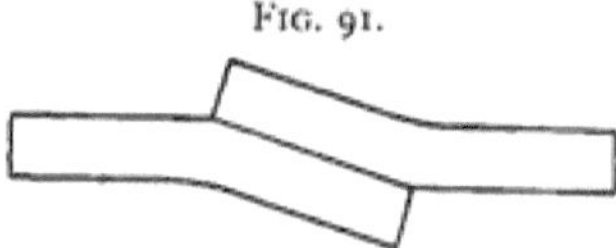

joint should be hammered or rolled to secure perfect contact through its whole length; the swell of the joint can be afterwards worked down to the thickness of the plate. This is a much better method than that shown in Fig. 90, in Bertram's experiments.

Lap-welded joints as shown in Fig. 90 are not recommended, except for thin plates, say ¼ inch and less, because they are weaker than scarf-welded, although there is no reduction of plate section through the joint. This has already been pointed out and illustrated in Fig. 89. The thicker the plates joined together the greater will be the distortion in the joint. This fact was clearly brought out in the test, page 99, which shows the ½-inch joint to be relatively weaker than the ⅜-inch joint in the same test.

Welding Bars.—If a weld is to be made in a brace or stay, the ends should be upset to about double the original diameter, afterwards bevelling to an angle of about 45°, as in Fig. 89. This form of joint is favorable to the escape of scale, flux, etc., out of the joint. After the parts are joined, the swelled portion of the joint can then be hammered down to the common diameter or size of the bar. Wherever possible, stays and braces should be made without welds.

Welding Plates.—The heating of two plates in a well-made, open fire is attended with greater risks than in the case of two bars of iron. The reasons are quite obvious: the ends of the bars are easily placed in the centre of the fire and entirely shut off from the injurious effects of free oxygen, if the fire is properly made. When a thick fire is built upon a tweer, the air passing up through it gives up its oxygen to the incandescent carbon, and carbonic acid gas is the product of this union. This gas in passing up through the bed of burning coal takes up another equivalent of carbon, and carbonic oxide gas is formed. Nitrogen is also present in the fire. But none of these gases have an injurious effect on iron, so far as welding is concerned. Therefore, the two bars of iron referred to above are in a highly heated chamber formed by the incandescent sides and cover of the fire. The included atmosphere being non-oxidizing, the bars may be readily brought to a welding-heat without fear of oxidation, for there is no excess of oxygen in the fire to come in contact with the iron. In the case of plates it is somewhat different, for the plates being hottest in the centre and of lower temperature towards the edges of the fire, it is not possible to confine the heated portion of the plates to a chamber of heated gases from which oxygen is excluded,—for no such chamber exists in an ordinary fire, and cannot from the nature of the case. Further, every movement of the plate brings the more or less highly heated portions in contact with the air: oxidation instantly occurs, forming an oxide of iron or hard cinder which prevents welding. There is at the same time a partial loss of iron; but this is not a serious matter in comparison with the bad effects resulting from the presence of oxide of iron in the weld.

In the manufacture of welded boilers as a business it would be necessary to construct a special heating apparatus, which would probably consist of an external and internal gas-furnace, operating on the principle of a blow-pipe, in which the flames of the burning gas would be directed against such portions of the joint as needed the greater heat. Such an apparatus could be made in which no free oxygen could reach the heated plates, and thus welds could be made without the use of a flux of any kind. The plates could be heated their whole length at one time, and when brought to the point of fusion could be welded by pressure instead of by hammering.

Localizing Heat in Welding.—Welding occurs only at the edges or ends of parts to be thus joined. Heating, therefore, should be confined to such portions only, and not allowed to extend over wide areas upon which no work is to be done. In the absence of special appliances no greater length of edge should be heated than can be conveniently and properly welded, because excessive temperatures occurring where no work is to be done, especially if in contact with the air, is fatal both to iron and steel plates. Without special heating appliances it is probable that not more than a few inches—say less than a foot—could be heated at any one time; and this heating might preferably begin at the centre of the length of the joint and work from there to either end, rather than begin welding at one end and work towards the other. Whichever method be adopted, there are likely to be strains set up in the plates thus joined, both in extension and compression, which can only be eliminated by proper annealing.

Bertram's Method.—Boiler-plates were welded at the Woolwich Dockyard in 1857. The edges were scarfed and placed together between two flames directed against either side, as shown in Fig. 92. These flames were obtained by the combustion of coal or coke, and were non-oxidizing in their character. When the two plates were raised to the welding temperature they were united by pressure or hammering in a special machine.

FIG. 92.

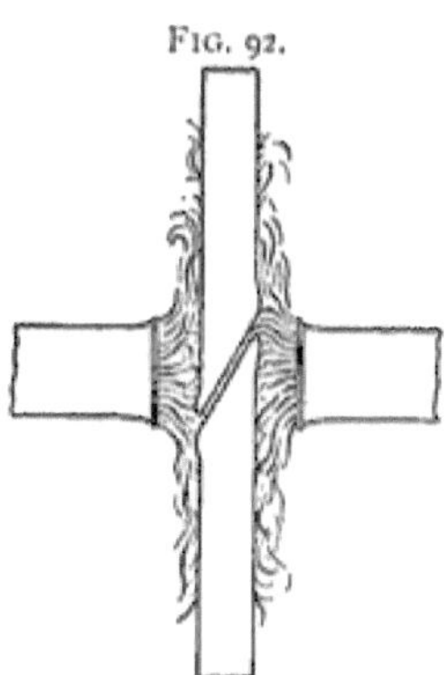

Tests made of these welded joints showed that the lap-welded test-pieces, as in Fig. 90, were inferior in strength to those scarf-welded, as in Fig. 89.

The specimens tested were 4 inches wide by 3/8, 7/16, and 1/2 inch in thickness. The lap of the joint was 1 1/4 inches, with results as follows:

The strength of a scarf-welded joint, Fig. 89, for the 1/2-inch plate was faulty; but for the 7/16- and 3/8-inch plates the welds were equal to that of the original plate.

The strength of the lap-welded joint, Fig. 90, was for the ½-inch plate 50 per cent. of the original plate, increasing to 69 per cent. in the $\frac{7}{16}$-inch plate, and 66 per cent. in the ⅜-inch plate.

From the above data it appears that the strength of joints united by lap-welding is scarcely greater than that secured by single riveting, the joint being about 40 per cent. weaker than the plates which compose it. Scarf-welding, on the contrary, equalled the strength of the plate. No doubt the shape of the joint under severe stress had much to do with the lowering of its strength in consequence of the indirect pull, as shown in Fig. 91.

Annealing Welded Joints.—It would be advantageous if, in the case of plate work, the whole structure after welding could be placed in an annealing furnace, properly heated, and allowed to cool gradually; but this is not always practicable. In lieu of this, a strip wide enough to cover any excessive heating on either side of the welded joint should be heated to a cherry-red and then allowed to cool gradually.

Practical Results.—Experiments on $\frac{7}{16}$-inch iron plates with welded joints, specimens taken from boiler-shells so as to test the efficiency of the welding, showed that of 23 tests 11 broke in the weld and 12 broke in the solid. The breaking strength of the solid plate was 46,368 pounds per square inch for the least strength, 57,792 pounds for the greatest,—an average of 52,865 pounds tensile strength for the whole series, showing that the iron was of good quality. The strength of the welded plates was 36,960 pounds per square inch for least strength, and 53,312 pounds for the greatest, or an average of 46,144 pounds. The efficiency of the welded joints on the total averages was 87.3 per cent. The efficiency of the weakest welded joint as compared with the average strength of the original plate was 68 per cent., showing throughout the series that the joints varied in efficiency from 68 to 87.3 per cent., which approximates that of good riveting.

The specifications for the United States protected cruisers "Columbia" and "Minneapolis" called for welded steel pipes, a contract executed by the Continental Iron-Works. These pipes varied from 10 to 20 inches inside diameter, and from $\frac{5}{16}$ to ⅜ of an inch thick; the maximum length was 16 feet, with flanges from ¾ to 1 inch thick after being faced. These pipes were to be made of plate steel, and were to comply with all the requirements called for in the specifications of the material used in the boilers, and be subjected to a water-test of 400 pounds per square inch.

The method of forming these flanged pipes was to scarf the edges, roll to the required diameter, and then weld, the length of each cylinder thus formed being about 8 inches shorter than the finished length of the pipe over the flanges. Steel disks of the diameter and thickness required to form the flanges were punched and flanged like a flue-head, with a cylindrical projection corresponding to that of the pipe and of the desired length, which, after scarfing in the lathe, was welded circumferentially to the pipe.

The experimental pipe was made of ¼-inch plate, as difficulty was expected in making the rivets in the head remain tight under such pressure as would be required to burst a vessel of the same diameter and ⅜ of an inch thick, and this proved to be true during the course of the experiment. The experimental pipe was 20 inches inside diameter and 42 inches long over the flanges. (Fig. 93.) The heads were ½ inch

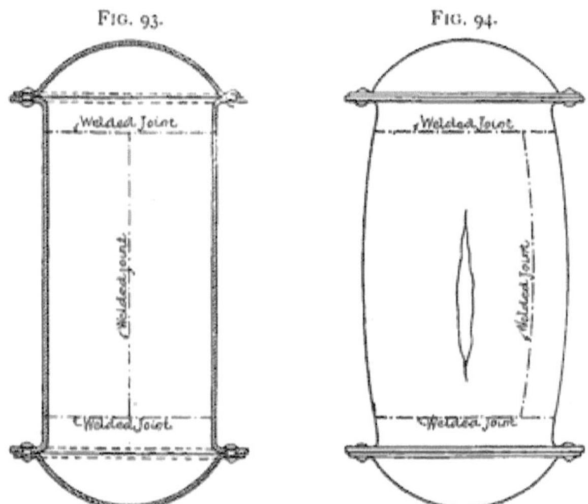

Fig. 93. Fig. 94.

thick, domed about 6 inches, and secured to the pipe by flanges ¾ of an inch thick. As the radius of the interior curve on the flange was 1 inch, the total area to strain the rivet was that of the diameter, or 22 inches. The heads were held in place by 34 drilled holes, fitted with rivets, machine-driven, $1\frac{1}{16}$ inches diameter, countersunk on each side. This experimental pipe had a water pressure applied up to 1700 pounds to the square inch, when it failed, as shown in Fig. 94, measurements of which are given in Fig. 95. It enlarged like a barrel, becoming over 24 inches in diameter at the middle of its length. The fracture occurred about 6 inches from the line of the longitudinal weld. This pressure is equivalent to a strain of about 68,000 pounds per square inch of section.

Three test-pieces were cut from the experimental pipe, as shown in Fig. 96, two of them being cut across the welded seam, leaving the welded part in the middle of the length, and the other from the unwelded part of the steel.

The test of these specimens shows a maximum tensile strength for the two having the weld of 58,230 pounds and 62,500 pounds respectively, and for the unwelded piece 61,470 pounds, the fracture in all the pieces occurring near the ends several inches from the weld.

Fig. 95.

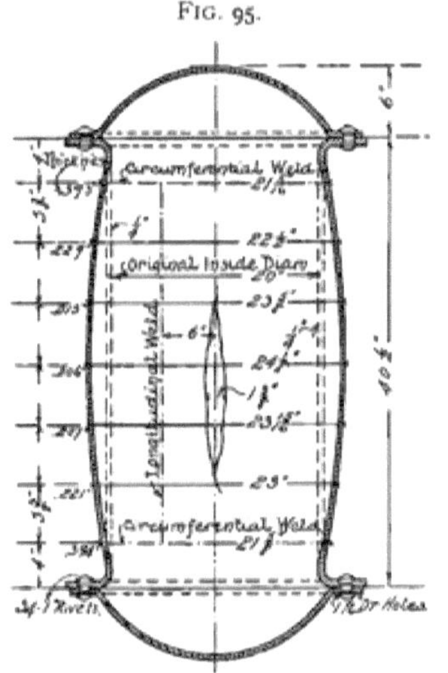

Fig. 96.

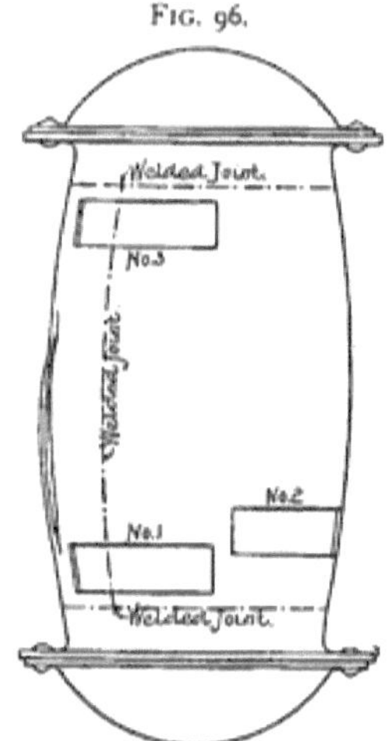

The specimens tested as follows:

Number of test-mark	1	2	3
Condition	Welded	Welded	No weld
Length of test-specimen, inches	8	8	8
Width, inches	1.105	1.008	1.010
Thickness, inches	.239	.222	.215
Area, square inches	.243	.224	.217
Final length, inches	9.50	9.28	9.28
Elongation, per cent.	18.75	16.00	16.00
Tensional stress on specimen, maximum pounds	14,150	14,000	13,340
Tensional stress per square inch, maximum pounds	58,230	62,500	61,470

At first sight there appears to be a discrepancy between the tensile strength of the metal, as indicated by the testing-machine, and that given from the gauge pressure. A careful examination of the outline of the ruptured vessel, Fig. 95, will, however, show that the stress on the metal at the time of rupture was much greater than would be obtained from a consideration of the original diameter and thickness of the metal.

The outside circumference from edge to edge of the rupture, of about the middle of its length, is $75\frac{11}{16}$ inches, corresponding to an outside diameter of 24.095 inches. The thickness of the metal at this point was 0.206 inches, making the internal diameter at the time of rupture $24.095 - 2 \times .206 = 23.683$ inches. Using these figures for thickness and diameter, the stress on the metal when the vessel burst becomes

$$f = \frac{1700 \times 23.683}{2 \times .206} = 97,720 \text{ pounds.}$$

Test-specimen No. 1 measured approximately 0.765 by 0.171 inch at point of fracture; or its area was 0.1308 square inch. The total stress on it at the time it broke was 14,150, and, consequently, the stress per square inch was

$$\frac{14,150}{.1308} = 108,200 \text{ pounds.}$$

Similarly, specimens Nos. 2 and 3 measured 0.775 by 0.165 and 0.783 by 0.169 inch respectively, giving stresses at the time of fracture of 109,500 and 100,800 pounds respectively, or an average for the three specimens of 106,167 pounds,—a result near enough to that obtained from the gauge pressure to leave little doubt of the correctness of the latter.

Furnace Flues.—In an internally fired boiler it is important that the main flue should be truly cylindrical, as the resistance to collapse depends largely upon this. Lap-joints prevent the plates forming a true circle; it has been the practice, therefore, among the best makers to employ in its construction a butt-riveted joint with the seam below the grates. The objections to this arrangement are, that it is impossible to perfectly calk such a seam when once in place; if the seam of rivets be along the bottom of the flue, the ready removal of ashes is prevented, and more or less of them will accumulate along the whole length of the furnace. Should there be a leaky joint, and this is not improbable, we may almost certainly count on an accumulation of hard-baked ashes and cinders, which lend themselves readily to surface corrosion. The best practice at this time is to make all such furnace flues with welded joints and flanged ends, placing the weld below the grate bars, so as to be away from the fire; the pressure, being wholly on the outside, tends to collapse, and thus to tighten the weld. An imperfect weld might, in such a flue, escape detection for a long time, but would soon make itself apparent in any case where internal pressures were employed.

Strength of Welded Joints.—Theoretically, the parts joined by welding should be equal in strength to the original bars or plates, and many tests have proven this to be the case. Were this true of a majority of welded joints, the claims of superiority for such joints over riveted joints would be realized; unfortunately, this is not the case in common practice.

The weakness of a welded joint is in part due to the unequal heating of the edges to be joined and the absence of immediate contact when the parts are brought together for welding, whether by hammers, rollers, or simple pressure. In the case of bars, metallic contact is usually had in welding, except in large pieces necessarily handled by a crane. The strength of welded bars of iron varies from 35 to 90 per cent., with an occasional bar showing full strength. In tests of iron chain cables in which a weld occurs in each link the failure almost invariably occurs in the weld, the best percentages of strength ranging anywhere from 70 to 90 per cent. of the original bar. A chain-cable link does not present the most favorable conditions for welding, but the workshop appliances and unusual skill on the part of the workman who is really a specialist make the average of chain-riveting equal to, if not superior to, ordinary forge-work where welding is not a continuous practice.

The uncertainty in regard to welded seams is equally shared by makers and users of steam boilers. So far as experimental tests have gone, welded seams do not average higher than single- or double-riveted joints, say from 60 to 70 per cent. of the strength of the original plate, which latter is, no doubt, injured by overheating at the time of making the weld. Test-sections cut from portions thus overheated show extreme brittleness and a reduction in tensile strength of the original plate of from 50 to 75 per cent.,—a very serious loss, some of which may be restored by judicious annealing; but furnaces large enough to admit the completed shell of a boiler for the purpose of annealing must be rare indeed, if they exist at all, in this country.

Welded-joint fractures do not always occur immediately in the joint, though such failure is known to be due to loss of strength occasioned by destructive treatment which the bar or plate receives in the operation of heating preparatory to welding. This loss of tensile strength, although occurring outside of the joint, is directly chargeable to the process of welding, even though the break did not occur in the weld itself. In a number of tests of welded bars, the break occurred both in and alongside of the weld, showing in each case from 15 to 40 per cent. reduction of strength. It is for reasons similar to these that thoroughly prepared boiler specifications require that stays and braces, whether of iron or of steel, intended for steam boilers shall not be welded if they are to be subjected to tensile stress. This exclusion of welded stays and braces accords with sound judgment, based upon a varied and sometimes disastrous experience.

Efficiency.—Welded seams in boiler-plates have not ordinarily yielded an efficiency higher than could have been supplied by double- or triple-riveted joints, or such a joint as a designer would have selected as suitable for the diameter of boiler, thickness of plate, and steam pressure to be carried.

Cost.—The relative cost of welding, as compared with punching and riveting, is probably in favor of welding when proper facilities are provided for heating, scarfing, hammering, annealing, etc.

Flanging.—The process of bending the edge of a plate so that a second plate may be fastened to it at an angle, such as a boiler-head, Fig. 97, or if a second plate is at a distance from the first plate, as in

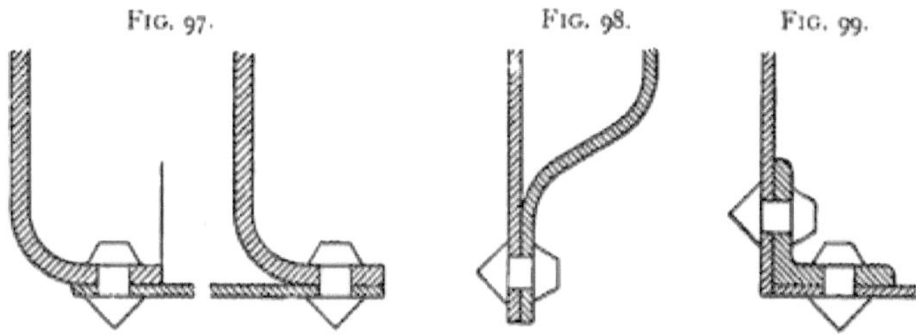

FIG. 97. FIG. 98. FIG. 99.

the bottom of a fire-box, Fig. 98, is called flanging. The advantages of a flanged joint over that in which the parts are dimensioned and secured by angle-iron corners, riveted as in Fig. 99, are that it is stronger, simpler, because consisting of fewer parts, and less liable to leakage, because there is one less riveted seam. In localities where flue boilers are popular, some very intricate flanging is required, many of these flues being as small as 6 inches, see Fig. 100, and few of them larger than 18 inches in diameter.

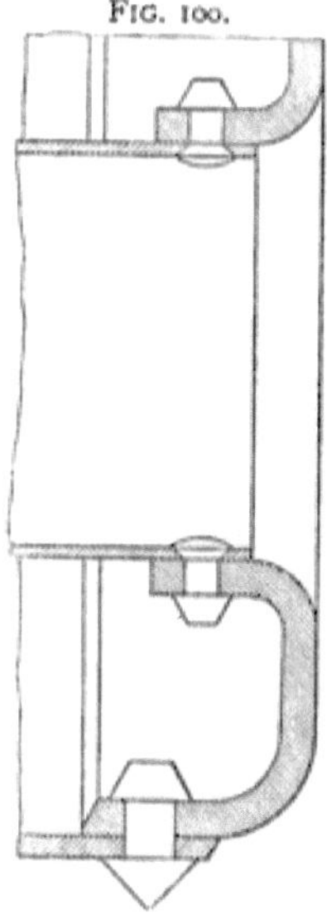

FIG. 100.

Heating of Plates.—Flanging should always be done at a bright-red heat, and work upon a plate should never be continued after it has cooled down below a dark cherry-red. Hand-flanging should always be performed over a cast-iron former with rounded corners. In heating a plate for hand-flanging, the heat should be confined to such portions of the edge of the plate as are to be flanged, and should not extend very far inward towards the centre. As much length of edge of the plate should be heated as can be conveniently flanged at one operation. This will require a special construction of fire, needing only a few bricks to give the necessary boundary to the enclosed fire and a proper arrangement of the fire itself, so that the entire edge can be inspected at any time during the progress of the heating of the plate. It seems almost needless to remark that the fire must be uniformly hot, absolutely clean, and free from ashes or clinkers, and must on no account have holes in it by which the blast can escape, or a burnt plate will surely result.

Hand-Flanging.—The flanging of an iron or steel plate should be done with wooden mauls, bending the plate over a cast-iron former. The blows should be quick, light, and distributed over as large a surface as possible in the shortest time, avoiding anything like short bends in turning the flange. The heating, when done in an ordinary open flange fire, must of necessity be local; there will be required, therefore, the greatest care in working. As the flanging approaches completion by successive stages of heating and hammering, care must be exercised that the plate, if of steel, is not ruined by splitting or cracking, which may be induced by internal strains.

The operation of hand-flanging is one requiring skill and judgment on the part of the workmen,—first, in the matter of heating, during which, if done in an open fire, as is usually the case, there is liability of overheating portions of the plate while other portions are not hot enough to insure the best working; second, in working down a hot sheet to the edge of the cast-iron former, the flange must be left true and accurate, free from lumps and wavy edges. Should the latter occur, the flange must be reheated and worked down with sledges and flatters until the entire edge is true, even, and accurately dimensioned. Attention must also be given to the flat portion of a flanged plate, to see that it is in presentable as well as in workable condition. Buckling is likely to occur, because the operation of flanging the edges subjects them to alternate strains of compression and elongation, much of which is transmitted to the centre portion of the plate.

Machine-Flanging.—There is a certain advantage in machine-flanging over hand-flanging in the one fact that the whole plate is heated in a special furnace, insuring a moderate and even temperature throughout. In hydraulic flanging the machine acts quickly by pressure, performing the entire operation upon a single plate in two or three minutes' time and without striking any blows, the metal flowing easily, naturally, and without abnormal strains into whatever shape the dies may give it. The centre of the plate is flat, and, on the whole, a much more satisfactory product than can be secured by hand.

Flanging-machines in which the edge of a circular plate is turned by means of rollers to a right angle turn out excellent work; but as such machines are confined to circular plates only, they are not adapted to general work, such as flanging irregular sheets, making flanged fire-door openings, flanged manholes, etc.

Hydraulic flanging-machines are more in favor at this time than any other, and, together with hydraulic riveting-machines, are now considered a necessary requisite in a first-class boiler-shop.

Hand- vs. Machine-Flanging.—Fig. 101 represents the thinning of the curve occasioned by the stretching of the plate over the cast-iron former in hand-flanging, the dotted line representing the normal curve and the middle line the actual thickness of metal. Fig. 102 is a representation of the thickening of the curve, taken from a roller-machine-flanged head. The normal curve, it will be noticed, falls considerably within the actual line of the metal. The advantages gained by the strengthening of the head at that particular point are quite obvious and are not likely to be underestimated.

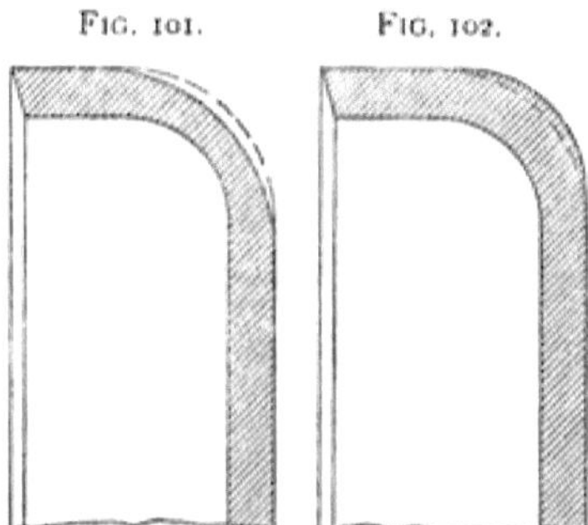

FIG. 101. FIG. 102.

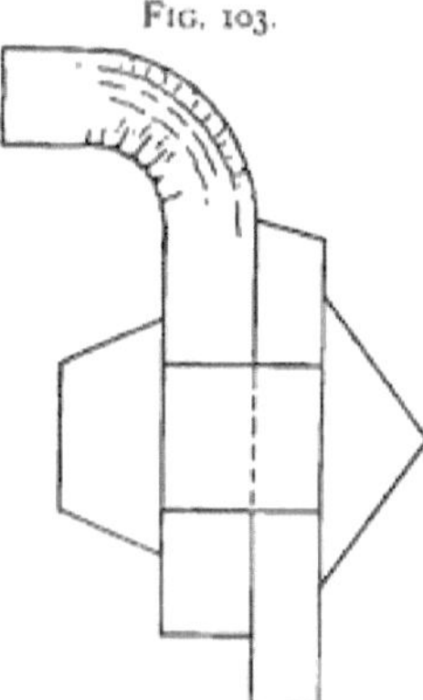

FIG. 103.

Flanging Iron Plates.—Iron plates are more severely tested by the act of flanging than by any other work done upon them. Iron, by reason of its fibrous structure, requires careful manipulation to prevent breaking in the bend, especially if the corner be too sharp, as is often the case. Mr. Allen called attention to this defect in flanging several years ago, supplemented by a sketch which is reproduced in Fig. 103, assigning the sharp curve and consequent distortion of metal in the bend as a direct cause of grooving.

Radius of Flange.—A defect more frequently met with in old than in later flanging is in the sharpness of the bend of the flange. Ordinary tubular boiler-heads, machine-flanged for the trade, have an inside radius of flange for heads $\frac{5}{16}$ to $\frac{3}{4}$ inch thick,—1 inch for the former to $1\frac{1}{2}$ inches for the latter,—nearly every mill that furnishes flanged heads having a radius of its own. Although there is no uniform standard for inside curves, they will generally be found to be ample for the thickness of the head and with no sharp corners. As the outer diameter of any tube is not likely to come nearer the outside of a flanged head than 3 inches, there is no reason for adhering to sharp curves; in fact, the large curve adds much to the stiffness of the head and reduces the area of flat surface to be fitted with stays.

When it is known that only the neutral axis of the plate does not change length in bending, that the outside surface must suffer extension and the inside surface compression, the advantage of a large radius in bending is made immediately apparent. Referring to Fig. 103, sketched from an actual specimen of flanged iron plate, it will be seen that the outer layers of metal separated and slid upon each other, as indicated by the transverse lines; the outer surface was filled with small cracks not unlike the season checks as seen in timber. The inside of the flange, as shown in the engraving, being in a state of undue compression, presents the appearance of a crushed and buckled-up mass of fibres, and it is particularly this disturbance of the fibres and the laminæ of the iron which renders it susceptible to the corrosive action of the acids present in the feed-water, which, together with the strains produced by expansion and contraction incident to the combined action of heat and pressure, result in corrosion, grooving or channelling of the flange, and possibly rupture. An examination of the engraving will show that the inner radius of the flange is but little more than the thickness of the plate.

Flanging Steel Plates.—Mild steel requires uniform heating, moderate curves, and gentle working to get the best results. This is best secured in machine-flanging, whether by rollers or by gently forcing the plate through a die by hydraulic pressure. The process of the flanging of mild steel need not be in any respect different from what has already been described relative to the subject in general.

Annealing Steel Plates.—After flanging a steel plate, whether by hand or machine, it should be immediately heated to a cherry-red to relieve it of all internal strains incident to working, allowing it to cool slowly, not disturbing it until entirely cold.

Thickness of Flanged Edges.—When a flat disk of metal is flanged, as in the case of a boiler-head, the edges of the flange, especially in hand-flanging, will be slightly thicker than the original plate. In machine-flanging, whether by rollers or by passing the plate through a die, the edges are not usually any thicker than other portions of the plate, except the excess due to the space allowance in the dies above the nominal thickness of the plate. This extra thickness is due to the longer circumference of the flat disk being compressed to the shorter circumference corresponding to the diameter of the finished head. The accumulated metal is worked down to the original thickness of plate, which causes an increase of depth of flange. Sexton gives the following practical instructions regarding flanging: "A plate during the process of flanging will gain twice its thickness in length of each flange. Thus, suppose you want to flange a circular plate to have a 3-inch flange all around, and to be 3 feet in diameter after being flanged and ⅜ inch thick; you must not add twice the width of the flange to the diameter, making 3 feet 6 inches, but twice the width of the flange, less four times the thickness,—making 3 feet 4½ inches. In marking the plate line out the exact diameter you want it to be after being flanged, then allow the width of the flange less twice the thickness, and when flanged the centre marks should be on the flange just where the curve joins the flat."

Flanging for flue-holes, manholes, etc., has just the opposite effect to that given in the preceding paragraph. Preparatory to flanging an opening in a plate, the whole depth of flange must be allowed when laying off the work. An inner hole is cut to the inside line of flange thus laid off, the plate heated, and the flange formed by forcing the metal outward from the plate in the desired direction,—as, for example, Fig. 100, which shows a flue-hole flanged to the inside of head, or as in Fig. 216, which shows a flange turned to the outside of head. As the diameter of the finished flange for flue-hole or manhole is worked from the diameter and thickness of the plate, it follows that the edge of the flange is thinner than that of the plate, the decrease in thickness depending upon the depth of flange to be formed.

CHAPTER V.

DETAILS AND STRENGTH OF CONSTRUCTION.

THE cylindrical shell of a steam boiler is commonly an aggregation of plates fastened together by riveted joints. These plates may be either wrought iron or mild steel, the thickness depending upon the diameter of the shell and the pressure to be carried. When boiler-shells were made of iron the plates were much narrower and shorter than can now be had in mild steel, because steel plates have no grain or fibre, but are homogeneous throughout, and can, therefore, be rolled lengthwise or crosswise as best suits the mill-man, which was not the case with wrought iron, as its fibrous nature had to be taken into account.

No difficulty is now experienced in getting mild steel plates of any size that may be economically handled or worked in a boiler-shop, thus effecting a reduction in the number of plates and number of joints in boiler-shells of at least one-half. Boiler-shells of 36 inches diameter by 8 feet in length can now be made of a single plate with one row of rivets if desired ; boilers 60 inches in diameter and less are commonly made of 2 sheets with 2 horizontal rows of riveted joints, forming the shell up to 16 or 18 feet in length ; boilers 72 inches in diameter are commonly made with a single sheet in the bottom and 2 or 3 sheets forming the upper half of the boiler for lengths up to 20 feet. In each of these arrangements no cross-seams are exposed to the fire, but in all cases where the upper and lower sheets of the boiler meet midway of the circumference there will be exposed to the action of the heat, but not to the direct action of the fire, 2 rows of riveted joints, unless the bottom sheet is made wide enough to extend around to the ⅔-heating-surface line, which is not a very good arrangement of spacing plates. Other shell designs are made up of 3 rings, each in a single piece, the length of the sheet reaching around the boiler and including the riveted joint ; the 3 widths thus riveted together make the required length of the boiler, usually not more than 20 feet.

Strength of Riveted Shell.—Wrought-iron boiler-plates should average 45,000 pounds and mild steel 55,000 pounds tensile strength per square inch of section ; but the gross strength of plate is lessened by the amount which has been taken out of it for the insertion of rivets, so that for single-riveted joints the net strength for plates $\frac{5}{16}$ inch thick is about 60 per cent. In double-riveted joints the net strength of the same thickness of plate is about 71 per cent. In triple-riveted joints for the same thickness of plate the net strength is about 79 per cent. A butt-joint triple-riveted, with outside and inside welt-strips, would have about 86 per cent. of the strength of the original plate.

Safe Working Pressure: Rule.—Multiply together the tensile strength of the plate, the thickness of the plate in parts of an inch, and the efficiency of the joint; divide the product thus obtained by one-half the diameter of the boiler multiplied by the factor of safety.

Example: What is the safe working pressure for the cylindrical shell of a boiler 72 inches in diameter, made of 55,000 pounds tensile strength steel plates, $\frac{7}{16}$ inch thick, butt-joints with outside and inside welt-strips, the efficiency of which is assumed to be 86 per cent. that of the original plate, factor of safety 5?

$$\frac{55,000 \times .4375 \times .86}{36 \times 5} = 114.97 \text{ pounds.}$$

Example 2: Shell 48 inches in diameter, $\frac{5}{16}$-inch plate, steel 55,000 pounds tensile strength, iron rivets, double-riveted lap-joint, assumed to have 67 per cent. of the efficiency of the original plate, factor of safety 5?

$$\frac{55,000 \times .3125 \times .67}{24 \times 5} = 95.96 \text{ pounds.}$$

In the first example, by referring to Table XXVIII. it will be seen that for steel plates, whether rivets are of iron or of steel, there is an excess of strength in the rivets: the joint would probably fail by tearing the plate through the line of rivet-holes, as in Fig. 68.

In the second example, by referring to Table XXV. it will be seen that for steel plates and iron rivets the percentage of joint is less in the proportion of 67 to 71 for the same spacing and diameter of rivet,—*i.e.*, $\frac{11}{16}$-inch rivets on 2⅝-inch centres. By referring to Table VII., page 59, it will be seen that the shearing strength of a $\frac{11}{16}$-inch iron rivet completely driven to fill a ¾-inch hole is 15,110 pounds. Turning now to Table X., page 60, the shearing strength of a steel rivet similarly dimensioned is found to be 19,715 pounds. If the joints were tested to rupture, the steel plate and iron rivets would probably fail by shearing the rivets, as in Fig. 58. If steel rivets were used, the failure would occur by tearing the plate across the line of rivet-holes, as in Fig. 59.

Tables of Working Pressures.—In the following tables, No. XXIX. gives the calculated working pressure for double-riveted lap-joints for iron shells and iron rivets, steel shells and iron rivets, steel shells with steel rivets, for the ordinary diameters of shell from 36 to 72 inches, and for the thickness of plates usually employed in such diameters, —*i.e.*, ¼ to ½ inch. A factor of safety of 5 is employed in this table.

Table XXX. contains calculated working pressures when similar cylindrical shells are triple-riveted. A factor of safety of 5 is employed in this table.

Table XXXI. contains working pressures calculated for diameters from 36 to 120 inches and for plates from ¼ inch to ¾ inch in thickness. These include iron and steel shells with iron and steel rivets; the longitudinal seams have butt-joints triple-riveted. The working pressure is based upon a factor of safety of 5.

TABLE XXIX.

WORKING PRESSURES FOR CYLINDRICAL SHELLS OF STEAM BOILERS, LAP-JOINTS, DOUBLE-RIVETED.

Factor of Safety, 5.

Diameter.	Thickness.	Iron Shell, Iron Rivets.	Steel Shell, Iron Rivets.	Steel Shell, Steel Rivets.
36	$\frac{1}{4}$ $\frac{5}{16}$	91 112	111 128	111 137
38	$\frac{1}{4}$ $\frac{5}{16}$	86 106	105 121	105 129
40	$\frac{1}{4}$ $\frac{5}{16}$	82 101	100 115	100 123
42	$\frac{1}{4}$ $\frac{5}{16}$	78 96	95 110	95 117
44	$\frac{1}{4}$ $\frac{5}{16}$	74 91	91 105	91 112
46	$\frac{1}{4}$ $\frac{5}{16}$	71 87	87 100	87 107
48	$\frac{5}{16}$ $\frac{3}{8}$	84 99	96 107	102 121
50	$\frac{5}{16}$ $\frac{3}{8}$	81 95	92 103	98 116
52	$\frac{5}{16}$ $\frac{3}{8}$	77 92	89 99	95 112
54	$\frac{5}{16}$ $\frac{3}{8}$	75 88	85 96	91 108
56	$\frac{5}{16}$ $\frac{3}{8}$	72 85	82 92	88 104
58	$\frac{5}{16}$ $\frac{3}{8}$	69 82	79 89	85 100
60	$\frac{5}{16}$ $\frac{3}{8}$	67 79	77 85	82 97
62	$\frac{3}{8}$ $\frac{7}{16}$	77 88	83 92	94 108
64	$\frac{3}{8}$ $\frac{7}{16}$	74 86	81 89	91 105
66	$\frac{3}{8}$ $\frac{7}{16}$	72 83	78 87	88 102
68	$\frac{3}{8}$ $\frac{7}{16}$	70 81	76 80	86 99
70	$\frac{3}{8}$ $\frac{7}{16}$	68 78	74 82	83 96
72	$\frac{3}{8}$ $\frac{7}{16}$ $\frac{1}{2}$	66 76 85	72 79 89	81 93 104

TABLE XXX.

WORKING PRESSURES FOR CYLINDRICAL SHELLS OF STEAM BOILERS, LAP-JOINTS, TRIPLE-RIVETED.

Factor of Safety, 5.

Diameter.	Thickness.	Iron Shell, Iron Rivets.	Steel Shell, Iron Rivets.	Steel Shell, Steel Rivets.
36	1/4 5/16	100 124	121 139	123 151
38	1/4 5/16	95 117	115 132	116 144
40	1/4 5/16	90 112	109 125	110 136
42	1/4 5/16	86 106	104 119	105 130
44	1/4 5/16	83 101	99 114	100 124
46	1/4 5/16	79 97	95 109	96 119
48	5/16 3/8	93 110	104 118	114 135
50	5/16 3/8	89 106	100 113	109 129
52	5/16 3/8	86 102	96 109	105 124
54	5/16 3/8	83 98	93 105	101 120
56	5/16 3/8	80 95	89 101	97 116
58	5/16 3/8	77 91	86 98	94 112
60	5/16 3/8	74 88	83 95	91 108
62	3/8 7/16	85 98	92 103	104 120
64	3/8 7/16	83 95	89 100	101 117
66	3/8 7/16	80 93	86 97	98 113
68	3/8 7/16	78 90	84 94	95 110
70	3/8 7/16	76 87	81 91	92 107
72	3/8 7/16 1/2	74 85 97	79 89 98	90 104 117

TABLE XXXI.

WORKING PRESSURES FOR CYLINDRICAL SHELLS OF STEAM BOILERS, BUTT-JOINTS, TRIPLE-RIVETED.

Factor of Safety, 5.

Diameter.	Thickness.	Iron Shell, Iron Rivets.	Steel Shell, Iron Rivets.	Steel Shell, Steel Rivets.
36	1/4	108	134	134
	5/16	135	165	165
	3/8	161	197	197
38	1/4	102	127	127
	5/16	128	156	156
	3/8	152	187	187
40	1/4	97	120	120
	5/16	121	148	148
	3/8	145	178	178
42	1/4	93	115	115
	5/16	116	141	141
	3/8	138	169	169
44	1/4	89	109	109
	5/16	110	135	135
	3/8	132	161	161
46	1/4	85	105	105
	5/16	106	129	129
	3/8	126	154	154
48	5/16	101	124	124
	3/8	121	148	148
	7/16	141	172	172
50	5/16	97	119	119
	3/8	116	142	142
	7/16	135	165	165
52	5/16	93	114	114
	3/8	111	137	137
	7/16	130	159	159
54	5/16	90	110	110
	3/8	107	132	132
	7/16	125	153	153
56	5/16	87	106	106
	3/8	103	127	127
	7/16	121	148	148
58	5/16	84	102	102
	3/8	100	123	123
	7/16	117	142	142
60	5/16	81	99	99
	3/8	97	118	118
	7/16	111	138	138
	1/2	128	157	157
62	3/8	93	115	115
	7/16	109	133	133
	1/2	124	152	152

TABLE XXXI.—*Continued.*

Diameter.	Thickness.	Iron Shell, Iron Rivets.	Steel Shell, Iron Rivets.	Steel Shell, Steel Rivets.
64	3/8	90	111	111
	7/16	106	129	129
	1/2	120	147	147
	9/16	135	165	165
66	3/8	88	108	108
	7/16	102	125	125
	1/2	117	143	143
	9/16	131	160	160
68	3/8	85	105	105
	7/16	99	121	121
	1/2	113	138	138
	9/16	127	155	155
70	3/8	83	102	102
	7/16	97	118	118
	1/2	110	134	134
	9/16	123	151	151
72	3/8	80	99	99
	7/16	94	115	115
	1/2	107	131	131
	9/16	120	147	147
	5/8	134	163	163
75	7/16	90	110	110
	1/2	102	125	125
	9/16	115	141	141
	5/8	128	157	157
78	7/16	87	106	106
	1/2	99	121	121
	9/16	111	135	135
	5/8	123	151	151
81	7/16	83	102	102
	1/2	95	116	116
	9/16	107	130	130
	5/8	119	145	145
84	1/2	92	112	112
	9/16	103	126	126
	5/8	115	140	140
	11/16	126	158	158
	3/4	137	167	167
87	1/2	89	108	108
	9/16	99	121	121
	5/8	111	135	135
	11/16	121	148	148
	3/4	132	162	162
90	1/2	86	105	105
	9/16	96	117	117
	5/8	107	131	131
	11/16	117	143	143
	3/4	128	156	156

TABLE XXXI.—*Continued.*

Diameter.	Thickness.	Iron Shell, Iron Rivets.	Steel Shell, Iron Rivets.	Steel Shell, Steel Rivets.
93	1/2	83	101	101
	9/16	93	114	114
	5/8	103	126	126
	11/16	114	139	139
	3/4	124	151	151
96	1/2	80	98	98
	9/16	90	110	110
	5/8	100	123	123
	11/16	110	134	134
	3/4	120	146	146
99	1/2	78	95	95
	9/16	87	107	107
	5/8	97	119	119
	11/16	107	130	130
	3/4	116	142	142
102	1/2	75	92	92
	9/16	85	104	104
	5/8	94	115	115
	11/16	104	127	127
	3/4	113	138	138
105	1/2	73	90	90
	9/16	82	101	101
	5/8	92	112	112
	11/16	101	123	123
	3/4	110	134	134
108	1/2	71	87	87
	9/16	80	98	98
	5/8	89	109	109
	11/16	98	120	120
	3/4	107	130	130
111	1/2	69	85	85
	9/16	78	95	95
	5/8	87	106	106
	11/16	95	116	116
	3/4	104	127	127
114	1/2	68	83	83
	9/16	76	93	93
	5/8	84	103	103
	11/16	93	113	113
	3/4	101	123	123
117	1/2	66	80	80
	9/16	74	90	90
	5/8	82	100	100
	11/16	90	110	110
	3/4	99	120	120
120	1/2	64	78	78
	9/16	71	88	88
	5/8	80	98	98
	11/16	88	108	108
	3/4	96	117	117

Philadelphia City Rules.—In estimating the strength of the longitudinal seams for rating maximum working pressure on cylindrical boiler-shells two rules should be applied :

Rule A.—From the pitch of the rivets in inches subtract the diameter of holes punched to receive the rivets ; divide the remainder by the pitch of the rivets. The quotient represents the percentage of strength of the solid part of the sheet.

Rule B.—Multiply the area of the hole filled by the rivet by the number of rows of rivets in the seam ; divide the product by the pitch of the rivets multiplied by the thickness of the sheet. This product, multiplied by the shearing strength of the rivet, divided by the tensile strength of the sheet, will give the percentage of the strength of the rivets in the seam as compared with the strength of the solid part of the sheet.

The shearing strength of a rivet in a composite joint made of iron rivets and steel plates shall not be considered in excess of 40,000 pounds. Take the lowest of the percentages as found by Rules A and B and apply that percentage as the value of the seam in the following rule (C), which determines the strength of the longitudinal seams.

Rule C.—Multiply the thickness of the boiler-plate in parts of an inch by the value of the seam as obtained by Rules A or B and by the ultimate tensile strength of the metal used in the plates ; divide this product by the internal radius of the boiler in inches multiplied by the factor of safety. The quotient will be the pressure per square inch at which the safety-valve may be set.

Boiler-Heads.—If the radius of the curvature of the convex head of the boiler be equal to the diameter of the shell of the boiler to which it is attached, then the metal in the head-sheet must be of the same thickness as the plates used in the shell or cylindrical part, and no bracing is necessary.

United States Rule for Boiler Pressure.—"Multiply one-sixth (1/6) of the lowest tensile strength found stamped on any plate in the cylindrical shell by the thickness—expressed in inches or parts of an inch—of the thinnest plate in the same cylindrical shell and divide by the radius, or half-diameter,—also expressed in inches,—and the sum will be the pressure allowable per square inch of surface for single riveting, to which add 20 per cent. for double riveting, when all the rivet-holes in the shell of such boiler have been 'fairly drilled' and no part of such hole has been punched.

"The hydrostatic pressure applied must be in proportion of 150 pounds to the square inch to 100 pounds to the square inch of the steam pressure allowed."

Factor of Safety.—This is a numerical term employed to indicate that proportion which the working pressure bears to the ultimate strength of a boiler. This proportion has long been fixed at 1/6 of the bursting pressure. For example, a riveted shell which is estimated to fail at the joints at a pressure of 570 pounds per square inch would be allowed a working pressure under a factor of safety of 6 as follows: $570 \div 6 = 95$ pounds.

Since mild steel has practically displaced wrought iron for boiler-shells a belief has gradually possessed the minds of designers of steam boilers that a factor of safety of 6 was too large; as a result, a factor of safety of 5 has been very generally adopted instead. The effect of this change is to increase the working pressure 20 per cent. over what was formerly allowed, taking the above example: $570 \div 5 = 114$ pounds, instead of 95 pounds.

The elastic limit of mild steel approximates ½ its tensile strength. The factor of safety has reference only to the ultimate strength. Working stress must always come within the elastic limit; if not, elongation and permanent set will occur. In other words, elastic limit is the yielding-point, and no stresses can be safely carried beyond that limit. Let us take Example 2, page 111, as a case in point. The whole strength of a 48-inch shell, $\frac{5}{16}$-inch steel, 55,000 pounds tensile strength, would be

$$\frac{55{,}000 \times .3125}{24} = 716 \text{ pounds.}$$

A riveted joint 67 per cent. as strong as the plate reduces the working-pressure limit as follows:

$$\frac{55{,}000 \times .3125 \times .67}{24} = 480 \text{ pounds.}$$

If this be divided by the factor of safety, we have, $480 \div 5 = 96$ pounds working pressure. The ratio which this bears to the elastic limit may be determined by substituting 27,500 pounds, or ½ the ultimate strength of the plate, thus:

$$\frac{27{,}500 \times .3125}{24} = 358 \text{ pounds per square inch,}$$

as the highest permissible pressure before the plates begin to stretch. The strength of the joint has to do with ultimate failure only. It also is affected by the elastic limit, for when stresses occur beyond this limit, stretching of plate occurs, as in Fig. 3: therefore,

$$\frac{27{,}500 \times .3125 \times .67}{24} = 240 \text{ pounds.}$$

The working pressure, as found above, was 96 pounds: then, $240 \div 96 = 2.5$ factor of safety with reference to elastic limit. Hydraulic tests seldom or never exceed 1½ times the working pressure. We have, then, $96 \times 1.5 = 144$ pounds test pressure; which is 96 pounds per square inch less than that necessary to reach the elastic limit.

Thickness of Boiler-Heads.—Externally fired cylindrical flue boilers under the United States Regulations require a thickness of material as follows: "For boilers having a diameter exceeding 32 inches and not exceeding 36 inches, not less than $\frac{1}{2}$ an inch; for boilers exceeding 36 inches in diameter and not exceeding 40 inches in diameter, not less than $\frac{9}{16}$ of an inch; for boilers exceeding 40 inches in diameter, not less than $\frac{1}{16}$ of an inch additional thickness for every 8 inches additional diameter required for boilers 40 inches in diameter.

"And the heads of steam- and mud-drums of such boilers shall have a thickness of material of not less than $\frac{1}{2}$ an inch."

Strains on a Boiler-Head.—The strains due to steam-pressure on a boiler-head are to be estimated as acting at right angles to the plate. Flat plates begin to bulge out at very low pressures, offering, as they do, very little resistance to bending. The common practice in boiler design is to transfer, by means of stays, the greater portion of the stress upon the flat surfaces of the heads to some portion of the cylindrical shell, and as this transference brings additional work upon the shell, the fastenings of the braces or stays must not be too much localized; otherwise distortion of the shell and consequent weakness may result.

In estimating the stress upon a boiler-head due to pressure alone, it is customary to make no account of the strength of the flat plate to be supported, but to transfer the whole of the estimated stress upon the stays. As a matter of fact, the whole stress is not thus transferred, but any difference between the two counts for that much additional to the factor of safety of the stay.

Unstayed Flat Heads.—The working pressure to be allowed when made of stamped materials on steam-drums or shells of boilers, when flanged and made of wrought iron or steel or of cast steel, shall, under the United States Regulations, be determined by the following rule:

"The thickness of plate in inches multiplied by $\frac{1}{6}$ of its tensile strength in pounds, which product divided by the area of the head in square inches multiplied by 0.09 will give pressure per square inch allowed. The material used in the construction of flat heads when tensile strength has not been officially determined shall be deemed to have a tensile strength of 45,000 pounds."

Bumped Heads.—The pressure allowed on bumped heads under the United States Regulations is determined by the following rule:

"Multiply the thickness of the plate by $\frac{1}{6}$* of the tensile strength, and divide by $\frac{1}{2}$ of the radius to which head is bumped, which will give the pressure per square inch of steam allowed."

---

* This means the factor of safety. If 5 be the factor employed in any boiler calculation relating to working pressure, then $\frac{1}{5}$ of the tensile strength is employed, instead of $\frac{1}{6}$ under the United States rule.

"*Example:* On a bumped head 54 inches in diameter, tensile strength 45,000 pounds, ¾ inch thick, bumped to a radius of 54 inches, pressure would be allowed as follows:

$$\frac{45,000 \times .75}{6 \times 27} = 208.33.$$

"Where the circumferential seam in such head is double-riveted to shell, which is entitled to an additional pressure, there will be allowed 20 per cent. additional pressure on said head."

Concave Heads.—The pressure allowed for concave heads of boilers under the United States Regulations is as follows:

"Multiply the pressure per square inch allowable for bumped heads attached to boilers or drums convexly by the constant 0.6, and the product will give the pressure per square inch allowable in concaved heads."

Boiler-Head Stays.—The end surfaces of steam boilers are stayed by means of braces extending from the heads of the shell or by longitudinal stay-bolts extending through from the front to the back heads. In all flue boilers in which the flues are riveted to the heads, the flues themselves act as stays, and commonly have strength enough to dispense with other stays below the water-line, except in very large boilers or for very high pressures. The holding power of wrought-iron tubes expanded in the heads is sufficient to withstand any working pressure occurring in that portion of stationary-engine boilers in which the tubes are located; but boilers of large diameter, such as marine boilers, are commonly fitted with stay-tubes in addition.

FIG. 104.

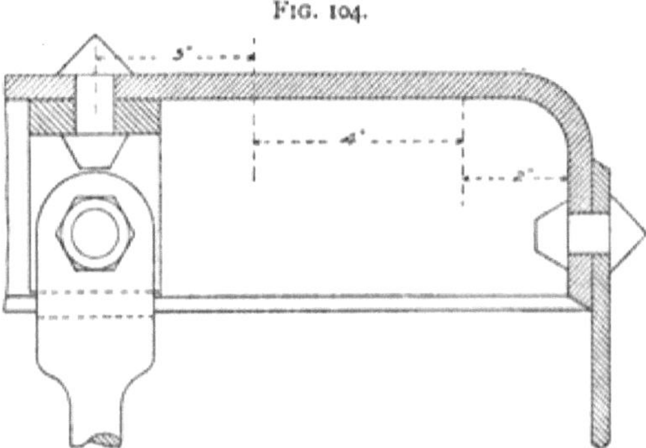

Flanging the edges of a boiler-head increases its stiffness along the outer edge. The radius of such a flange will average not far from 1½ inches for large sizes; 2 inches of this outer flanged surface of the head may be left to take care of itself. In addition to this, the influence of the flange extends inward, and no braces need be located within 4 inches of the flange radius for pressures less than 100 pounds per square inch, as shown in Fig. 104; and unless in the case of curve intersections and odd spaces which require a brace, the regular line of braces for pressures not exceeding the above need not be closer than 3 inches, also shown in the illustration.

The holding power of the tubes imparts sufficient stiffness to the boiler-head as not to require braces nearer than 4 inches, so that in all ordinary calculations the area to be supported would be represented by a segment of a circle, as shown in Fig. 105, of 6 inches less radius than the boiler-head and its base-line 4 inches above the tubes.

FIG. 105.

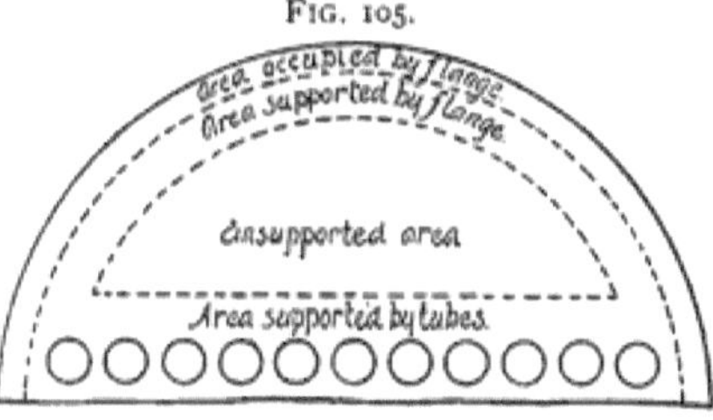

The location of stay-centres is not easily worked out except on the drawing-board. This may, perhaps, be best shown by the illustration, Fig. 106, which represents a boiler-head 60 inches in diameter. The tube-line, $\frac{3}{5}$ from the bottom, is 36 inches, and to this is added the 4 inches supported by the tubes, making 40 inches in all; this leaves 20 inches to the top of the head, from which 6 inches are deducted, this being the area supported by the flange. We have, therefore, a segment of a circle with a chord of 43 inches and a height of 14 inches, as shown. Let us assume that each brace will have 1 square inch of area—this corresponds to 1⅛ inches diameter—and that the steam-pressure to be carried is 100 pounds per square inch. We have then as a practical limit 60 inches of surface for each stay to support; it will be a near enough approximation to call it 8 inches square. Whatever the system of spacing, this distance of 8 inches ought not to be greatly exceeded, even though a corresponding reduction from centre to centre be made in the other direction. The chord of 43 inches, when laid off in 8-inch sections, as shown, will require 5 stays. The height of the segment, being 14 inches, will require 2 stays. By simply intersecting the lines passing through these centres, the number and location of stays is had. These centres may be considerably changed without in the least affecting the strength of the boiler-head, so long as the work thrown upon each stay is not more than

FIG. 106.

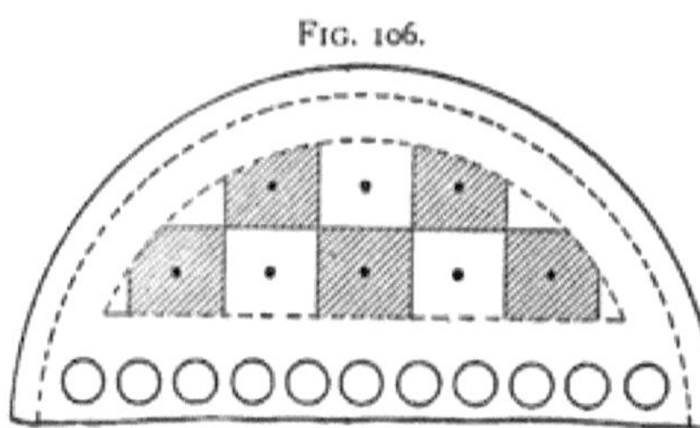

6000 pounds per square inch of section. The area to be covered by any one brace may be determined, when the steam pressure is known, by the use of Table XXXII.

Another method is to calculate the area of the segment, which in this case is 418 inches; divide it by the number of braces to be used (8 in this case); this will give the area to be supported by each: thus 418 ÷ 8 = 52.25 square inches, corresponding to a diameter of 8⅛ inches. These circles transferred to the drawing as in Fig. 107 approximate very closely those previously described.

FIG. 107.

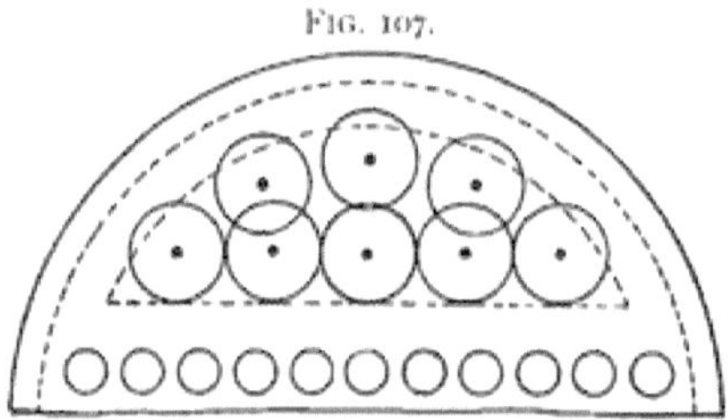

The working strength assumes an ultimate strength of 6000 pounds per square inch of section.

TABLE XXXII.

DIRECT BRACES FOR STEAM BOILERS.

For diagonal braces, see page 129.

BRACE.	WROUGHT-IRON STAYS.		INCHES SQUARE EACH BRACE WILL SUPPORT FOR PRESSURE PER SQUARE INCH.			
Diameter, Inches.	Area, Square Inch.	Working Strength, Pounds.	75 Pounds.	100 Pounds.	125 Pounds.	150 Pounds.
⅞	.60	3,600	7.	6.	5.4	4.9
1	.78	4,712	7.9	6.9	6.1	5.6
1⅛	.99	5,964	8.9	7.7	6.9	6.4
1¼	1.23	7,362	9.9	8.6	7.7	7.0
1⅜	1.48	8,880	10.7	9.5	8.5	7.7
1½	1.77	10,620	11.9	10.4	9.2	8.5

The United States Regulations are: "No braces or stays hereafter employed in the construction of boilers shall be allowed a greater strain than 6000 pounds per square inch of section. Braces must be put in sufficiently thick that the area in inches which each has to support, multiplied by the pressures per square inch, will not exceed 6000 when divided by the cross-sectional area of the brace or stay.

"Steel stay-bolts exceeding a diameter of 1¼ inches and not exceeding a diameter of 2½ inches at the bottom of the thread may be allowed a strain not exceeding 8000 pounds per square inch of cross-section; steel stay-bolts exceeding a diameter of 2½ inches at bottom of thread may be allowed a strain not exceeding 9000 pounds per square inch of cross-section; but no forged or welded steel stays will be allowed.

"The ends of such stays may be upset to a sufficient thickness to allow for truing up and including the depth of the thread.

"And all such stays after being upset shall be thoroughly annealed."

Lloyd's Rule.—The working pressure allowed on flat surfaces supported by stays is according to Lloyd's rule as follows:

$$\frac{C \times T^2}{p^2} = \text{working pressure in pounds per square inch.}$$

Where T = thickness of plate in $\frac{1}{16}$ of an inch.

p = greatest pitch in inches.

C = 90 for $\frac{7}{16}$ plate and thinner fitted with screw stays with riveted heads.

C = 100 for plates thicker than $\frac{7}{16}$ fitted with screw stays with riveted heads.

C = 110 for $\frac{7}{16}$ plate and thinner fitted with screw stays and nuts.

C = 120 for plates thicker than $\frac{7}{16}$ fitted with screw stays and nuts.

C = 140 for plates fitted with stays with double nuts.

C = 160 for plates fitted with stays with double nuts and washers, at least $\frac{1}{2}$ thickness of plates and a diameter of $\frac{2}{5}$ of the pitch, riveted to the plates.

The United States Regulations differ slightly from Lloyd's rule in the value of constants employed. The working pressure allowed on flat surfaces fitted with screw stay-bolts and nuts, or plain bolt with single nut and socket, or riveted head and socket, will be determined by the following rule:

"When plates $\frac{7}{16}$ inch thick and under are used in the construction of marine boilers, using 112 as a constant, multiply this by the square of the thickness of plate in 16ths of an inch. Divide this product by the square of the pitch or distance from centre to centre of stay-bolt."

Example: A plate $\frac{7}{16}$ inch thick, with stays placed on 6-inch centres, using the constant 112 as above, would yield

$$\frac{7 \times 7 \times 112}{6 \times 6} = 152.44 \text{ pounds}$$

as the working pressure allowed, provided the strain on stay or bolt does not exceed 6000 pounds per square inch of section.

"Plates above $\frac{7}{16}$ inch thick, the pressure will be determined by the same rule, excepting the constant will be 120. Then, a plate ½ inch thick, stays spaced 7 inches from centre, would be as follows: 120, the constant, multiplied by 64, the square of thickness in 16ths of an inch, equals 7680; which, divided by the square of 7 inches (distance from centre to centre of stays), which is 49, would give 156 pounds working pressure."

On other flat surfaces there may be used stay-bolts with the ends threaded, having nuts on same, both on the outside and inside of plates. The working pressure allowed would be as follows:

"A constant 140, multiplied by the square of the thickness of plate in 16ths of an inch. This product divided by the pitch or distance of bolts from centre to centre, squared, gives the working pressure."

Example: A plate ¾ inch thick, supported by bolts 14-inch centres, would be

$$\frac{140 \times 144}{196} = 102 \text{ pounds working pressure.}$$

Same thickness of plate, with bolts 12-inch centres, would be

$$\frac{140 \times 144}{144} = 140 \text{ pounds working pressure.}$$

"Flat part of boiler-head plates, when braced with bolts having double nuts and a washer at least ½ the thickness of head, where washers are riveted to the outside of the head and of a size equal to ⅞ of the pitch of stay-bolts, or where heads have a stiffening plate covering the area braced, will equal the thickness of the head and washers, the head and stiffening plate being riveted together, with rivets spaced and of sufficient sectional area of rivets, shall be allowed a constant of 200, rivets to be spaced by thickness of washer on the stiffening plate. Boiler-heads so reinforced will be allowed a thickness to compute pressure allowed of 80 per cent. of the combined thickness of head and washer or head and stiffening plate."

Example: A boiler-head plate ¾ inch thick, with washers ⅜ inch thick and 12¼ inches square, supported by bolts 14-inch centres, would be allowed a working steam pressure as follows:

Thickness of plate and washers equals $\frac{3}{4} + \frac{3}{8} = \frac{9}{8}$ inch; 80 per cent. of which combined thickness equals $\frac{80}{100} \times \frac{9}{8}$ inch = .9 inch = 14.4 16ths of an inch. Then, by rule,

$$\frac{200 \times 14.4^2}{14^2} \quad \frac{200 \times 207.36}{196} = 211 \text{ pounds working pressure.}$$

"Plates fitted with double angle-iron and riveted to plate with leaf at least ⅔ thickness of plate and depth at least ¼ of the pitch would be allowed the same pressure as determined by formula for plate with washer riveted on."

"*Example:* A boiler-head plate ¾ inch thick, supported by angle-iron ½ inch thick and 3½ inches depth of leaf, and with bolts 14-inch centres, would be allowed a working steam pressure as follows:

"Thickness of head and leaf of angle-iron equals $\frac{3}{4} + \frac{1}{2} = \frac{5}{4}$; 80 per cent. of which combined thickness equals $\frac{80}{100} \times \frac{5}{4}$ inch = 1 inch = 16 16ths of an inch. Then, by rule,

$$\frac{200 \times 16^2}{14^2} \quad \frac{200 \times 256}{196} = 261 \text{ pounds working pressure.}$$

"But no flat surface shall be unsupported at a greater distance in any case than 16 inches, and such flat surfaces shall not be of less strength than the shell of the boiler, and able to resist the same strain and pressure to the square inch. In allowing the strain on a screw stay-bolt, the diameter of the same shall be determined by the diameter at the bottom of the thread."

Details of Stays and Braces.—The stays and braces in steam boilers working under high pressures require careful working out, that they do not of themselves become an element of weakness. A less number of strong stays is to be preferred to a larger number of weaker ones. Too many braces, especially below the water-line, interfere with the circulation; the interior of the boiler above the water-line would also be more or less inaccessible, and thus prevent proper inspection, as well as interfere in the facility of making repairs.

The spacing of braces and the kind best adapted for any given area will depend upon the extent of that area, the thickness of the plate, and the steam pressure. No particular form of stay is recommended to the exclusion of others, as several kinds may be included in the same boiler to good advantage.

FIG. 108.

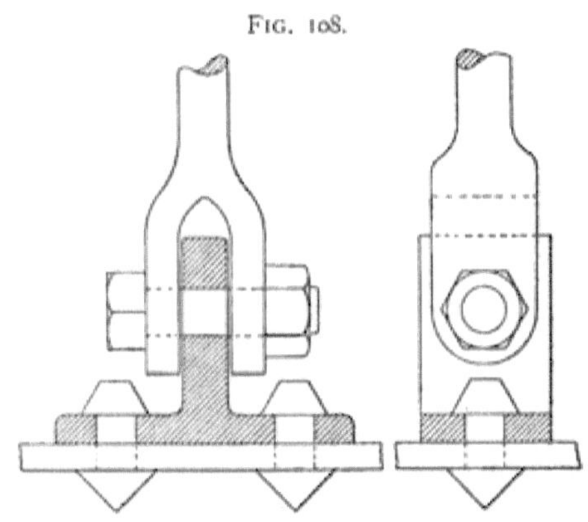

A crowfoot is a link-joint secured to wrought-iron lugs, ⊥-shaped, riveted to crown-sheets, heads, or other flat surfaces, and adapted for the fastening of a brace or stay by means of a bolt or split pin, as shown in Fig. 108.

Bolts and nuts are to be preferred to split pins, because they give less chance for the spreading of the eye of the brace. One or other of the ends of the split pins are liable to break off when spreading, or closing them for removal. Difficulty is sometimes experienced in getting a bolt back into place after removal by reason of the springing of the boiler-plate. In such a case a mortised drift-pin, as in Fig. 109, may be used and left in place, after the insertion of the spring cotter to prevent its working back and out of place.

FIG. 109.

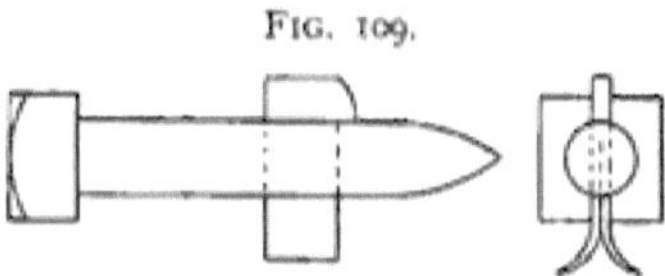

In case two points of support are required for a single rod, the end of the stay should be made preferably of two links, as shown in Fig.

110. Wherever practicable such forgings should be made without welding. No objections exist to cutting them out of odd pieces of plate of proper thickness.

Fig. 110.

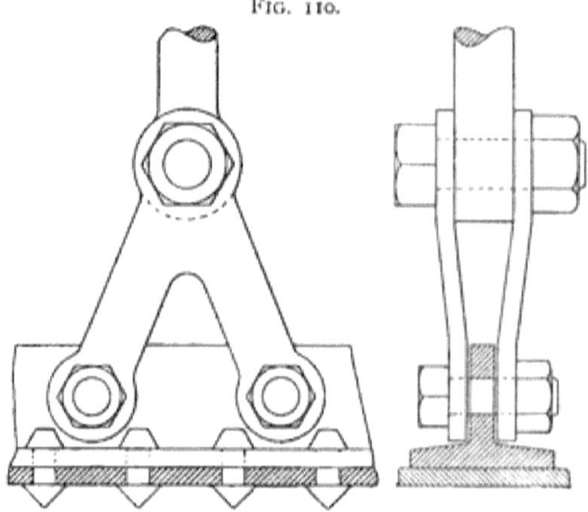

When fitting the joints or other fastenings of stay-bolts the holes should be drilled. The pins or bolts should be sufficiently near the size of the hole to make a shearing instead of a bending stress. A not uncommon practice of bending a piece of square iron over the horn of an anvil and welding it, making an end similar to *A*, Fig. 111, is not first-class work. It is much better to make the end as at *B* and drill the hole not more than $\frac{1}{16}$ larger than the bolt.

Fig. 111.

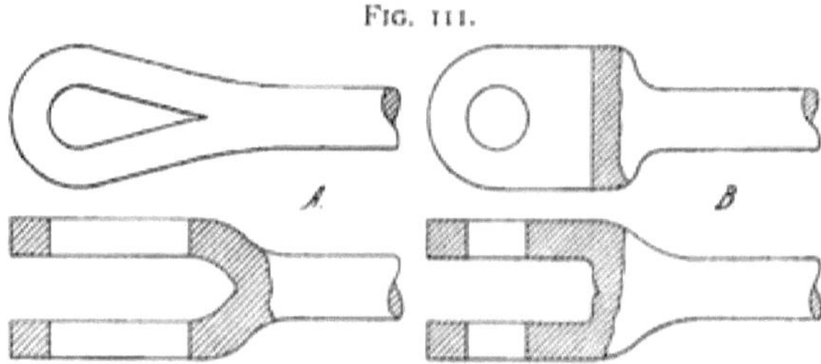

A gusset-stay is an iron plate forming a diagonal brace, one end of which is fitted with a flanged foot for riveting to the head and the other end flanged for riveting to the shell, as shown in Fig. 112.

Gusset-stays are usually cut out of plate metal, with two pieces of angle-iron riveted to each end, as shown in Fig. 112, *B*. Less frequently the end of the plate itself is flanged and a piece of angle-iron riveted to it, as shown at *A*.

A larger factor of safety must be given gusset-stays than for ordinary oblique stays, because the tension on a stay of this kind is not uniform across its area, but is greater near one edge. Its sectional area should be at least four times that calculated for a round or square stay for the same pressure.

FIG. 112.

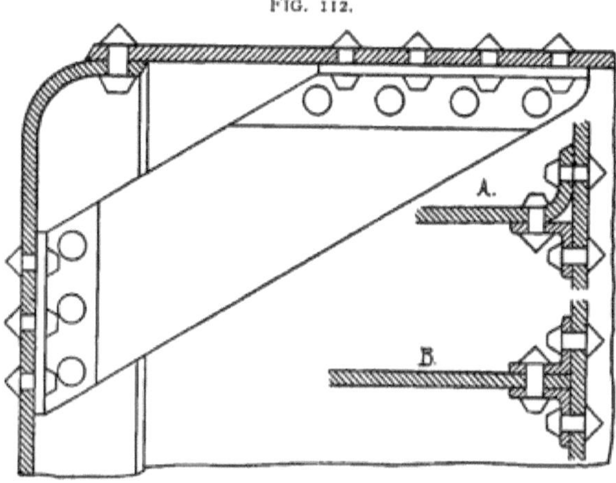

Longitudinal stays are rods extending through from one end of a boiler to the other. These rods are commonly fitted with nuts and washers, as shown in Fig. 326, and further in detail for the same boiler in Fig. 113. When such stays are of considerable length, say 18 or 20 feet, they should have a central support to prevent drooping in the

FIG. 113.

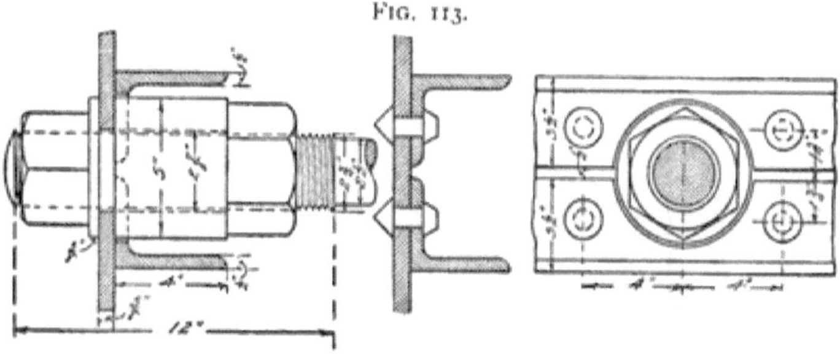

centre. Whenever possible longitudinal stays should be made without welding ; but if welding be a necessity, the material should then be iron, and not steel. Upsetting the ends to get a larger diameter of screw-threads should be confined to iron only. Stays are sometimes necessary in places where for reasons they cannot pass through the plates, as in Fig. 113; in such cases the ends have been fitted as in Fig. 114. After the stay is in place the distance piece at *A* is fitted to the exact length between the flat plate and the stay, after which the tap-bolt is inserted and screwed up tight. The latter arrangement is not as satisfactory as the former one, as well as being more expensive to make. Its use is confined to those places in which no other form of stay can be used.

FIG. 114.

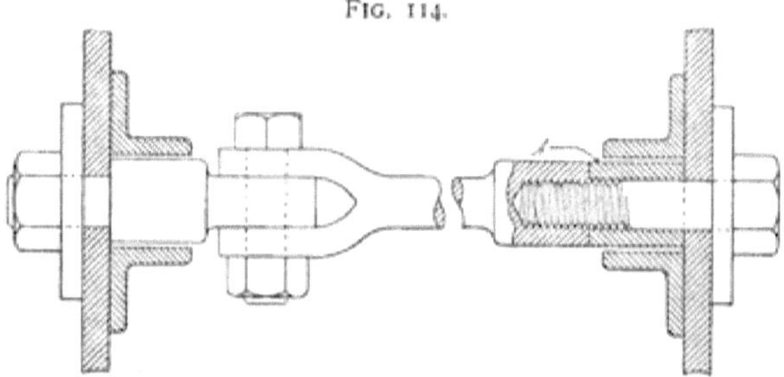

Diagonal Stays.—It is always desirable that stays and tie-rods lead at right angles to and from the surfaces to be stayed, but this is a condition not always practicable, and diagonal stays become a necessity.

Knowing the area of a direct stay required for any working pressure, the area of a diagonal stay may be found for the same pressure thus: Multiply the length of the proposed diagonal stay in inches by the proper area in square inches required for a direct stay; divide the product thus obtained by the length in inches of a right-angle line drawn from the face of the surface to be stayed to the centre of the first

FIG. 115.

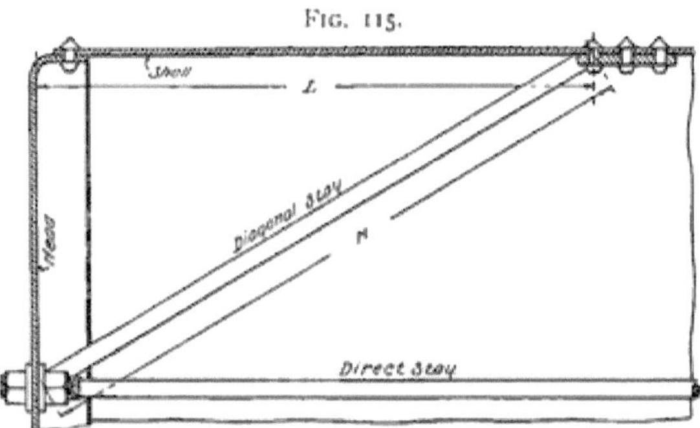

rivet in the proposed diagonal stay: the quotient will give the area of the smallest part of the diagonal stay.

Let the diagram (Fig. 115) represent a portion of the boiler-head to be stayed, and from a common centre imagine a direct stay leading to the opposite head of the boiler, also a diagonal stay, *H*, attaching to the shell of the boiler at a distance, *L*. The direct stay is the best, but if for any reason it cannot be used, the enlarged area of a diagonal stay to carry the same load may be found in the use of the above rule, as follows:

Example: A direct stay of 1¼ inches diameter or 1.227 square inches area is sufficient for a given area of surface and steam pressure. Allowing 6000 pounds per square inch of area of stay, what must be the area of a diagonal stay having a distance, *H*, of 48 inches and a length, *L*, of 42 inches? Area of direct stay = 1.227 square inches × 48 = 58.896 ÷ 42 = 1.402 square inches area, corresponding to a diameter of nearly 1⅜ inches.

Each diagonal stay must be separately calculated. The following formulæ may be useful in this connection. Let

A = surface to be supported in square inches.
B = working pressure in pounds.
H = length of diagonal stay in inches.
L = length of line drawn at a right angle from the surface to be supported to the end of the diagonal stay in inches.
S = working stress per square inch on stay in pounds.
a = area required for direct stay in square inches.
a_1 = area of diagonal stay in square inches.
d = diameter of diagonal stay in inches.

Then $a_1 = \frac{a \times H}{L}$.

$$H = \frac{a_1 \times L}{a}.$$

$$d = \sqrt{\frac{a_1}{.7854}}, \text{ or } d = \sqrt{\frac{a \times H}{.7854 \times L}}, \text{ or } d = \sqrt{\frac{A \times B \times H}{.7854\ S \times L}}.$$

$$B = \frac{.7854 \times d^2 \times S \times L}{A \times H}.$$

The stress upon a diagonal stay is equal to the stress which a perpendicular stay supporting a like surface would sustain, divided by the cosine of the angle which it forms with the perpendicular to that surface requiring to be supported.

Let us suppose an area of 64 square inches to be supported under 100 pounds pressure and the angle of the stay to be 30°, as in Fig. 116, what would be the stress upon such a diagonal stay?

The cosine of 30° is 0.866. Then

$$\frac{100 \times 64}{.866} = 7390 \text{ pounds,}$$

as against 100 × 64 = 6400 pounds if a direct stay-rod could be used.

Two forms of diagonal stays are shown in Fig. 117. The one recommended is that having a T-head with a rivet on either side of the rod

FIG. 116.

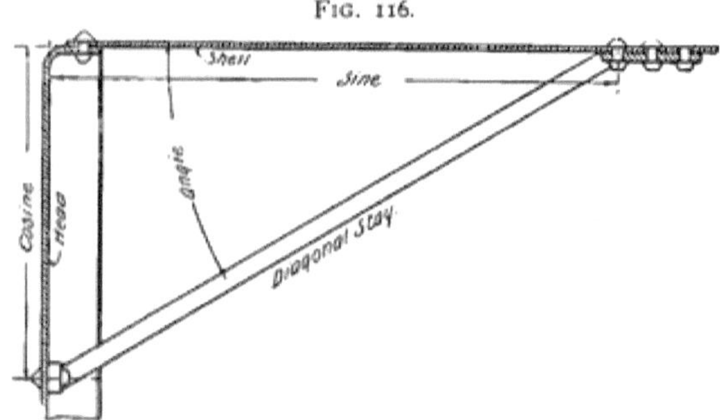

connecting with the shell. The flat stay with bent end is not as rigid as the former, because there is more or less yielding at the joint inside the head, unless the stay is quite thick, which is usually not the case.

FIG. 117.

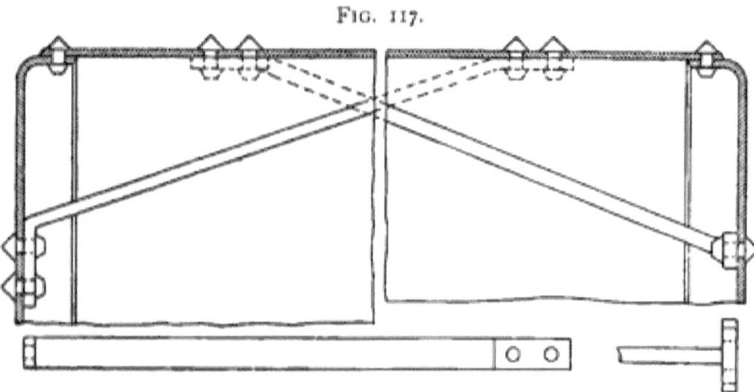

The diagonal brace (Fig. 118), by the Lukens Iron and Steel Company, is made of plate steel bent to the form shown in the engraving. It will be seen that the bend which occurs at the end fitted for staying

FIG. 118.

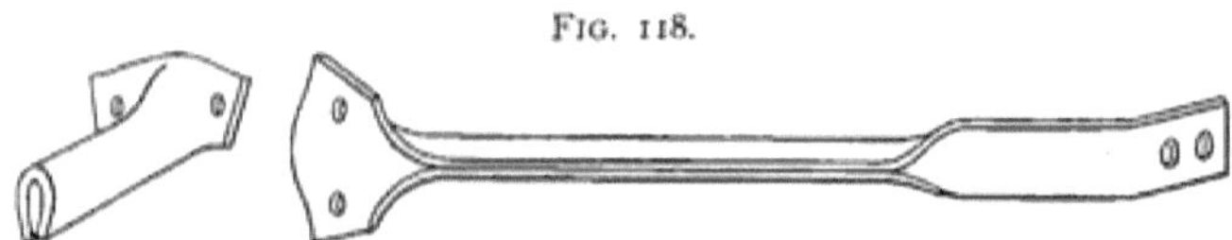

the head is braced down to the lower edge of the plate by a peculiar curve for securing the necessary rigidity.

Longitudinal or other stays are sometimes made of two lengths, to be connected in the centre, in which case a turnbuckle, as in Fig. 119, is used; but this detail had better be omitted if it is possible to make a single-link connection with pin-joints at either or both ends, as screw-threads are apt to waste away inside of a steam boiler.

FIG. 119.

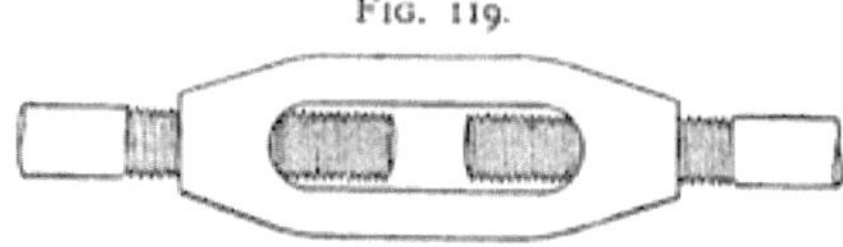

FIG. 120.

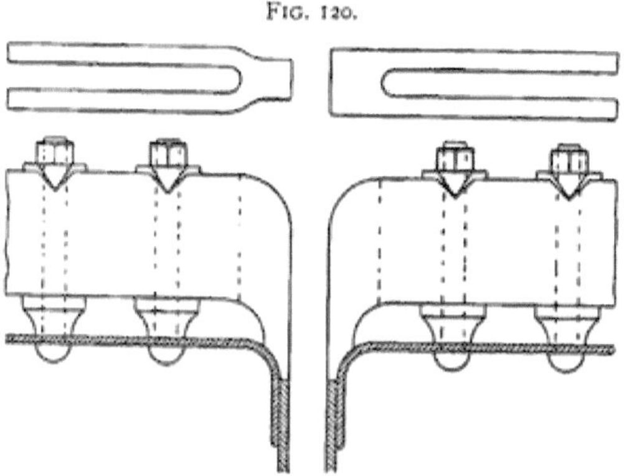

Crown-Bars.—Flat crown-sheets for boilers of the locomotive type are commonly supported by crown-bars extending across the furnace; less frequently they extend lengthwise of the furnace. Crown-bars are generally made up of two wrought-iron bars welded at the ends, as

FIG. 121.

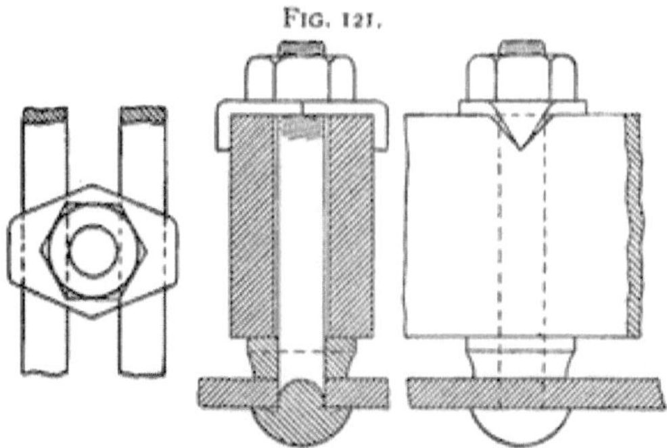

shown in Fig. 120. The depth will vary from 4 to 6 inches; the thickness of the bars will average not far from ¾ of an inch, though ⅞-inch metal is sometimes used; the distance apart is 1 inch or 1⅛ inches, depending on whether ⅞- or 1-inch stay-bolts are used. A clear water-way between the crown-sheet and the crown-bars of 1 to 1½ inches is allowed; the latter distance is preferable. The ends of the crown-bars should be carefully fitted by chipping and filing, so that a good bearing is had on the end of the side sheets not only, but upon the flanged curve of the crown-sheet as well. The crown stay-bolts seldom vary from the two diameters already referred to,—viz., ⅞ inch or 1 inch,—and these are placed on about 4½-inch centres across the crown-sheet. The details of the head of such stay-bolts vary somewhat; for example, the stays experimented upon by Mr. Cole (Figs. 133 to 147) are all in use. The illustration, Fig. 121, represents a practice of the Pennsylvania Railroad, while Fig. 122 represents a stay-bolt head and neck recommended by J. G. A. Meyer in his "Modern Locomotive Construction." This bolt has a slight taper under the head; the crown-sheet is reamed out to fit this taper. The advantage claimed for this form is, that should leaks occur the bolts may be readily taken out, the tapered part extended by turning a small portion off the head and refitting the bolt in the crown-sheet. Cast-iron washers or distance-pieces are placed between the crown-sheet and the crown-bar for tightening the stay-bolts without springing the crown-sheet. These washers contribute also to the stiffness of the crown-bars, the effect being that of deepening the crown-bar by including the crown-sheet, which now acts as a bottom flange to the girder spanning the fire-box.

FIG. 122.

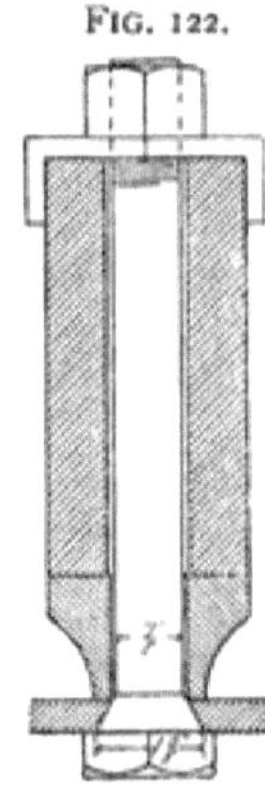

The strength of a crown-bar is usually determined by the formula used when a beam is supported at both ends and uniformly loaded, in which

W = load in pounds.
b = breadth of crown-bar in inches.
d = depth of crown-bar in inches.
l = length of crown-bar in inches.
16,000 = a constant.

When a crown-bar is made of two pieces, as shown in the illustrations, the breadth b includes both,—that is, if the crown-bar is ⅝-inch thick, b = .625 × 2 = 1.25 inches.

Example: A crown-bar is 38 inches long, 4½ inches deep by ¾ inch in thickness, welded in pairs as in Fig. 120. The sustaining power if uniformly loaded would be

$$\frac{16{,}000 \times d^2 \times b}{l} \text{ or } \frac{16{,}000 \times 4.5^2 \times 1.5}{38} = 12{,}789 \text{ pounds.}$$

If these crown-bars are to be placed on 5-inch centres, the area supported would be 38 × 5 = 190 square inches. Then 12,789 ÷ 190 =

67.31 pounds per square inch, a pressure much too low for the modern boiler of the locomotive type. Let us assume a steam pressure of 140 pounds per square inch to be necessary for the work. We have then $190 \times 140 = 26,600$ pounds to be supported, or $26,600 - 12,789 = 13,811$ pounds for each girder in excess of what the girder itself can safely carry. The best way to provide for this is to connect the crown-bars by means of braces with the roof-sheet overhead, as shown in Fig. 123. If two braces be used, the stress to be borne by each will be $13,811 \div 2 = 6906$ pounds. If we allow 6000 pounds per square inch for the brace, we have $6906 \div 6000 = 1.151$ area of cross-section at the smallest part, which approximates a diameter of 1⅜ inches.

FIG. 123.

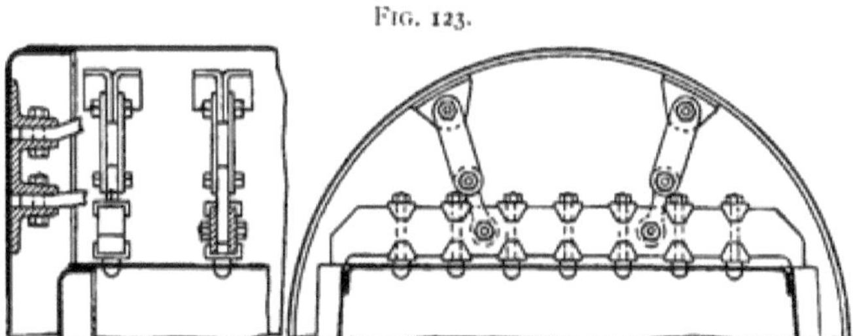

The crown-bars referred to in the above example are placed crosswise of the furnace, but in some boilers they are arranged to extend lengthwise of the furnace. The latter arrangement affords in locomotives a better circulation over the crown-sheet; on the other hand, it adds to the length of the crown-bar, which must be made deeper and will require more roof-stays to prevent deflection at its centre. The present tendency in locomotive designs is against crown-bars altogether and towards that of radial stays, or the employment of a Belpaire fire-box.

Radial Stays are employed in curved crown-sheets of internally fired boilers. These extend from the crown-sheet to the outside of the boiler, as shown in Fig. 124. This arrangement of stays has much to commend it. The pressure of steam tends to burst the shell and to collapse the fire-box. But by a correctly laid out scheme of radial stays these pressures are made to assist in counteracting each other by bringing the stress of both the outer and inner sheets upon the intervening stays. These stays are commonly $\frac{15}{16}$ inch in diameter, 12 threads per inch, with one end enlarged to $1\frac{1}{16}$ inches, so that the smaller screw shall pass easily through the larger hole. The sketch, Fig. 125, illustrates average practice in locomotive work. The plates are represented as being further strengthened by nuts on the outside, a practice by no means universal, the commonest method being to cut the stay-bolts to length in place and then rivet the ends without nuts.

FIG. 124.

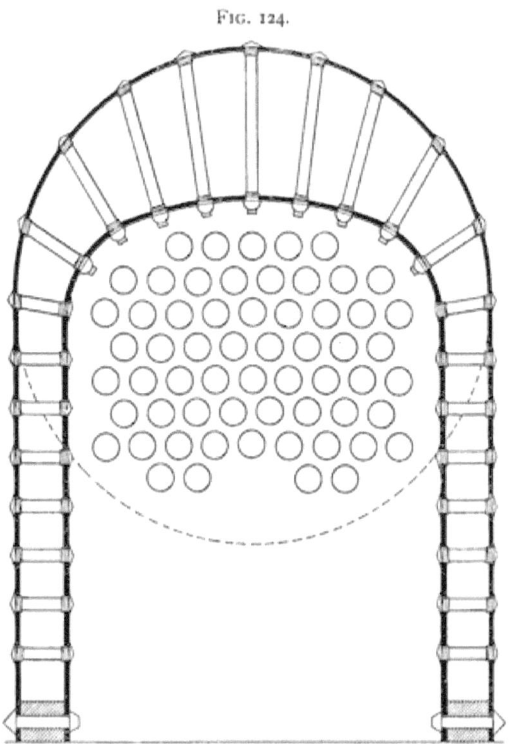

FIG. 125.

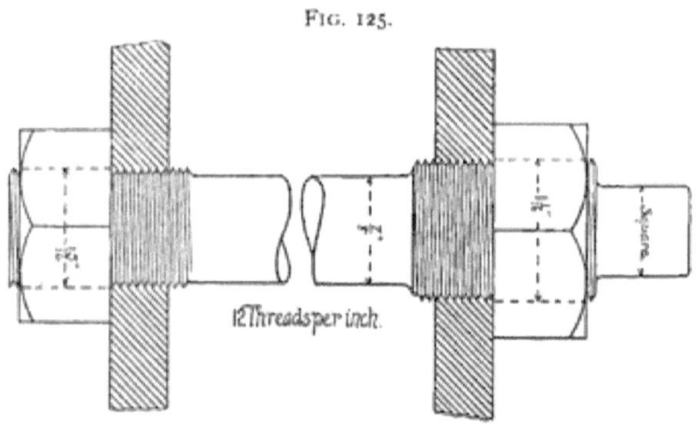

Stay-bolts for flat surfaces under high pressures, such as locomotive fire-boxes, are commonly placed on 4-inch centres, as shown in Fig. 126. Each stay-bolt is required, therefore, to take the pressure of a surface of 4 inches square, as indicated by the dotted lines, less its own area, which is commonly omitted, to be on the safe side. Suppose the boiler pressure to be 175 pounds per square inch. We then have $4 \times 4 \times 175 = 2800$ pounds tension upon each stay-bolt. If the stay-bolts have 1 inch threads and are ⅞-inch diameter between the threads, it is the area, 0.601 inch, due to this latter diameter which must take the load. Assuming 50,000 pounds as the tensile strength of the wrought iron, we have $50{,}000 \times .601 = 30{,}500$ pounds as the breaking strength of the bolt. Therefore,

FIG. 126.

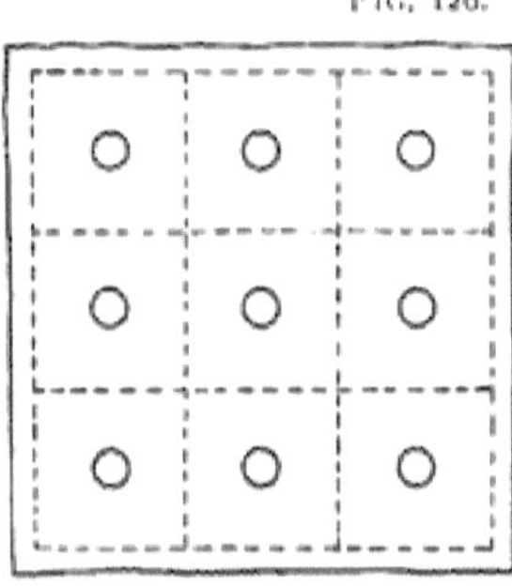

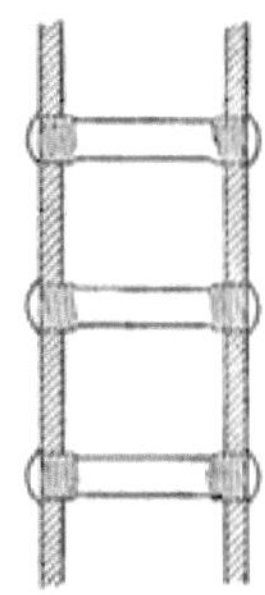

$$\frac{30{,}500}{2800} = 10.9 \text{ the factor of safety,}$$

which for bolts under bending as well as tensile strain is the least admissible limit. A 1-inch bolt is recommended instead, which will give under similar conditions a factor of safety of 14. No stay-bolt less than ¾ inch diameter should be used in flat fire-box sheets, no matter how low the pressure.

Table XXXIII. shows the proper spacing from centre to centre of stay-bolts for flat surfaces for pressures of 50 to 150 pounds per square inch, thicknesses of plate from ¼ to ½ inch for stay-bolts ¾ to 1¼ inches in diameter.

The Strains upon a Screwed Stay-Bolt are not the same as upon a rivet, or calculations regarding them would be very simple. Unfortunately, such calculations are at best of little value, because stay-bolts seldom or never fail in mere tensile strength. It is their inability to withstand the bending stresses which centre immediately inside of the outside sheet; and it is here that almost all stay-bolt failures occur. Railway master mechanics, who have had a larger experience in such matters than any one else, have generally reached the conclusion, based entirely upon experience, that no screwed stay-bolts less than ⅞ inch in diameter shall be used at all in locomotive fire-boxes, and then only for pressures up to 150 pounds per square inch. For pressures greater than that, 1 inch screw stay-bolts of the best quality refined iron, placed on 4-

inch centres, are recommended. Such staying may be counted on for at least five years' service for pressures up to 175 pounds per square inch. The screw threads are finer than those of the United States standard. A common pitch for screw stay-bolts is 12 threads per inch.

TABLE XXXIII.

PROPORTIONS FOR STAY-BOLTS FOR FLAT SURFACES.

Pressure per Square Inch.	Centre to Centre of Stay-Bolts in Inches.				
	¼-inch Plate.	$\frac{5}{16}$-inch Plate.	⅜-inch Plate.	$\frac{7}{16}$-inch Plate.	½-inch Plate.
	¾-inch Stay.	¾-inch Stay.	⅞-inch Stay.	1-inch Stay.	1⅛-inch Stay.
50	6	7	8	9	10
60	5⅜	6⅜	7¼	8⅛	9
70	5	5¾	6⅝	7½	8⅜
80	4⅝	5½	6¼	7⅛	7⅞
90	4⅜	5⅛	5⅞	6⅝	7⅜
100	4¼	4¾	5½	6¼	7
110	4	4⅝	5¼	5⅞	6⅝
120	3⅞	4⅜	5	5¾	6⅜
130	3¾	4¼	4⅞	5½	6⅛
140	3⅝	4⅛	4⅝	5¼	6
150	3½	4	4⅜	5	5⅝

Stay-bolts with a drilled hole, as in Fig. 127, are quite common, especially in boilers of the locomotive type. The hole may be say $\frac{3}{16}$

Fig. 127.

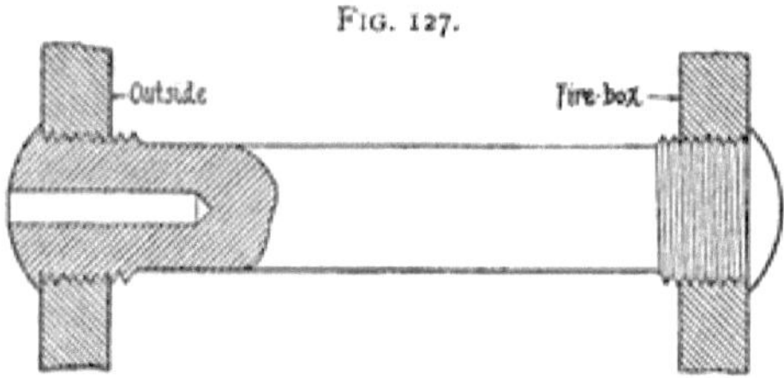

inch in diameter and extend inward perhaps an inch or a little more. Inasmuch as breakages almost always occur at the outside sheets, the drilled hole should always be at the outside end of the stay-bolt. In the event of breakage occurring immediately inside of the outer sheet, warning is given by the escape of water through the central hole.

Fig. 128.

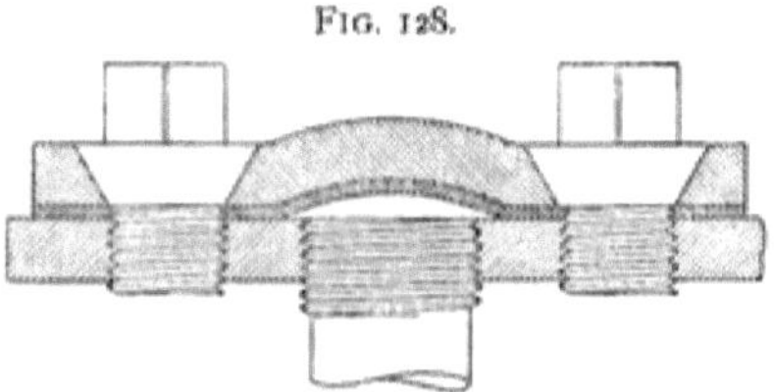

It sometimes happens that leakage occurs around a stay-bolt, not serious enough to shut down the boiler for repairs. A soft patch may be put over it with tap-bolts, as shown in Fig. 128, at any time the pressure is off.

Flexible Stay-Bolts.—The outer shell and the fire-box of an internally fired boiler, such as a locomotive boiler, never expand alike; and it is the greater expansion of one plate over another which causes the breakage of stay-bolts. To obviate this, flexible stay-bolts have been designed, one of which is shown in Fig. 129. In this illustration a T-iron is riveted to the outer shell. Depending from this at fixed intervals are duplex hangers, on either side of which are stay-bolts passing through the crown-sheet and securely fastened by thread and nut. The proper tension for the stays can be had by nuts at the hangers, which are afterwards held in place by the split pins above.

FIG. 129.

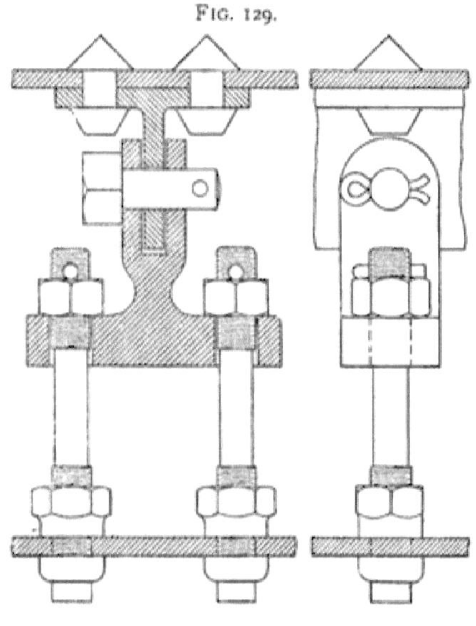

One objection to the design in the preceding paragraph is the space occupied, which prevents its use in places other than above crown-sheets. The staying of flat surfaces along the water-leg of a boiler requires different designs, one of which is shown in Fig. 130. This stay is screwed into the fire-box sheet, making a joint against the button-head, as shown in the engraving. At the outer end of this stay-bolt is fitted a spherical washer and nut. This washer fits into a seating screwed into the outer sheet. As such an arrangement is not likely to remain tight under varying pressures and conditions of expansion, a cap is screwed over the nut and washer, making a water-tight joint. The arrangement shown in Fig. 131 is the same, except that, instead of a conical washer, the under side of the bolt-head is turned spherical, adapting it to a seating as in the previous example. This bolt is screwed into the fire-box sheet from the outside and then riveted over. In the event that either of these two caps should occupy more space on the outside of the boiler than can be given to them, another form of seating is shown in Fig. 132, which diminishes the outer distance about one-half. It will be seen in these three examples that considerable movement may occur before the sides of the stay-bolt touch the seating through which they pass.

Cole's Experiments.—Investigations into the holding power at different temperatures of various styles of locomotive fire-box crown-stays was made by Mr. Francis J. Cole, the results of which were presented in a communication to the American Society of Mechanical Engineers, 1897. The tests were made as nearly as possible under the same conditions as in actual service.

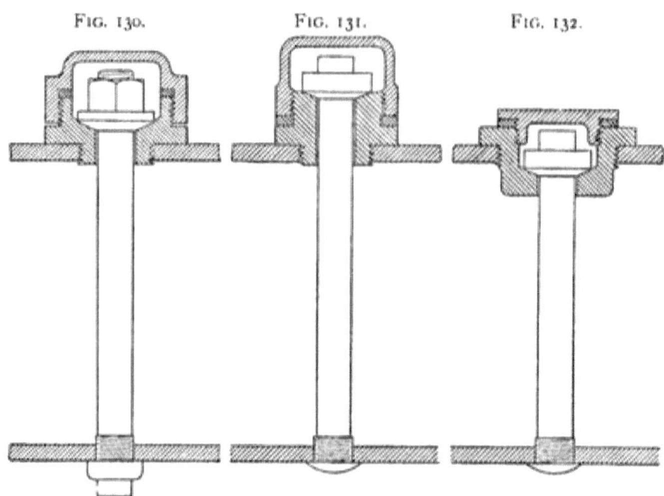

FIG. 130. FIG. 131. FIG. 132.

The material used to represent the stays was 1-inch round mild steel, 58,390 pounds tensile strength, an elastic limit of 38,900 pounds, and an elongation of 30.25 per cent. in 8 inches. The sheets were mild steel ⅜ inch thick, mostly cut from the same plate; the tensile strength was 59,150 pounds, elastic limit 28,800 pounds, elongation 31.75 per cent. in 4 inches when tested lengthwise of the grain; 58,400 pounds tensile strength, 28,040 pounds elastic limit, elongation of 28 per cent. in 4 inches when tested crosswise of the grain.

In all these tests it is assumed that the bolts are spaced 4 x 4 inches from centre to centre, supporting an area of 16 square inches. The bagging down characteristic of an overheated crown-sheet caused by low water was imitated by heating the specimens to a bright red and the use of a bearing-plate of ½-inch steel, 8 inches square, with a hole 4½ inches in diameter bored through its centre. The area of this hole is 15.9 square inches.

Specimen, Fig. 133, represents a 1-inch stay, with head ⅛ inch above the sheet riveted over. This specimen developed under cold test an elastic limit of 12,400 pounds, the tensile strength being 16,700 pounds, yielding as shown in Fig. 134. The effect of the hot test is shown in Fig. 135. The sheet was at a bright red heat after parting, the stay-bolt pulling through the sheet. The tensile strength was 3570 pounds.

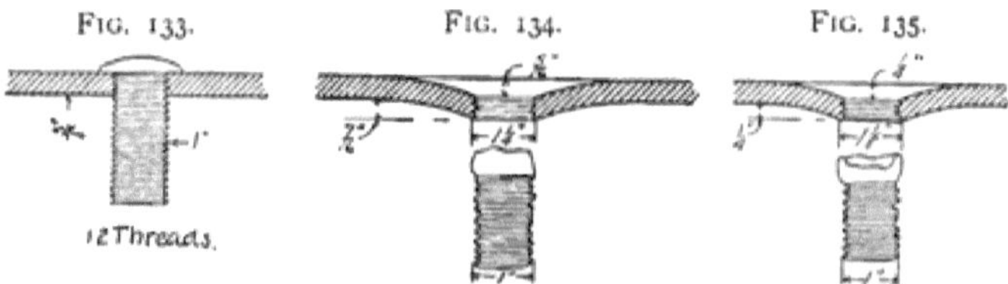

FIG. 133. FIG. 134. FIG. 135.

Specimen, Fig. 136, represents a 1-inch stay-bolt with a ⅞ standard nut tapped out to 1 inch, 12 threads, and riveted over the top of the nut, the end projecting about ⅝ inch for that purpose. This specimen under cold test had an elastic limit of 28,000 pounds and a tensile strength of 43,100 pounds. The effect of the test was to break the stay, as shown in Fig. 137. The hot test consisted in heating the specimen to a dull red, which then showed an elastic limit of 14,500 pounds and a tensile strength of 21,500 pounds at the time of yielding. The final failure is represented in Fig. 138. The plate was almost black

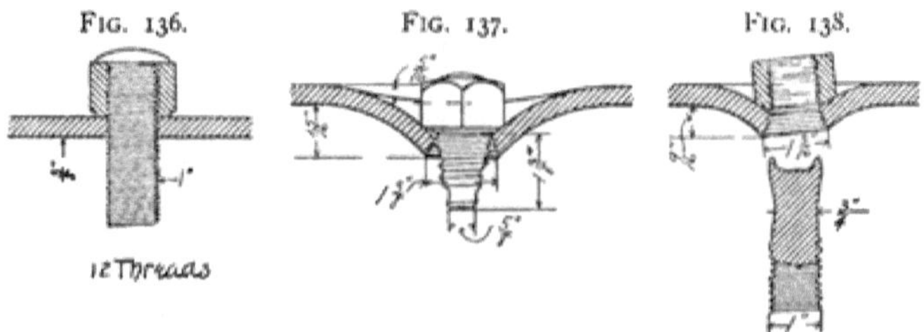

FIG. 136. FIG. 137. FIG. 138.

after parting. The holding power of a stay provided with a nut is considerably increased, when red hot, by countersinking the nut and riveting the bolt end into it.

Specimen, Fig. 139, represents a 1-inch stay with button head, no groove under the head, the plate slightly countersunk to fit a corresponding projection under the head. This specimen under cold test broke the bolt midway, as shown in Fig. 140, the elastic limit being 27,000 pounds, the tensile strength 39,800 pounds. When heated to a bright red the bolt parted under the head, as shown in Fig. 141, the tensile strength being 8000 pounds.

The specimen, Fig. 142, was an unthreaded stay-bolt with a button-head fitted into a 1$\frac{1}{16}$ reamed hole. Under cold test this specimen pulled through the sheet, as shown in Fig. 143. The elastic limit was 32,500 pounds, tensile strength 43,100 pounds. It will be observed that the effect of the cold test was to rupture the sheet as well as to distort the head. When heated to a bright red the stay-bolt pulled through the sheet, as shown in Fig. 144, exhibiting a tensile strength of 9700 pounds. The bolt and plate parted while bright red.

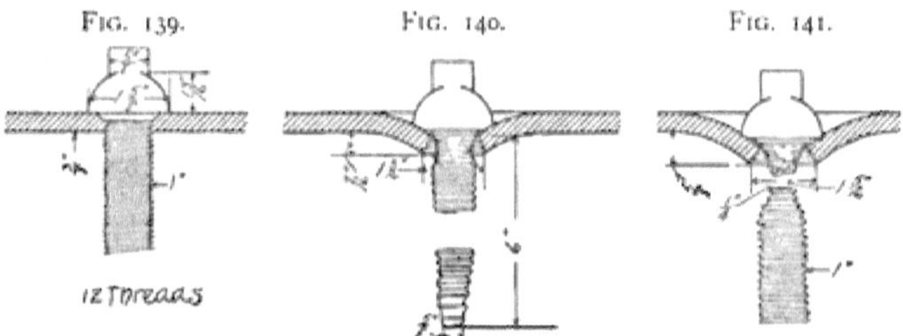

Fig. 139. Fig. 140. Fig. 141.

The above tests seem to indicate that the best riveted head which can be formed cold, made in the usual conical shape, has a holding power, hot and cold, much less than the solid head; but the objection to the use of solid heads is the liability of injury when screwed into a fire-box where the holes are not tapped at right angles to the sheet, and where the surface of the sheet is curved, but this objection can easily be removed by properly seating the head.

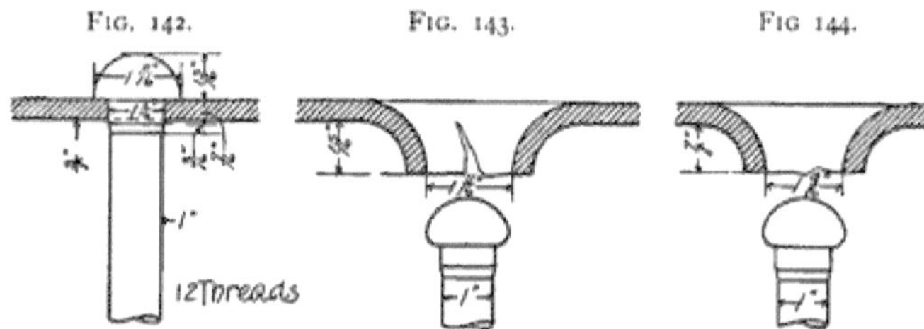

Fig. 142. Fig. 143. Fig 144.

Specimen, Fig. 145, was a button-headed stay-bolt with 1¼-inch tapered reamed hole, 3-inch thimble and nut. Under cold test the effect of the pull was to break the bolt 7 inches from the plate, as shown in Fig. 146. The elastic limit was 22,300 pounds, the tensile strength 40,300 pounds. Under hot test, at a bright red, the stay-bolt pulled through the sheet, as shown in Fig. 147, showing a tensile strength of 9660 pounds, the parts being still at a bright red after their separation.

The results of the whole series of tests, of which only a few are here given, seem to indicate that the average holding power of the usual form of stay-bolt, at a dull red or almost black heat, would be decreased from its strength when cold by about 50 per cent.; at a bright red, to about ⅕ of its original strength, except in specimens, Figs. 139, 142, and 145, which are decreased to about ¼ of their original strength. In the case of specimens 142 and 145 their holding power would be very much increased by the use of a thicker crown-sheet, as they mostly fail, both hot and cold, by the head pulling through the sheet.

Mr. Cole's conclusions are, that the centre rows of the crown-stays in a locomotive boiler should be provided with solid button-heads like Fig.

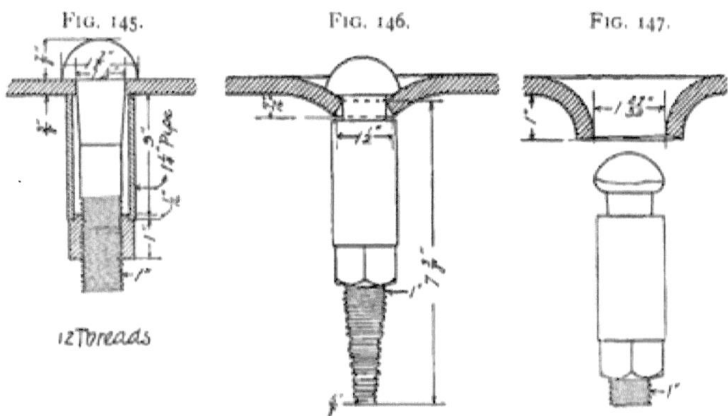

FIG. 145. FIG. 146. FIG. 147.

139, or with nuts having a countersunk cavity into which the stay-bolt is riveted, to prevent pulling through in case the crown-sheet is overheated.

Material for Stay-Bolts.—Thus far the best quality of refined iron has been most satisfactory. Good stay-bolt iron will equal mild steel in strength, and from its fibrous character will better withstand the strains of alternate heating, cooling, and bending than steel appears to do. Material for stay-bolts or braces should have an ultimate tensile strength of not more than 65,000 pounds per square inch, nor less than 50,000 pounds; it should show 20 per cent. elongation, and not more than 35 per cent. of reduction in area at point of fracture. The objection brought against the employment of steel for stay-bolts by master mechanics of railroads is, that they break off in a few months' use, and this has been especially the case in the throat-sheet and the front parts of the side sheets. The reason assigned for this breakage is the crystalline structure of the steel and the repeated bendings to which the stays are subjected, owing to the difference in expansion between the inner and outer fire-box sheets that they brace together.

Boiler-Tubes.—These are made of wrought iron or mild steel; charcoal-iron lap-welded tubes are commonly called for in boiler specifications. The standard dimensions of such tubes to 21 inches diameter are given in Table XXXIV.

Tubes are always measured by outside diameter, and are commonly true to gauge, so that heads, tube-plates, etc., can be bored or otherwise fitted without taking measurements directly from the tubes themselves. The bursting and collapsing pressures of solid-drawn iron tubes are calculated to be, according to Clark, as in Table XXXV.

TABLE XXXIV.

STANDARD BOILER-TUBES, LAP-WELDED, WROUGHT IRON.

Outside.		Thickness, Inches.	Weight per Foot, Pounds.	Heating Surface, 1 Foot in Length.		Area of Opening.	
Diameter, Inches.	Circumference, Inches.			Outside, Square Feet.	Inside, Square Feet.	Square Feet.	Square Inches.
1½	4.71	.08	1.25	.393	.349	.0097	1.40
1¾	5.50	.10	1.67	.458	.408	.0133	1.91
2	6.28	.10	1.98	.524	.472	.0177	2.56
2¼	7.07	.10	2.34	.589	.540	.0230	3.31
2½	7.85	.11	2.76	.655	.598	.0284	4.09
2¾	8.64	.11	3.05	.720	.663	.0350	5.04
3	9.43	.11	3.33	.785	.729	.0422	6.08
3¼	10.21	.12	3.96	.851	.789	.0495	7.12
3½	11.00	.12	4.27	.916	.854	.0580	8.36
3¾	11.78	.12	4.59	.982	.919	.0673	9.69
4	12.57	.13	5.32	1.047	.979	.0763	10.99
4½	14.14	.13	6.01	1.178	1.110	.0981	14.13
5	15.71	.14	7.23	1.309	1.234	.1215	17.50
6	18.85	.15	9.35	1.571	1.492	.1771	25.51
7	21.99	.17	12.44	1.833	1.743	.2417	34.81
8	25.13	.18	15.11	2.094	1.998	.3180	45.80
9	28.27	.19	18.00	2.356	2.254	.4048	58.29
10	31.42	.21	22.19	2.618	2.506	.4998	71.98
11	34.56	.22	25.49	2.880	2.764	.6075	87.48
12	37.70	.23	28.52	3.142	3.022	.7205	103.75
13	40.84	.24	32.21	3.403	3.279	.8554	123.19
14	43.98	.25	36.27	3.665	3.534	.9943	143.19
15	47.12	.26	40.61	3.927	3.791	1.1438	164.72
16	50.27	.27	45.20	4.189	4.047	1.3032	187.67
17	53.41	.28	49.90	4.451	4.305	1.4738	212.23
18	56.55	.29	54.82	4.712	4.560	1.6543	238.22
19	59.69	.30	59.48	4.974	4.817	1.8465	265.90
20	62.83	.32	66.77	5.219	5.068	2.0443	294.37
21	65.97	.34	73.40	5.498	5.320	2.2522	324.31

TABLE XXXV.

BURSTING AND COLLAPSING PRESSURES (CALCULATED) FOR SOLID-DRAWN WROUGHT-IRON TUBES.

Diameter.	Thickness.	Bursting Pressures.	Collapsing Pressures.
Inch.	Inch.	Pounds.	Pounds.
1½	.083	6200	5200
1¾	.083	5300	4300
2	.083	4500	3700
2¼	.095	4600	3600
2½	.109	4800	3600
2¾	.109	4300	3100
3	.120	4400	3000
3¼	.120	4000	2700
3½	.134	4200	2700
3¾	.134	3900	2400
4	.134	3600	2100
4½	.134	3200	1700
5	.134	2800	1400
6	.148	2600	1000

Expanders.—A thin tube can be expanded in a bored hole so as to make a steam- and water-tight joint; this practice, however, is confined to tubes having a less diameter than 5 inches. Tubes 6 inches in diameter and larger are commonly riveted to flanged heads, as shown in Fig. 148. Two forms of expanders are in use: the Prosser expander, shown in Fig. 149, which consists of a number of steel segments with radial joints held together by an external steel spring band, the whole being so arranged that the expander when collapsed is of less diameter, and may thus be inserted in the end of the tube to be fitted. A tapered steel pin passes through the centre of these steel pieces, and by driving on the end of this pin these segments are forced out radially against the tube. By successive operations of driving and slacking, turning the expander slightly after each such expansion, the end of the tube is stretched until it accurately fills the hole. The expander is made partially concave near the end, the length of this groove approximating the thickness of the head and about three times the thickness of the tube. Tubes put in by this method, being expanded on both sides of the tube-plate in one operation, serve as braces, and tend greatly to stiffen the head. Before the tubes are put in place they should be carefully cut to length before expanding, as the chipping off the end of the tube in place is not only unworkmanlike, but there is danger of splitting the tube.

FIG. 148.

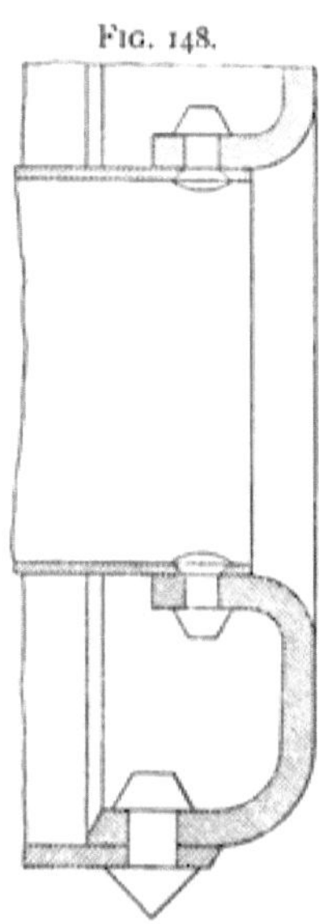

The Dudgeon expander, shown in Fig. 150, is designed to expand a tube by means of a continuous rotary pressure. It consists of a hollow cylinder provided with openings to receive three or more steel rollers; these rollers engage the inner diameter of the tube on the outside, and

FIG. 149.

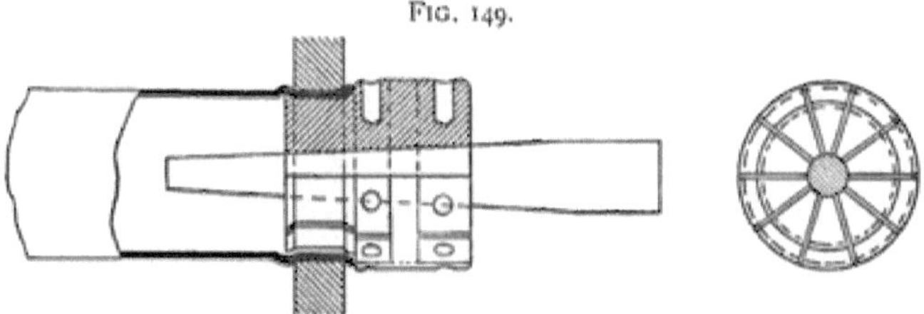

rest upon a conical mandrel on the inside. A guide-sleeve is provided which bears against the tube-sheet; this is secured to the hollow cylinder by a set screw. By shifting the position of this guide-sleeve the different thicknesses of tube-plate are provided for, and thus the expander will answer for several thicknesses of tube-sheet. By revolving the expander and at the same time gently forcing in the conical mandrel, the rollers are forced gently outward against the inner circumference of the tube, enlarging its diameter until it completely fills the bored hole.

FIG. 150.

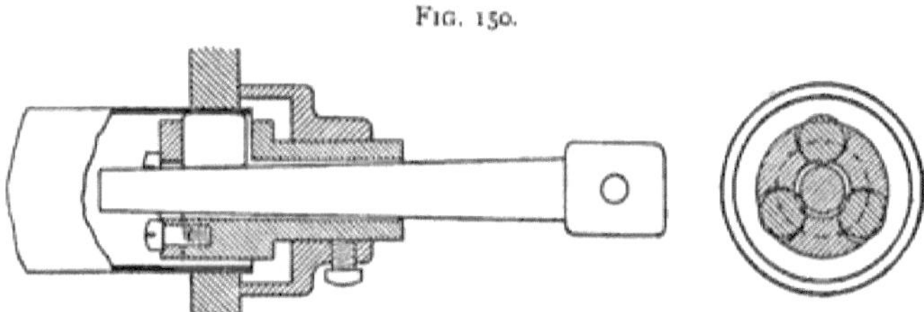

Shock's observations on the Dudgeon expander indicate that it might become a dangerous instrument in the hands of an inexperienced or careless person, since the operation of rolling out the metal may be continued until the tube-ends are entirely cut off without giving warning. To prevent this, the taper mandrel often carries a loose collar, which may be secured in any position by means of a set screw, and thus limit the distance which the mandrel may enter the tool and force out the rollers.

Prosser and Dudgeon Expanders Compared.—The difference in effect by the use of the two expanders is shown in Fig. 151, in which *A* represents a tube expanded by the Prosser, and *B* represents work done by the Dudgeon expander. Some twenty years ago a series of experiments were made at the Washington Navy Yard to determine

FIG. 151.

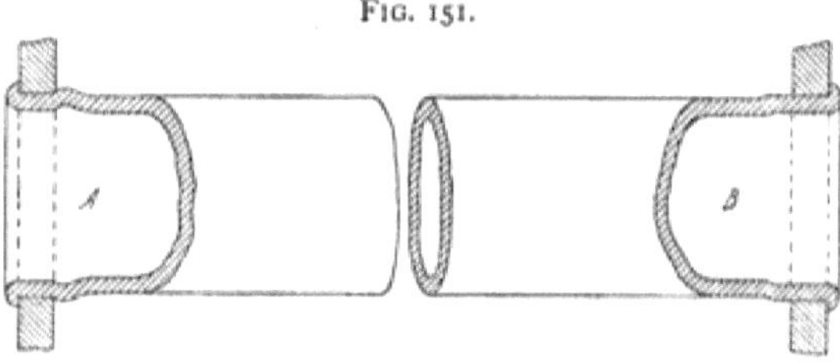

the holding power of boiler-tubes secured by various methods. So far as relates to tube-expanders, the following general conclusions were reached, viz.: (1) The tubes fixed by the Dudgeon expander and beaded over have a considerably stronger hold of the tube-plates than those fixed by the Prosser expander, particularly with thin tube-plates; (2) that if the tubes were not beaded over, the hold afforded by the

Dudgeon is less than that afforded by the Prosser system of fixing; (3) that with both expanders the introduction of ferrules, Fig. 152, adds very materially to the holding power of the tubes; (4) that on the whole the effect of ferrules is with the Dudgeon expander proportionately greater in thick than in thin tube-plates, while in the case of the Prosser expander the proportionate increase of resistance afforded by the introduction of ferrules is not materially affected by the thickness of the tube-plates.

FIG. 152.

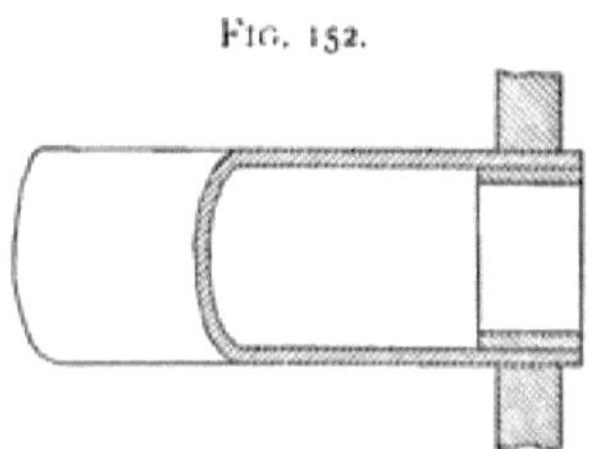

The distortion of flue-sheets by the action of expanders was investigated by Mr. Brown in the Dubuque shops of the Chicago, Milwaukee and St. Paul Railway, in which two engines were undergoing repairs, one receiving a new flue-sheet, the other a new fire-box. After the flues in one engine had been set in the usual manner by the use of a Prosser sectional expander, two flues were removed from the region of the upper corners and the hole in the sheet calipered. It was found that the hole was $\frac{3}{32}$ inch out of round. The same thing was done with the other boiler, and the holes in it were found to be $\frac{1}{8}$ inch out of round. This was before the boiler had been fired up. It was then determined to make accurate measurements of a fire-box entirely new. Accurate measurements taken before and after the flues were set and expanded by a Prosser expander showed that the sheet was expanded upward $\frac{3}{16}$ inch and sideway $\frac{1}{16}$ inch. Experiments of the same kind were then made with another boiler, only a Dudgeon roller was employed to expand the flues. There was no distortion of sheets caused by the roller expander. To amplify this test, they took a discarded flue-sheet, reamed the holes true, and rolled pieces of flues in the holes. Some of the pieces were rolled until they were as thin as a piece of paper, and in no case was it found that the hole was distorted by the action of the roller.

The manufacturers of the Dudgeon expander supplemented these experiments by taking a wrought-iron tube 4 inches in diameter and $\frac{3}{16}$ inch thick, and fitted a ring $\frac{1}{2}$ inch thick on one end, and then applied all the power they could upon a flue-roller. The pressure was so great that it made the metal flow outward, but it left the holes perfectly true.

It appears that the obvious lesson of these experiments is, abandon the sectional flue-expander. Fire-boxes are known to have had the middle of the flue-sheet forced up almost an inch above its original level; this commonly has been attributed to expansion due to the action of heating and cooling. There is good reason for believing that the expander was the real cause of the distortion.

Holding Power of Tubes.—The Hartford Steam-Boiler Inspection and Insurance Company had tests made for them to determine the holding power of 3-inch standard wrought-iron tubes. The tubes were rolled into ⅜-inch plate in the ordinary way, without any expanding other than that produced by a Dudgeon expander: ⅜-inch plate is thinner than is usually used for heads of boilers of ordinary dimensions. A thicker head or tube-sheet would give more frictional surface and, consequently, more holding power. The test consisted in determining the stress necessary to draw the tubes out of the plates. Fig. 151, *B*, represents the end of the three tubes experimented upon, which we designate as *a*, *b*, *c*.

The greatest observed stress sustained without the tube yielding in the plate was,—

Specimen *a*	6000 pounds.
Specimen *b*	4500 pounds.
Specimen *c*	7000 pounds.

The observed stress which occasioned yielding was,—

Specimen *a*	6500 pounds.
Specimen *b*	5000 pounds.
Specimen *c*	7500 pounds.

To ascertain the holding power of tubes in an ordinary tubular boiler, multiply the holding power of one tube by the number of tubes.

Riveting over the ends of tubes is quite generally practised, and when well done makes a very strong joint; but those who are familiar with this kind of work know that in many cases the ends of the tubes are frayed out and split, and until the thumb-tool is brought to bear on the job it has a very unpromising look. Such work yields readily to the action of the heated gases, and after a time the riveting or beading fractures and crumbles off and very little strength remains, a result due to want of proper annealing of the ends of the tubes, but quite as often to bad workmanship.

Another method of fastening tubes into the tube-sheet is to adjust the tubes so that they shall project slightly beyond the tube-sheet, roll them in with a Dudgeon expander, and then flare or expand the ends with a suitable tool or set. These experiments extend to 3-inch tubes in ⅜-inch plates thus expanded with results as below.

The tube was fastened in the plate by being expanded, and the end of the tube, which projected $\frac{3}{16}$ inch beyond the plate, was flared so that the external diameter of the extreme end was 3.2 inches, while the diameter of the tube where it entered the plate was expanded to 3.1 inches diameter.

The test was made by observing the stress required to draw the tube out of the plate, but the tube was not wholly removed from the plate in the specimen *e*.

The stress which was sustained without the tube yielding in the plate was,—

For specimen *d* . . . . 20,000 pounds.
For specimen *e* . . . . 18,500 pounds.

The observed stress which first produced yielding was,—

For specimen *d* . . . . 20,500 pounds.
For specimen *e* . . . . 19,000 pounds.

And the observed stress which occasioned failure was,—

For specimen *d* . . . . 21,000 pounds.
For specimen *e* . . . . 19,500 pounds.

From the foregoing it will be seen that the observed stress which first produced yielding was 20,500 pounds and 19,000 pounds. To ascertain the holding power of the tubes in an ordinary tubular boiler we multiply the holding power of one tube by the number of tubes. The above company, in their publication the "Locomotive," from which these figures are taken, state that they have had boilers with tubes set in this way under their care for some years, and have seen nothing to lead them to apprehend any trouble. The ends have given little or no trouble by being subjected to the heated gases, but a projection of the tube beyond the sheet of more than ⅛ inch before expanding is not recommended.

Stay-Tubes.—These are seldom used in stationary boiler practice, and in marine practice not as much as formerly; their use originated at a time when the holding power of expanded tubes had not been experimentally determined. It is now known that such holding power is more than equal to any pressure occurring in the spaces between tubes in any ordinary tube-head.

A tube of extra thickness, threaded and fitted with a nut outside the tube-head, as shown in Fig. 153, is perhaps the simplest method of fitting a stay-tube. Inasmuch as the outside thread over which a nut must screw easily cannot have its diameter changed, such a tube cannot be expanded in place, being simply a hollow stay in which all the stress comes upon the outside nut.

Fig. 153.

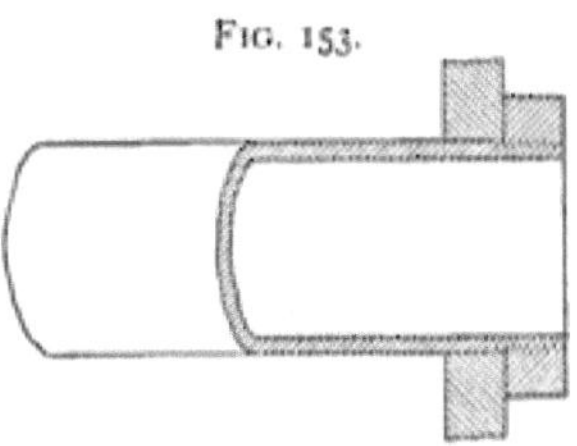

Upsetting each end of an extra thick tube and threading both the tube and the tube-sheet, screwing the former into the latter, as in Fig. 154, and afterwards expanding the tube so as to make a tight joint, is a method that has been used for staying tube-heads, but is not now in common use. The tube-end shown in Fig. 155, in which a nut is added, giving the tube additional thread surface, which by reason of the fineness of the threads is not needed, provides a lock-nut against the tube-head, which assists in making a tight joint.

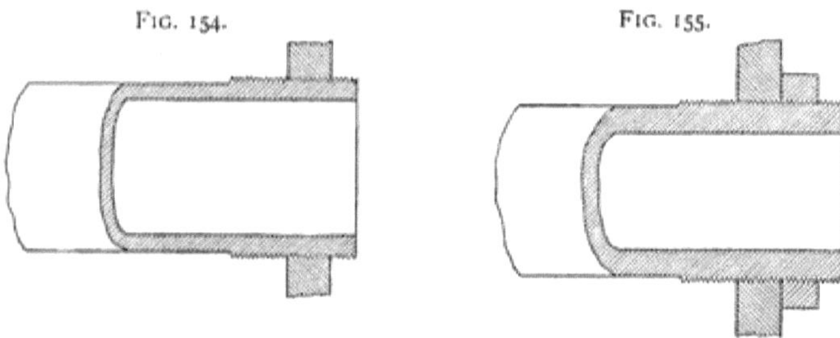

FIG. 154. FIG. 155.

Furnace-Flues.—Flues subjected to external pressure must always be kept perfectly cylindrical. When such flues are riveted the joints should be butt-riveted, and not lap-riveted. Welding should be practised whenever practicable. In the designing of furnace-flues, large diameters and short grates are to be preferred to small furnaces and long grates, because with the larger diameter of furnace a better combustion of fuel is had, and this contributes to higher evaporative efficiency. Fairbairn's experiments showed that the strength of a plain tube under collapsing pressure varied inversely with the length; this led to the making of tubes in short sections with flanged ends, as in Fig. 156, known as Adamson's flanged seam. This method of construction makes an excellent flue. It is very elastic and permits of free expansion. The Adamson flue has sufficient strength if riveted flange to flange without the intermediate ring shown in the engraving, but this ring is used in order to give a calking edge on each side of the lap. This design was a great advance over the plain furnace: it is sufficiently flexible longitudinally to prevent destructive strains within the boiler, and the flanged rings make it much stronger to resist collapse.

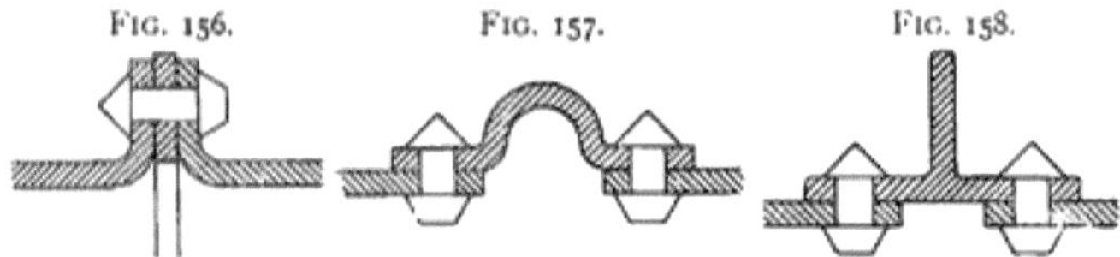

FIG. 156. FIG. 157. FIG. 158.

The Bowling hoop is shown in Fig. 157. This is a weldless hoop, made in either wrought iron or steel. It has been largely used for strengthening the flues of internally fired boilers.

The T-iron ring, shown in Fig. 158, is also a weldless hoop; it was the first form of strengthening ring employed for furnace-flues. The objection to it is that it holds the flue too rigidly and does not permit of free expansion and contraction. The Bowling hoop is superior in permitting such movement, but both kinds are faulty in exposing a double thickness of plates and two rows of rivets to the flames from the furnace.

Corrugated Flues.—These flues are now in common use and are deservedly popular. The corrugations afford a resistance to collapse sufficient for all pressures now used in steam engineering. The corrugations render such flues longitudinally elastic and thereby reduce the local strains within a boiler to a minimum. The material of such flue should be of equal thickness, that the expansion be equal throughout its length. The plates should be of such thickness and the corrugations of such size and form as to prevent sagging in the middle of its length. Corrugated furnaces range from $\frac{7}{16}$ to $\frac{5}{8}$ inch in thickness. A 3 feet 6 inch flue, $\frac{5}{8}$ inch thick, with the Morison corrugations, will carry 198 pounds working pressure, in which example the desirable limit of thickness and the practical limit of steam pressure seem to meet.

The Fox corrugated furnace flue, shown in Fig. 159, was introduced about twenty years ago. Its merits were quickly appreciated, and its

Fig. 159.

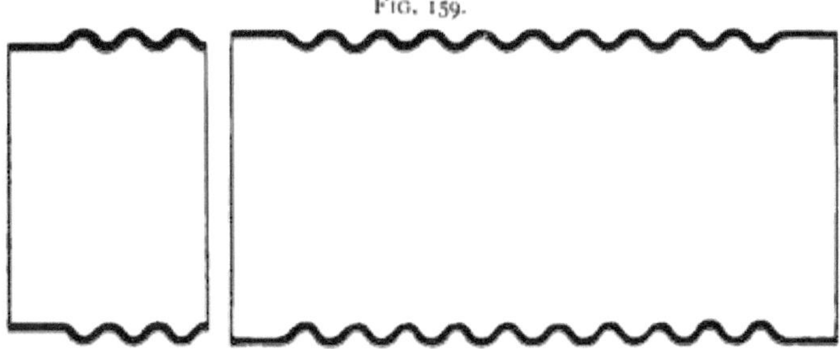

almost universal adoption in marine-boiler construction followed because the corrugated flue was stronger than the other flues which preceded it for the same weight of material. It had the further merit of readily accommodating itself longitudinally to the varying stresses incident to wide changes of temperature, but it has been pointed out that the extreme longitudinal elasticity of this form of corrugation is unnecessarily great, and detracts from the strength to resist deflection or other deformation.

The Morison suspension furnace shown in Fig. 160 possesses the same general characteristics as the Fox corrugated furnace, but in a more pronounced and, it is claimed, more perfect development. Referring to the engraving, it will be seen that this furnace consists of a series of long curves projecting inward towards the fire. The strengthening projections are outward or towards the water space, and are thus protected from the fire, each curve being approximately a catenary. The distance between centres of ridge arch supporters and the general proportions as adopted have been experimentally determined and offer great resistance to distortion or collapse. This form of corrugation presents a crown surface in which the facilities for lodgment of scale is reduced to a minimum, coupled with a maximum convenience for readily removing the same when formed.

FIG. 160.

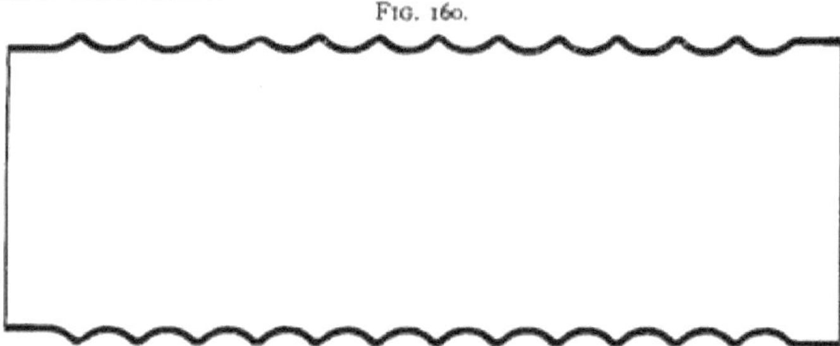

The peculiar form of the long suspension curve is emphasized as that feature of the furnace which has gained for it a practical success over the Fox corrugation, inasmuch as the tension on the material is more uniformly distributed in the Morison curves than in the Fox section, with its series of semicircles. Recent experiments have shown the Morison furnace-flue has a slightly less longitudinal elasticity than the Fox, as might have been expected from the shape of the curves; but experience has proved, so it is claimed, that the Morison has a greater tendency to preserve its original circular form under work, due, no doubt, to the uniform stiffening effect of the ridges, which gives a little less elasticity, and corrects to an appreciable extent the disposition to sag under severe conditions of work.

Strength of Flues and Tubes to Resist Collapse.—The strength of a cylinder resisting internal pressure is not affected by its length. The reverse is true of cylinders subjected to external pressure, for such cylinders are liable to collapse through want of conformity to a true circle, and this liability to collapse increases with the length of the cylinder. It is for this reason that, for the same thickness of metal, a lap-welded tube will withstand a higher pressure for the same length than would a lap-riveted tube, because the latter must of necessity be out of round the thickness of metal at the lap-joint.

Horizontal flue boilers externally fired, when made of lap-welded flues, commonly range in diameter of flue from 6 to 16 inches. Boilers of this kind are rarely more than 20 feet long. The standard lap-welded flues in such boilers withstand collapse under ordinary working pressures which probably do not average more than 75 to 90 pounds. In the case of longer boiler-shells the flues must be lengthened also, the ordinary practice being for flues 12 inches in diameter and larger to make riveted flues in short sections, say 26 to 48 inches; the circular seams, presenting two thicknesses of metal, are favorable to resist collapse. Such flues are not commonly larger than 18 inches in diameter.

Under the United States Regulations, lap-welded flues for horizontal flue-boilers, externally fired for Western river steamboats, over 6 inches in diameter, and not exceeding 16 inches in diameter, and not longer than 18 feet, the steam-pressure not exceeding 60 pounds per square inch, are not required to be made in sections. If the steam pressure is more than 60 pounds and does not exceed 120 pounds per square inch, such flues may be allowed if made in sections not exceeding 5 feet in length, and properly fitted one into the other, and substantially riveted.

Furnace Flues.—The external pressure allowable on flues not more than 42 inches in diameter, such flues being used as furnaces in boilers, under the United States Regulations may be determined by the following formula:

$$\frac{89{,}600 \times T^2}{L \times D} = P,$$

in which D = diameter of flue in inches.
89,600 = a constant.
T = thickness of flue in decimals of an inch.
L = length of flue in feet, not to exceed 8 feet.
P = pressure of steam allowable in pounds.

Example: Given a flue 40 inches in diameter, 7 feet in length, and ½ inch in thickness, required working pressure to be allowed.

Substituting values in the formula, and performing the operation indicated, we have

$$P = \frac{89{,}600 \times T^2}{L \times D} = \frac{89{,}600 \times .25}{7 \times 40} = \frac{22{,}400}{280} = 80 \text{ pounds pressure.}$$

Provided, that when such flues are made in sections of less than 8 feet in length and flanged to a depth of not less than 2½ inches, and

FIG. 161.

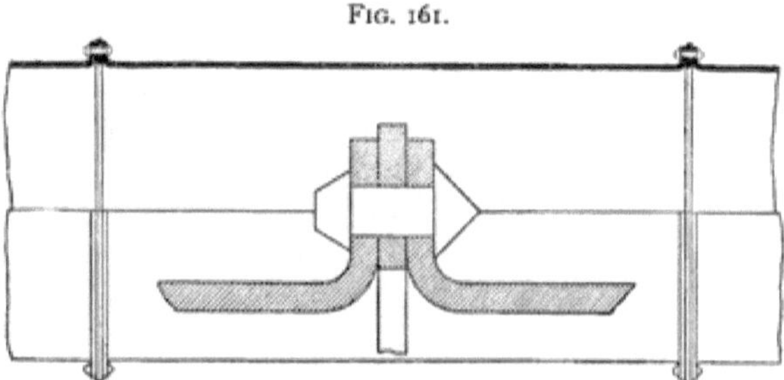

are substantially riveted together with wrought-iron rings between such flanges, such rings having a thickness of not less than ½ inch and a width of not less than 2½ inches, see Fig. 161; or, in lieu thereof, angle-iron rings are employed, such rings having a thickness of material of not less than double the thickness of the material in the flue and a depth of not less than 2½ inches, and substantially riveted in position with wrought-iron thimbles between the inner surface of such ring and the outer surface of the flue, at a distance from the flue not to exceed 2 inches, with rivets having a diameter of not less than 1½ times the thickness of the material in the flue, and placed apart at a

FIG. 162.

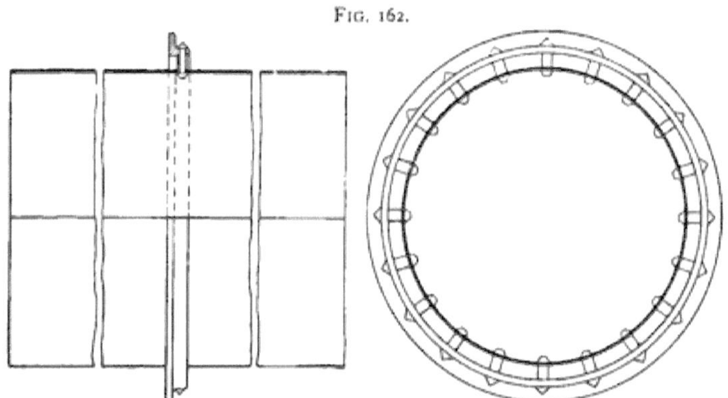

distance not to exceed 6 inches from centre to centre at the outer surface of the flue, see Fig. 162, the distance between the flanges, or the distance between such angle-iron rings, shall be taken as the length of the flue in determining the pressure allowable.

Example: Given a flue 40 inches in diameter, 8 feet long, and ½ inch in thickness, having one ring at the middle of its length, required the pressure allowable by the inspectors:

Substituting values in the formula, and performing the operation, we have

$$P = \frac{89{,}600 \times T^2}{L \times D} = \frac{89{,}600 \times .25}{4 \times 40} = \frac{22{,}400}{160} = 140 \text{ pounds.}$$

To determine the thickness of material required, multiply the diameter of the flue in inches by the length of the flue in feet, and multiply the product by the pressure per square inch in pounds, and divide the last product by the constant, 89,600; then extract the square root of the quotient. The answer will give the thickness of material required.

Example: Let

40 = diameter of flue or furnace in inches.
4 = length of flue or length of furnace in feet.
140 = pressure per square inch in pounds.
89,600 = a constant.

Then we have

$$\sqrt{\frac{40 \times 4 \times 140}{89,600}} = .5, \text{ thickness of material in decimals of an inch.}$$

Corrugated Furnace-Flues.—The strength of corrugated flues, see Figs. 159 and 160, when used for furnaces (corrugation not less than 1½ inches deep), and provided that the plain parts at the ends do not exceed 6 inches in length, and the plates are not less than $\frac{5}{16}$ inch thick when new, corrugated and practically true circles, may be calculated from the following formula :

$$\frac{14,000}{D} \times T = \text{working pressure in pounds per square inch.}$$

14,000 = a constant.
T = thickness in inches.
D = mean diameter in inches.

Example: Given a corrugated flue 40 inches mean diameter, ½ inch thick, required the pressure allowed :

$$\frac{14,000}{40} \times .5 = 175 \text{ pounds working pressure.}$$

To find the thickness of metal for a corrugated furnace-flue, the formula

$$\frac{P \times D}{14,000} = \text{thickness}$$

may be used.

Example: Given a furnace 40 inches mean diameter, to carry 175 pounds working pressure, required the thickness of metal :

$$\frac{175 \times 40}{14,000} = .5 \text{ inch thickness of metal.}$$

Manholes.—Any steam boiler having either steam-space or water-space large enough to admit a man should be provided with a manhole.

FIG. 163.

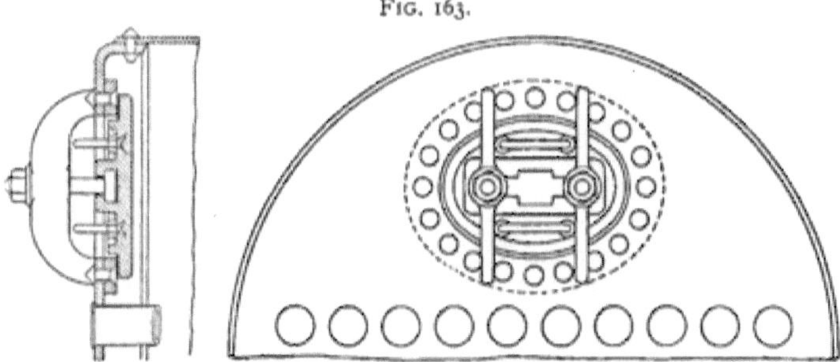

Men are not all the same size, but the standard dimensions of manholes, so far as such a detail can be standardized, is 11 x 15 inches, which will admit a good-sized man. Smaller openings, such as 9 x 14½ inches and 8 x 12 inches, and so on down to small handholes, serve a useful purpose in case of examination, cleaning, and repairs.

The location of a manhole should be such that a man can enter the boiler easily. In horizontal tubular and flue boilers the best place for a manhole is in one of the heads, as shown in Fig. 163; but if for any reason the manhole should not be placed there, it may then be located in the shell, as shown in Fig. 164. This arrangement is not a good one, except for large boilers, because of the difficulty a man has in straightening himself out in the small height occurring between the top of the tubes and the inside of the shell.

FIG. 164.

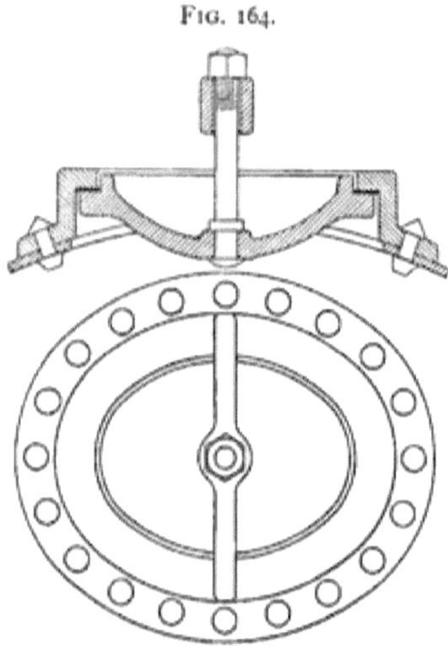

Wrought-iron strengthening rings of not less than 2 x 1 inches should be riveted around manholes when cut through a flat plate, as shown in Fig. 163, to restore as much as possible the loss of strength occasioned by cutting an opening in the plate, a recommendation which also applies to a steam dome-head when made of plate metal. An additional strengthening plate is shown in Fig. 165. It will be observed that the inside plate is double-riveted and the outer ring is single-riveted. The inside of a strengthening ring serves also as a bearing surface for the joint between the boiler-head and manhole-plate, which necessitates countersinking the rivet-heads.

FIG. 165.

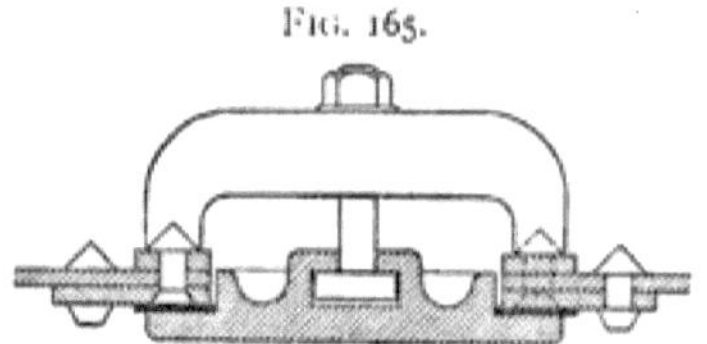

A flanged opening, shown in Fig. 166, has the merit of sufficiently strengthening the plate without the necessity of a separate ring and the insertion of a row of rivets, common to other methods of reinforcement. This is an excellent method of making a manhole, and nearly all manufacturers of boiler-plate which furnish flanged heads have the necessary machinery for forming flanged manholes. The additional price charged for this extra work is reasonable.

A manhole in the shell of a boiler needs to have special precautions taken to insure its being perfectly safe, for it must be remembered that the strain on the circumferential shell of a boiler is much greater than upon the heads. Two forms of cast-iron frames for manhole openings

FIG. 166.

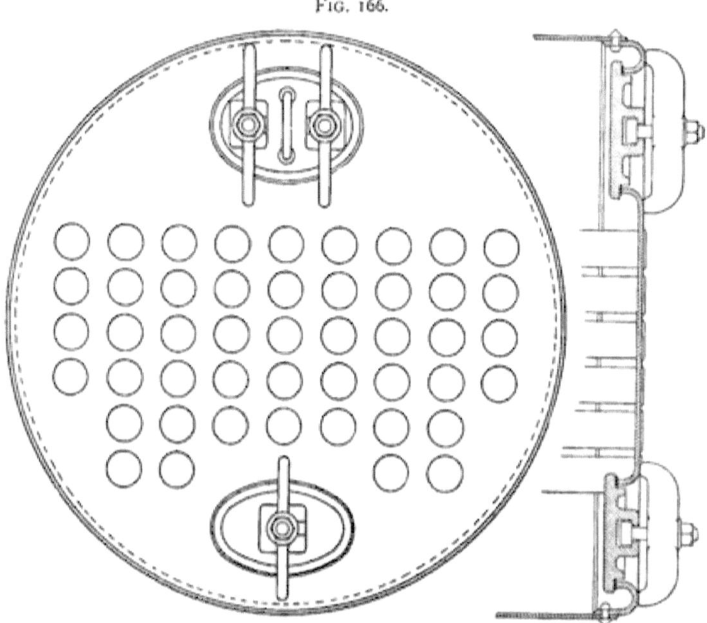

are in use, the one shown in Fig. 164, known as an outside frame, and Fig. 167, an inside frame. Of these two the latter is to be preferred, because the strains upon the frame are those of compression and not those of extension, as in the case of the former illustration. It is important—and this detail should never be overlooked—that the long diameter of the manhole should be placed across the boiler and never lengthwise of it, because the strength of the circumferential shell is affected more by longitudinal openings than by transverse ones. If, now, a standard 11 x 15 manhole opening is cut in the shell, its long diameter parallel to that of the centre line of the boiler, 15 inches of the solid plate would be affected, but if the manhole be placed across the boiler, only 11 inches would be affected.

FIG. 167.

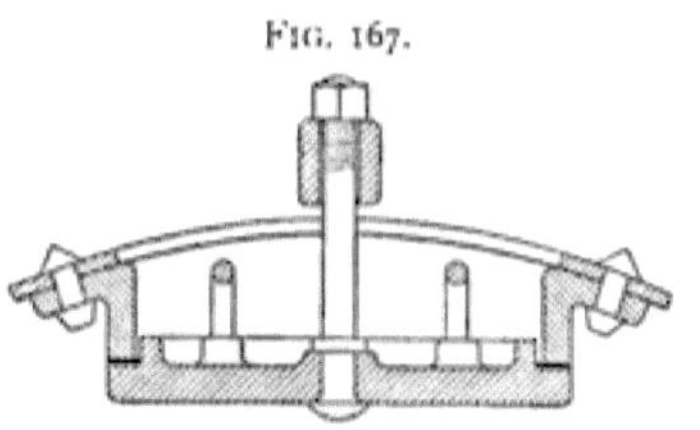

The strength of the frame surrounding the opening must make good the loss of plate removed for the manhole. It is important that the frame, especially if an outside one like Fig. 164, be made of some more tenacious material than cast iron. It is recommended that a steel casting be used instead, or, if possible, a wrought-steel frame, shown in Fig. 168. This latter is one of several details published in this volume relating to the Galloway boiler by the Edgemoor Iron Company. These manhole frames are for the present imported from England. The inside frame, Fig. 167, should also be made in steel casting in preference to cast iron.

FIG. 168.

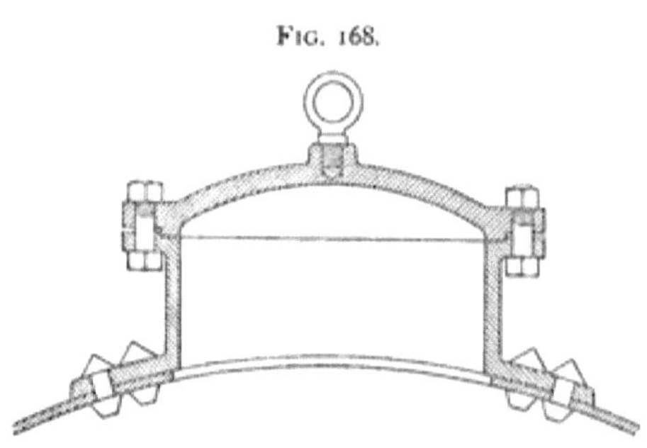

The Eclipse manhole, a wrought-steel frame, shown in Fig. 169, is recommended as a satisfactory device for the purpose. The fact that it is pressed out of plate metal having the same physical properties as that of the shell to which it is attached is much in its favor.

Whatever form of frame be chosen for a manhole in the shell of a boiler, its flange must be of sufficient area of cross-section to restore any loss of strength caused by cutting the opening in the boiler.

FIG. 169.

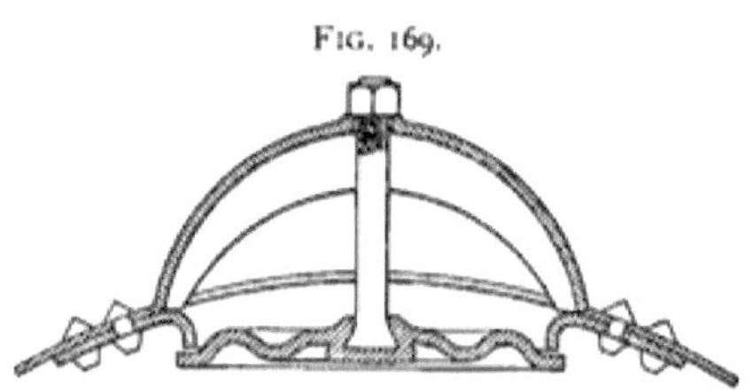

The placing of a manhole in the shell of a boiler having been decided upon, its location should be, if possible, in the centre of a wide sheet and not close to a circumferential riveted seam. No other opening, such as steam-nozzle, etc., should be included in the same plate as that which includes the manhole.

Manhole plates or covers are commonly made of cast iron and in general detail as shown in Fig. 167. The pressure of the steam forces the plate against the seat; there is required nothing but some form of gasket not affected by heat and moisture to complete the joint. The wrought-iron loop handles are usually included in the casting, and are convenient when placing manhole plates in position. The bolts may be placed in cored pockets, as shown in Fig. 165. This is a preferable arrangement to drilling through the plate and riveting, as shown in Fig. 167. For covering openings in frames, as in Fig. 168, cast and wrought plates are both in use; it will be observed that the pressure is all taken up by the bolts, as in the case of a cylinder-head.

Plate-iron or steel manhole covers are now furnished with wrought metal frames, as shown in Fig. 169, which represents a design by Roe. This cover is pressed in shape by suitable dies, the object being to secure lightness as well as strength and furnish a reliable cover at a low price. In the illustration the yoke, which is also of plate metal pressed into shape, is shown lengthwise of the manhead instead of at right angles to it, a position which it occupies in use. The object in showing the yoke as above was to save the additional engraving necessary to show it in another position.

Swinging Manhead.—This device, furnished with the Cahill boiler and shown in Fig. 170, is very simple, and is intended to save the labor of taking manhole plates out of boilers and lifting them to a

FIG. 170.

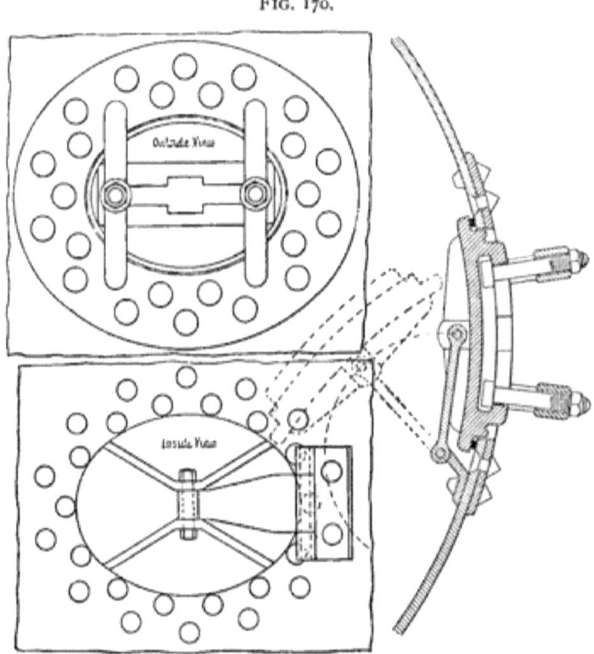

place of security, and then going through the annoyance of putting them back in their places again after the work in the boiler is finished. By loosening the nuts on this manhead, a slight push swings the head in as though it were a door, and it fits back in place against the drum without occupying any appreciable amount of room, and when the time arrives to again close it, it is pulled back to its place. Being hinged, the seats come together in the same place. The joint having been once properly fitted is tight for all time, needing only a gasket to complete it.

FIG. 171.

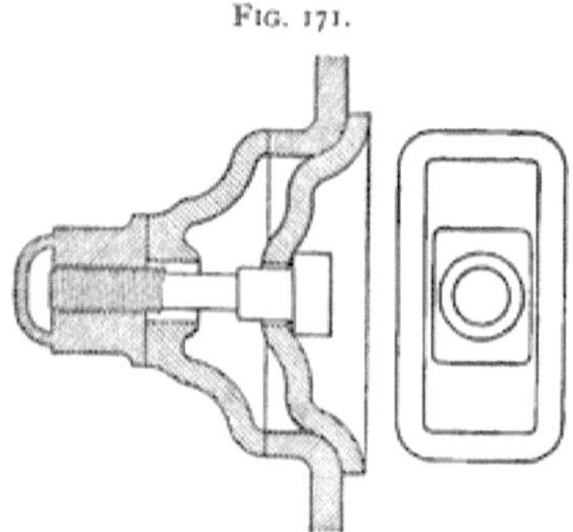

Handholes are to be treated in the same manner as manholes if the opening be greater than 4 x 6 inches. Handhole plates should by preference be inside of the boiler, as in Fig. 166. An example of a handhole plate outside of the boiler is shown in Fig. 171, a detail belonging to a water-tube boiler. In this case the stress is upon the bolt, which must be large enough to carry the pressure due to the entire opening.

Supporting Boilers in the Furnace.—Horizontal tubular boilers are commonly furnished with cast-iron wings riveted on both sides of a boiler and near each end, as in Fig. 172. These wings are from 1 inch to 1½ inches thick, depending on the size of the boiler, 1¼ inches being a fair average. The location of wings as regards height is commonly such as to give three-fifths of the shell to the furnace for heating

FIG. 172.

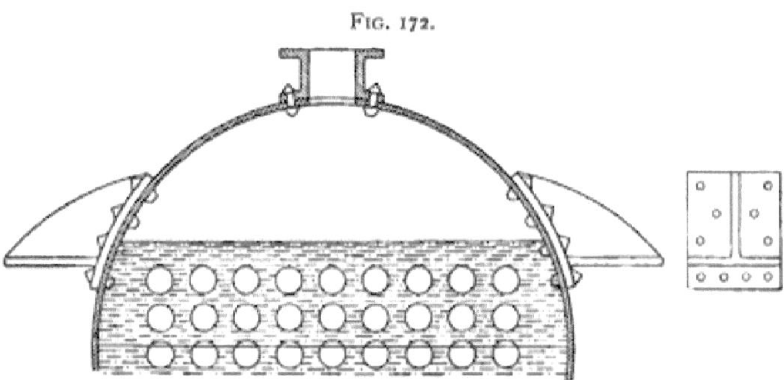

surface. It is not always practicable to carry the wings up to that height, in which case the furnace-walls are carried up higher between the supporting piers than the lower face of the wings.

Detachable wings are sometimes furnished boilers for narrow openings which prevent a boiler passing through with wings as above, in which case a base is securely riveted to the boiler, as in Fig. 173, the face of which does not project beyond the diameter of the boiler. A wing is then bolted, as shown in both elevation and plan. A lug being

Fig. 173. Fig. 174.

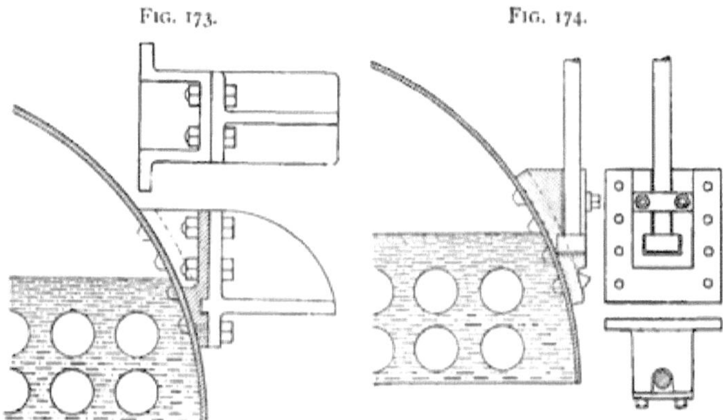

included in the wing and a corresponding recess in the base brings all the load upon the cast portions and entirely relieves the bolts from shearing strains. Another form of suspension is shown in Fig. 174, in which a special head is forged on the end of a hanging bolt which fits into a corresponding recess in the casting riveted to the boiler-shell. A plate of iron with two tap-bolts prevents the hanging bolt working out of place. This form of hanger is also within the diameter of the boiler.

Fig. 175.

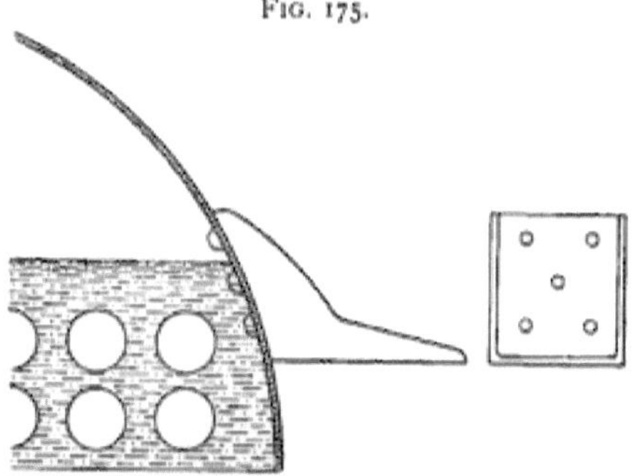

A stamped metal bracket or supporting wing, shown in Fig. 175, made of a single plate of boiler steel by the Lukens Iron and Steel Company, is now being placed upon the market. It is well formed, light, and strong. Being made of mild steel, it is not likely to fail by any of the methods which characterize the failure of cast iron.

A hook-strap riveted to the shell, as in Fig. 176, with an eye-bolt suspended from an overhead beam, is occasionally employed in horizontal-boiler settings, but its use is not at all common.

A loop-strap riveted to the shell, as in Fig. 177, has long been in use for suspending boilers of small diameter. The rivets are not in shearing stress, but in tension, all the strain coming upon the heads of the rivets. Ordinarily these are forged out of bar iron, but Fig. 178 shows a loop-strap forged out of plate steel. These are not in very common use, having been but recently introduced.

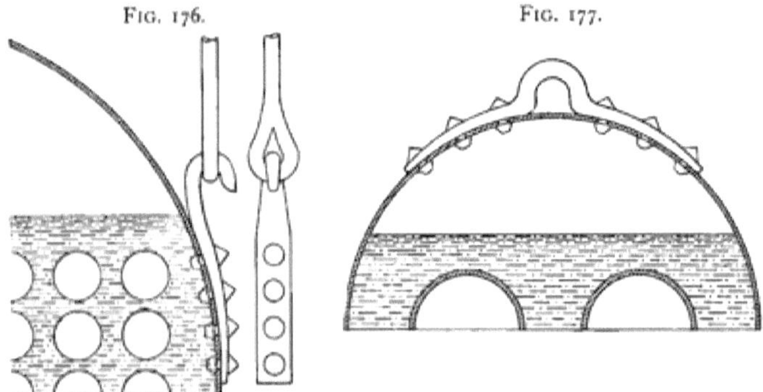

Fig. 176. Fig. 177.

Boilers suspended by links or bolts from overhead are to be provided with a girder extending across the boiler. Fig. 179 represents one made of cast iron suspending a two-flue boiler. Fig. 180 is a representation of a tubular boiler suspended from two channel-beams placed back to back, as shown in the central section. This latter method of using channel-beams is recommended.

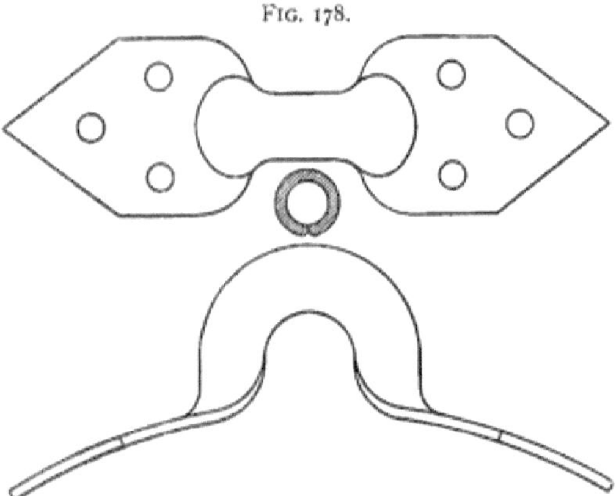

Fig. 178.

A bracket with suspension-rod fitted with a pin in double shear, as shown in Fig. 181, is a device which enables the boiler to swing free in the furnace. The nuts at the four corners can be tightened up so as to entirely relieve the boiler of twisting strains occasioned by walls being out of level, which sometimes occurs when using plain wings resting upon brick walls or piers.

The hook bracket and link shown in Fig. 182 allows a little more freedom of movement in expansion and contraction than is the case with any of the preceding methods of hanging.

FIG. 179.

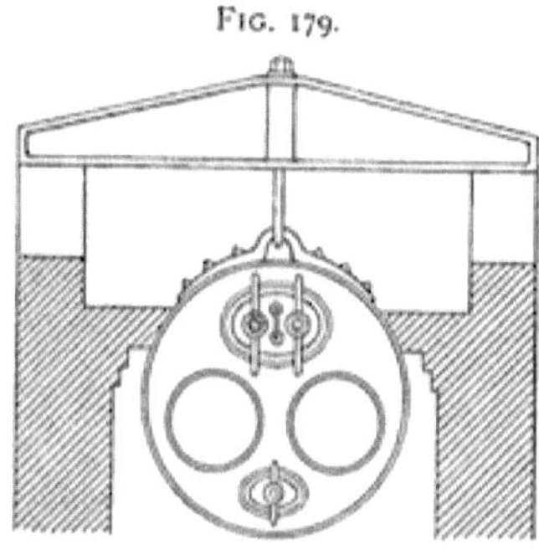

The expansion of a horizontal boiler will be in the direction of least resistance; and as it is quite undesirable that the expansion shall occur from the rear of a boiler to the front, it is customary to have the front wings or brackets of a boiler rest upon the brickwork and provide the rear wings or brackets with a plate and rollers, as shown in Fig. 183. This will secure a fixed distance for the front end of the boiler, and all variations in length due to expansion or contraction occur at the rear end of the boiler.

The expansion of a boiler, in case the front end of it rests upon a cast-iron fire-front, is often provided for by building up a brick pier near the rear end and adapting to it a cast-iron top, cylindrical rollers, and a cradle for the boiler to rest upon, as shown in Fig. 184. Another

FIG. 180.

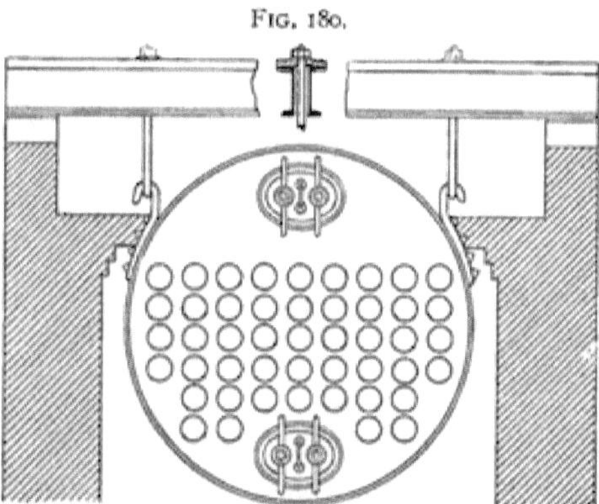

method of accomplishing the same thing is shown in Fig. 185, which consists of a single roller, curved to fit the boiler and adapted to roll upon a cast-iron block of the same radius as that of the boiler.

Expansion from the rear of the boiler to the front, if for any reason

FIG. 181. FIG. 182.

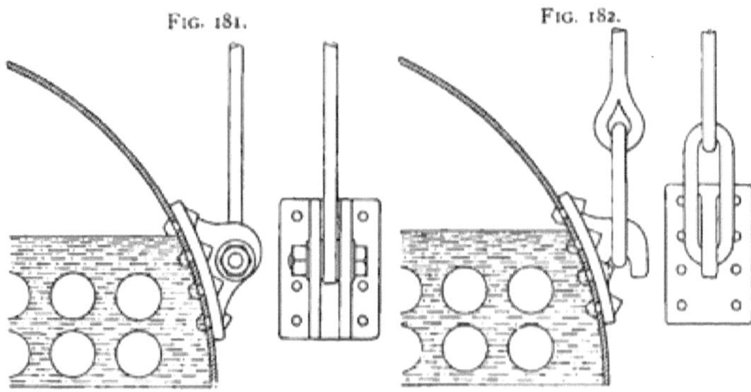

this is desirable, can be had by putting wings resting upon the brickwork at the rear end only and suspending the front by links or bars, as shown in Fig. 181 or Fig. 182, at the discretion of the designer.

FIG. 183.

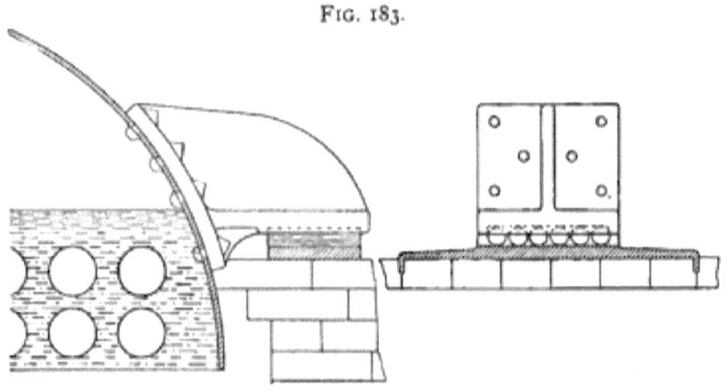

Boilers are sometimes set with one end of the boiler resting upon the fire-front, the rear end resting in a cast-iron stand, as shown in Fig.

FIG. 184.

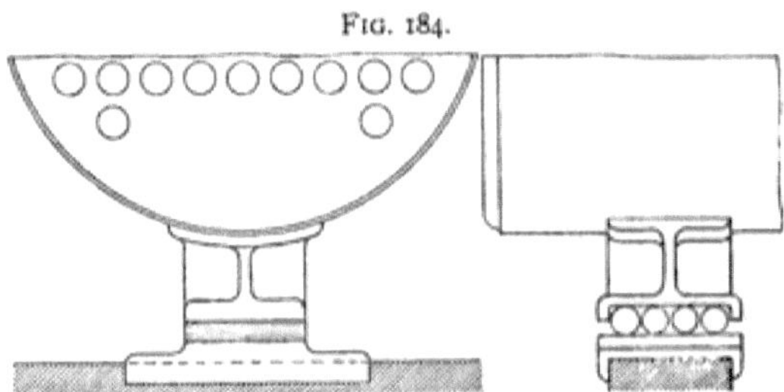

186, without any special provision for expansion; but this is not good practice, and if employed at all should be confined to small boilers.

An expansion-rocker for a large boiler is shown in section in Fig. 313 and in elevation in Fig. 349.

Fire-Door Openings.—Boilers of the locomotive type and vertical tubular boilers,—in fact, all boilers in which there is an outer and inner sheet with a water-space between,—will require either a ring around the fire-door opening or a flanging of the plates to effect a closure between the two plates.

FIG. 185.

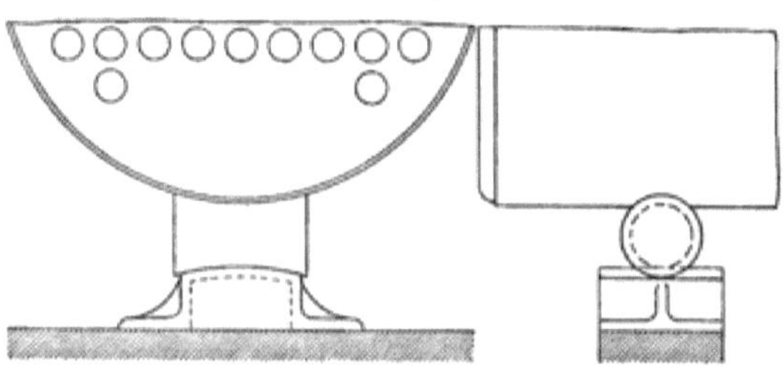

These openings are usually made oval for stationary boilers, ranging from 8 x 12 inches for portable engine boilers to 16 x 20 for large stationary boilers. In locomotive practice they are quite as often made round, and commonly 16 inches, but occasionally as large as 18 inches in diameter.

For portable boilers a ring such as shown in Fig. 187 is used more than anything else. The distance between the outside and the inside sheets is commonly from 2½ to 3 inches. These rings are frequently

FIG. 186.

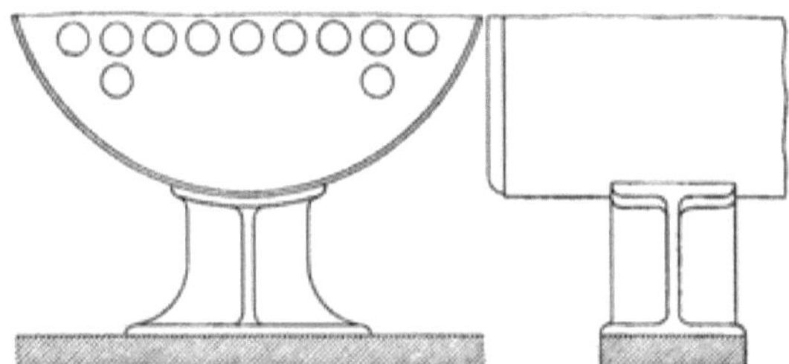

made of cast iron, a practice to be condemned, because of the brittle nature of the material, the frequency of rivet-holes, and the unequal expansion to which it is subjected. It is a much better plan to use steel castings, or to make them of wrought iron with a welded joint, afterwards shaping the ring to fit the surfaces to be joined and drilling the holes to suit the pitch of the rivets.

A fire-door opening made of a Z-bar, such as shown in Fig. 188, is seldom used. It is not an easy ring to make, because the outer and inner flanges subjected to compression and extension make it difficult to bend in circular form. It offers a convenience in riveting which would be advantageous in case the ring was formed in a die in a forging-machine, which is really the only practical way of producing such rings at a low cost.

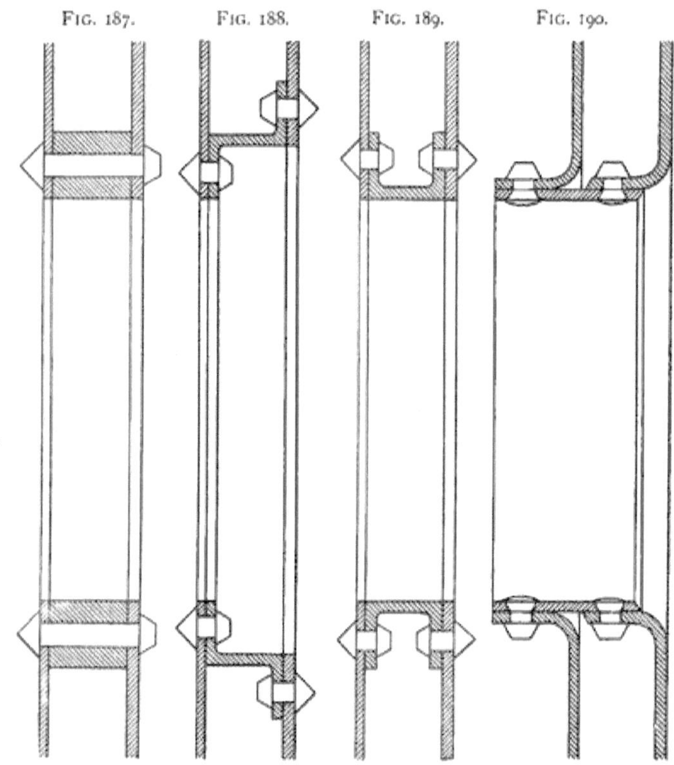

Fig. 187. Fig. 188. Fig. 189. Fig. 190.

The double-flanged ring shown in Fig. 189 is not unlike a channel-bar in its cross-section. This ring should preferably be made with a welded joint, the flanges being turned in a flanging-machine to such dimensions as will admit the rivets, not, as shown in the engraving, directly opposite each other, but, to lessen the distance between the two sheets, the rivets should be staggered, placing them at half-pitch on opposite sides.

The flanging of a fire-box sheet and the head-sheet outward with a welded ring, joining the two as shown in Fig. 190, is much used in locomotive practice and makes a very substantial joint. To save wear at the bottom of the fire-door opening, it is recommended that a piece of mild steel (say 2 inches in width by ½ inch thick) be riveted for at least a third of the distance around the lower part of the opening. This will form an excellent leverage upon which the fire-tools can be used and save wear on the plates.

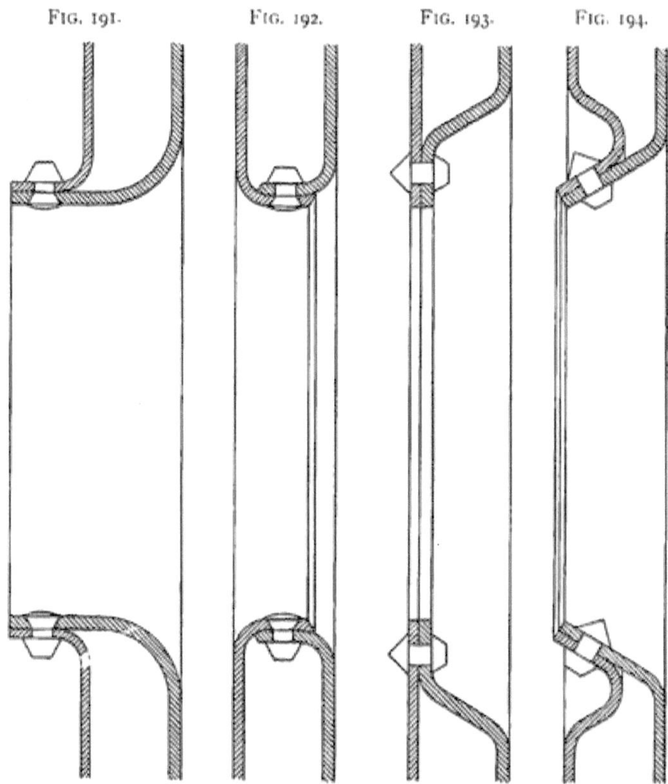
FIG. 191. FIG. 192. FIG. 193. FIG. 194.

A fire-door opening in which the outer and inner sheets are flanged one inside of the other and held by a single row of rivets is shown in Fig. 191. This method of flanging is open to the objection regarding the wear on the lower half of the opening by fire-tools already referred to, so that a bar of the same dimensions as indicated above should be riveted on the lower third or half of this opening. The thickness of the flange on the outside of the inner plate will be considerably reduced by reason of the stretch to which it is subjected to bring it in line with the shorter flange of the outside sheet. The designer will, of course, take this into account and adapt the diameter and spacing of the rivets to the thinner flange.

Flanging the fire-box sheet and the outside sheet each towards the other, as shown in Fig. 192, is an excellent way of joining a fire-door opening. This design is largely used in locomotive practice.

For vertical tubular boilers the outside sheet is usually left cylindrical and the fire-box sheet flanged outward to meet it, as shown in Fig. 193. This method of flanging is not often practised and has no advantages over the wrought-iron ring shown in Fig. 187.

The flanging of the outer and the fire-box sheet as shown in Fig. 194 is occasionally met with in locomotive practice. This flanging is accomplished by the use of dies in a flanging-machine, and makes a very good opening. An error which crept into the drawing unobserved will be readily noticed as such,—the lower rivet is shown just the reverse of what it should have been.

Steam-Dome.—This is a small reservoir attached by a base-flange to the shell of a boiler, as shown in Fig. 195, for the purpose of increasing the steam-room, both in quantity and, especially, in height. Some makers place the upper manhole in the dome instead of the boiler-head; in such cases the dome-head is not infrequently made of cast iron, as in Fig. 196. This arrangement for entering the boiler above the tubes requires that the shell have an opening large enough for a man to enter the main steam-space. The weakening of the main shell incident to cutting out so large a portion of the plate is made good by flanging the shell into the dome, and sometimes further strengthening the shell by riveting around it a heavy wrought-iron ring, the size of which and the proportion of the rivet area must equal the strength of the other riveted joints.

Fig. 195.

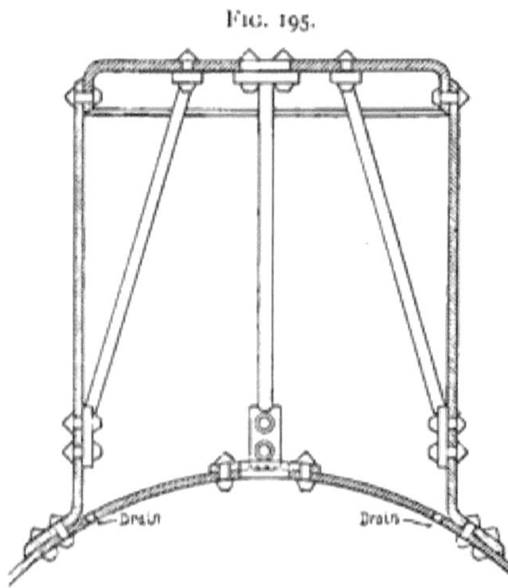

Steam-domes of large diameter when made similar to Fig. 195 affect the working strength of the main shell, because the area covered by the dome has a pressure on both sides of the plate. This counter pressure neutralizes the action of the steam pressure over the area thus covered. The pressure exerted by the steam acting upon the main shell on either side of the dome tends to flatten the main shell within the dome. It is for this reason that large domes with flat heads usually have stays reaching from the dome-head to the shell, as in Fig. 197, in order to overcome any tendency to flattening at that point.

FIG. 196.

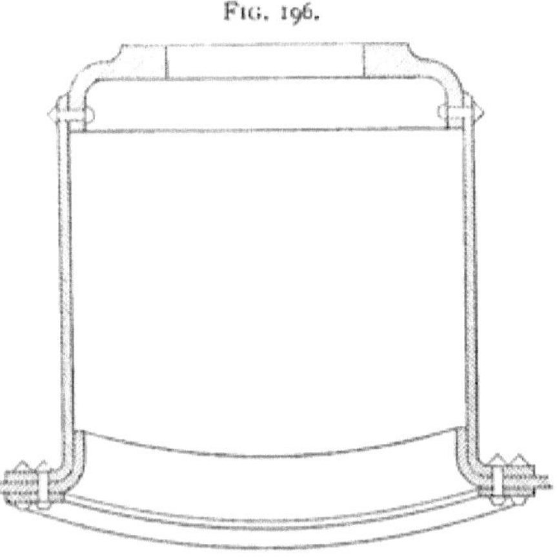

A drainage-hole must be provided at the lowest point on each side of the dome to conduct any water of condensation or that due to priming back into the boiler, as shown in Fig. 195. The proportions for steam-domes vary according to the fancy of the builder, but approximate one-half the diameter of the boiler-shell. A close approximation to average practice may be had by consulting Table XXXVI.

FIG. 197.

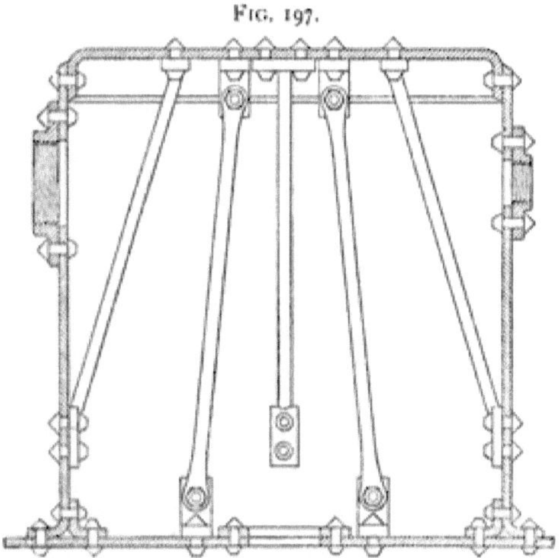

TABLE XXXVI.

PROPORTIONS FOR STEAM-DOMES. FOR 100 POUNDS PRESSURE, DOUBLE-RIVETED, STEEL SHELL AND HEAD, IRON RIVETS.

Diameter of Boiler.	Size of Dome.		Thickness of Metal.	
	Diameter.	Height.	Shell.	Head.
Inches.	Inches.	Inches.	Inch.	Inch.
36	20	22	1/4	5/16
38	20	22	1/4	5/16
40	22	24	1/4	5/16
42	22	24	1/4	5/16
44	24	26	1/4	5/16
46	24	26	1/4	5/16
48	26	28	1/4	5/16
50	26	28	5/16	3/8
52	28	30	5/16	3/8
54	28	30	5/16	3/8
56	30	32	5/16	3/8
58	30	32	5/16	3/8
60	32	34	5/16	3/8
62	32	34	5/16	3/8
64	34	36	5/16	3/8
66	34	36	3/8	7/16
68	36	38	3/8	7/16
70	36	38	3/8	7/16
72	36	40	3/8	7/16

The thicknesses given in the table for the shell of the dome are those for the top, where the shell joins the head by a riveted joint. Plates can be rolled on special order $\frac{1}{16}$ inch thicker at the bottom for turning a flange, but the thickness given in the table is ample for 100 pounds pressure without additional thickness for flanging. The flanging at the bottom of a steam-dome should be wide enough for a double row of rivets, because the shell of a boiler ought to have additional stiffness around the base of the dome to assist in counteracting the strains incident to cutting through the shell, as well as the flattening tendency that occurs when steam acts on both sides of a curved sheet. An outside and inside flange, as shown in Fig. 197, is not largely employed in boiler-making;

Fig. 198.

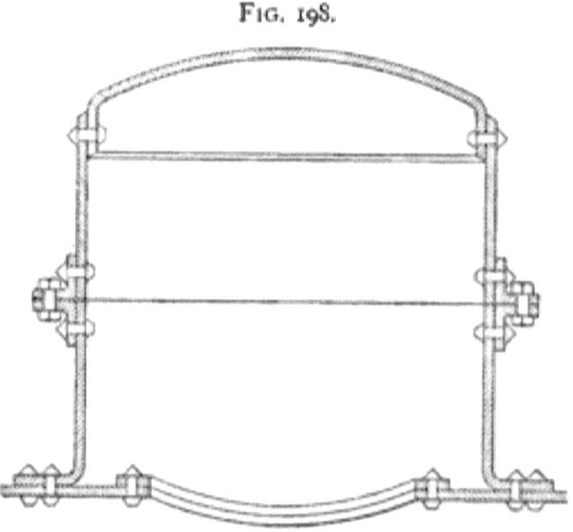

it is more expensive to make than simply extending the width of the flange for two rows of rivets, and offers no additional advantages for extra cost.

Stiffening rings riveted around an opening into a steam-dome are shown in Figs. 195 and 197. These add to the strength of the shell by restoring a portion of that lost by cutting through the plate.

A steam-dome with a flanged joint and removable top for a locomotive boiler is shown in Fig. 198. The bottom is joined to the boiler-shell by a double-riveted joint. The boiler-shell has a large opening reinforced by a ring riveted to it. This dome is made in halves, that ready access may be had to the throttle-valve located within it, an arrangement more costly and no more efficient than the current American practice of making dome-heads of cast iron with an opening large enough to insert, examine, or remove the throttle-valve, the cover to the opening containing the safety-valves, Fig. 196. A flanged opening in the shell of a boiler to fit the interior of a steam-dome, as shown in Fig. 199, is seldom practised in stationary boiler-work; it is, however, largely employed in locomotive practice, but the bumped head shown in the engraving is now largely employed, as it needs no bracing.

FIG. 199.

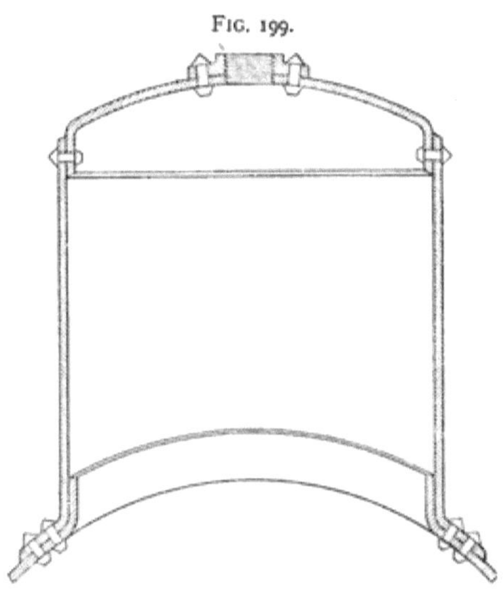

Steam-Drum.—This appendage to a steam boiler has for its object the increasing of the steam-room of a boiler in a separate vessel, connected with the boiler by a small nozzle. When a steam-drum extends across a single boiler, as in Fig. 200, a common proportion is to make the drum one-half the diameter of the boiler, the length of the drum to be twice its own diameter. A manhole should be provided for the purpose of internal examination. Corrosion is quite as likely to occur in a steam-drum as in the boiler.

Nozzles for steam-drums may be of cast iron, riveted to the shells of both drum and boiler, the flanges faced for making a joint with a suitable gasket. The interior diameter of nozzles may be for boilers of 36 to

Fig. 200.

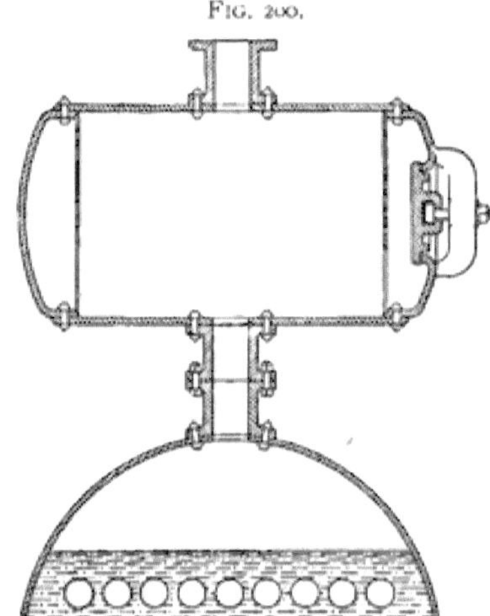

44 inches diameter, 4 inches; of 46 to 54 inches diameter, 5 inches; of 56 to 72 inches diameter, 6 inches. The metal in nozzles should be about an inch thick,—not for strength to resist the pressure of steam, but the better to withstand the effects of rough handling, etc. The thickness of shells for steam-drums may be ¼ inch for all sizes up to 36 inches diameter. The longitudinal seams for diameters 24 inches and larger should be double-riveted. It is recommended that bumped heads be used rather than flat ones.

A cross steam-drum to connect two or more boilers is in very common use in connection with horizontal tubular and flue boilers, as shown in Fig. 201. The diameter may be one-half that of the boiler

Fig. 201.

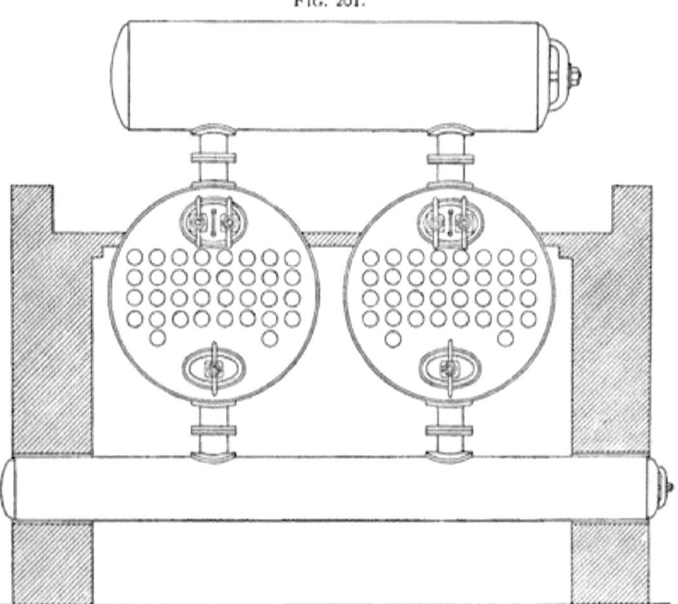

with which it is connected; the length may be that which the boilers measure from outside to outside of shells, as indicated in the engraving. Where two or more boilers are included within the same furnace, no other fittings in the steam connections are necessary than are shown, but if the boilers are separately set, as in Fig. 202, so that any one of

FIG. 202.

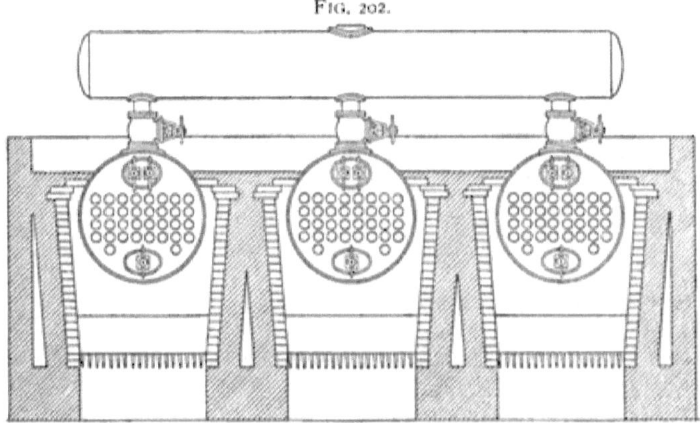

them can be withdrawn from service, a stop-valve should be placed between each of the boiler and steam-drum nozzles; if so, the safety-valve must not be placed on the steam-drum, but on a separate nozzle on the shell of the boiler.

Longitudinal steam-drums are often used on horizontal tubular boilers when the latter are 5 feet and larger in diameter. It is a common practice to fit such drums with cast-iron connecting-nozzles, as shown in Fig. 203. This arrangement is not without objection; the principal one

FIG. 203.

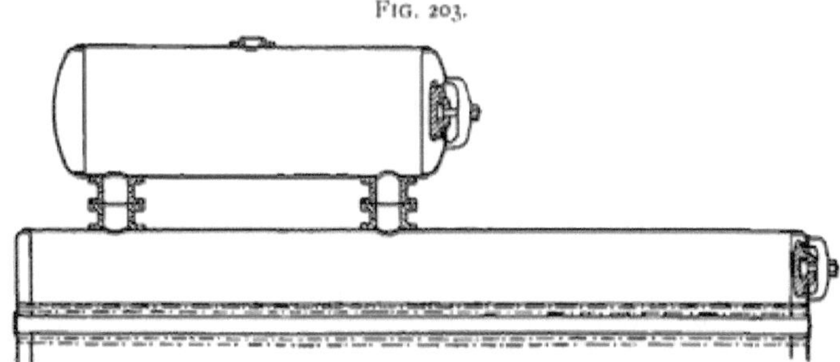

urged against it is, that the expansion of the boiler and drum are not likely to be coincident, and thus unequal strains occur at the riveted intersections, the boiler-joint frequently becoming leaky, and a leak thus caused is difficult, if not impossible, to stop. For this reason some designers use one steam-nozzle only, the other being simply a stand to carry the weight of that end of the drum. When one nozzle only is used it is not infrequently made of plate metal, and is commonly larger in diameter than tabulated above, and riveted to both boiler and drum.

Nozzles.—The three sizes of nozzles referred to above are shown in Fig. 204. The writer is not partial to either steam-domes or steam-

FIG. 204.

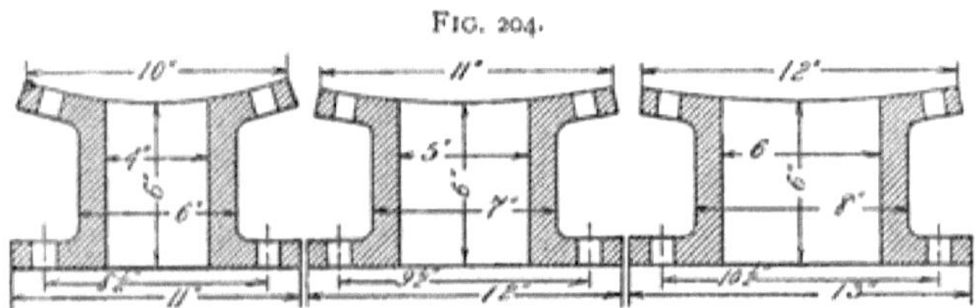

drums when they exist for no other reason than to furnish additional steam-room; if, however, either seem desirable, it is better to use a drum, because the opening in the boiler is less, there are fewer rivet-holes, and these are confined to a smaller diameter. The sizes of nozzles as given above are ample for any steam service that will ever be required of them.

Mud-Drums.—This appendage to a steam boiler is located underneath and commonly at the rear end, as shown in Fig. 205, which represents the drum attached to a horizontal tubular boiler, the end of the drum passing through the rear end of the furnace-walls. The boiler must be suspended in such a manner that no portion of its weight comes upon the mud-drum. In building a wall around a mud-drum, no portion of the brickwork should be in contact with it. A good method is to make the opening in the rear wall about ½ inch larger all around than the mud-drum, and after completion drive in a piece of asbestos packing-rope or plaited gasket to fill the hole: this will be at once tight, flexible, and enduring.

A cross mud-drum, as shown in Fig. 206, must be free from the weight of the boiler, as already referred to. Instead of providing for expansion at the rear end of the boiler, the better plan is to put expansion-rollers, shown in Fig. 183, under the front wings of the boiler, so as to throw all the expansion towards the front, the rear end having no free movement. Where one mud-drum serves for two boilers attached, as shown in Fig. 201, this same remark regarding expansion applies, if anything, with added emphasis.

The size of mud-drums is not fixed by any definite rules. The following proportions agree closely with average practice: The diameter of mud-drum for a 36-inch boiler may be 12 inches; for 38- to 52-inch boilers, the mud-drum may be 14 inches; for 54- to 66-inch boilers, 16-

FIG. 205.

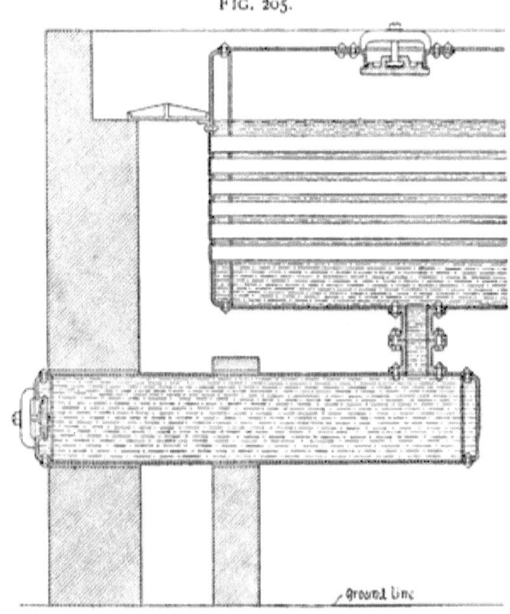

FIG. 206.

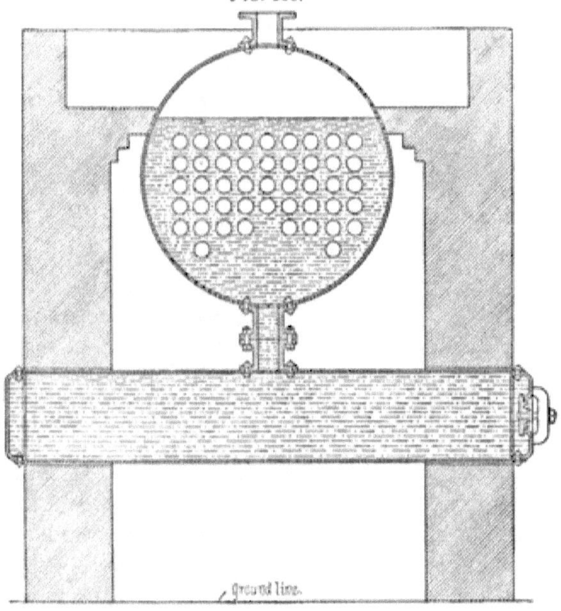

inch mud-drums; and for 72-inch boilers and larger, 18-inch mud-drums. The length of the mud-drum will depend upon whether it leads through the rear wall, Fig. 205, or crosswise of the furnace-setting and into the two side-walls, Fig. 206, and whether or not two or more boilers are attached to the same mud-drum. These lengths will be fixed by the draughtsman preparing the designs.

Material for Mud-Drums.—In ordinary horizontal tubular- and flue-boiler installations, the mud-drum is commonly made of the same material as the boiler to which it is attached. In water-tube boilers the mud-drum is almost invariably made of cast iron, this material being less injuriously acted upon by corrosive gases, water, etc., than is the case with wrought plates, whether of iron or mild steel.

Functions of a Mud-Drum.—It was formerly, and is to some extent at present, the practice to have the feed-water enter the boiler through the mud-drum, in which case the latter performs the functions of a heater. The feed-water entering the boiler through the mud-drum does so at a much less velocity than if the feed-pipe led directly into the boiler. This slow movement of the water through the mud-drum, in contact with the heated gases, brings its temperature up to that within the boiler or nearly so; it thus relieves the latter in part from the bad effects of cold water localized within the boiler, such as leaky tubes and riveted joints when occasioned by the contraction of metal in the vicinity of the water-inlet. This relief, if any, is simply due to a higher temperature of feed-water. The proper function of a mud-drum is not to heat the feed-water, but if the water contains mud or any other foreign substance which may be precipitated by heating the water, the expectation is that much of this foreign matter will be lodged by a downward circulation in the mud-drum and blown out from there.

FIG. 207.

To accomplish this to the best advantage, the mud-drum should be protected from the atmosphere of heated gases which surrounds it. One method of doing this in connection with a horizontal tubular boiler is shown in Fig. 207, where a brick arch is thrown over the mud-drum, thus wholly removing it from the direct action of the hot gases. In the illustration of the Stirling boiler, Fig. 378, the construction of furnace to protect the mud-drum from the direct action of the heat is especially noticeable.

Water-Surface.—In designing steam boilers of whatever type there should be provided a large water-surface for the disengagement of steam. Such a surface secures a steadiness of water-level and prevents foaming. Boilers having a large water-surface for the liberation of steam furnish dryer steam than those boilers in which there is a deficiency of liberating surface. This is notably the case in small vertical tubular boilers, in which the liberating surface is small as compared with the total heating surface.

Steam-Room.—In horizontal tubular and flue boilers the steam-room is nearly constant for a given diameter. The upper line of the top row of tubes is commonly two-thirds of the diameter of the boiler from the bottom; over this is from 2 to 4 inches of water, depending on the size of the boiler; all above this is steam-room. In water-tube boilers the water is carried up to about one-half the diameter of the combined water- and steam-drum. In ascertaining the steam-room in any cylindrical shell, it is the area of the segment above the water multiplied into the length of the steam-space. Knowing the diameter of the boiler and the height of the segment, the area of the latter may be calculated by means of Table XXXVII.

TABLE XXXVII.

AREA OF CIRCULAR SEGMENTS; DIAMETER OF CIRCLE, 1.

Height.	Area.	Height.	Area.	Height.	Area.	Height.	Area.
.100	.0409	.205	.1158	.310	.2074	.415	.3081
.105	.0439	.210	.1199	.315	.2120	.420	.3130
.110	.0470	.215	.1240	.320	.2167	.425	.3180
.115	.0502	.220	.1281	.325	.2213	.430	.3229
.120	.0534	.225	.1323	.330	.2260	.435	.3279
.125	.0567	.230	.1365	.335	.2307	.440	.3328
.130	.0600	.235	.1407	.340	.2355	.445	.3378
.135	.0634	.240	.1449	.345	.2403	.450	.3428
.140	.0668	.245	.1492	.350	.2450	.455	.3478
.145	.0703	.250	.1536	.355	.2498	.460	.3527
.150	.0739	.255	.1579	.360	.2546	.465	.3577
.155	.0775	.260	.1623	.365	.2594	.470	.3627
.160	.0811	.265	.1667	.370	.2642	.475	.3677
.165	.0848	.270	.1711	.375	.2690	.480	.3727
.170	.0885	.275	.1755	.380	.2739	.485	.3777
.175	.0923	.280	.1800	.385	.2787	.490	.3827
.180	.0961	.285	.1845	.390	.2836	.495	.3877
.185	.1000	.290	.1891	.395	.2885	.500	.3927
.190	.1039	295	.1936	.400	.2934		
.195	.1078	.300	.1982	.405	.2983		
.200	.1118	.305	.2028	.410	.3032		

To use the table, divide the height of the segment by the diameter of the boiler; find the quotient in the table opposite to which is the corresponding area for a circle whose diameter is 1; square the diameter of the boiler and multiply it by the area found as above: the product will be the area of the segment.

FIG. 208.

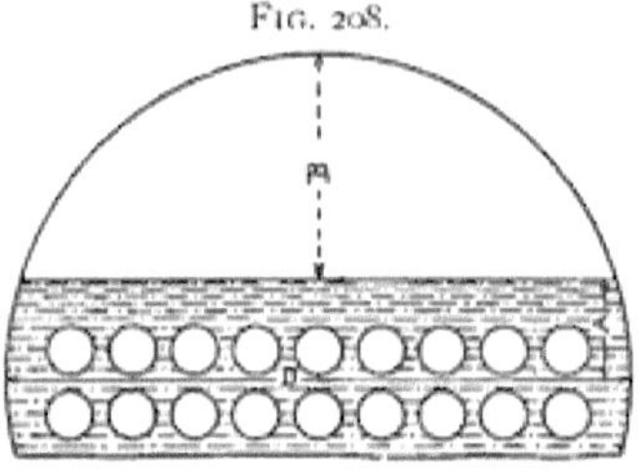

Example: Let Fig. 208 represent a boiler in which the diameter D = 48 inches and the height of the segment B = 14 inches. What will be the contents of the steam-room in cubic feet if the boiler be 14 feet long?

$$\frac{14}{48} = .292.$$

The nearest number in the tabular heights is 0.290, and its corresponding area is 0.1891. The total area of the segment is 48 × 48 × .1891 = 435.69 square inches. Length, 14 feet = 14 × 12 = 168 inches. Then

$$\frac{168 \times 435.69}{1728} = 42.36 \text{ cubic feet.}$$

Heating Surface.—All that portion of a boiler subject to the direct action of the heat of the furnace, and that traversed by the heated products of combustion after leaving the furnace, is reckoned as water-heating surface when the opposite side of such a surface is in contact with water, and superheating surface where such surfaces pass through the steam-room of a boiler. Unless otherwise stated, heating surface means water-heating surface.

The most effective heating surface is that of the fire-box in internally fired boilers and the exposed shell surface in externally fired boilers. Flue- and tube-surfaces are less effective than either of the above types, since the limiting figure for one boiler might not be suitable for another.

Experiments made with a locomotive boiler on the Northern Railway of France, the boiler being divided into several sections, showed a rapid decrease in the rate of evaporation as the distance from the furnace increased, so that the tubes at their extreme ends scarcely evaporated any water at moderate rates of combustion. The furnace-sheets and a small portion of the contiguous tube-surface forming the first section evaporated at the rate of 44.6 pounds of water hourly per square foot of heating surface.

Horse-Power of Boilers.—There is no standard for measuring boiler power by extent of heating surface, nor can there be, because the evaporating power for similar areas throughout the boiler is very unequal. By experiment the efficiency of certain types of boilers has been ascertained with considerable accuracy, so that it is a well-established commercial practice to sell steam boilers by extent of heating surface as a basis of horse-power.

The common rating of horizontal tubular boilers is 15 square feet of heating surface per horse-power. The heating surface includes two-thirds of the shell and all of the tube-surface.

Flue boilers are similarly rated at 12 square feet.

Vertical tubular boilers are rated at 12 square feet.

Cylinder boilers are rated at 9 square feet per horse-power.

For horizontal tubular boilers the usual allowance for each horse-power is,—

Steam for heating, etc.	15 square feet heating surface.
For plain throttle engine	15 square feet heating surface.
For single Corliss engine	12 square feet heating surface.
For compound Corliss condensing	10 square feet heating surface.

Hence a boiler for furnishing steam for

Plain slide-engine, with 1500 square feet surface	100 horse-power.
Simple Corliss engine, same boiler	125 horse-power.
Compound engine, same boiler	150 horse-power.

The best method is to compare boilers by their evaporative efficiency and not by heating surface.

The following is an approximate consumption of steam per indicated horse-power per hour for engine :

Plain slide engine	60 to 70 pounds,
High-speed automatic engine	30 to 50 pounds,
Simple Corliss engine	25 to 35 pounds,
Compound Corliss engine	15 to 20 pounds,
Triple-expansion engines	13 to 17 pounds,

depending upon the horse-power, steam pressure, condition of engine, load, etc.

The Centennial Standard.—The Centennial Exhibition occurred in 1876. The Committee of Judges to whom was committed the trials of competing boilers adopted as a unit of horse-power "30 pounds of water evaporated into dry steam per hour from feed-water at 100° Fahr., and under a pressure of 70 pounds per square inch above the atmosphere." These conditions were considered by the committee to fairly represent average practice.

The American Society of Mechanical Engineers' standard is the Centennial standard in another form,—viz., "34½ pounds of water evaporated from a feed-water temperature of 212° Fahr. into steam of the same temperature."

CHAPTER VI.

EXTERNALLY FIRED BOILERS.

EXTERNALLY fired boilers are usually cylindrical shells with or without flues or tubes and set in brickwork, the furnace being located underneath one end of the shell of the boiler.

A Cylinder Boiler consists of a plain cylindrical shell with closed ends, no flues or tubes being used. Cylinder boilers are seldom used singly, but commonly in batteries of from two to four, and occasionally six, with one furnace common to all. The diameters range from 26 to 40 inches, having lengths of from 10 to 15 diameters, according to location, service, etc. They are not as economical of fuel as either flue or tubular boilers; their use is restricted, therefore, to the lumber and coal regions, or such other localities where the feed-water is bad and the fuel abundant and cheap.

A cross-section and foundation plan of a battery of three cylinder boilers, 36 inches in diameter by 36 feet long, by the Jeanesville Iron Works, for an anthracite mine in Pennsylvania, is shown in Fig. 209. These boilers have two points of support placed at 8 feet 3 inches from each end, a cast-iron suspension girder extending across the furnace walls, shown in Fig. 210, carrying one end of the three boilers. Cylinder boilers are commonly fitted with cast-iron heads, and this is the case in the present example. This head is detailed in Fig. 211. The lower nozzle is for the feed-water, the upper one for the steam connection; the piping may be at either the front or rear of the boiler, as may be most convenient. The feed-pipe may extend some distance back from its entrance and upward, towards the centre of the boiler, to prevent local contraction, which would occur if the cold water came directly upon the plates at either end. The chimney is situated at the rear of the boilers, suitable flues being constructed in brickwork, as shown in plan in Fig. 209 and in sectional elevation in Fig. 212.

One objection to this type of boiler is the excessive ground-room required as compared with other types when large heating surfaces are necessary to furnish a given quantity of steam. In the boiler illustrated above, as well as cylinder boilers generally, only one-half the circumference is available for the transmission of heat to the water. We have, then: $\frac{1}{2}$ circumference of 3 feet = 4.71 feet; length, 36 feet × 4.71 feet = 169.56 square feet for one boiler, or 169.56 × 3 = 508.68 square feet total heating surface. The ground-room occupied, apart from the chimney and flues leading thereto, is 472.29 square feet. Allowing 9 square feet of heating surface per horse-power, we have

$$\frac{508.68}{9} = 56.52 \text{ horse-power.}$$

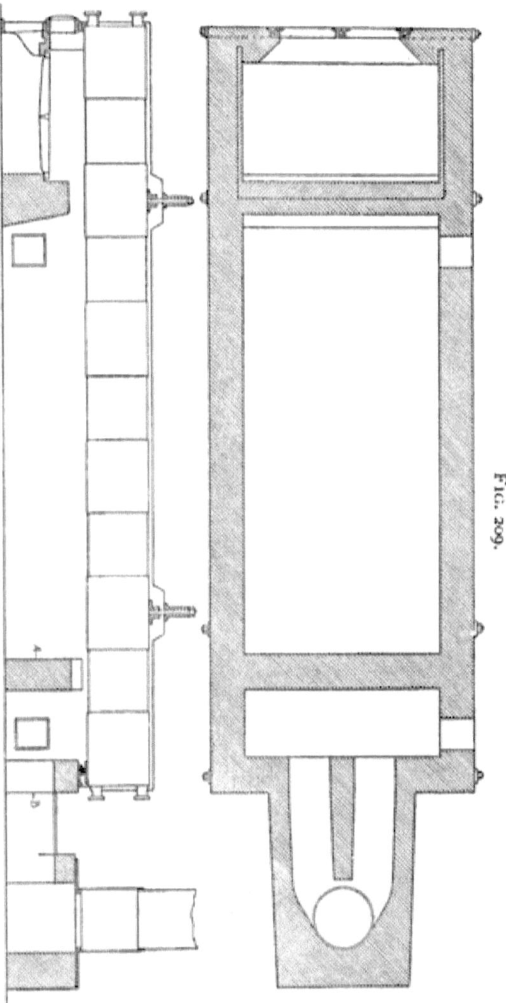

FIG. 209.

This same power could have been furnished by a horizontal tubular boiler 60 inches diameter by 14 feet long on the basis of 15 square feet per horse-power, which would have required a ground space of 9 x 19 feet = 171 square feet, the former being 2.76 times as much.

FIG. 210.

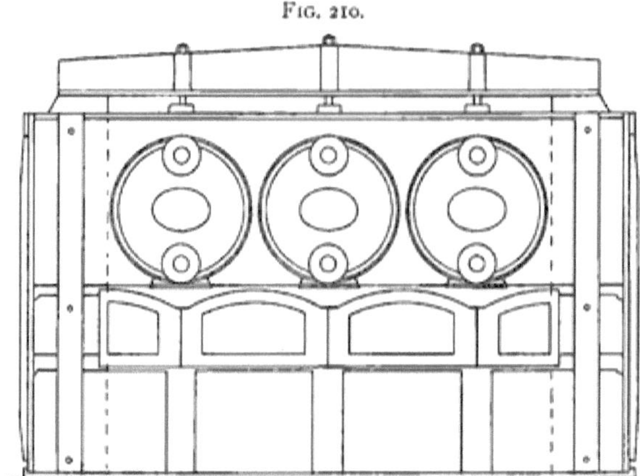

Another objection to this type of boiler is the incomplete utilization of heat as compared with other types, the direct escape of gases from the furnace to the chimney resulting in a large waste of heat, which,

FIG. 211.

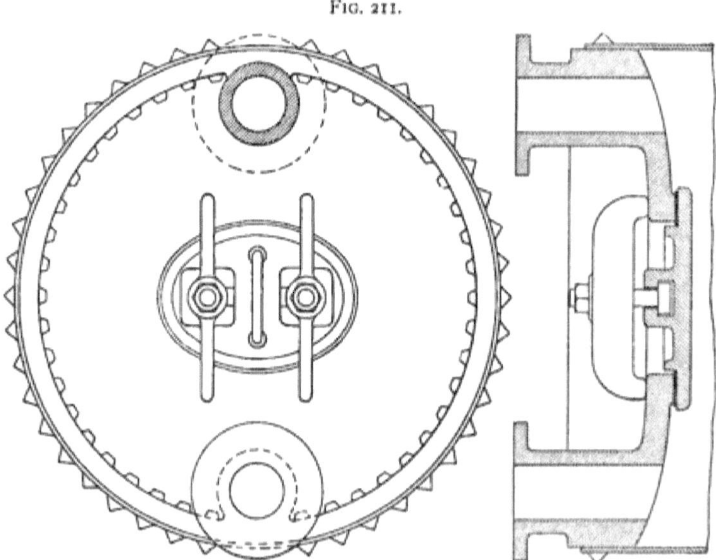

going on continuously, could not be borne under any circumstances which involved purchase of fuel. The temperature of escaping gases when the fires are forced is not infrequently over 800° Fahr., showing a wasteful expenditure of fuel.

FIG. 212.

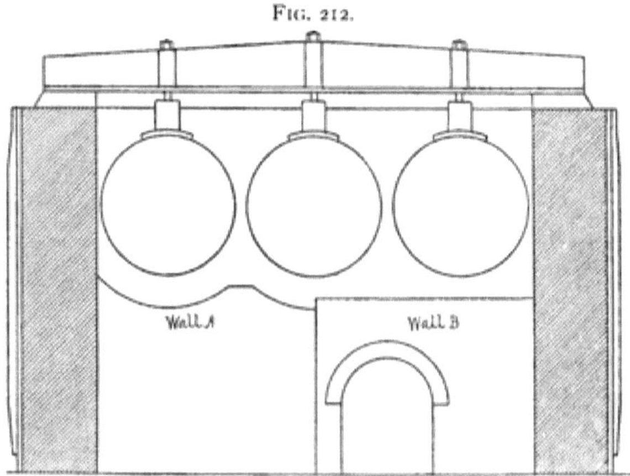

Double-Deck Cylinder Boilers.—The lack of sufficient heating surface in a cylinder boiler to bring the temperature of the escaping gases down to that of a flue or tubular boiler, when the boiler is worked up to its capacity, has led to the placing of an additional cylinder under the main boiler, back of the bridge-wall, as shown in Fig. 213. The lower shell is in reality an exaggerated mud-drum and heater, its office being to absorb heat from the outgoing gases. Whatever waste heat can thus be reclaimed is gain. Such boilers are not in common use: the difference in temperature of water in the upper and lower cylinders produces unequal expansion, and this brings additional and variable stresses upon the joints at the necks connecting the two shells, resulting in troublesome leaks and corrosion induced by unusual strains.

The French Boiler, or Elephant Boiler, is one large cylinder with two or three smaller cylinders underneath connected by suitable nozzles, as shown in Fig. 214. The lower cylinders are filled with water and almost wholly surrounded by heated gases. The furnace being under these small cylinders, the products of combustion act first upon them throughout their length, then return to the front end of the boiler along one side of the main cylinder, and finally pass to the chimney along the other side of the main cylinder. The particular boiler here illustrated is one in which the famous Mulhouse experiments were made. The principal dimensions are:

Main cylinder, 3 feet 9 inches diameter by 20 feet 6⅜ inches long; three lower cylinders, 19.7 inches diameter by 32 feet 9¾ inches long; grate surface, 20 square feet; total heating surface, 607.6 square feet.

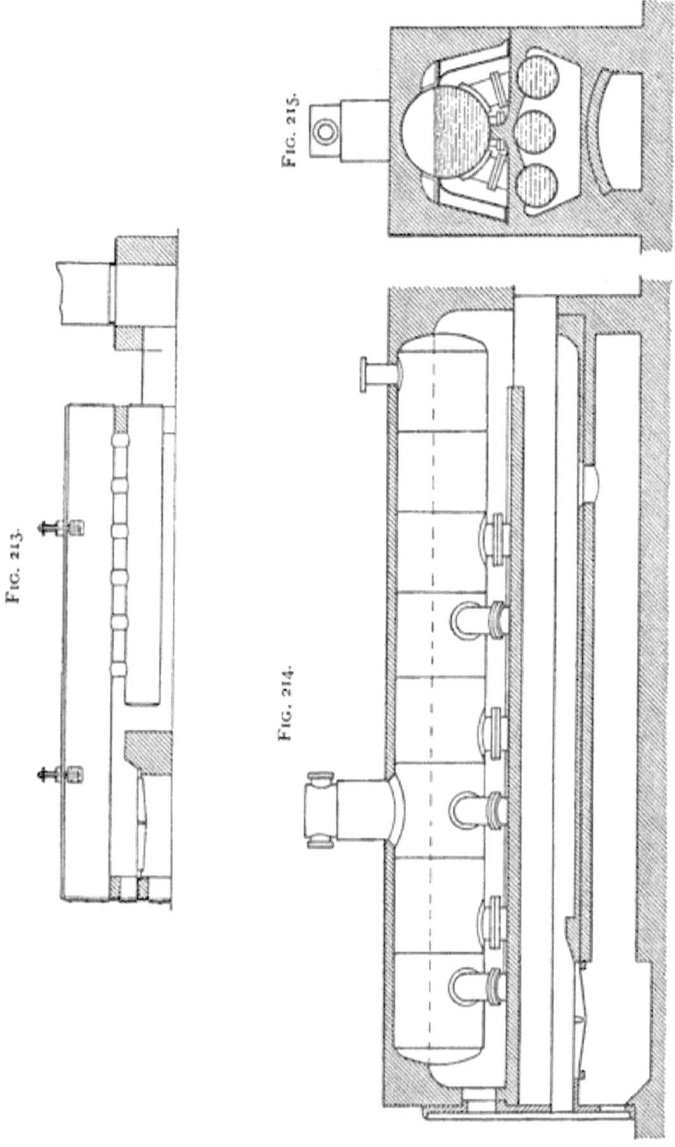

Fig. 213.

Fig. 214.

Fig. 215.

RESULTS OBTAINED.

Equivalent evaporation from and at 212° Fahr. per pound of coal	8.97 pounds.
Equivalent evaporation from and at 212° Fahr. per pound of net combustible	10.37 pounds.
Equivalent evaporation from and at 212° Fahr. per hour per square foot of heating surface	3.28 pounds.
Weight of air supplied per pound of coal consumed	14.89 pounds.
Mean temperature of gases entering chimney	425° Fahr.

This shows a good rate of evaporation, but it does not surpass that of a good tubular boiler, which can be bought for much less money.

Two-Flue Boilers.—These boilers are extensively used in the Western States, especially in the river towns and coal-mining districts. Many of the light-draught steamboats which navigate the Western rivers use boilers of this design.

In diameter such boilers range from 36 to 48 inches, and from 16 to 24 feet in length, with an occasional increase for the larger diameters to 30 feet. The flue diameter is commonly 12 inches for 36- and 38-inch shells, 13 inches for 40-inch shells, 14 inches for 42-inch shells, 15 inches for 44-inch shells, 16 inches for 46- and 48-inch shells. For ordinary land service the pressures are not much above 75 pounds per square inch, and flues are made ¼ inch thick for all diameters. Lap-welded tubes can be had in either steel or iron up to 20 feet in length; but the advantages accruing from a double thickness of plate at riveted circumferential seams to resist collapse keeps the riveted flue still in favor for the larger diameters. Lap-welded flues can be cut in lengths and made up with riveted circumferential seams to accommodate any length of shell. The United States Regulations fix the greatest length of such sections for diameters of 12 to 18 inches at 3 feet. The thickness of flues from 12 to 15 inches diameter may be ¼ inch; 16 to 18 inches diameter, $\frac{9}{32}$ inch. Such 12- and 14-inch flues will pass United States inspection for 150 pounds working pressure. For 16- to 18-inch flues,—the 16 x $\frac{9}{32}$-inch flues will be passed for 150 pounds working pressure, the 17-inch flues for 141 pounds, and the 18-inch flues for 134 pounds.

FIG. 216.

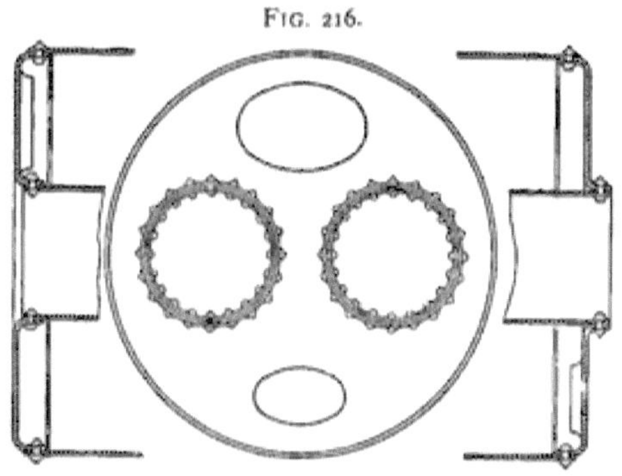

The flanging of two-flue boiler-heads is commonly as shown in Fig. 216, the illustration representing a lap-welded flue. The rear head is flanged in, and that end of the flue is first riveted in place. The front head is flanged out and flue riveted on the outside.

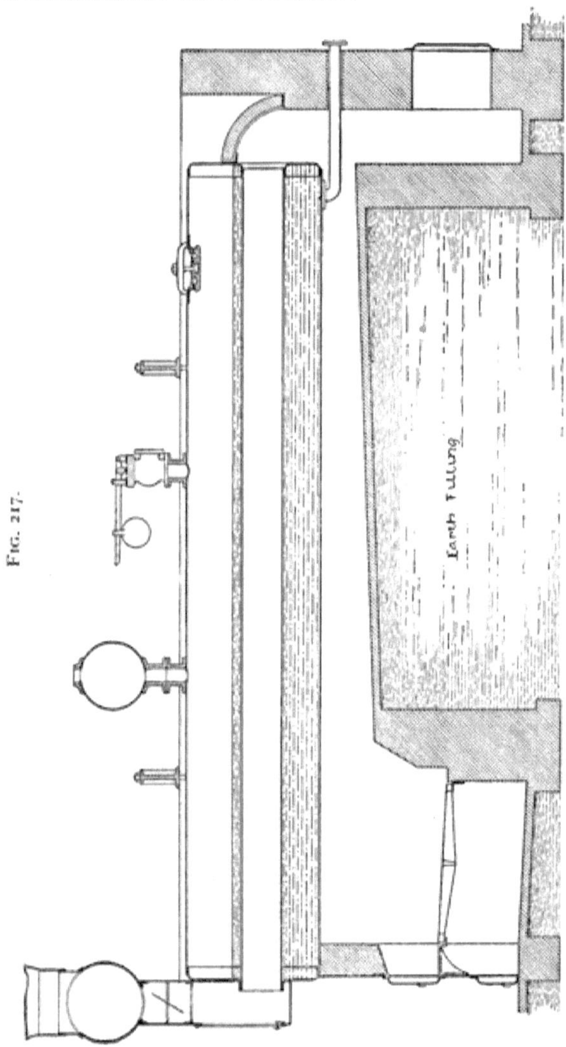

FIG. 217.

A longitudinal elevation of a boiler of 40 inches diameter with 2 flues of 13 inches diameter, each boiler 20 feet long, is shown in Fig. 217. A cross-sectional elevation showing two such boilers set in the same furnace is given in Fig. 218. The grate-bars, adapted for burning bituminous coal, are placed 36 inches below the bottom of the boilers, thus affording a roomy combustion-chamber above the fuel.

Fig. 218.

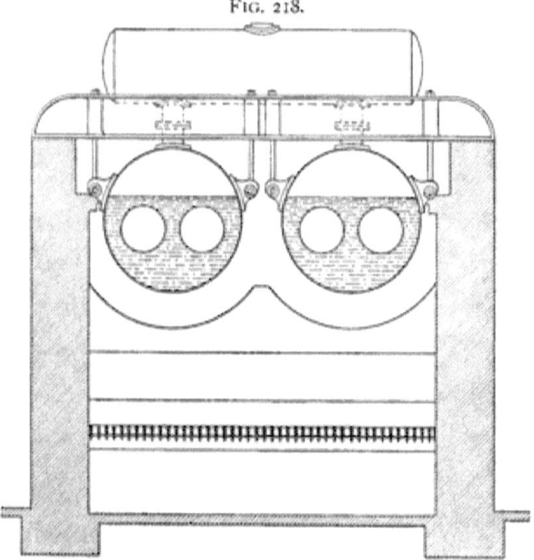

Five-Flue Boilers.—The power of resistance to prevent collapse of a boiler-flue decreases with its diameter. For this reason several smaller flues are preferred to two large ones. When more than two flues are wanted, the next practical number is five. Three flues are placed in the upper row and two in the space underneath.

In some localities the flue-diameters vary for a given boiler; for example, a

44-inch boiler may have 3 8-inch and 2 10-inch flues.
46-inch boiler may have 2 8-inch, 2 9-inch, and 1 13-inch flues.
48-inch boiler may have 2 8-inch, 2 10-inch, and 1 12-inch flues.
50-inch boiler may have 2 8-inch, 2 10-inch, and 1 14-inch flues.

This arrangement of flue-diameter is not known to possess any advantage over the simpler one in having all the flues of the same diameter. There is a disadvantage in the fact that escaping gases flow in the direction of least resistance, the larger flues robbing the smaller ones of their proportion of heated gases. The same remarks apply here regarding details of construction and collapsing resistance of flues as were given in the section on two-flue boilers. If the flues are all of the same diameter, 7-inch flues will answer for 36- and 38-inch shells, 8-inch flues for 40-inch shells, 9-inch flues for 42-inch shells, and 10-inch flues for 44- to

Fig. 219.

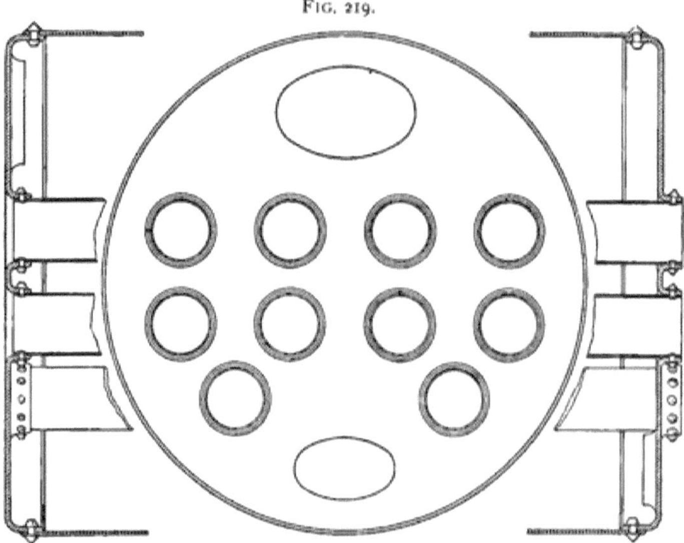

48-inch shells. Boilers of this type are commercially rated at 12 square feet of heating surface per horse-power.

Fig. 220.

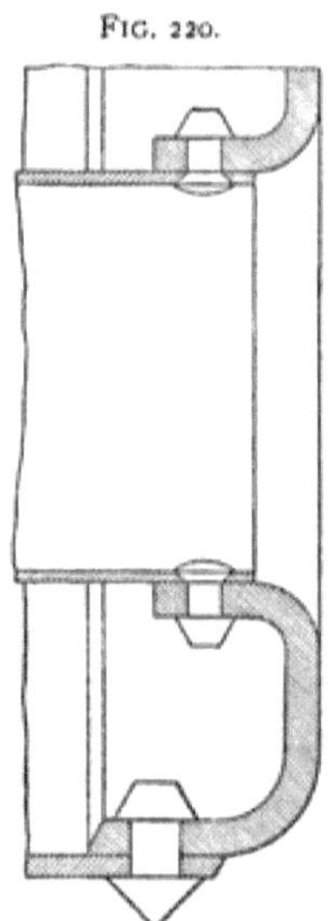

Six-Inch-Flue Boilers.—This design of boiler has long been a favorite one in the Western States; and when a choice is made away from a two-flue boiler it usually passes over to this, because a larger heating surface is had than in two-flue boilers, and there are larger spaces between the flues than is the case in tubular boilers, making it a compromise boiler well adapted to water heavily charged with scale-making impurities. The tubes are lap welded and riveted into the heads, as shown in Fig. 219. The rear head is flanged in and the front head flanged out. A detail of a riveted flue and section of head is shown in Fig. 220. A few special tools are needed for holding the rivets in place, and perhaps a few special riveting-hammers. The rivets should be ½ inch diameter for 6-inch flues. The holes in the head should be countersunk, as shown in Fig. 220, so that the rivet-point shall project into the flue as little as possible. When putting the flues in the boiler, begin with the bottom ones, insert the rivets from the inside of the boiler, and rivet them on the inside of the flue. This can be done with little difficulty from the outside of the boiler.

The number of 6-inch flues that can be gotten into a given diameter of head depends entirely upon the number of flanged openings which can be made in it. In a properly constructed machine there is no difficulty in flanging them within 3 inches of each other and keeping the face of the boiler-head flat. The number of 6-inch lap-welded flues for each diameter of boiler may be as follows:

Diameter of boiler	40	44	48	54	60	66	72
Number of 6-inch flues . .	6	8	10	12	18	22	26

The commercial rating of a 6-inch-flue boiler is 12 square feet of heating surface per horse-power.

TABLE XXXVIII.

PRINCIPAL DIMENSIONS OF SIX-INCH-FLUE BOILERS.

Commercial Rating.	Shell. Diameter.	Shell. Length.	Number of Flues, all 6 inches in diameter.	Thickness. Shell.	Thickness. Heads.	Dome. Diameter.	Dome. Height.	Dome. Thickness. Shell.	Dome. Thickness. Head.	Heating Surface.	Weight, including Full Front Fixtures.*
H.-P.	Inch.	Feet.		Inch.	Inch.	Inch.	Inch.	Inch.	Inch.	Sq. Feet.	Pounds.
40	48	16	10	5/16	7/16	26	28	5/16	3/8	380	12,065
45	48	18	10	5/16	7/16	26	28	5/16	3/8	426	12,465
50	54	16	12	5/16	7/16	30	34	5/16	3/8	454	14,025
55	54	18	12	5/16	7/16	30	34	5/16	3/8	508	14,725
60	54	20	12	5/16	7/16	30	34	5/16	3/8	566	15,825
65	60	16	18	11/32	7/16	32	36	5/16	3/8	624	17,120
75	60	18	18	11/32	7/16	32	36	5/16	3/8	702	18,370
80	60	20	18	11/32	7/16	32	36	5/16	3/8	780	19,700
85	66	18	22	3/8	7/16	36	40	3/8	7/16	827	21,800
95	66	20	22	3/8	7/16	36	40	3/8	7/16	919	23,370
90	72	16	26	7/16	1/2	36	40	3/8	7/16	854	22,800
100	72	18	26	7/16	1/2	36	40	3/8	7/16	961	24,050
110	72	20	26	7/16	1/2	36	40	3/8	7/16	1067	25,750

Tubular Boilers.—The diameter of tubes in an ordinary horizontal tubular boiler ranges from 3 to 4 inches, seldom above or below these diameters; the ordinary diameters for boiler shells lie between 36 and 72 inches. For 36-inch boilers up to 48 inches diameter, 3-inch tubes are commonly employed; from 48 to 60 inches diameter of shell, the tubes are commonly 3½ inches; 4-inch tubular boilers usually range

* Fixtures comprise: Full front with anchors, coking-plate and stack-plate, grate- and bearing-bars, flue-plate or arch-bars, wall-binder bars and rods, expansion-plates and rollers to go under boiler-brackets, rear ash-door and frame, and smoke-stack with bands and guys.

from 54 to 72 inches diameter. The dividing line for changing tube-diameter is a variable one and rests largely upon the judgment of the designer, but the above covers the usual practice.

Fig. 221.

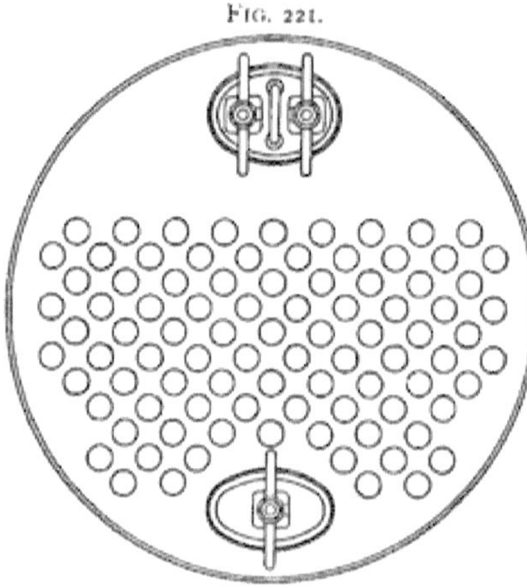

The number of tubes for a given head requires correct judgment and liberal spacing. A common defect in the construction of horizontal tubular boilers is the insertion of too many tubes, and for large boilers the tubes are sometimes too small in diameter. An example is here given in Fig. 221, which represents a 60-inch boiler fitted with 91 tubes 3 inches in diameter, arranged zigzag. This boiler-head ought not to have contained more than 76 tubes 3 inches in diameter, and preferably the tubes should have been placed in vertical rows. A much better arrangement would have been to select tubes 3½ inches in diameter, which would have permitted the insertion of 56 when placed in vertical rows. Overcrowding of tubes prevents a proper circulation of water and increases the resistance to the free escape of steam by impeding the upward movement of the heated water from the shell to the steam-space above. The spacing of tubes to insure a proper circulation of water in the boiler ought to bear some relation to the diameter of the tube, and in carefully prepared specifications this is usually the case.

Fig. 222.

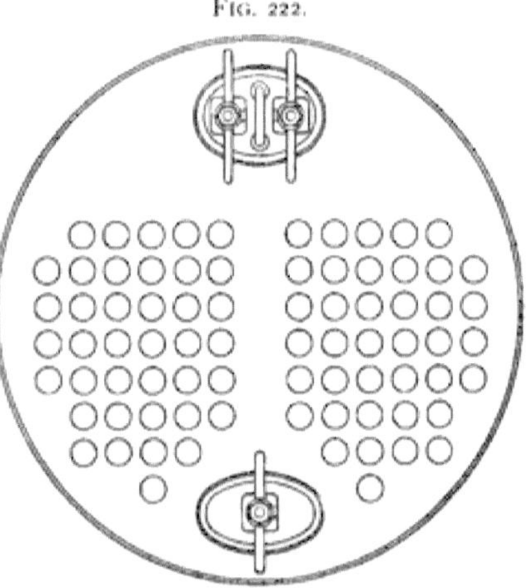

A central water-space is sometimes provided in the arrangement of tubes by separating them into two groups, as shown in Fig. 222.

This method of tube distribution was formerly believed to have peculiar merit, because it permitted a vertical movement of the water in the centre of the boiler, rising in the centre and descending at the sides; but it is quite probable that the reverse of this is true. In any event, what is wanted is not a circulation of water around a group of tubes, but a circulation between them, and as water, like every other fluid, will flow in the direction of least resistance, these circular currents will be set up within the boiler, one on either side, to the practical exclusion of water-currents through the mass of tubes. The objection to such an arrangement is, that scale is likely to form around the tubes, and this will, by reason of its granular surface, retain such floating matter as may be in mechanical suspension in the water, thus accumulating a coating much thicker in a given time than would be the case if there were a freer circulation of water between the tubes themselves. The recommendation is, to dispense with this central space and set the tubes wider apart in their horizontal centres.

The distance from the side of a tube to the inside of the boiler-shell should not in the case of a 36-inch boiler be less than 2 inches, for a 48-inch boiler it should be not less than 2½ inches, and for 50 inch boilers and larger the distance should be not less than 3 inches, to secure good circulation of water.

Fig. 223.

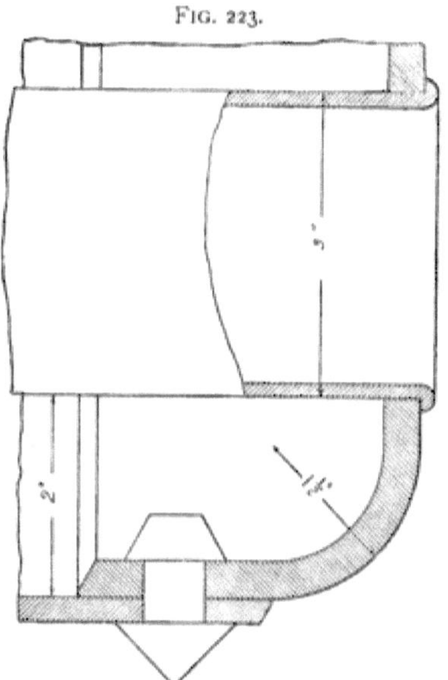

The distance from the side of a tube to the shell of a boiler is limited by the radius of the flange of the head, as shown in Fig. 223, which represents a 36-inch machine-flanged head ⅜ inch thick, with an outside radius of 1½ inches and a 3-inch tube. As there is more or less distortion of the head where the flange-curve merges into the flat plate, the nearest distance at which it is advisable to locate a tube is ½ inch from the centre of the flange-curve; this makes the distance from the inside of the shell to the tube 2 inches, as shown. The flange of a 72-inch head, ½ inch thick, with a 4-inch tube, is shown in Fig. 224, in which the outside radius is 2 inches, and the distance from the inside of the shell to the side of the tube is 2½ inches, which is the least practical limit in locating the tube. These limiting facts govern the location of tubes with reference to the shell; but the matter of circulation must not be overlooked, and a further allowance of clear waterway will greatly facilitate the circulation at the point where it is particularly needed to accommodate the ascending currents caused by the contact of water with the hot shell, and such increased allowance will add sufficiently to the efficiency of the boiler to fully compensate for any loss of heating surface occasioned by the omission of a few tubes which, if retained, might prove a positive detriment to the boiler.

FIG. 224.

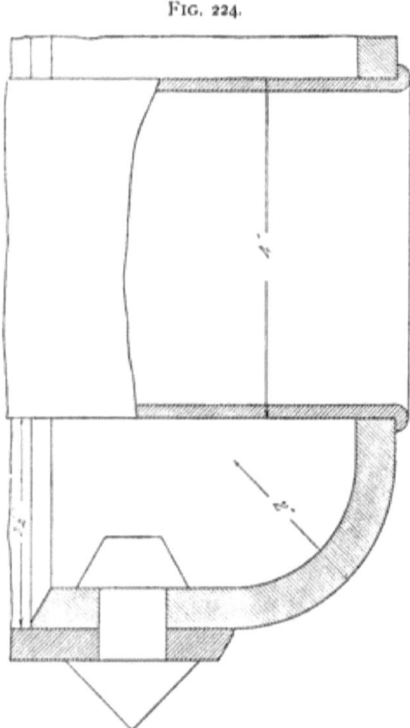

The horizontal distance between tubes should not in any case be less than one-third the diameter of the tube for the smallest diameter of shell in which tubes are to be used; and this distance should increase in proportion to the depth of water below the top row of tubes,—in other words, the distance increases with the diameter of the shell. If tubes are too close together horizontally, the circulation is interfered with and the efficiency of the boiler lowered. The writer has in mind a boiler-shell which bagged no less than three times in as many weeks under moderately heavy firing, the reason for which was afterwards traced directly to a want of proper circulation caused by the tubes being too close together to begin with, and this meagre allowance of space was further lessened by a slight accumulation of scale, which so interfered with the water circulation that the boilers utterly failed at a much less power-rating than could have been secured had there been a proper allotment of space between the tubes themselves, and between the outside of the tube and the inside of the shell.

The horizontal spacing of tubes in the accompanying tables is based upon many years of observation, experiment, and trial. It is recommended that for 3-inch tubes the horizontal centres be 4 inches for shells 36 to 40 inches diameter ; 4¼-inch centres for shells 42 to 52 inches diameter ; 4½-inch centres for shells 54 to 60 inches diameter. See Table XXXIX.

For 3½-inch tubes the horizontal centres should be 5 inches for 48- to 56-inch shells, 5¼-inch centres for shells 58 to 62 inches diameter, 5½-inch centres for shells 64 and 66 inches diameter. See Table XLI.

For 4-inch tubes the horizontal centres may be 6 inches for all diameters from 54 to 72 inches inclusive. See Table XLIII.

The vertical spacing of tubes is of less importance than the horizontal spacing. In the tables above referred to this distance is uniformly 1 inch between the outside diameters of the tubes, which is as little metal as should intervene between them.

The upper limit of tubes varies somewhat, and may be said to lie between three-fifths and two-thirds of the height from the bottom in the best practice. The drawings from which the accompanying tables were prepared had a uniform limit of two-thirds, the only variation being in one or two cases where the tubes were carried about one-half inch above this line in order to get a better arrangement of tubes at the bottom of the boiler. One of these was the 70-inch boiler with 4-inch tubes, a size of boiler quite unusual in trade and not likely to be called for in any engineering specifications. The two-thirds limit ought not to be encroached upon, as the steam-room is none too large after allowing say three inches for depth of water above the tubes.

FIG. 225.

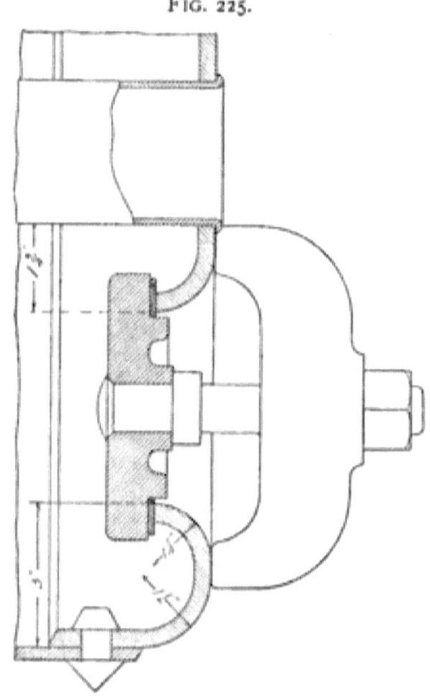

The location of a tube with reference to a flanged handhole or manhole will depend upon the radius of the curve if flanged as in Fig. 225. The illustration represents a 4 x 6 handhole in a 36-inch head, the latter being ⅜ inch thick. The radius for such a flanged opening would probably be 1¼ inches, as shown, to which an allowance of ½ inch is added, making the nearest distance from the inside of the opening to the side of the tube 1¾ inches, to which an

TABLE XXXIX.

3-INCH TUBULAR BOILER HEADS.

FIG. 226.

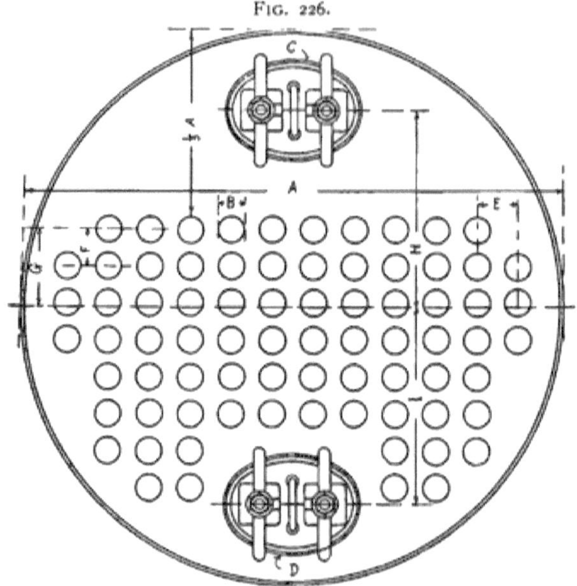

Reference Number.	Diameter. Head. A.	Diameter. Tube. B.	Manhole. Upper. C.	Manhole. Lower. D.	Tubes Centre to Centre. Hori-zontal. E.	Tubes Centre to Centre. Verti-cal. F.	Centre of Boiler to First Row of Tubes. G.	Centre of Boiler to Centre of Manholes. Upper. H.	Centre of Boiler to Centre of Manholes. Lower. I.	Number of Tubes.	Length of Boiler.
	Ins.	Ins.	Ins.	Ins.	Ins.	Ins.	Ins.	Ins.	Ins.		Feet.
1	36	3	6 x 10	4 x 6	4	4	4½	12	13	28	8 to 12
2	38	3	6 x 10	6 x 10	4	4	5	13	13	32	8 to 12
3	40	3	9 x 14½	6 x 10	4	4	5¼	12½	14	34	8 to 12
4	42	3	9 x 14½	6 x 10	4¼	4	5½	13½	15	36	8 to 12
5	44	3	9 x 14½	9 x 14½	4¼	4	5¾	14½	14½	38	10 to 12
6	46	3	9 x 14½	9 x 14½	4¼	4	6	15½	15½	42	10 to 12
7	48	3	11 x 15	9 x 14½	4¼	4	6½	15½	16½	46	10 to 12
8	50	3	11 x 15	9 x 14½	4¼	4	6¾	16½	17½	52	10 to 12
9	52	3	11 x 15	9 x 14½	4¼	4	7¼	17½	18½	54	10 to 12
10	54	3	11 x 15	9 x 14½	4½	4	7½	18½	19½	60	10 to 12
11	56	3	11 x 15	11 x 15	4½	4	8	19½	19½	64	10 to 12
12	58	3	11 x 15	11 x 15	4½	4	8¼	20½	20½	70	10 to 12
13	60	3	11 x 15	11 x 15	4½	4	8½	21½	21½	76	10 to 12

manhole,
in a larger

ZE OF GRATE.		COMMERCIAL HORSE-POWER.		
Length.		15 Feet.	12 Feet.	10 Feet.
Calculated Feet.	Proposed Feet and Inches.			
2.75′ @ 7′	3′ 0″	14.74	18.43	22.11
3.15′ @ 8′	3′ 3″	18.43	23.03	27.64
3.54′ @ 9′	3′ 6″	22.11	27.64	33.17
2.98′ @ 7′	3′ 0″	16.59	20.74	24.88
3.41′ @ 8′	3′ 6″	20.74	25.92	31.10
3.83′ @ 9′	4′ 0″	24.88	31.10	37.32
3.01′ @ 7′	3′ 0″	17.59	21.99	26.39
3.44′ @ 8′	3′ 6″	21.99	27.49	32.98
3.86′ @ 9′	4′ 0″	26.39	32.98	39.58
3.04′ @ 7′	3′ 0″	23.25	29.06	34.87
3.47′ @ 8′	3′ 6″	27.90	34.87	41.85
3.91′ @ 9′	4′ 0″	32.55	40.69	48.82
3.05′ @ 7′	3′ 0″	24.50	30.63	36.76
3.49′ @ 8′	3′ 6″	29.40	36.76	44.11
3.92′ @ 9′	4′ 0″	34.30	42.88	51.46
3.23′ @ 7′	3′ 3″	25.81	36.51	40.22
3.69′ @ 8′	3′ 9″	32.17	40.22	48.26
4.16′ @ 9′	4′ 3″	37.54	46.92	56.30
3.40′ @ 7′	3′ 6″	29.11	39.39	43.67
3.88′ @ 8′	4′ 0″	34.93	43.67	52.40
4.37′ @ 9′	4′ 6″	40.76	50.95	61.14
3.68′ @ 7′	3′ 9″	32.46	40.58	48.69
4.20′ @ 8′	4′ 3″	38.95	48.69	58.43
4.73′ @ 9′	4′ 9″	45.44	56.81	68.17
3.69′ @ 7′	3′ 9″	33.72	42.15	50.58
4.21′ @ 8′	4′ 3″	40.47	50.58	60.70
4.74′ @ 9′	4′ 9″	47.21	59.01	70.81
3.94′ @ 7′	4′ 0″	37.07	46.34	55.60
4.50′ @ 8′	4′ 6″	44.48	55.60	66.73
5.06′ @ 9′	5′ 0″	51.90	64.87	77.85
4.05′ @ 7′	4′ 0″	39.38	49.22	59.07
4.63′ @ 8′	4′ 9″	47.25	59.07	70.88
5.20′ @ 9′	5′ 3″	55.13	68.91	82.69
4.28′ @ 7′	4′ 3″	42.73	53.41	64.09
4.89′ @ 8′	5′ 0″	51.27	64.09	76.91
5.50′ @ 9′	5′ 6″	59.82	74.77	89.72
4.49′ @ 7′	4′ 3″	46.07	57.59	69.11
5.14′ @ 8′	5′ 2″	55.29	66.11	82.93
5.78′ @ 9′	5′ 9″	64.50	80.63	96.76

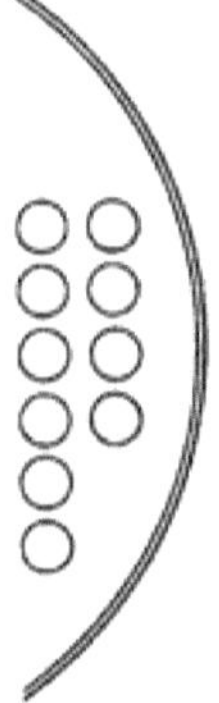

ɔituminous
diameters;
ɔd modern
ɔrced draft
onger tube
ın is ordi-
ith natural

l arrange-
ater-tubes,
over each
g. 227, or
as in Fig.
ctically set-
he former,
e the latter
ɔyed. The
probably
belief that
l current
ıe water to
ıeat on its
, as would

TABLE XL.

HORIZONTAL TUBULAR BOILERS, FITTED WITH 3-INCH TUBES.

Reference Number.	Boiler Shell: Diameter	Boiler Shell: Length	Number of 3-inch Tubes	Internal Area of Whole Number of Tubes	Heating Surface: Tubes	Heating Surface: Shell	Heating Surface: Total	Grate Area: Proportion to Tube Area: 7 Times	8 Times	9 Times	Ratio of Grate Surface to Total Heating Surface when Grate Area is 7, 8, and 9 times that of the Tube Area: 7 Times	8 Times	9 Times	Size of Grate: Width in Feet	Length: Calculated Feet	Length: Proposed Feet and Inches	Commercial Horse-Power: 15 Feet	[illegible] Feet	20 Feet
	Inches	Feet		Sq. Feet	Sq. Feet	Sq. Feet	Sq. Feet	Sq. Feet	Sq. Feet	Sq. Feet									
	[illegible]	[illegible]	[illegible]	[illegible]	[illegible]	[illegible]	[illegible]	[illegible]	[illegible]	[illegible]	[illegible]	[illegible]	[illegible]	[illegible]	[illegible]	[illegible]	[illegible]	[illegible]	[illegible]
1	[illegible]	[illegible]	[illegible]	[illegible]	[illegible]	[illegible]	[illegible]	[illegible]	[illegible]	[illegible]	[illegible]	[illegible]	[illegible]	[illegible]	[illegible]	[illegible]	[illegible]	[illegible]	[illegible]
	[illegible]	[illegible]	[illegible]	[illegible]	[illegible]	[illegible]	[illegible]	[illegible]	[illegible]	[illegible]	[illegible]	[illegible]	[illegible]		[illegible]	[illegible]	[illegible]	[illegible]	[illegible]
	[illegible]	[illegible]	[illegible]	[illegible]	[illegible]	[illegible]	[illegible]	[illegible]	[illegible]	[illegible]	[illegible]	[illegible]	[illegible]	[illegible]	[illegible]	[illegible]	[illegible]	[illegible]	[illegible]
2	[illegible]	[illegible]	[illegible]	[illegible]	[illegible]	[illegible]	[illegible]	[illegible]	[illegible]	[illegible]	[illegible]	[illegible]	[illegible]	[illegible]	[illegible]	[illegible]	[illegible]	[illegible]	[illegible]
	[illegible]	[illegible]	[illegible]	[illegible]	[illegible]	[illegible]	[illegible]	[illegible]	[illegible]	[illegible]	[illegible]	[illegible]	[illegible]		[illegible]	[illegible]	[illegible]	[illegible]	[illegible]
	[illegible]	[illegible]	[illegible]	[illegible]	[illegible]	[illegible]	[illegible]	[illegible]	[illegible]	[illegible]	[illegible]	[illegible]	[illegible]	[illegible]	[illegible]	[illegible]	[illegible]	[illegible]	[illegible]
3	[illegible]	[illegible]	[illegible]	[illegible]	[illegible]	[illegible]	[illegible]	[illegible]	[illegible]	[illegible]	[illegible]	[illegible]	[illegible]	[illegible]	[illegible]	[illegible]	[illegible]	[illegible]	[illegible]
	[illegible]	[illegible]	[illegible]	[illegible]	[illegible]	[illegible]	[illegible]	[illegible]	[illegible]	[illegible]	[illegible]	[illegible]	[illegible]		[illegible]	[illegible]	[illegible]	[illegible]	[illegible]
	[illegible]	[illegible]	[illegible]	[illegible]	[illegible]	[illegible]	[illegible]	[illegible]	[illegible]	[illegible]	[illegible]	[illegible]	[illegible]	[illegible]	[illegible]	[illegible]	[illegible]	[illegible]	[illegible]
4	[illegible]	[illegible]	[illegible]	[illegible]	[illegible]	[illegible]	[illegible]	[illegible]	[illegible]	[illegible]	[illegible]	[illegible]	[illegible]	[illegible]	[illegible]	[illegible]	[illegible]	[illegible]	[illegible]
	[illegible]	[illegible]	[illegible]	[illegible]	[illegible]	[illegible]	[illegible]	[illegible]	[illegible]	[illegible]	[illegible]	[illegible]	[illegible]		[illegible]	[illegible]	[illegible]	[illegible]	[illegible]
	[illegible]	[illegible]	[illegible]	[illegible]	[illegible]	[illegible]	[illegible]	[illegible]	[illegible]	[illegible]	[illegible]	[illegible]	[illegible]	[illegible]	[illegible]	[illegible]	[illegible]	[illegible]	[illegible]
5	[illegible]	[illegible]	[illegible]	[illegible]	[illegible]	[illegible]	[illegible]	[illegible]	[illegible]	[illegible]	[illegible]	[illegible]	[illegible]	[illegible]	[illegible]	[illegible]	[illegible]	[illegible]	[illegible]
	[illegible]	[illegible]	[illegible]	[illegible]	[illegible]	[illegible]	[illegible]	[illegible]	[illegible]	[illegible]	[illegible]	[illegible]	[illegible]		[illegible]	[illegible]	[illegible]	[illegible]	[illegible]
	[illegible]	[illegible]	[illegible]	[illegible]	[illegible]	[illegible]	[illegible]	[illegible]	[illegible]	[illegible]	[illegible]	[illegible]	[illegible]	[illegible]	[illegible]	[illegible]	[illegible]	[illegible]	[illegible]
6	[illegible]	[illegible]	[illegible]	[illegible]	[illegible]	[illegible]	[illegible]	[illegible]	[illegible]	[illegible]	[illegible]	[illegible]	[illegible]	[illegible]	[illegible]	[illegible]	[illegible]	[illegible]	[illegible]
	[illegible]	[illegible]	[illegible]	[illegible]	[illegible]	[illegible]	[illegible]	[illegible]	[illegible]	[illegible]	[illegible]	[illegible]	[illegible]		[illegible]	[illegible]	[illegible]	[illegible]	[illegible]
	[illegible]	[illegible]	[illegible]	[illegible]	[illegible]	[illegible]	[illegible]	[illegible]	[illegible]	[illegible]	[illegible]	[illegible]	[illegible]	[illegible]	[illegible]	[illegible]	[illegible]	[illegible]	[illegible]
7	[illegible]	[illegible]	[illegible]	[illegible]	[illegible]	[illegible]	[illegible]	[illegible]	[illegible]	[illegible]	[illegible]	[illegible]	[illegible]	[illegible]	[illegible]	[illegible]	[illegible]	[illegible]	[illegible]
	[illegible]	[illegible]	[illegible]	[illegible]	[illegible]	[illegible]	[illegible]	[illegible]	[illegible]	[illegible]	[illegible]	[illegible]	[illegible]		[illegible]	[illegible]	[illegible]	[illegible]	[illegible]
	[illegible]	[illegible]	[illegible]	[illegible]	[illegible]	[illegible]	[illegible]	[illegible]	[illegible]	[illegible]	[illegible]	[illegible]	[illegible]	[illegible]	[illegible]	[illegible]	[illegible]	[illegible]	[illegible]
8	[illegible]	[illegible]	[illegible]	[illegible]	[illegible]	[illegible]	[illegible]	[illegible]	[illegible]	[illegible]	[illegible]	[illegible]	[illegible]	[illegible]	[illegible]	[illegible]	[illegible]	[illegible]	[illegible]
	[illegible]	[illegible]	[illegible]	[illegible]	[illegible]	[illegible]	[illegible]	[illegible]	[illegible]	[illegible]	[illegible]	[illegible]	[illegible]		[illegible]	[illegible]	[illegible]	[illegible]	[illegible]
	[illegible]	[illegible]	[illegible]	[illegible]	[illegible]	[illegible]	[illegible]	[illegible]	[illegible]	[illegible]	[illegible]	[illegible]	[illegible]	[illegible]	[illegible]	[illegible]	[illegible]	[illegible]	[illegible]
9	[illegible]	[illegible]	[illegible]	[illegible]	[illegible]	[illegible]	[illegible]	[illegible]	[illegible]	[illegible]	[illegible]	[illegible]	[illegible]	[illegible]	[illegible]	[illegible]	[illegible]	[illegible]	[illegible]
	[illegible]	[illegible]	[illegible]	[illegible]	[illegible]	[illegible]	[illegible]	[illegible]	[illegible]	[illegible]	[illegible]	[illegible]	[illegible]		[illegible]	[illegible]	[illegible]	[illegible]	[illegible]
	[illegible]	[illegible]	[illegible]	[illegible]	[illegible]	[illegible]	[illegible]	[illegible]	[illegible]	[illegible]	[illegible]	[illegible]	[illegible]	[illegible]	[illegible]	[illegible]	[illegible]	[illegible]	[illegible]
10	[illegible]	[illegible]	[illegible]	[illegible]	[illegible]	[illegible]	[illegible]	[illegible]	[illegible]	[illegible]	[illegible]	[illegible]	[illegible]	[illegible]	[illegible]	[illegible]	[illegible]	[illegible]	[illegible]
	[illegible]	[illegible]	[illegible]	[illegible]	[illegible]	[illegible]	[illegible]	[illegible]	[illegible]	[illegible]	[illegible]	[illegible]	[illegible]		[illegible]	[illegible]	[illegible]	[illegible]	[illegible]
	[illegible]	[illegible]	[illegible]	[illegible]	[illegible]	[illegible]	[illegible]	[illegible]	[illegible]	[illegible]	[illegible]	[illegible]	[illegible]	[illegible]	[illegible]	[illegible]	[illegible]	[illegible]	[illegible]
11	[illegible]	[illegible]	[illegible]	[illegible]	[illegible]	[illegible]	[illegible]	[illegible]	[illegible]	[illegible]	[illegible]	[illegible]	[illegible]	[illegible]	[illegible]	[illegible]	[illegible]	[illegible]	[illegible]
	[illegible]	[illegible]	[illegible]	[illegible]	[illegible]	[illegible]	[illegible]	[illegible]	[illegible]	[illegible]	[illegible]	[illegible]	[illegible]		[illegible]	[illegible]	[illegible]	[illegible]	[illegible]
	[illegible]	[illegible]	[illegible]	[illegible]	[illegible]	[illegible]	[illegible]	[illegible]	[illegible]	[illegible]	[illegible]	[illegible]	[illegible]	[illegible]	[illegible]	[illegible]	[illegible]	[illegible]	[illegible]
12	[illegible]	[illegible]	[illegible]	[illegible]	[illegible]	[illegible]	[illegible]	[illegible]	[illegible]	[illegible]	[illegible]	[illegible]	[illegible]	[illegible]	[illegible]	[illegible]	[illegible]	[illegible]	[illegible]
	[illegible]	[illegible]	[illegible]	[illegible]	[illegible]	[illegible]	[illegible]	[illegible]	[illegible]	[illegible]	[illegible]	[illegible]	[illegible]		[illegible]	[illegible]	[illegible]	[illegible]	[illegible]
	[illegible]	[illegible]	[illegible]	[illegible]	[illegible]	[illegible]	[illegible]	[illegible]	[illegible]	[illegible]	[illegible]	[illegible]	[illegible]	[illegible]	[illegible]	[illegible]	[illegible]	[illegible]	[illegible]
13	[illegible]	[illegible]	[illegible]	[illegible]	[illegible]	[illegible]	[illegible]	[illegible]	[illegible]	[illegible]	[illegible]	[illegible]	[illegible]	[illegible]	[illegible]	[illegible]	[illegible]	[illegible]	[illegible]
	[illegible]	[illegible]	[illegible]	[illegible]	[illegible]	[illegible]	[illegible]	[illegible]	[illegible]	[illegible]	[illegible]	[illegible]	[illegible]		[illegible]	[illegible]	[illegible]	[illegible]	[illegible]

Reference Number.	Diameter.	
	Head.	Tube.
	A.	B.
	Ins.	Ins.
1	36	3
2	38	3
3	40	3
4	42	3
5	44	3
6	46	3
7	48	3
8	50	3
9	52	3
10	54	3
11	56	3
12	58	3
13	60	3

extra ¼ or ½ inch might be added if practicable. If a large manhole, such as the standard 11 x 15 inches, be employed, it will be in a larger boiler with a thicker head, say ½ inch or more in thickness, the outer radius of the curve will be greater and may vary from 1½ to 2 inches, depending upon the dies used, so that it is altogether probable that no less distance could be safely made than 2½ inches instead of 1¾, as shown in Fig. 225.

Fig. 227.

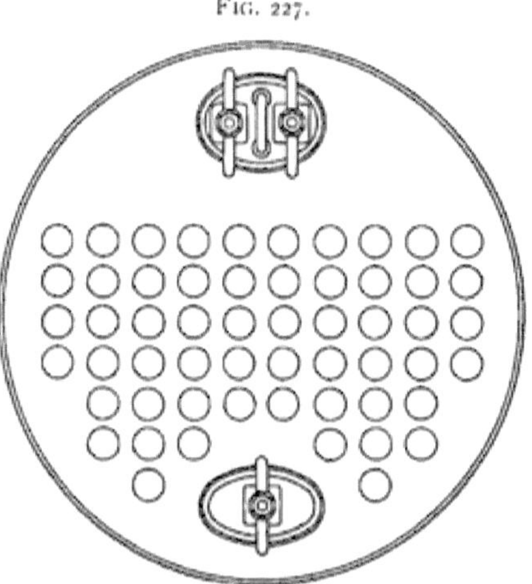

The length of tube is governed somewhat by its diameter, and for the three sizes in common use, viz., 3, 3½, and 4 inches, this length approximates 50 diameters, if to be used in connection with bituminous coal. For anthracite coal the length may be increased to 60 diameters; and these two proportions, with slight variations, represent good modern practice. If a forced draft be employed a longer tube can be used than is ordinarily the case with natural draft.

Fig. 228.

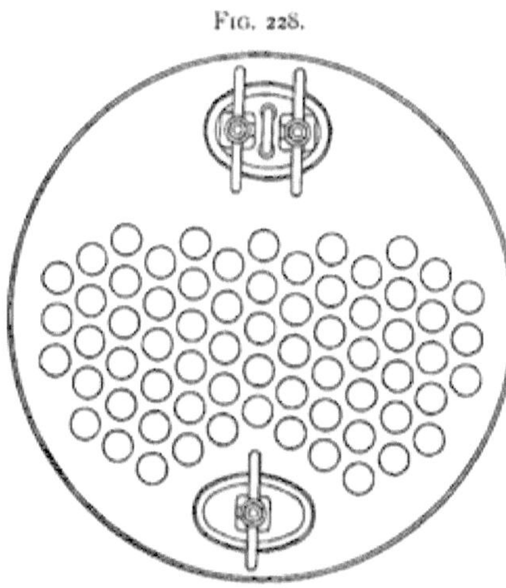

The vertical arrangement of the water-tubes, whether directly over each other, as in Fig. 227, or placed zigzag, as in Fig. 228, is now practically settled in favor of the former, though for a time the latter was much employed. The zigzag spacing probably originated in a belief that an intercepted current would enable the water to take up more heat on its way upward than if it proceeded directly to the water-surface, as would be the case when the tubes are placed in vertical rows.

TABLE XLI.

3½-INCH TUBULAR BOILERS.

FIG. 229.

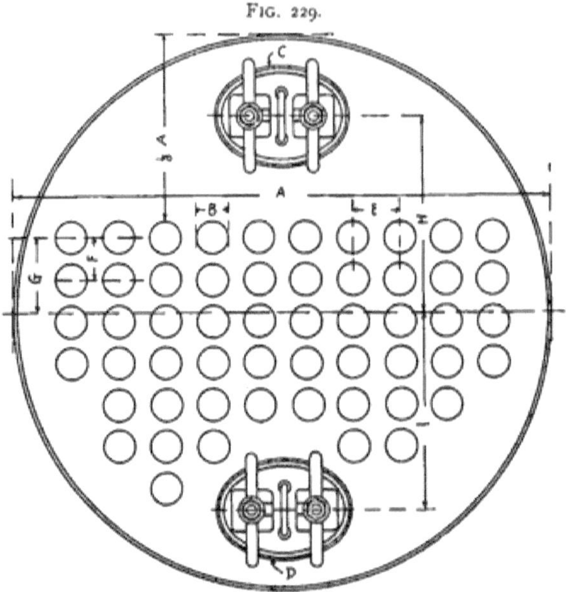

Reference Number.	Diameter.		Manhole.		Tubes Centre to Centre.		Centre of Boiler to First Row of Tubes.	Centre of Boiler to Centre of Manholes.		Number of Tubes.	Length of Boiler.
	Head.	Tube.	Upper.	Lower.	Hori-zontal.	Verti-cal.		Upper.	Lower.		
	A.	B.	C.	D.	E.	F.	G.	H.	I.		
	Ins.	Ins.	Ins.	Ins.	Ins.	Ins.	Ins.	Ins.	Ins.		Feet.
14	48	3½	11 x 15	9 x 14½	5	4½	6¼	15¼	16½	34	12 to 14
15	50	3½	11 x 15	9 x 14½	5	4½	6½	16½	17½	38	12 to 14
16	52	3½	11 x 15	9 x 14½	5	4½	7¼	17½	18½	46	12 to 14
17	54	3½	11 x 15	9 x 14½	5	4½	7¼	18½	19½	47	12 to 14
18	56	3½	11 x 15	11 x 15	5	4½	7½	19½	19½	50	12 to 14
19	58	3½	11 x 15	11 x 15	5¼	4½	7¾	20½	20½	52	12 to 14
20	60	3½	11 x 15	11 x 15	5¼	4½	8¼	21½	21½	56	14 to 16
21	62	3½	11 x 15	11 x 15	5¼	4½	8½	22½	22½	60	14 to 16
22	64	3½	11 x 15	11 x 15	5½	4½	9	23½	23½	64	14 to 16
23	66	3½	11 x 15	11 x 15	5½	4½	9¼	24½	24½	70	14 to 16

Reference Number.	Boiler-Shell.		IZE OF GRATE.		Commercial Horse-Power.		
	Diameter.	Length.	Length.		15 Feet.	12 Feet.	10 Feet.
			Calculated Feet.	Proposed Feet and Inches.			
	Inches.	Feet.					
	48	12	3.45′ @ 7′	3′ 6″	30.95	38.68	46.42
11	48	14	3.94′ @ 8′	4′ 0″	36.11	45.13	54.16
	48	16	4.43′ @ 9′	4′ 6″	41.26	51.58	61.89
	50	12	3.69′ @ 7′	3′ 9″	34.13	42.66	51.19
15	50	14	4.22′ @ 8′	4′ 3″	39.81	49.77	59.72
	50	16	4.75′ @ 9′	4′ 9″	45.50	56.88	68.25
	52	12	4.32′ @ 7′	4′ 6″	40.24	50.31	60.37
16	52	14	4.93′ @ 8′	5′ 0″	46.95	58.69	70.43
	52	16	5.55′ @ 9′	5′ 6″	53.66	67.08	80.49
	54	12	4.25′ @ 7′	4′ 3″	41.23	51.53	61.84
17	54	14	4.85′ @ 8′	5′ 0″	48.10	60.12	72.15
	54	16	5.46′ @ 9′	5′ 6″	54.97	68.71	82.45
	56	12	4.35′ @ 7′	4′ 6″	43.68	54.60	65.52
18	56	14	4.97′ @ 8′	5′ 0″	50.96	63.70	76.44
	56	16	5.59′ @ 9′	5′ 6″	58.24	72.80	87.36
	58	12	4.38′ @ 7′	4′ 6″	45.39	56.74	68.09
19	58	14	5.00′ @ 8′	5′ 0″	52.96	66.20	79.44
	58	16	5.63′ @ 9′	5′ 9″	60.52	75.66	90.79
	60	12	4.55′ @ 7′	4′ 6″	48.57	60.72	72.86
20	60	14	5.20′ @ 8′	5′ 3″	56.67	70.84	85.00
	60	16	5.85′ @ 9′	6′ 0″	64.78	80.96	97.15
	62	12	4.71′ @ 7′	4′ 9″	51.76	64.70	77.64
21	62	14	5.38′ @ 8′	5′ 6″	60.39	75.48	90.58
	62	16	6.06′ @ 9′	6′ 0″	69.01	86.27	103.52
	64	12	4.86′ @ 7′	5′ 0″	54.94	68.67	82.41
22	64	14	5.57′ @ 8′	5′ 6″	64.10	80.12	96.14
	64	16	6.20′ @ 9′	6′ 3″	73.26	91.57	109.88
	66	12	5.17′ @ 7′	5′ 3″	59.59	74.49	89.39
23	66	14	5.91′ @ 8′	6′ 0″	69.52	86.91	104.29
	66	16	6.64′ @ 9′	6′ 9″	79.46	99.32	119.18

TABLE XLII.

HORIZONTAL TUBULAR BOILERS, FITTED WITH 3½-INCH TUBES.

Reference Number	Boiler Shell		Number of 3½-inch Tubes	Internal Area of Whole Number of Tubes	Heating Surface			Grate Area						Size of Grate			Commercial Horse Power		
	Diameter	Length			Tubes	⅔ Shell	Total	Proportional to Tube Area			Ratio of Grate Surface to Total Heating Surface when Grate Area is 7, 8, and 9 times that of the Tube Area			Width in Feet	Length		[illegible] Feet	[illegible] Feet	[illegible] Feet
								7 Times	8 Times	9 Times	7 Times	8 Times	9 Times		Calculated Feet	Proposed Feet and Inches			
	Inches	Feet		Sq. Feet	Sq. Feet	Sq. Feet	Sq. Feet	Sq. Feet	Sq. Feet	Sq. Feet									
	[illegible]	[illegible]	[illegible]	[illegible]	[illegible]	[illegible]	[illegible]	[illegible]	[illegible]	[illegible]	[illegible]	[illegible]	[illegible]	[illegible]	[illegible]	[illegible]	[illegible]	[illegible]	[illegible]
14	[illegible]	[illegible]	[illegible]	[illegible]	[illegible]	[illegible]	[illegible]	[illegible]	[illegible]	[illegible]	[illegible]	[illegible]	[illegible]	[illegible]	[illegible]	[illegible]	[illegible]	[illegible]	[illegible]
	[illegible]	[illegible]	[illegible]	[illegible]	[illegible]	[illegible]	[illegible]	[illegible]	[illegible]	[illegible]	[illegible]	[illegible]	[illegible]	[illegible]	[illegible]	[illegible]	[illegible]	[illegible]	[illegible]
	[illegible]	[illegible]	[illegible]	[illegible]	[illegible]	[illegible]	[illegible]	[illegible]	[illegible]	[illegible]	[illegible]	[illegible]	[illegible]	[illegible]	[illegible]	[illegible]	[illegible]	[illegible]	[illegible]
15	[illegible]	[illegible]	[illegible]	[illegible]	[illegible]	[illegible]	[illegible]	[illegible]	[illegible]	[illegible]	[illegible]	[illegible]	[illegible]	[illegible]	[illegible]	[illegible]	[illegible]	[illegible]	[illegible]
	[illegible]	[illegible]	[illegible]	[illegible]	[illegible]	[illegible]	[illegible]	[illegible]	[illegible]	[illegible]	[illegible]	[illegible]	[illegible]	[illegible]	[illegible]	[illegible]	[illegible]	[illegible]	[illegible]
	[illegible]	[illegible]	[illegible]	[illegible]	[illegible]	[illegible]	[illegible]	[illegible]	[illegible]	[illegible]	[illegible]	[illegible]	[illegible]	[illegible]	[illegible]	[illegible]	[illegible]	[illegible]	[illegible]
16	[illegible]	[illegible]	[illegible]	[illegible]	[illegible]	[illegible]	[illegible]	[illegible]	[illegible]	[illegible]	[illegible]	[illegible]	[illegible]	[illegible]	[illegible]	[illegible]	[illegible]	[illegible]	[illegible]
	[illegible]	[illegible]	[illegible]	[illegible]	[illegible]	[illegible]	[illegible]	[illegible]	[illegible]	[illegible]	[illegible]	[illegible]	[illegible]	[illegible]	[illegible]	[illegible]	[illegible]	[illegible]	[illegible]
	[illegible]	[illegible]	[illegible]	[illegible]	[illegible]	[illegible]	[illegible]	[illegible]	[illegible]	[illegible]	[illegible]	[illegible]	[illegible]	[illegible]	[illegible]	[illegible]	[illegible]	[illegible]	[illegible]
17	[illegible]	[illegible]	[illegible]	[illegible]	[illegible]	[illegible]	[illegible]	[illegible]	[illegible]	[illegible]	[illegible]	[illegible]	[illegible]	[illegible]	[illegible]	[illegible]	[illegible]	[illegible]	[illegible]
	[illegible]	[illegible]	[illegible]	[illegible]	[illegible]	[illegible]	[illegible]	[illegible]	[illegible]	[illegible]	[illegible]	[illegible]	[illegible]	[illegible]	[illegible]	[illegible]	[illegible]	[illegible]	[illegible]
	[illegible]	[illegible]	[illegible]	[illegible]	[illegible]	[illegible]	[illegible]	[illegible]	[illegible]	[illegible]	[illegible]	[illegible]	[illegible]	[illegible]	[illegible]	[illegible]	[illegible]	[illegible]	[illegible]
18	[illegible]	[illegible]	[illegible]	[illegible]	[illegible]	[illegible]	[illegible]	[illegible]	[illegible]	[illegible]	[illegible]	[illegible]	[illegible]	[illegible]	[illegible]	[illegible]	[illegible]	[illegible]	[illegible]
	[illegible]	[illegible]	[illegible]	[illegible]	[illegible]	[illegible]	[illegible]	[illegible]	[illegible]	[illegible]	[illegible]	[illegible]	[illegible]	[illegible]	[illegible]	[illegible]	[illegible]	[illegible]	[illegible]
	[illegible]	[illegible]	[illegible]	[illegible]	[illegible]	[illegible]	[illegible]	[illegible]	[illegible]	[illegible]	[illegible]	[illegible]	[illegible]	[illegible]	[illegible]	[illegible]	[illegible]	[illegible]	[illegible]
19	[illegible]	[illegible]	[illegible]	[illegible]	[illegible]	[illegible]	[illegible]	[illegible]	[illegible]	[illegible]	[illegible]	[illegible]	[illegible]	[illegible]	[illegible]	[illegible]	[illegible]	[illegible]	[illegible]
	[illegible]	[illegible]	[illegible]	[illegible]	[illegible]	[illegible]	[illegible]	[illegible]	[illegible]	[illegible]	[illegible]	[illegible]	[illegible]	[illegible]	[illegible]	[illegible]	[illegible]	[illegible]	[illegible]
	[illegible]	[illegible]	[illegible]	[illegible]	[illegible]	[illegible]	[illegible]	[illegible]	[illegible]	[illegible]	[illegible]	[illegible]	[illegible]	[illegible]	[illegible]	[illegible]	[illegible]	[illegible]	[illegible]
20	[illegible]	[illegible]	[illegible]	[illegible]	[illegible]	[illegible]	[illegible]	[illegible]	[illegible]	[illegible]	[illegible]	[illegible]	[illegible]	[illegible]	[illegible]	[illegible]	[illegible]	[illegible]	[illegible]
	[illegible]	[illegible]	[illegible]	[illegible]	[illegible]	[illegible]	[illegible]	[illegible]	[illegible]	[illegible]	[illegible]	[illegible]	[illegible]	[illegible]	[illegible]	[illegible]	[illegible]	[illegible]	[illegible]
	[illegible]	[illegible]	[illegible]	[illegible]	[illegible]	[illegible]	[illegible]	[illegible]	[illegible]	[illegible]	[illegible]	[illegible]	[illegible]	[illegible]	[illegible]	[illegible]	[illegible]	[illegible]	[illegible]
21	[illegible]	[illegible]	[illegible]	[illegible]	[illegible]	[illegible]	[illegible]	[illegible]	[illegible]	[illegible]	[illegible]	[illegible]	[illegible]	[illegible]	[illegible]	[illegible]	[illegible]	[illegible]	[illegible]
	[illegible]	[illegible]	[illegible]	[illegible]	[illegible]	[illegible]	[illegible]	[illegible]	[illegible]	[illegible]	[illegible]	[illegible]	[illegible]	[illegible]	[illegible]	[illegible]	[illegible]	[illegible]	[illegible]
	[illegible]	[illegible]	[illegible]	[illegible]	[illegible]	[illegible]	[illegible]	[illegible]	[illegible]	[illegible]	[illegible]	[illegible]	[illegible]	[illegible]	[illegible]	[illegible]	[illegible]	[illegible]	[illegible]
22	[illegible]	[illegible]	[illegible]	[illegible]	[illegible]	[illegible]	[illegible]	[illegible]	[illegible]	[illegible]	[illegible]	[illegible]	[illegible]	[illegible]	[illegible]	[illegible]	[illegible]	[illegible]	[illegible]
	[illegible]	[illegible]	[illegible]	[illegible]	[illegible]	[illegible]	[illegible]	[illegible]	[illegible]	[illegible]	[illegible]	[illegible]	[illegible]	[illegible]	[illegible]	[illegible]	[illegible]	[illegible]	[illegible]
	[illegible]	[illegible]	[illegible]	[illegible]	[illegible]	[illegible]	[illegible]	[illegible]	[illegible]	[illegible]	[illegible]	[illegible]	[illegible]	[illegible]	[illegible]	[illegible]	[illegible]	[illegible]	[illegible]
23	[illegible]	[illegible]	[illegible]	[illegible]	[illegible]	[illegible]	[illegible]	[illegible]	[illegible]	[illegible]	[illegible]	[illegible]	[illegible]	[illegible]	[illegible]	[illegible]	[illegible]	[illegible]	[illegible]
	[illegible]	[illegible]	[illegible]	[illegible]	[illegible]	[illegible]	[illegible]	[illegible]	[illegible]	[illegible]	[illegible]	[illegible]	[illegible]	[illegible]	[illegible]	[illegible]	[illegible]	[illegible]	[illegible]

Ratio of Tube Area to Grate Area.—This area is often referred to in essays on steam boilers as if the tube area in a boiler was to be fixed after the grate surface had been determined upon; as a matter of fact, this is not the method by which this ratio is obtained. For a given diameter of boiler a certain diameter of tube is decided upon. The number of tubes to be placed in the head will be the greatest that the upper limit of water-space and distance from boiler-shell will allow, provision being made for the proper manhole or handhole under the tubes.

It was found at an early date that boilers built with a certain ratio of tube area to grate surface did better work than others of different proportions, and that with a certain fixed relation of tube area to grate surface the evaporative efficiency of a boiler could be approximately forecast.

If a tubular boiler has been designed upon extent of heating surface alone, the tube area is, of course, known in advance. The proportionate grate area for such tube area will depend somewhat upon the kind of fuel to be used and upon its rate of combustion. For anthracite coal the grate area may be 8 times the sectional area of tubes for ordinary draft. For bituminous coal the grate area may be 7 times, or under exceptional conditions as little as 6 times, the sectional area of tubes. If the fuel be of indifferent quality or the draft be sluggish, the proportion may be 9 of grate to 1 of tube area. A very good average is 8 of grate to 1 of tube area.

Heating Surface and Grate Area.—The total heating surface of a boiler, as commonly determined, is two-thirds the superficial area of the shell, to which is added the superficial area of all the tubes. If less than two-thirds the circumference of the boiler is exposed to the action of the heated gases, then the shell heating surface will be in whatever proportion of circumference of shell is thus exposed multiplied by the length of the boiler. All heating surfaces are expressed in square feet.

The proportion of heating surface to grate area will vary according to the kind of boiler. A two-flue boiler 48 inches by 26 feet, for example, would have a ratio probably of 14 feet of heating surface to 1 of grate; a 60-inch by 24-feet boiler with 18 6-inch tubes would have a ratio of heating surface to grate area of about 32 to 1; a 72-inch by 20-feet boiler with 68 4-inch tubes would have a ratio of heating surface to grate area of about 54 to 1. All the above are taken from actual examples. Neglecting the first, we have a proportion varying from 32 to 1 to 54 to 1, and between these two limits most of the tubular boilers in use are likely to be found.

The coal burnt per hour per square foot of grate in each of the above boilers was: For the two-flue boilers, 14.5 pounds; for the 6-inch tubular boiler, 41.38 pounds; for the 4-inch tubular boiler, 35 pounds. Bituminous coal was used in each of the above.

The coal burnt per hour per square foot of heating surface was: For the two-flue boiler, 0.934 pounds; for the 6-inch tubular boilers, 1.277 pounds; for the 4-inch tubular boiler, 0.650 pounds.

The temperature of the escaping gases was: For the two-flue boiler, 735° Fahr.; for the 6-inch tubular boiler, 542° Fahr.; for the 4-inch tubular boiler, 585° Fahr. Of these, the temperature of the first is too high; the second approaches closely that temperature at which the chimney gives its best working results; the third is rather too high for boilers of this type for best economy.

The heating surface in square feet required to develop 1 horse-power (34½ pounds of water evaporated per hour from and at 212° Fahr.) was: For the two-flue boiler, 3.34; for the 6-inch tubular boiler, 4.74; for the 4-inch tubular boiler, 6.61.

For small tubular boilers, the ratio of heating surface to grate area will be found to vary from 25 to 1 up to 40 to 1. All things considered, a ratio of from 30 to 35 square feet of heating surface to 1 square foot of grate surface will be found to give good results. Much will depend upon the intensity of the draught, because this determines how many pounds of coal shall be burned per square foot of grate in a given time. The higher the rate of combustion the greater proportionally may be the extent or ratio of heating surface. Excess of heating surface may have the effect of reducing the temperature of gases to a point so low that the chimney will not give sufficient draft to get the most economical results of the furnace.

The ratio of tube area and heating surface to grate surface in the case of a 48-inch boiler fitted with 46 3-inch tubes: if the proportion of the tube area to that of the grate be 7 to 1, the grate area would contain 13.58 square feet; if 8 to 1, 15.52 square feet; and if 9 to 1, 17.46 square feet. As 3-inch tubular boilers vary in length, we will select three,—viz., 10, 12, and 14 feet. There must be a ratio of grate to total heating surface most economical for a given draft. For example, the ratio of heating surface to grate surface when the latter is 7 times that of the tube area, would be 32.14 to 1 of grate for a 10-foot boiler; for a 12-foot boiler this ratio is increased to 38.65 to 1; and for a 14-foot boiler to 45.03 to 1.

If the grate area be 8 times that of the tube area, then the ratio of heating surface to grate surface would be: For a boiler 10 feet long, 28.12 to 1 of grate; if 12 feet long the ratio would be 33.82 to 1; and if 14 feet long the ratio would be 39.40 to 1.

If the ratio of grate area to that of tube area be 9 to 1, the ratio of heating surface to grate area would be: For the 10-foot boiler, 25 to 1; for the 12-foot boiler, 30.06 to 1; and for the 14-foot boiler, 35.03 to 1.

An examination of these three lengths and three proportions of tube area to grate area show that good average results will be secured if the boiler is 12 feet long and the ratio of grate to tube area 8 to 1.

A 3½-inch tubular boiler, 60 inches diameter, for example, will show nearly the same results,—that is: If the ratio of grate to tube area is 7 to 1, the best length approximates 12 feet; if 8 to 1, the best length approximates 14 feet; and if 9 to 1, the best length approximates 16 feet. Each of these is a little less than 35 feet of heating surface to 1 of grate. In case the ratio of grate be 8 to 1 of tube area, the length might be 16 feet, provided the draught is good.

Referring now to the 4-inch tubular boilers: If a diameter of 66 inches be selected, this contains 56 4-inch tubes. If the proportion of the grate area to tube area be 7 to 1, the greatest length approximates 16 feet, and if 8 to 1, the length might be increased to 18 feet, a ratio of 36.08 to 1, and if the ratio be 9 to 1, a length of 20 feet would be required to secure a ratio of 35.91 square feet of heating surface to 1 of grate surface. By an examination of the accompanying table, in which the ratio of total heating surface to grate surface is approximately 35 to 1, it will be seen that the length of tube approximates very closely 18 feet, which has been found in practice to be a very satisfactory length for boilers from 5 to 6 feet in diameter fitted with 4-inch tubes.

Tables.—The leading properties of horizontal tubular boilers from 36 to 72 inches diameter are conveniently arranged for reference in the annexed tables. The strength of shells for corresponding and larger diameters, together with the several kinds of riveting best suited to the strength of plates used, has already been given in Tables XXIX., XXX., and XXXI. The number and arrangement of tubes as given in Tables XXXIX. to XLIV. accords with what has been said in this chapter regarding horizontal spacing and the distance of top of upper row of tubes from the bottom of the boilers. Regarding the latter, it may be said that two-thirds is the highest permissible limit; some designers place it at three-fifths instead, in order to get a lower water level. Between these two limits lies the best modern practice. The tube area, total heating surface, and grate area are all expressed in square feet. The ratio of tube area to grate surface is given as 7, 8, and 9, these being the ordinary ratios in practice. The width of grate is in all cases that of the diameter of the boiler; the length is calculated for the fixed ratios of 7, 8, and 9 feet. As these present fractional dimensions, a proposed length of grate-bars for common use is appended.

Retarders for Fire-Tubes.—A retarder is a thin strip of sheet iron fitting loosely in and running the whole length of a tube; this iron is twisted one or more spiral convolutions in its whole length. See Fig. 231.

The insertion of retarders in fire-tubes or flues is but rarely practised in this country. The selection of the word "retarder" is not a happy one, though, of course, it does obstruct the flow of gas through a tube, as any obstruction in a tube will retard the flow of the gases through

TABLE XLIII.

4-INCH TUBULAR BOILERS.

FIG. 230.

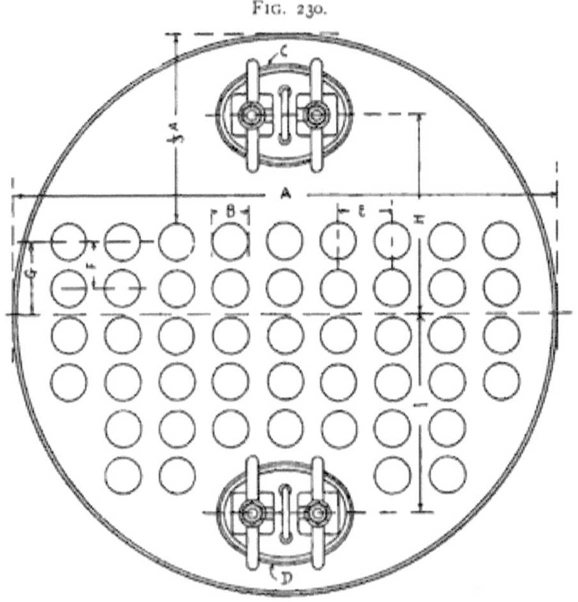

Reference Number.	Diameter.		Manhole.		Tubes Centre to Centre.		Centre of Boiler to First Row of Tubes.	Centre of Boiler to Centre of Manholes.		Number of Tubes.	Length of Boiler.
	Head.	Tube.	Upper.	Lower.	Horizontal.	Vertical.		Upper.	Lower.		
	A.	B.	C.	D.	E.	F.	G.	H.	I.		
	Ins.	Ins.	Ins.	Ins.	Ins.	Ins.	Ins.	Ins.	Ins.		Feet.
24	54	4	11 x 15	11 x 15	6	5	7	18½	18½	36	16 to 20
25	56	4	11 x 15	11 x 15	6	5	7¼	19½	19½	38	16 to 20
26	58	4	11 x 15	11 x 15	6	5	7¾	20½	20½	40	16 to 20
27	60	4	11 x 15	11 x 15	6	5	8	21½	21½	47	16 to 20
28	62	4	11 x 15	11 x 15	6	5	8¼	22½	22½	49	16 to 20
29	64	4	11 x 15	11 x 15	6	5	8¾	23½	23½	51	16 to 20
30	66	4	11 x 15	11 x 15	6	5	9	24½	24½	56	16 to 20
31	68	4	11 x 15	11 x 15	6	5	9½	25½	25½	62	16 to 20
32	70	4	11 x 15	11 x 15	6	5	9¾	26½	26½	64	16 to 20
33	72	4	11 x 15	11 x 15	6	5	10¼	27½	27½	74	16 to 20

Reference Number.	Boiler-Shell.		ze of Grate.		Commercial Horse-Power.		
	Diameter.	Length.	Length.		15 Feet.	12 Feet.	10 Feet.
			Calculated Feet.	Proposed Feet and Inches.			
	Inches.	Feet.					
	54	16	4.28′ @ 7′	4′ 3″	49.28	61.56	73.88
24	54	18	4.89′ @ 8′	5′ 0″	55.41	69.26	83.11
	54	20	5.48′ @ 9′	5′ 6″	61.56	76.95	92.34
	56	16	4.35′ @ 7′	4′ 6″	51.83	64.78	77.74
25	56	18	4.97′ @ 8′	5′ 0″	58.30	72.89	87.46
	56	20	5.59′ @ 9′	5′ 6″	64.78	80.98	97.17
	58	16	4.42′ @ 7′	4′ 6″	54.39	67.99	81.58
26	58	18	5.05′ @ 8′	5′ 0″	61.19	76.49	91.78
	58	20	5.68′ @ 9′	5′ 9″	67.99	84.98	101.98
	60	16	5.01′ @ 7′	5′ 0″	62.54	78.17	93.81
27	60	18	5.73′ @ 8′	5′ 9″	70.35	87.94	105.53
	60	20	6.44′ @ 9′	6′ 6″	78.17	97.72	117.26
	62	16	5.05′ @ 7′	5′ 0″	65.11	81.38	97.67
28	62	18	5.79′ @ 8′	5′ 9″	73.25	91.56	109.85
	62	20	6.51′ @ 9′	6′ 6″	81.39	101.74	122.09
	64	16	5.11′ @ 7′	5′ 3″	67.68	84.60	101.52
29	64	18	5.84′ @ 8′	6′ 0″	76.15	95.17	114.21
	64	20	6.57′ @ 9′	6′ 6″	84.60	105.75	126.89
	66	16	5.43′ @ 7′	5′ 6″	73.60	92.00	110.40
30	66	18	6.21′ @ 8′	6′ 3″	82.80	103.50	124.20
	66	20	6.99′ @ 9′	7′ 0″	92.00	115.00	138.00
	68	16	5.84′ @ 7′	6′ 0″	80.63	100.79	120.95
31	68	18	6.67′ @ 8′	6′ 9″	90.71	113.39	136.07
	68	20	7.51′ @ 9′	7′ 6″	100.79	125.99	151.19
	70	16	5.86′ @ 7′	6′ 0″	83.21	104.01	124.81
32	70	18	6.70′ @ 8′	6′ 9″	93.61	117.01	140.41
	70	20	7.53′ @ 9′	7′ 6″	104.01	130.01	156.02
	72	16	6.58′ @ 7′	6′ 6″	94.71	119.22	142.06
33	72	18	7.52′ @ 8′	7′ 6″	106.55	133.18	159.82
	72	20	8.46′ @ 9′	8′ 6″	118.38	147.98	177.58

TABLE XLIV.

HORIZONTAL TUBULAR BOILERS, FITTED WITH 4-INCH TUBES.

Reference Number.	Boiler-Shell.		Number of 4-inch Tubes.	Internal Area of Whole Number of Tubes.	Heating Surface.			Grate Area.						Size of Grate.			Commercial Horse-Power.		
								Proportional to Tube Area.			Ratio of Grate Surface to Total Heating Surface when Grate Area is 7, 8, and 9 times that of the Tube Area.				Length.				
	Diameter.	Length.			Tubes.	Shell.	Total.	7 Times.	8 Times.	9 Times.	7 Times.	8 Times.	9 Times.	Width in Feet.	Calculated Feet.	Proposed Feet and Inches.	15 Feet.	14 Feet.	10 Feet.
	Inches.	Feet.		Sq. Feet.	Sq. Feet.	Sq. Feet.	Sq. Feet.	Sq. Feet.	Sq. Feet.	Sq. Feet.									
[illegible]	[illegible]	[illegible]	[illegible]	[illegible]	[illegible]	[illegible]	[illegible]	[illegible]	[illegible]	[illegible]	[illegible]	[illegible]	[illegible]	[illegible]	[illegible]	[illegible]	[illegible]	[illegible]	[illegible]

it; but the object of a retarder is a wholly different one, and has nothing to do with the flow of the gases.

Retarders* are intended to increase the amount of heat transmitted to the tube surface from the hot gases, and it does it in two ways: first, by a mixing action upon the gas in the tubes. The friction upon the surface of the retarder aids in stirring up the gas in its passage through the tube and in mixing the hot gas at the centre with the cold film next the surface of the tube. Also in every horizontal tube there is a tendency for the gases to be cooler in the upper part of the tube and hotter in the lower, for the upper part of the tube extracts heat far more readily from the gases than the lower half. The twist of the retarder has the effect of repeatedly turning over the gas in the tube as it flows along.

In the second place, the retarder acts by direct radiation of heat to the tube surface. While this action may not be apparent at first sight, it is of such importance that it should be clearly understood. To this end the fact should first be realized that the temperature of the tube surface exposed to the fire in any steam boiler is practically the same as that of water in contact with it, no matter what the temperature of the gas on the other side, supposing, of course, the tube surface to be clean. The reason is that water absorbs heat many times as rapidly as gas. Now suppose we place in the tube any solid body, of any shape whatever. Manifestly, as it is surrounded and bathed on all sides by gases at a temperature of say 1000° Fahr., it will, if it loses no heat, soon become of the same temperature as the hot gases. Suppose the surface of the tube is of a temperature of 300° Fahr., the hot body at the centre of the tube will energetically radiate heat to the walls of the tube and will materially increase the amount of heat transmitted to the water.

FIG. 231.

WHITHAM'S EXPERIMENTS.—A trial of retarders on a 100 horse-power horizontal tubular boiler was conducted by J. M. Whitham, to ascertain under what conditions, if any, they would add to the efficiency of the boiler. The boiler was 60 inches diameter by 20 feet long, fitted with 44 tubes 4 inches in diameter. The water-heating surface was 1137 square feet. Ratio of heating surface to grate surface was 42.6 to 1.

* Jay M. Whitham and C. W. Baker, "Transactions American Society of Mechanical Engineers," vol. xvii.

The conclusions reached were:

Retarders in fire-tubes of a boiler interpose a resistance varying with the rate of combustion.

Retarders result in reducing the temperature of the waste gases and in increasing the effectiveness of the heating surface of the tubes.

Retarders show an economic advantage when the boiler is pushed, varying in the tests from 3 to 18 per cent.

Retarders should not be used when boilers are used very gently and when the stack draught is small.

It is probable that retarders can be used with advantage in plants using a fan- or steam-blast under the fire, or a strong natural or induced chimney draught, when burning either anthracite or bituminous coals.

Fig. 232.

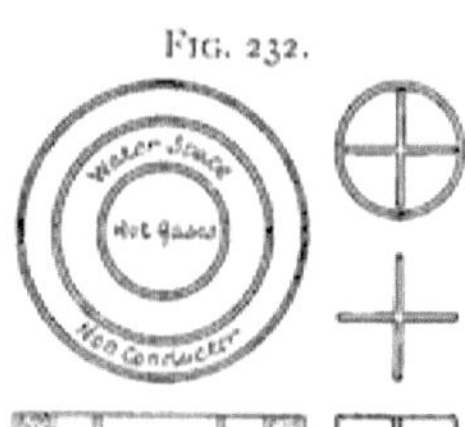

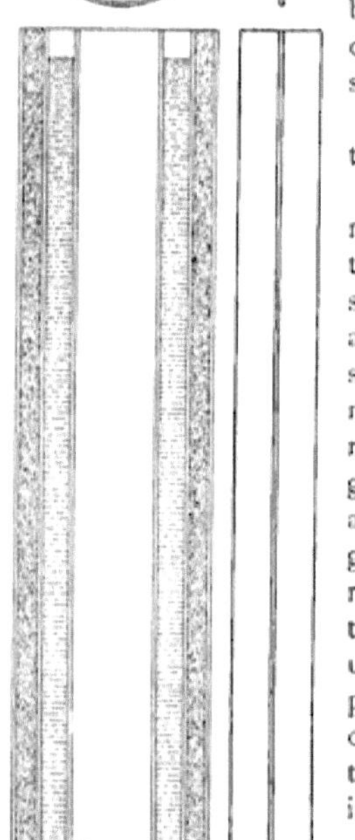

Retarders may often prove to be, for an existing plant, as economical as an economizer, and will not, in general, interpose as much resistance to the draught.

The economic results obtained on the boiler tested are ideal, showing that it was clean, the coal good in quality, and the firing skilful.

With retarders, the tubes are more effectively cleaned than without their use.

Baker's Experiments.—These were made to test the efficiency of radiators with the apparatus shown in the accompanying sketch, Fig. 232. It represents a section of a single tube of a vertical boiler. The water space surrounding it is well protected by non-conductors. Through the tube a current of hot gas is caused to flow from a lamp, gas-jet, or other suitable source, and the amount of heat transmitted to the water in a given time is measured. The test is then repeated under identical conditions, except that a radiator of the form shown at the upper part and at the side of Fig. 232 is placed in the tube. The increased amount of heat transmitted to the water is taken as the amount due to the radiation from the internal piece.

Experiments with the apparatus showed the following general results:

That the percentage of increase in heat transmitted due to radiation increases with increase in temperature of gases passing through the tube.

That the percentage of increase in heat transmitted due to radiation is larger in vertical tubes than in horizontal, on account of the fact that a given area of heating surface in a horizontal tube absorbs heat faster by direct contact with the gases than the same area in a vertical tube.

Experiments on actual boilers indicate that either device is most useful on boilers with short tubes of not too small diameter and with an abundance of draught.

Fig. 233.

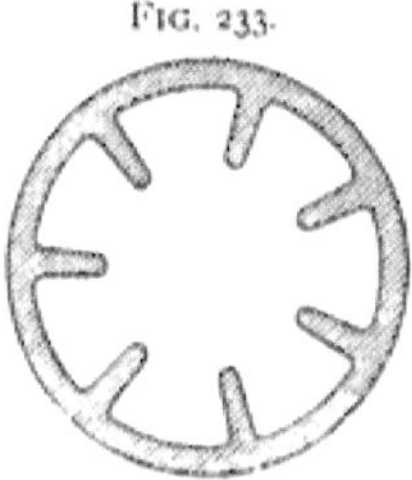

With either device the tube surface must be kept clean, otherwise the increased efficiency will soon disappear, as is the case with the Serve ribbed tube, an illustration of which is given in Fig. 233, when care is not taken in this respect.

The economic gain by the use of either radiators or retarders depends entirely upon the temperature at which the boiler is discharging its hot gases. It may be assumed that every 100° Fahr. reduction in the temperature of the waste gases represents from 5 to 10 per cent. saving in fuel.

In general it will not usually be found worth while to introduce either retarders or radiators in the tubes of any boiler, unless the thermometer shows its hot gases to be discharging at a temperature of over 550° Fahr.

Double-Deck Boilers.—These boilers consist of an upper and lower horizontal cylindrical shell, as shown in Fig. 234. These two

Fig. 234.

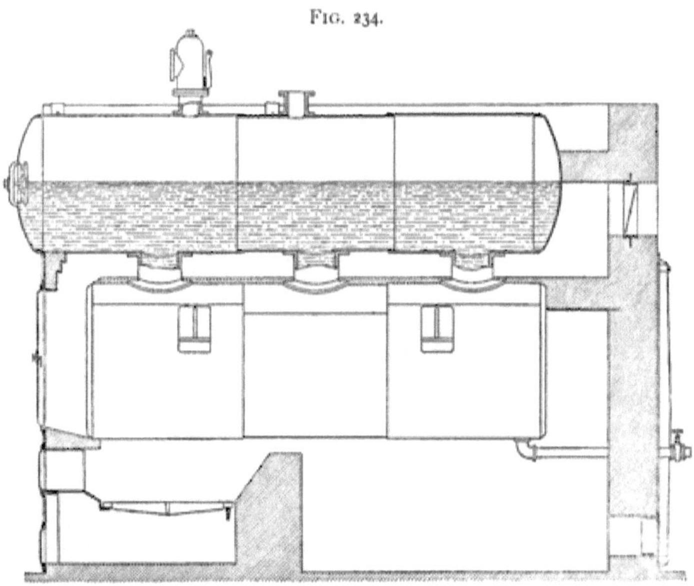

shells are connected by two or more necks; the present illustration shows three necks, which is the common practice. The lower shell is fitted with as many tubes as can be conveniently arranged in it, having due regard for the manholes necessary for internal examination and cleaning. The upper shell is fitted with no tubes, and is intended to be about half filled with water; these details are shown in the half-sectional elevation, Fig. 235. The connecting necks are usually made of flange steel, and are as large in diameter as can be conveniently adapted to the two shells: the larger the diameter the better the circulation; but however large these necks may be, the circulation in a boiler of this type is not as good as in the ordinary horizontal tubular type, in which the steam-space is included in the upper part of the boiler. The only apparent reason for constructing a boiler in this manner is that of increasing the total heating surface by carrying the fire-line nearer the top of the outer shell, as well as increasing the number of tubes in a boiler of a given diameter. The circuit of the products of combustion from the furnace is underneath and around the sides of the lower shell to the rear end, returning through the fire-tubes to the front, thence under the upper drum to the rear of the boiler; the gases then pass off to the chimney. A test of a boiler of this type, in which the lower shell was 54 inches in diameter by 12 feet in length, having 118 tubes 3 inches in diameter, the upper drum being 32 inches in diameter and of the same length, measured 1281 square feet of total heating surface. This boiler evaporated 10 pounds of water from and at 212° Fahr. per pound of coal, showing that boilers thus constructed have not sufficient advantages over the ordinary horizontal tubular boilers to pay for their extra cost.

FIG. 235.

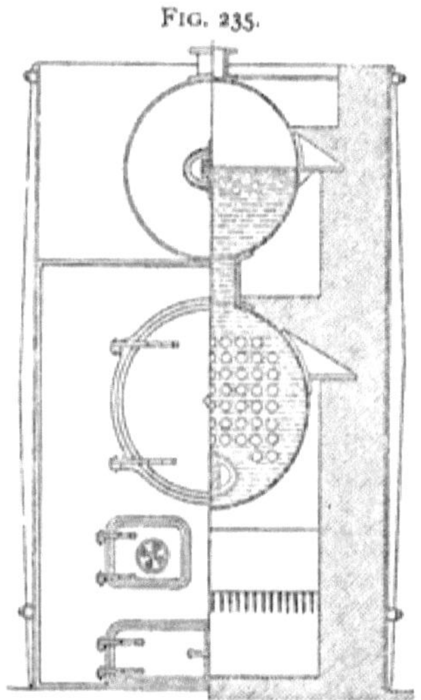

Triplex Boiler.—This boiler differs from the one previously described in having two lower shells mounted side by side, and with liberal connections from these two lower shells upward into a combined water- and steam-drum above, as shown in Figs. 236 and 237. The circulation of the water is upward through the front connections, returning downward to the shells by similar rear connections, these being arranged, as shown in the drawings, to induce and maintain a longitudinal circulation in a circuit through the upper drum and lower shells. To avoid the difficulties ordinarily met in supporting boilers of this type,

Fig. 236.

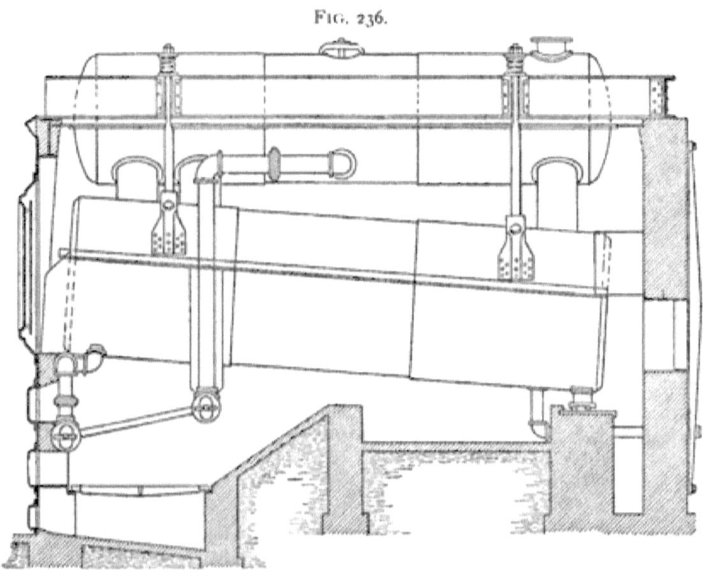

Fig. 237.

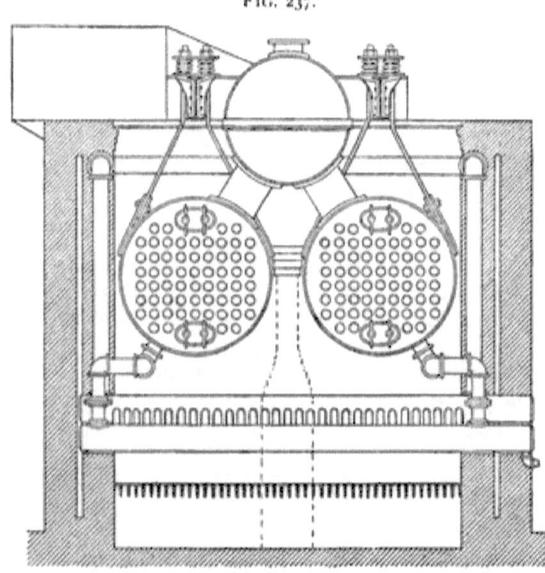

and to allow for expansion and contraction, the front ends of the tubular shells are held in a fixed position, the rear ends being free to move backward and forward. The weight of the boiler and contents is sustained at the back partly by rollers underneath, and partly by slings which suspend the weight from points overhead. The slings are equipped with springs which, while maintaining a nearly uniform strain on the slings and girders, allow by their elasticity for imperfections in workmanship and adjustment, as well as for settling or other changes which may occur; they provide also for vertical expansion and contraction of the parts by changes of temperature. The slings are so adjusted that they will support about three-fourths of the weight, leaving the other quarter to run on the rollers.

The principal dimensions of the boiler illustrated, from designs by J. T. Fanning, are here given:

Diameter of each lower shell	58 inches.
Length between heads and length of tubes	16 feet 9 inches.
Number of tubes in each shell, 4 inches outside diameter	62.
Diameter of steam-drum	48 inches.
Length of drum	16 feet.
Diameter of necks connecting shell and drums	15 inches.
Area of heating surface, one boiler	2705 square feet.
Area of grate surface, upper	41.8 square feet.
Area of grate surface, lower	49.5 square feet.
Total area of grate surface	91.3 square feet.
Ratio of heating surface to total grate surface	29.6 to 1.

The boilers were set with the Hawley down-draft furnace, described in the next chapter. These boilers were erected under a guarantee by the contractor that each boiler should develop 250 standard horse-power, and evaporate 10½ pounds of water per pound of Youghiogheny coal. The guarantees of the Hawley Down-Draft Furnace Company were that each boiler should develop 300 horse-power when the chimney-draught was 0.6 of an inch, evaporate 10.5 pounds of water per pound of combustible from and at 212° Fahr. when burning Pennsylvania (bituminous) coal, and consume 95 per cent. of the smoke.

The results of tests made by G. H. Barrus, based on dry coal, are given in the table of comparative performance of modern boilers. From this it appears that the evaporation per pound of coal from and at 212° Fahr. was 10.846 pounds, and the evaporation per pound of combustible from and at 212° Fahr. was 11.551 pounds, meeting the guarantees of both contractors in respect to economy. The boiler developed 315.2 horse-power with a chimney-draught of 0.25 of an inch water pressure for 1½ hours, meeting the conditions of contract relating to capacity,—no smoke whatever escaping from the chimney except at times of firing, or when the bed of coal was disturbed by the use of the poker; even then the quantity was exceedingly small and within the guarantee, and its color barely perceptible. Assuming that the boiler develops 250 horse-power and operates 24 hours per day, the quantity of lump coal consumed on the basis of this test would be 10 tons per day, and the quantity of slack 11.11 tons per day. At $3.65 per ton for lump coal and $2.60 per ton for slack coal, the cost of fuel for a day's run of 24 hours on each boiler when developing 250 horse-power would be $36.50 for the lump coal and $28.89 for the slack,—the difference between the two being $7.51 in favor of the slack coal.

FIG. 238.

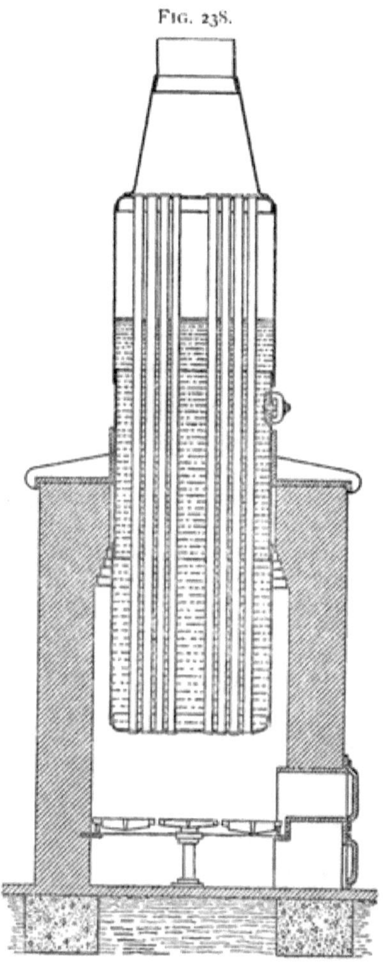

A vertical tubular boiler externally fired, as shown in Fig. 238, is not often met with. It presents some good features, however: one of which is, that an equal amount of heating surface is provided within the combustion-chamber as would have been provided and without the complications involved in the design of a fire-box. This is simply a cylindrical shell fitted with tubes of a diameter and in number suited to the diameter of the boiler. Around the outer shell of the boiler is riveted a band of iron of sufficient thickness and depth to support the boiler in place, and without bringing undue strain upon the rivets by which this band is fastened to the boiler. The furnace consists of a circular grate, as shown in the engraving, the height of the furnace being proportioned to the length of the shell, which in this case is a little less than one-half its height. A cast-iron cap covers the top of the furnace walls, through which, and upon an annular ring included in the cap, the boiler is suspended. The performance of this boiler is the same as that of horizontal tubular boilers having the same heating surface.

CHAPTER VII.

BOILER FURNACES AND SETTINGS.

BEFORE entering upon the design of a boiler furnace it is necessary to know among other things what kind of a boiler is to be used, its size, the kind of fuel to be used, and the rate of combustion. It is needless to say that many boiler plants throughout the country are developing less power than they ought because of defective designs for the furnace and poor construction.

Batteries of Boilers.—If we select the horizontal tubular boiler as an example, we shall have probably the most popular boiler in use at this time. Such boilers are practically limited as to diameter from 36 to 72 inches, in length from 10 to 20 feet, there being, of course, occasional variations outside these dimensions. For large powers the question is raised at the outset whether two or more boilers shall be set in a single furnace as a "battery" or whether each boiler shall be set singly. Both methods are practised; both have reasons for and against. Let us take the case of a steam plant requiring six boilers to do the work, with spare boilers for cleaning and repairs. If the boilers are set singly, one spare boiler will be enough, or seven boilers in all; if the boilers are set in pairs, eight boilers will be required; if three boilers are set in a single furnace, nine boilers will be required. The settings for the single boilers will be most expensive, the boilers in pairs less so, and for the three boilers in battery the cheapest. Whatever is saved in furnace walls can be applied towards the cost of the additional boilers. Two or three cylinder and two-flue boilers, which are always of small diameter relatively to tubular boilers, may preferably be set in a single furnace. So also tubular boilers up to 48 inches diameter may be set in pairs; for larger diameters the writer prefers that they be set singly.

The Size of the Grate.—The ratio of heating surface to grate area will vary with the kind of boiler: a cylinder boiler, for example, will require less grate area than one presenting a larger heating surface, if economic evaporation only be considered; but such boilers are not run along economic lines so much as that of capacity, and this necessitates a larger grate area. In the battery of cylinder boilers, Fig. 212, which represents actual practice, there are three boilers, each 3 x 36 feet. The heating surface at $\frac{2}{3}$ of the shell is 610 square feet for the three; the grate surface is 9 feet 8 inches wide by 6 feet long,—say 58 square feet,—a ratio of heating surface to grate area of 10.87 to 1. One reason for the large grate area is that non-merchantable refuse is commonly burnt at coal mines, saw-mills, etc., where cylinder boilers are mostly in use. If refuse coal is burnt, the rate of combustion is much less than is the case with a better grade of fuel, and this requires a larger grate area. One important fact must not be overlooked in this connection, shell heating surface is much more effective than tube surface. A tubular boiler having a ratio of heating surface to grate area of 32.6 to 1 would not be three times as effective as the cylinder boilers now under consideration.

If a change be made from cylinder to two-flue boilers to do the same work, there would be required three boilers 42 inches in diameter by 20 feet long, each boiler fitted with two 14-inch flues. The combined heating surface is 882 square feet, or 272 square feet more than the cylinder boilers. The grate surface for the two-flue boilers in a battery of three would be about 12 feet wide by 4 feet long,—say 48 square feet; then $882 \div 48 = 18.37$ to 1 is the ratio of heating surface to grate area, about 70 per cent. greater than in the case of cylinder boilers. The estimated steam-producing capacity of the two batteries of boilers is practically alike.

To do the same work with a horizontal tubular boiler would require one boiler say 66 inches in diameter by 16 feet long, with 56 4-inch tubes, or a tube area of 4.27 square feet, the total heating surface of which would be 1104 square feet. Taking into account the inferior quality of fuel, as was the case with the cylinder boiler, the ratio of grate area to tube area may be 9 to 1; that is, the tube area being 4.27 square feet, the grate area would be 38.43 square feet. If the length of the grate be fixed at 6 feet as a maximum, we have $38.43 \div 6 = 6.4$ feet as its width,—say 6 feet 5 inches. The ratio of total heating surface to grate area would be $1104 \div 38.43 = 28.74$ to 1. With a better quality of fuel and good draft the ratio of grate area to tube area might be made 7 to 1, in which case the grate area would be $4.27 \times 7 = 29.9$ square feet. If the length of grate be 5½ feet, we have $29.9 \div 5.5 = 5.43$,—say 5 feet 6 inches square. The ratio of heating surface to grate area would be $1104 \div 29.9 = 38.6$ to 1.

To recapitulate: We have for the cylinder boiler 1 square foot of grate for each 10.87 feet of heating surface; two-flue boilers, 1 square foot of grate for each 18.37 feet of heating surface; tubular boiler, 1 square foot of grate for each 28.74 feet of heating surface, on the basis of 9 square feet of grate to 1 of tube area, or 1 square foot of grate for each 38.60 square feet of heating surface, on the basis of 7 square feet of grate to 1 of tube area. Taking the first three results, we have for a given evaporation, or, as commonly stated, a given horse-power, with three different sizes of grates, viz.: 10.87, 18.37, and 28.74 square feet. The explanation is that the last figure will yield both high economy and capacity, the middle figure yields less capacity and less economy, and the first figure slightly less capacity and very much less economy; the temperature of the products of combustion being highest with the cylinder boiler, perhaps 800° Fahr., lower with the two-flue boiler, perhaps 600° Fahr., and least with the tubular boiler, perhaps 500° Fahr.

The rate of combustion will affect the proportions of grate relatively to the total heating surface. Bituminous coal with natural draught will burn from 10 to 45 pounds per hour per square foot of grate, a difference wide enough to show that for ordinary steam-boiler furnaces it is not worth while to enter upon this problem without knowing all the conditions which affect combustion, and these cannot usually be known in advance.

The practical considerations which affect the size of the grate are the diameter of the boiler, which determines the width of the furnace, and the impossibility of keeping the fire clean at the bridge-wall if the grates exceed 7 feet, which establishes the maximum limit of length in hand-firing, and this ought to be shortened to 6 feet to get the best results. For horizontal tubular boilers it is customary to set the side walls 4 inches away from the boiler, as shown in Fig. 272. If this distance be brought vertically downward the greatest ordinary width of furnace will be had; the least width is seldom less than the diameter of the boiler, see Fig. 271. One-thirty-fifth of the total heating surface will make a good average area for the grate surface, or eight times the combined area of the tubes will approximate good working conditions.

Grate-Bars.—These are usually made of cast iron, and consist of alternate supports for the fuel to be burned and spaces for supplying the air needed for combustion. Every portion of the entire area of the grate surface should be made up of these alternate "lands" and spaces. The top surface of the grate should be perfectly level, otherwise it will be difficult to properly clean the fire with ordinary hand-tools. The grates must rest upon proper supports, and these must be of such form, dimensions, and so located under the grates as to prevent their warping or getting out of shape by the action of the radiant heat from the fire. The grates should have at least one end free for expansion; otherwise they will get out of shape.

FIG. 239.

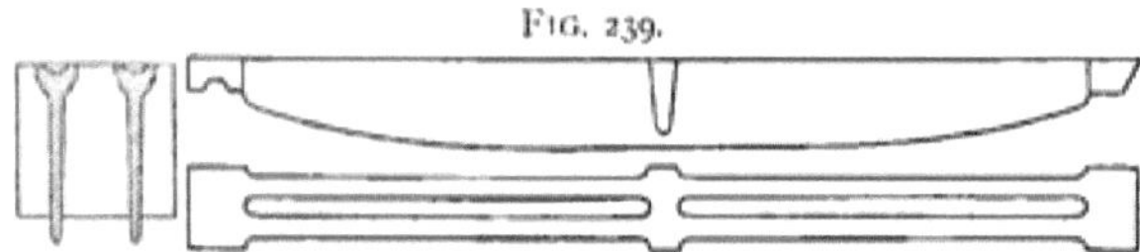

The clear opening through the grates is usually as large and as much subdivided as possible, that each portion of the fire shall have its proper supply of air. They commonly have from 3/8- to 5/8-inch openings, with only as much lateral obstruction at the ends and centre of the bars as is necessary to enable them to keep their shape. Such a grate is shown in Fig. 239, the metal and spaces being about equally divided. Mr. Barrus's observations are, that "Grates with 50 per cent. air-space gave 3.2 per cent. better results based on coal and 1.7 per cent. better results based on other combustibles than grates with 60 per cent. air-space.

"The smaller the rate of combustion, the smaller should be the opening for draught through the grates. The gain probably comes about by preventing the introduction of too great an excess of air over that required for combustion."

The practical spacing of grate-bars for the admission of air will depend on the kind of fuel to be burned. For bituminous coal it may be ½ or ⅝ inch, but for anthracite coal—and especially when burning the finer grades, such as rice, buckwheat, and pea coal—the distance must be less, and may approximate ⅜ inch in width. The larger the area of opening, the less will be the resistance of the air passing into the fire; but if the draught is good, this is a matter of less moment than the resistance of the air passing through the burning fuel. In the case of fine anthracite this resistance becomes so great that it is necessary to employ a forced draught in order to burn the fuel at a sufficiently rapid rate to develop the full power of a boiler.

The distance for the level of the grates above the fire-room floor will vary from 24 to 30 inches. The first is probably the best distance, all things considered; and nearly all fire-fronts approximate this height.

The slope of grate-bars is usually from the furnace-front towards the bridge-wall in the proportion of about one inch in twenty of the length of the bar. No advantage accrues in this; and for stationary boilers, externally fired, as good results may be had if the grates are set level. Inclining grates to the rear probably originated in marine practice, in order to protect the fireman from live coals rolling out of the furnace when stoking in rough weather.

FIG. 240.

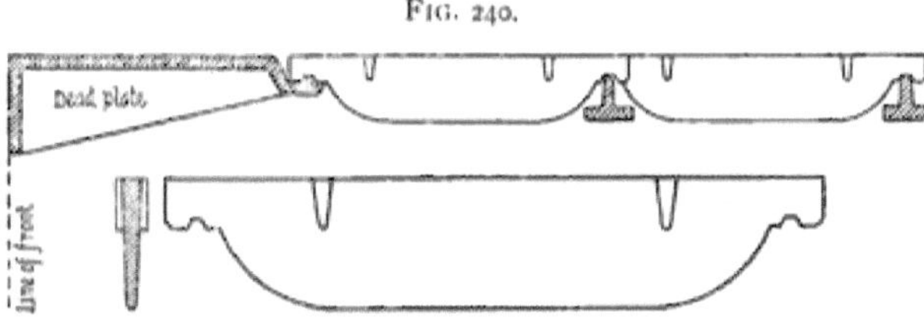

Grate-bars are commonly cast in pairs, as this adds much to their stiffness over single bars, and are thus better able to resist bending and twisting. Cast-iron bars should have a groove on the top along their whole length, as in Fig. 239. This groove, filling with ashes, prevents in a measure the clinkers adhering to the grate. Short bars suffer less in distortion by overheating than long ones. Thin and deep grate-bars, such as shown in Fig. 240, give excellent results. The bottom edge should be no thicker than will insure a sound casting. Such bars are usually cast singly and not in pairs. No difference need be made in the construction of grates between anthracite and bituminous coals.

Fig. 241.

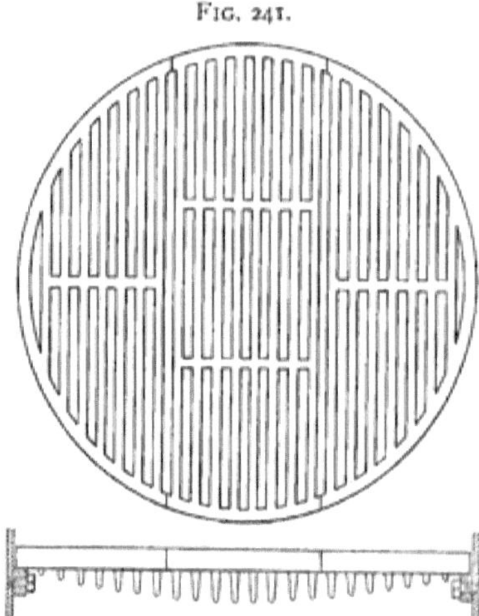

A grate for a circular fire-box is shown in Fig. 241 which does not differ essentially from the one referred to above, except that instead of being confined to two or three bars a larger number are included in a single casting. The engraving shows the grate to be made up of three sections. It is obvious, however, that the number of sections will depend upon the diameter of the fire-box. The width of the sections ought not to greatly exceed 12 inches.

A stationary grate with tilting section at the end next the bridge-wall is shown in Fig. 242. This is a very convenient arrangement for dropping ashes, clinkers, etc., into the ash-pit underneath. Grates as long as 6 feet should be provided with some device similar to this to prevent ashes from accumulating at the bridge-wall, which accumulation has the effect of shortening the furnace and diminishing the grate area.

A herring-bone grate, shown in Fig. 243, is extensively used in this country, its popularity not being confined to any one locality. The angular spacing readily permits expansion and contraction to take place

Fig. 242.

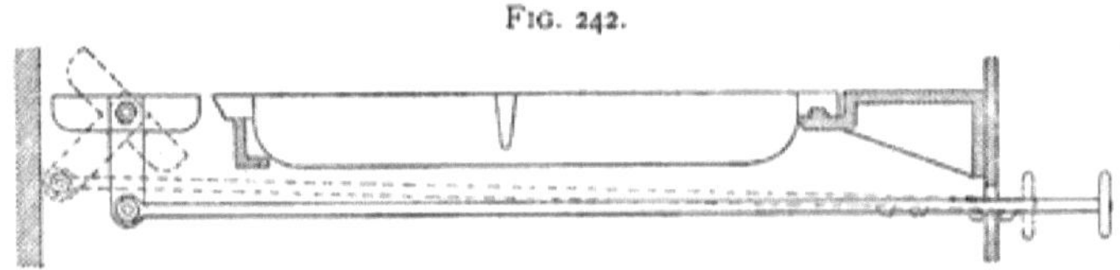

without breakage. It withstands the effects of repeated heating and cooling quite successfully. An adaptation of this grate for a circular fire-box is shown in Fig. 244.

A revolving grate is shown in Fig. 245. This is a very convenient arrangement for shaking the ashes out of the fire, for breaking up masses of clinker, and for dumping the contents of the furnace into the ash-pit underneath. The two middle grates in the upper row show how this is accomplished; the side bars show how the grates are kept level by locking upon one side wall and then upon each other.

FIG. 243.

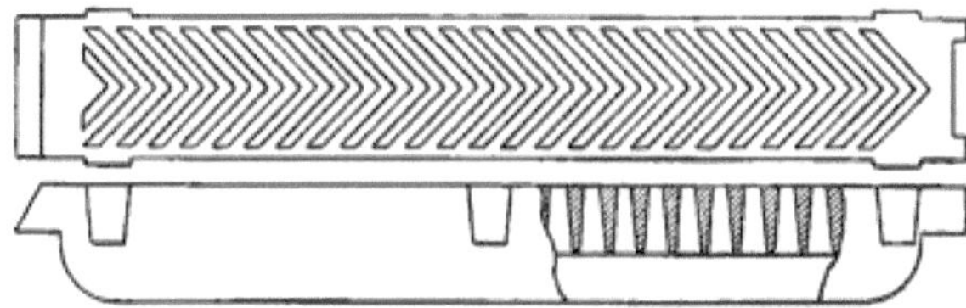

Shaking- or moving-grates are designed to facilitate cleaning the fires and, without opening the fire-doors, the removal of loose ashes and clinkers which would ordinarily fall into the ash-pit by the use of a slice-bar, also to break up and loosen the bed of burning fuel if the latter has a caking tendency.

FIG. 244.

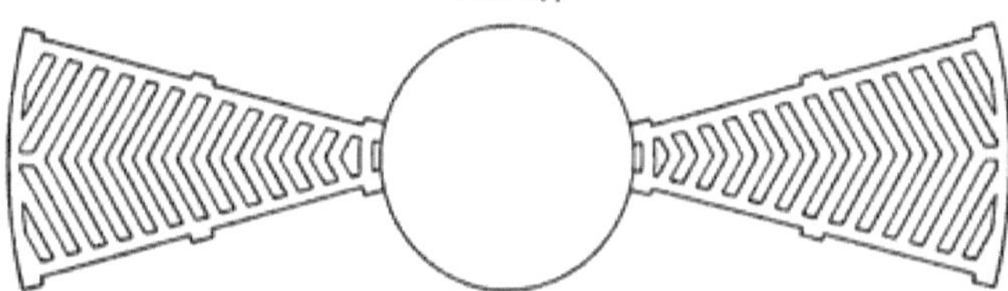

Some very absurd claims have been put forth regarding the great saving in fuel by the use of such grates, but the following extract from a letter puts the question in a way too often overlooked, and that is in reference to the protection afforded the fireman himself, who asks: "If

FIG. 245.

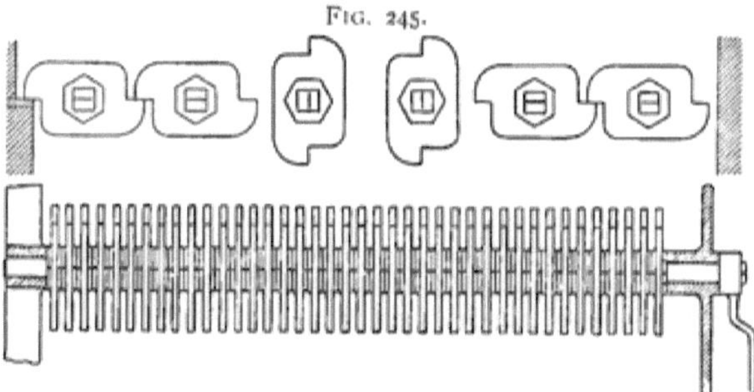

the doors have to be opened to clean the fires, is there not a loss in the admission of cold air to the boiler, to say nothing regarding the injury to the boiler itself? Again, can any fireman with a slice-bar, no matter how dexterous he may be, thoroughly stir out the ashes from every part of the grate, and if not, is not the supply of air to that extent cut off? If a fireman can, with comparatively little effort and without exposure to intense heat, thoroughly clean every inch of his grate by using the shaking-grate, why, in the interest of humanity and economy, should he not be furnished one?"

Shaking-grates will break up a bed of non-caking coal or anthracite coal very satisfactorily, but they do not always succeed in breaking up large masses of coal which may have fused together because of its caking quality. The fireman who relies upon a shaking-grate to break up such masses of caked coal will waste a great deal of fine coke in process of burning, which will drop into the ash-pit through the openings occasioned by the tilting of the grate-bars.

The Butman grate, shown in Fig. 246, has long been used in the Western States in furnaces burning bituminous coal. The grate is ex-

Fig. 246.

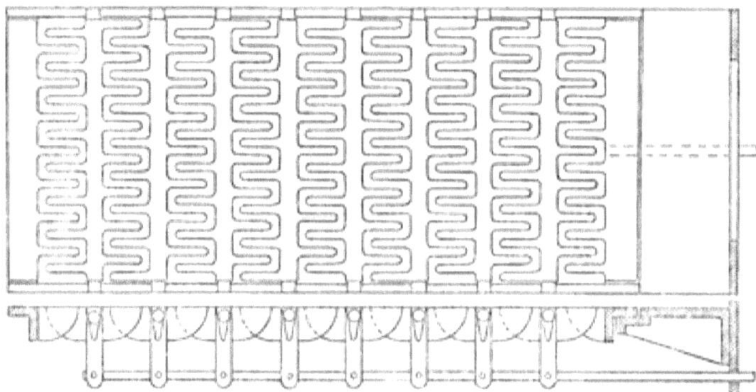

ceedingly simple and very effective; it consists of two side-bars notched to receive the grates, which are provided on their upper surface with oblique cutting edges and interlocking fingers, which also form parallel cutting edges. The cross-bars to which these interlocking fingers are attached are corrugated in the direction of their length, and taper from the top to the bottom. The spacing of the interlocking fingers is such that a series of irregular openings occur over the whole surface, and these openings are so dimensioned that a full supply of air is had for the fuel. The fingers, having semicircular ends, always preserve the same distance whether stationary or being rocked, this detail of construction preventing the loss of small fuel by its falling through into the ash-pit when cleaning the fire. Each rocking-bar has an arm projecting downward, attached to a connecting bar operated from the front of the furnace. When the fire is being cleaned by this rocking motion the whole surface is broken up at the same time, any accumulation of clinker being prevented by the cutting action of the locking fingers.

A shaking-grate known as the Ætna is shown in Fig. 247. Upon ordinary bearing-bars, common to all furnaces, are laid two or more stationary bars, *A* (depending upon the width of the furnace). Near either end is an opening, shown in one of the detail drawings, in which are placed two rockers. These spring freely like a scale bearing upon pivoted edges. Upon these rockers are placed the moving bars, *B* and *C*. When at rest the grate-bars present the same level surface as an ordinary grate. Attached to the front rocker is a socket, shown in the two upper drawings, which reaches almost to the ash-pit door. A lever put in this socket and worked imparts to both sets of bars at the same time a vertical and horizontal movement over the entire surface. By this action the entire surface of the fire is cleaned and opened, a condition which can be maintained with little effort by one or two movements of the lever at periodic intervals.

Fig. 247.

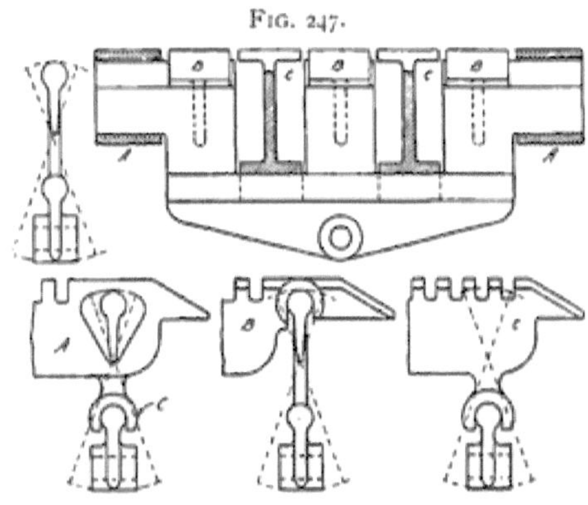

Fig. 248.

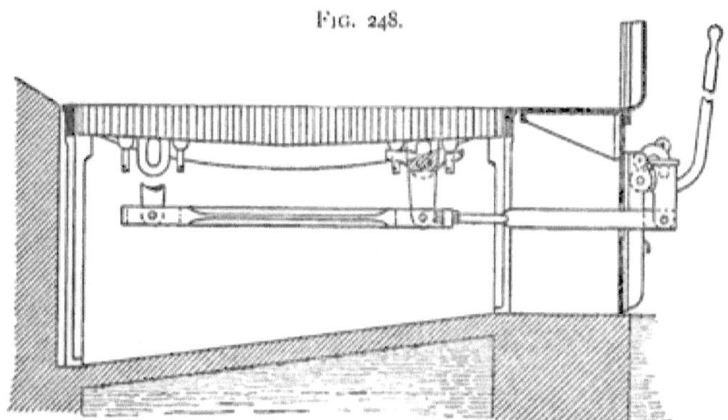

The Rose grate is shown in sectional elevation in Fig. 248. It differs from the preceding designs in that a vertical movement is given each alternate bar by rocking-levers located underneath the grate.

Each bar has V-shaped corrugations along its entire length. The bars are of unusual depth and taper from top to bottom, and are well calculated, therefore, to resist the action of fire. A plan of a set of grates is shown in Fig. 249. A centre-bar extends the length of the furnace, dividing it into two sections. This centre-bar locks into the two end-pieces by a dovetail joint, and thus serves to hold the frame together. Whatever number of bars are assembled to make a grate, usually from two to three feet in width, are set in a frame which is entirely independent of the boiler or brick-work. Two rocking-levers are suspended in each section of the frame and coupled together, as shown in Fig. 248; additional to this coupling-bar, there is still another, extending out through the fire-front and there connected to a bell-crank provided with a socket for the insertion of a hand-lever for operating or shaking the grates. A locking-dog hinged to the front, the free end dropping into a socket in the bell-crank, holds the locking levers in neutral position, but when the grates are to be shaken it is lifted out of its socket and thrown back against the front; a to-and-fro movement of the hand-lever will raise or lower and at the same time give a slight end movement to each alternate grate-bar for each movement of the hand. The grate-bars in their up-and-down movement pass each other at a sufficient distance to sift out all ashes and dirt accumulated under the fire, but not far enough to allow coal to fall through into the ash-pit. By the same movement of the grates the fire-surface is lifted, allowing air to pass more freely through the bed of fire than is common to the ordinary form of stationary grates.

FIG. 249.

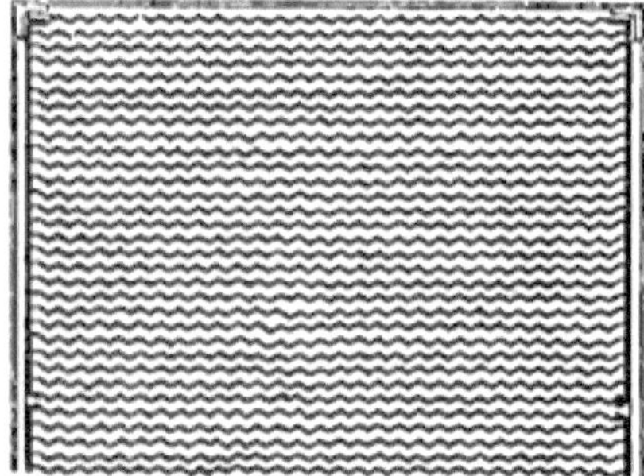

Deterioration in Grates.—The principal cause which contributes to the rapid burning out of grate-bars is the action of the furnace heat, which will in time destroy any set of grates, but the want of a proper flow of air through the grates will cause overheating, whether it occurs through too little air-space in the grates themselves, or by these spaces becoming obstructed through any cause, thus preventing the cooling effect of the air on its passage to the fire; but a more common reason is found in the impurities of the coal, and especially in the chemical combinations of sulphur and iron, which abound in more or less quantity in all red-ash coals. Any coal which forms an easily fused clinker will soon injuriously affect the grates.

Effect of Water-Pan in Ash-Pit.—There is no advantage in the use of a water-pan in stationary boilers, and its use is not recommended. The common notion that the vapor rising from the surface of the water and passing upward through the fire is decomposed and assists combustion is a mistake. Action and reaction are equal, and just as much energy is expended in disassociating the gases as is afterwards obtained by their recombination. The ash-pans underneath locomotive fire-boxes and marine boilers of the locomotive pattern have water-pans as security against fire, which may be caused by live coals dropping down upon wood-work.

Stationary boiler ash-pits should be kept dry and reasonably clean. As all the heat of the cinders and ashes is taken up by the air on its way to the grates, it will be seen that a hot, dry ash-pit is the best.

Mechanical Stokers.—The earliest mechanical stoker for steam-boilers was probably that designed by James Watt (1785), which was simply a device to push the coal after it was coked at the front end of the grate back towards the bridge. The suggestion of Watt was followed by numerous patents in Europe for mechanical stokers, but none of these have been adopted to any considerable extent in this country because of their supposed lack of adaptability to American boiler practice. Their non-adoption here was partly because the mechanism was of too complicated a nature to be entrusted to the average fireman, or the rate of speed at which the machine would be required to run was such that it would be wanting in durability. Another objection was, that in many cases the grates were so placed as to be almost inaccessible, either for examining or altering the condition of the fire by hand or for the renewing of worn-out parts.

The mechanical stokers in use in this country are almost entirely of American design, and are well adapted for the service to which they are applied, whether for burning graded small coals or mine refuse, such as screenings or other unmerchantable coal. A mechanical stoker is particularly suited to the burning of such refuse material, and instances are plentiful in which more steam was supplied when using bituminous slack or refuse coal in a furnace provided with an automatic stoker than the ordinary grate-surface supplied when using lump coal and when hand-fired, the boilers and all other accessories being similar, this increase in performance being due to the constant supply of coal to the fire without any disturbing influences. The labor required for boiler attendance with a mechanical stoker is much less than that required for the ordinary flat grate of like capacity and burning similar fuel. The time consumed in cleaning a fire under a boiler provided with such a machine is claimed to be less than one-tenth that required for similar work with ordinary furnaces.

Fuel economy is one of the claims persistently urged by makers of mechanical stokers,—not that a furnace fed by machine will accomplish results different than if fed by hand under the best conditions, but hand-firing occurs at irregular intervals, and usually more coal is put on the fire than is necessary. A mechanical stoker with continuous feed will supply the coal as needed without disturbing the progress of the fire, not cooling it by an excess of coal, nor by the admission of cold air in the furnace by the opening of the fire-door.

The prevention of smoke by the use of a mechanical stoker is largely brought about by a uniform delivery of fuel to the grates, maintaining a high temperature at all times in the furnace and combustion-chamber. This constant fuel supply, especially in the case of fine coal, permits the gradual distillation of the gases from the coal as it enters the furnace, and the automatic movement of the grates keeps the air-spaces open, insuring the combination of the oxygen of the air with the hydrocarbon gases in the furnace, and the complete utilization of the latter in heat-making.

The Murphy Furnace.—This furnace was specially designed for the use of small-sized bituminous coal and slack. A cross-sectional elevation is shown of it in Fig. 250, where the magazines into which the coal is put are at the sides of the combustion-chamber. At the bottom

FIG. 250.

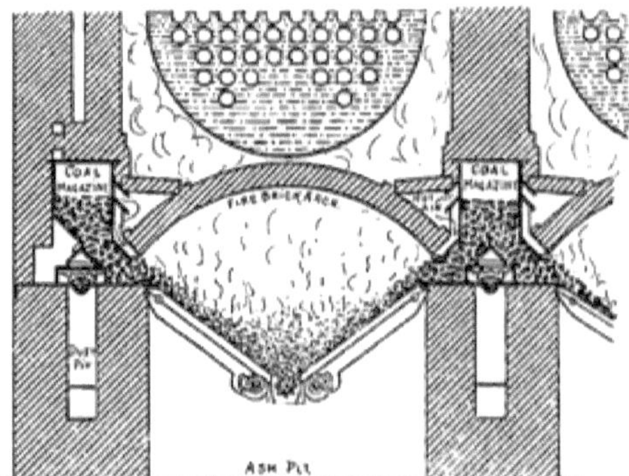

of each magazine is a casting, part of which is used as a coking-plate; the inclined grates rest against it at their upper ends. On the central part of this plate is an inverted open box (called the stoker-box) with a rack under each end. A shaft worked from the exterior has a pinion geared to each rack, whereby the stoker-box is moved back and forth on the coking-plate as rapidly or as slowly as may be required by the rate of combustion. The stoker-boxes, through the motion thus communicated, push the coal from the edge of the coking-plates to the grates below.

Immediately over the coking-plate is the arch-plate. At the line where the fire-brick comes in contact with the arch-plate are ribs an inch apart; these form the skewback on which the arch rests; the spaces thus formed between the ribs are air-ducts. The admission of air is regulated by a register at the front. The air passes through the flues, up and over the arch, taking up heat from the front, arch, and arch-plate; it then passes down through the small openings in the arch-plate and supplies the coking fuel with the proper quantity of air and at a high temperature. This secures the immediate combustion of the uncombined gases. A longitudinal sectional elevation of this furnace is shown in Fig. 251.

This furnace has sufficient coking capacity to feed 50 pounds of coal per square foot of grate per hour, and slowly enough to give ample time for the complete expulsion of the volatile gases. The air, which is delivered in numerous jets, is so distributed as to thoroughly mix with the liberated gases, resulting in the complete prevention of smoke escaping from the furnace. By the time the coal is fairly on the grates the gaseous part of the fuel is consumed; what remains is coke, and this is not smoke-producing. The coke gets its needed supply of air through the grates and is burned in the usual way. The grates are kept constantly in motion. The incoming air to the furnace passes between the bars and thus supplies the lower bed of fuel. The incandescent coke as it burns is moved down towards the centre of the furnace, at the bottom of which is a clinker-breaker for grinding any clinker and refuse to be deposited in the ash-pit below.

Fig. 251.

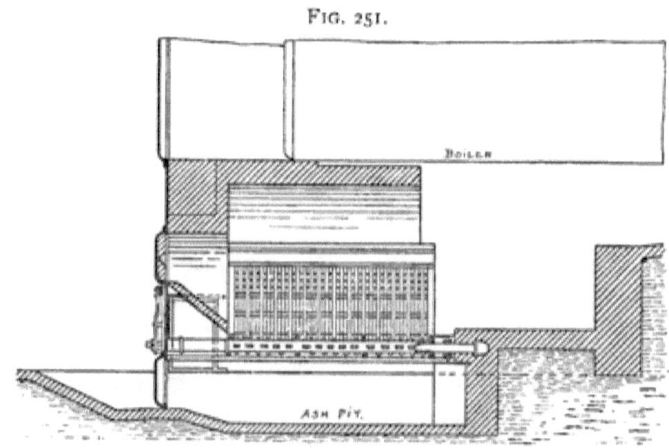

A single or double engine with automatic governor and proper gearing is placed on one side of a battery of boilers for operating a reciprocating bar across the outside of the entire front, and to which all the working parts are attached by links which may be detached or attached at the will of the operator. Any one or all of the movable parts may be detached from the automatic bar, and the furnaces may be successfully operated by hand, the fuel being fed in the same manner as when controlled by the engine.

The Roney Mechanical Stoker is shown in sectional elevation in Fig. 252. A hopper for the coal is attached to the boiler front; in the lower part of the hopper is a pusher, to which is attached, by a flexible connection, the feed-plate forming the bottom of the hopper. The pusher, by a vibratory motion, carrying with it the feed-plate, gradually forces the fuel over the dead-plate to the grates below. These grates

Fig. 252.

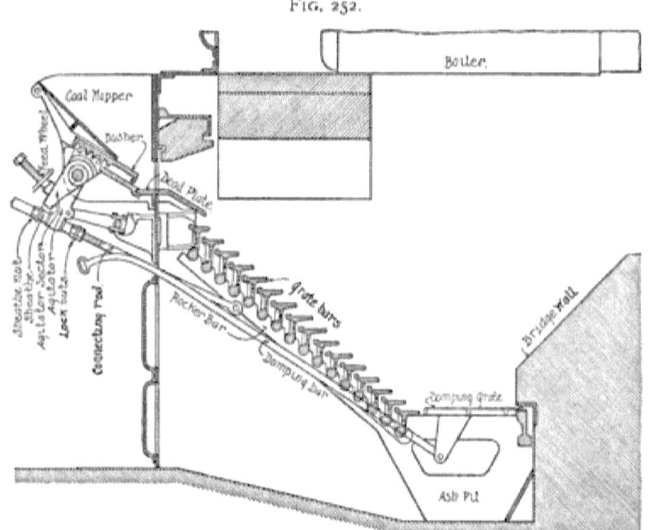

are horizontal flat-surfaced bars running from side to side of the furnace; they are carried on inclined side bearers extending from the throat of the hopper to the rear and bottom of the ash-pit. The grates, therefore, in their normal condition form a series of steps, upon the top step of which coal is fed from the dead-plate. These steps at the inclination given, however, prevent the free descent of the coal, but each bar rests in a concave seat in the bearer and is capable of a rocking motion through an adjustable angle. All the grate-bars are coupled together by a rocker-bar, the notches of which engage with a lug on the lower rib of each grate-bar, pin connections being made with two of the grate-bars only for the purpose of holding the rocker-bar in position. A variable back-and-forth motion being given to the rocker-bar through a connecting rod, the grate-bars necessarily rock in unison, now forming a series of steps and now approximating to an inclined plane, with the grates partly overlapping. The depending webs of the grate-bars are perforated with longitudinal slots, so placed that the condition of the fire can be seen at all times and free access be had to all parts of the grate to assist, when necessary, the removal of clinker. The slots also serve an important purpose in furnishing an abundant supply of air for combustion.

Assuming the grates to be covered by a bed of coal and fresh fuel being fed in at the top, it is obvious that when the grates rock forward the fire will tend to work down in a body. But before the coal can move too far the bars rock back to the stepped position, checking the downward motion, breaking up the bed of fuel over the whole surface, and admitting a free volume of air through the fire. The rocking motion is slow, being from seven to ten strokes per minute, according to the kind of coal. This alternate starting and checking motion, being continuous, keeps the fire constantly stirred and broken up from underneath and finally lands the cinder and ash on the dumping-grate below. By releasing the dumping-rod the dumping-grate tilts forward, throwing the cinder into the ash-pit, after which it is again closed, ready for further operation. The dumping-grate is made in two parts, so that each half can be dumped separately.

The actuating mechanism is simple. All motion is taken from one driving-shaft. In a single stoker this shaft may either be driven through a worm gear from a small engine attached to the boiler front, or it may be driven by a link-belt from any convenient point of the nearest shaft. In large batteries of boilers the driving-shaft is extended across all the boiler fronts, delivering power to each stoker; this, with the coal elevators and conveyers, is driven by a small independent engine. The largest stoker can easily be turned over by hand, indicating the small amount of power necessary to operate it. The worm-gear shaft carries a disk and wrist-pin from which a link couples to the agitator, shown attached to the boiler front, underneath the dead-plate. Through the eye of the agitator passes a stud screwed into the pusher, on which stud is a feed-wheel by which the stroke of the pusher, and consequently the amount of feed, is regulated. The agitator having a fixed stroke, it is apparent that if the feed-wheel is run down against it in the position shown in the engraving, the pusher will be given its full traverse and the greatest feed. If run back to clear the travel of the agitator, the pusher will, of course, have no motion, and the feed will stop. Between these extremes any desired rate of feed can be given.

In a like manner the rock of the grate-bars can be adjusted between any limiting angles, and over a range of motion from no movement to full throw, by means of the sheath-nut and jam-nuts on the connecting rod. By these two adjustments the whole action of the stoker is controlled, the fires forced, checked, or banked at will. There are poker-doors in the front, on each side of the hopper, through which the whole grate can be seen and the condition of the cinder on the dumping-grate determined. If the location is convenient, a cleaning-door may be introduced into the side of the furnace to facilitate examination of the boiler.

The Wilkinson Mechanical Stoker is shown in sectional elevation in Fig. 253 and in front elevation in Fig. 254. The grate-bars are a series of hollow castings approximating a rectangular cross-section, placed side by side, and inclined towards the bottom of the furnace at an

FIG. 253.

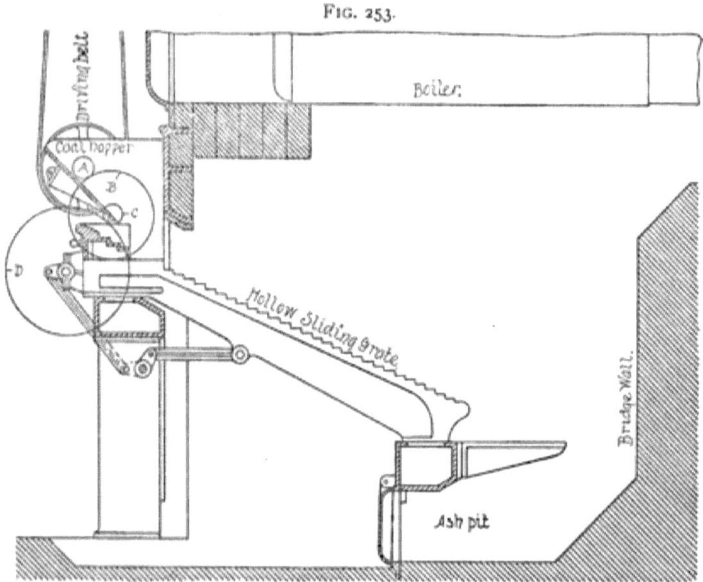

angle of about twenty-five degrees with the horizontal. The upper end is open to admit the blast-pipe; it projects through, and is supported by the stoker front, the lower ends sliding on and supported by a hollow cast-iron box, as shown. This lower box, or bearing-bar, has finger-grates about 15 inches long secured to its rear face The bars are 4-inch centres, so that, practically, the air-openings are restricted to the risers. Throughout the inclined length and on the face of the bar is cast a succession of steps. Through the rise of each step a vent of about 1/4 by 3 inches is provided to admit air through the fire to the combustion-chamber.

The feeding is accomplished by the motion of the grates. The pusher shown in the bottom of the hopper and resting on the grate-bars is secured to each alternate bar by a dowel-pin and moves with them, feeding the fuel in measured quantities from the hopper to the upper end of the grate. The continuous back-and-forth motion of the grate-bar is for the purpose of maintaining a uniform thickness of fire by a gradual descent of the fuel from the top to the bottom of the grate, depositing the clinker and ash on the stationary grate shown projecting from the bearer-bar at the ash-pit. The accumulated ash is pushed off this stationary grate into the ash-pit by the reciprocating motion of the bars, to be removed in the usual manner or by special appliance. The mechanism for effecting the entire operation of the stoker consists of a pulley, compound gearing, toggle-shaft, and quadrant, all of which are shown in the two engravings above referred to.

FIG. 254.

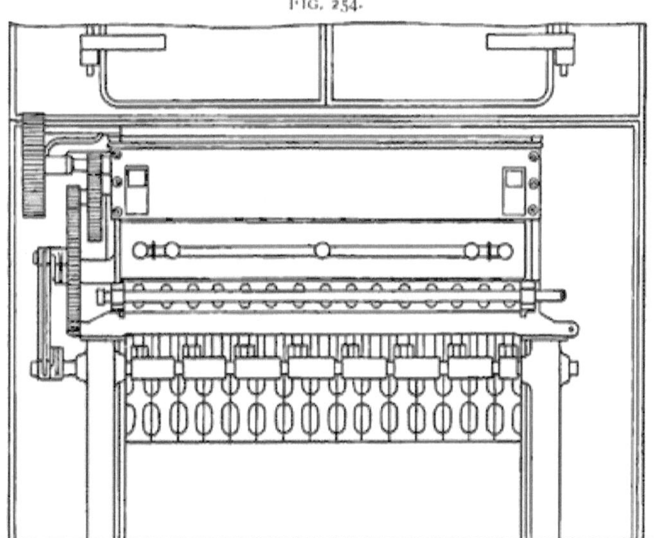

The blast is saturated steam, through a nozzle of $\frac{1}{16}$-inch opening, giving an induced current of air controlled by a regulating valve. This method of injecting the air into a hollow bar, from which there is no escape except into and through the fire, is set forth as one of the merits of this machine. The steam is decomposed by the incandescent fuel and fills the combustion-chamber with burning gases, resulting, according to Whitham, in a more uniform distribution of the effective heating surface of the boiler, reduces local injury, gives a more uniform expansion to the parts of the boiler, but is apt to cause a loss of heat in the stack if not properly controlled.

The cost of the steam-blast is from 5 to 11 per cent. of the steam generated by the boilers, estimated at $1000 per year for a 1000 horse-power plant, on a 10-hour basis, when fuel is $2.00 per ton.

An automatic damper-regulator should not be used with this stoker; simply observe the best position of damper suited to the work and keep it there. When a strong natural draft exists, it may be well to partly close the damper at the chimney and force the steam-jets harder.

When running too slow, the fire will burn out before reaching the bottom of the grate. When running too fast, live coal will be pushed out with the ash. Between these two extremes a speed may be found best suited to the requirements of any boiler plant.

The object of blowing steam into the grate is in part to assist the combustion and prolong the life of grate-bars. Most anthracite coals of inferior sizes clinker badly if the fire is forced; such clinkers "freeze" to the sides of the furnace and to the grates; the effect of a steam-blast is to chill the grates and non-combustible material against them, so that a clinker cannot form. It is at all times necessary with this machine to have some steam passing through the jets when fired up, therefore the steam-jets should all be open and blowing alike. When the furnace is in operation, the damper and steam-jets should never be entirely closed.

The Playford Mechanical Stoker is shown in longitudinal sectional elevation in Fig. 255 and in cross-sectional elevation in Fig. 256. As will be seen, it consists of an endless revolving chain of grate-bars, operated by power, preferably by a small engine taking steam from the boiler. A fire-brick arch is sprung across the boiler front under which this machine is located when in use; a similar arch is sprung across the rear end of the furnace adjoining and attached to the bridge-wall. The grate-bars are attached to each other by links, shown in perspective in Fig. 257. The sprocket-wheels engage between the links, giving the whole grate surface a movement inward, the rapidity of which is controlled by the attendant to suit the needs of the furnace. The grates, bearing-shaft, etc., are mounted on a frame provided with wheels resting upon rails, shown in both engravings. This permits withdrawal of the whole grate surface and bed of fire from the furnace at any time.

A coal-hopper is provided at the boiler front; an adjustable gauge regulates the amount of coal which may pass into the furnace from the hopper. As the movement of the grates is very slow, the volatile matter, in the case of bituminous coal, is driven off by the intense heat of the front arch, and these gases are not only here brought to the point of ignition, but are consumed in the combustion-chamber immediately adjoining. The incandescent carbon remaining on the grate is burnt during the interval required for the passage of the grates up to the rear arch, under which the grates begin their return to the front, the ashes falling off into the ash-pit underneath, and being brought forward to the front of the boiler or wherever else desired by the spiral ash-conveyer, shown in Fig. 255.

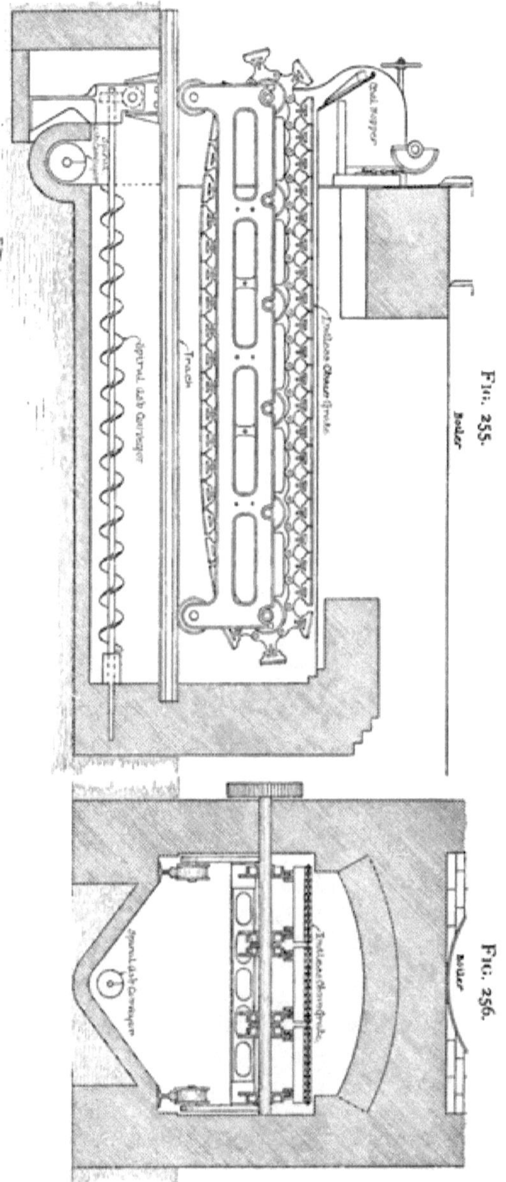

Fig. 255.

Fig. 256.

FIG. 257.

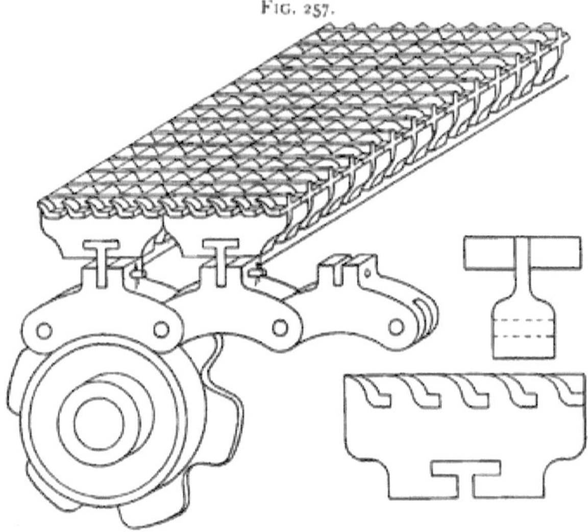

The Coxe Mechanical Stoker is shown in sectional elevation through a furnace in Fig. 258. It consists of an endless travelling grate which receives the fuel at one end, burns it as it moves slowly along,

FIG. 258.

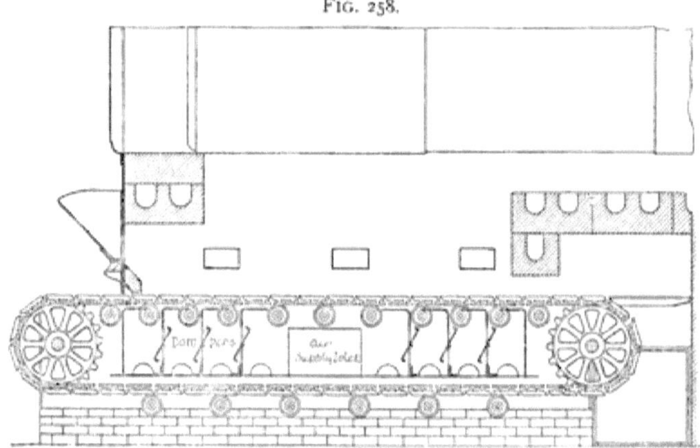

and deposits the ashes at the other end. It was originally designed to burn the finest sizes of anthracite coal and is shown arranged for this grade of fuel, but it is capable of handling bituminous coal as well.

A coal-hopper, into which the coal is placed by the fireman, is located immediately in front of the boiler. The depth of coal upon the grates is regulated by a sliding gate, which can be raised or lowered as a thicker or thinner bed of fire is required.

After the coal passes under the regulating gate, it flows over fire-brick "ignition blocks" set on an incline immediately inside the fire-front and above the grates, which blocks are used only with anthracite coal; these blocks, becoming incandescent, or nearly so, retain sufficient heat to insure ignition of the incoming supply of fresh coal. In burning bituminous coal these ignition blocks are omitted and a hopper is put on that admits of the fuel being deposited directly on the grates, as no extra care is necessary to insure perfect ignition of bituminous coal.

The entire grate is set travelling at the rate of four to six feet an hour, or at any other like speed, according to the demands upon the boiler, etc., and thus the fuel which has been fed in from the hopper at the forward end of the furnace is burned completely during its slow passage to the dumping end, where the ashes are left, while the grate reverses underneath and comes back, as shown, to the front end of the furnace, to pass in again like an endless belt.

The smaller sizes of anthracite coal pack down closely and cannot be burned unless a forced draught is used, and as this draught should vary in intensity according to the state of the fire, all the air that enters the fire is conducted into the largest chamber at the centre of the frame, see Fig. 258; and the only air that can enter the adjacent chamber must pass through dampers in the partitions separating these chambers, which are closed at the bottom, so that the supply can be regulated to suit

FIG. 259.

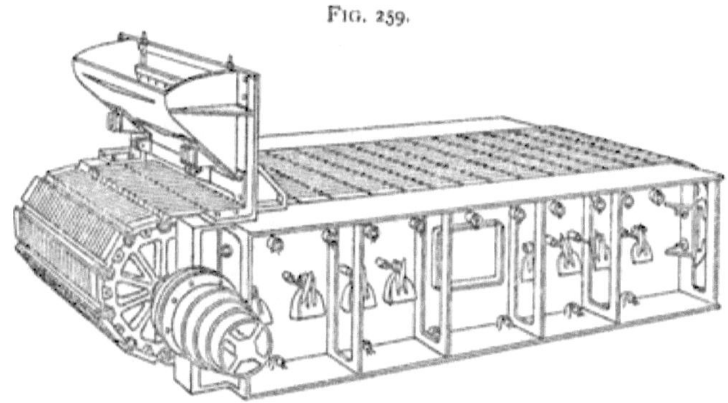

existing conditions. In practice, the maximum amount of air is supplied at the centre, and this is diminished to a minimum at each end, the dampers being regulated so that the pressure is diminished in each succeeding chamber. The handles which regulate these dampers are shown in Fig. 259 projecting from the sides of the frame. Preference is given to a dry-air- or fan-blast, because with the fan-blast the rate of combustion per square foot per hour is greater than with a steam-jet. The percentage of carbon left in the ash is less, and as it is not necessary to prevent the formation of clinker, which is the chief reason for using steam, the loss and waste due to the steam-jet are avoided. It is claimed for this furnace that the finer coals are burned with a less excess of air than is common, and the results with all fuels are above the average.

The American Stoker is shown in sectional elevation in Fig. 260. The coal-hopper is placed at the front of the boiler, in common with other

FIG. 260.

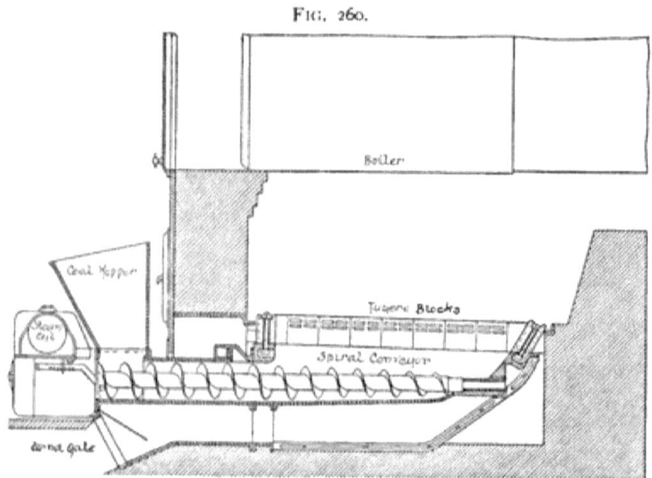

stokers. Underneath this coal-hopper is a spiral conveyer, located in a trough with a half-round bottom, shown in Fig. 261, which extends nearly the entire length of the magazine. Immediately beneath is located the wind-box, to which is connected the piping for air-blast ; the other end of the wind-box opens into the air-space between the magazine and outer casing. The upper end of the magazine is surrounded by tuyeres, or air-blocks, these being provided with openings for the discharge of air-blast.

The air is delivered in the approximate proportions of 150 cubic feet of air to each pound of coal fed, and at a pressure ranging from ½ ounce to 1 ounce at the tuyeres. This pressure is only such as to admit of the thorough mixing of the air with the coal, and must not be confused with the ordinary forced draught. A wind-gate controlled by a lever, shown underneath the conveyer, or hopper, Fig. 260, enables the operator to regulate the supply of air to suit the amount of coal fed. Being thus independent of natural draught in the air supply, the supply of coal

Fig. 261.

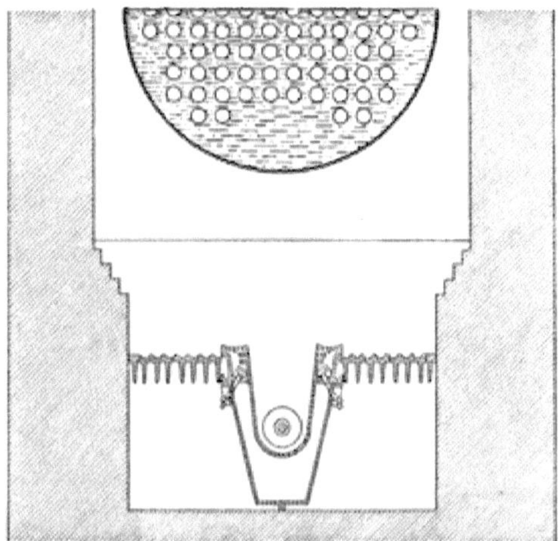

being also under complete control, the fire can be forced at a moment's notice, or it can be as quickly reduced.

The operating mechanism, apart from the fan for furnishing the blast, is located beneath and in front of the hopper. A steam motor for operating the stoker is shown in end elevation in Fig. 262. The only moving parts of this motor are the reciprocating piston and rod, the valve movement being similar in its construction and operation to that of a single steam-pump. The piston-rod carries a cross-head, which by means of suitable connecting links operates a pawl mechanism, which in turn actuates the ratchet-wheel mounted on the conveyer-shaft.

Fig. 262.

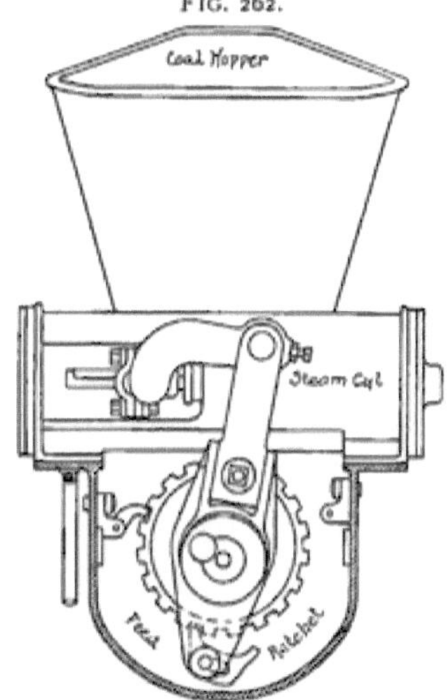

The space on each side of the stoker, between the tuyere blocks and the side-walls of the furnace, is occupied by dead plates or air-tight grates.

The tuyere blocks shown in Fig. 260 are substantial castings. The air passes through the blocks, protecting them to a considerable degree; but in case a block is damaged by the action of the heat it can be easily and inexpensively replaced. These blocks are practically the only part of the stoker subject to renewal.

The operation of this stoker is as follows: The coal is fed into the hopper by hand or by any convenient mechanism; from there it is carried by the conveyer into the magazine, and then forced upward by the action of the spiral conveyer until it overflows on both sides of the tuyeres, spreading upon and over the dead-grates the entire width of the furnace. The entire mass of the coal above the tuyeres and all of that above the dead-grates is in active combustion, consisting usually of a bed of burning coke from 14 to 18 inches in depth. The feeding of the coal is continuous, thus keeping the incandescent body of fuel slightly agitated and preventing the formation of large clinkers; this agitation also allows the air to thoroughly mix through the burning coal. The amount of coal fed is regulated by the speed at which the motor is run.

In cleaning the fires, the non-combustible is usually found in the shape of a vitrified clinker deposited upon the two sides of the furnace, and is removed through the ordinary feed-doors. By the use of a slice-bar this clinker can be raised from the grate and afterwards pulled out with a hook. The central portion of the fire is not disturbed.

The Jones Under-Feed Mechanical Stoker is shown in Fig. 263. It consists of a hopper to receive the coal; underneath is a cylin-

FIG. 263.

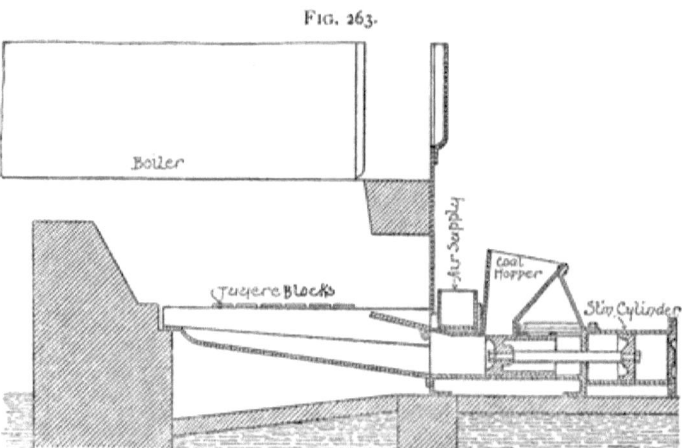

der, in which is fitted a plunger or ram. A steam cylinder immediately adjoining is furnished with a piston, the rod of which, passing through a stuffing-box, attaches and gives motion to the plunger or ram under the coal-hopper. A slide-valve, operated by hand, admits steam on either side of the piston for the forcing of coal into the furnace or the withdrawal of the plunger for a fresh charge. The central trough, or retort, shown in Fig. 264, into which the coal is forced, is inclined upward at the angle which gives a nearly uniform thickness or spread of coal throughout the length of the furnace. On either side of this retort are tuyere pipes or openings, shown in the cross-sectional engraving, Fig. 264. Ordinary grate-bars on either side of the retort fill out the width of the furnace, the length and width of the furnace being suited to the size of the boiler and rate of combustion.

Fig. 264.

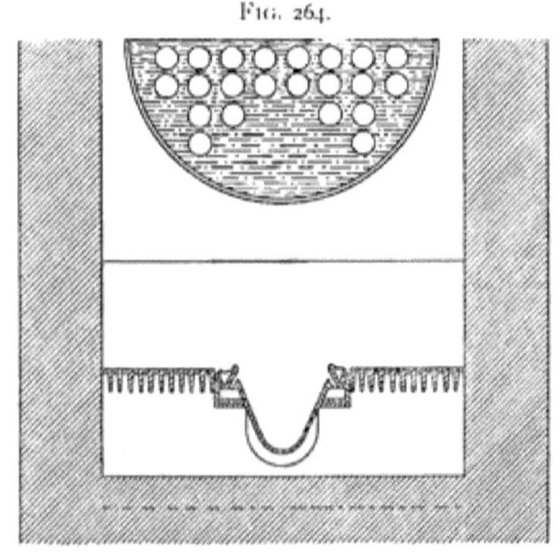

Additional to the stoker, the equipment comprises a blower and engine, steam and air piping, and usually a new grate-bearing bar. A blower must be used that will hold a 4-ounce pressure at the tuyeres at all times, delivering 150 cubic feet of air for each pound of coal burned, estimating upon the greatest amount of coal to be burned per hour. The engine for furnishing power to maintain the requisite amount and pressure of air should preferably be independent. The blower should not be driven from the main shaft, because air cannot be supplied to the stoker in starting, nor at any time, except the main engine be running. Suppose the furnace is to fired and with no steam in the boiler. The retort is first filled with coal level with the top of the tuyere pipes. Fire is then started on the side grates, as usual, until steam is raised. The ash-pit doors that admit air to the side grates are then closed. Then the coal is placed in the hopper outside of the boiler-front. The steam ram is then withdrawn by shifting the lever. The desired quantity of coal then falls from the hopper in front of the ram, and upon its return stroke is forced into the retort. Air under pressure is then admitted into the tuyere pipes. The air issues through the slots shown in Fig. 264, over the top of the fuel in the retort, but under and through the burning fuel. The result is that the heat from the burning fuel over the retort slowly liberates the gas from the green fuel in the retort.

The gas, being thoroughly mixed with the incoming air before it passes the burning fuel above, results in a bright, clear fire, free from smoke, and the complete consumption of all the heat-producing elements in the fuel. The retort being practically air tight from below, and the fuel being in a compact mass in the retort, the air will find its way in the direction of least resistance, which is upward; consequently combustion takes place only above the air-slots. Hence the castings of the retort are always cool and not subject to the action of the fire. The incoming fresh fuel from the retort forces the resulting ash and clinker over the top of the tuyere pipes on the side grates, from whence they may be removed at any time without interfering with the fire in the centre of the furnace.

The furnace should not be charged with more than 100 pounds of coal at a time. It is better to put in small charges at regular intervals, giving the coal ample time to coke before forcing it over the air-slots and into the fire.

Hawley Down-Draft Furnace.—This furnace consists essentially of two water-drums, one located immediately inside the boiler-front and

Fig. 265.

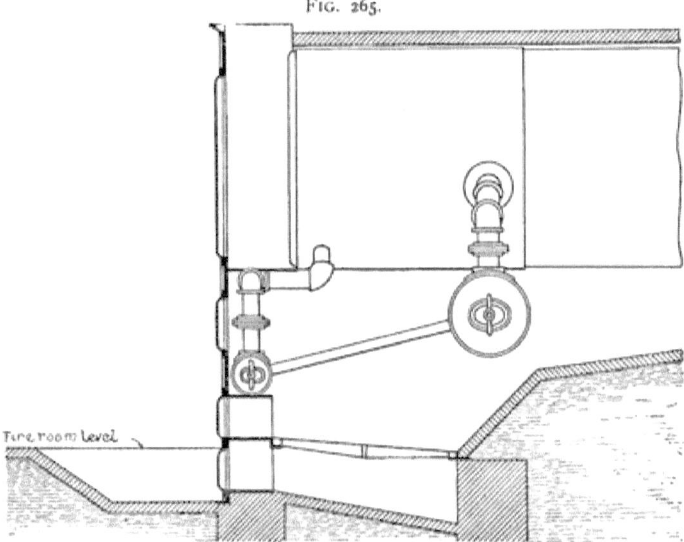

another back the proper depth of the furnace. These two water-drums are connected with each other by wrought-iron pipes, forming a water-grate, as shown in Fig. 265. The forward drum is in water communication with the bottom of the boiler through a wrought-iron pipe connection. The rear drum is similarly connected, but the pipe connection to the boiler has its opening higher up, see Fig. 266. This arrangement of pipes, drums, and water-grates is conducive to water circulation and adds to the efficiency of the boiler.

Fig. 266.

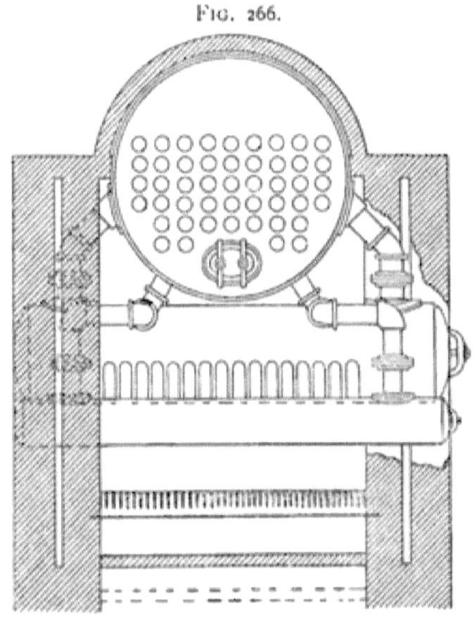

The wrought-iron pipes forming the water-grates are sometimes arranged in a zigzag row in the drums and sometimes in a single row. It matters little which arrangement is employed, as the results are practically the same for the same amount of tube heating surface thus employed.

The front water-drum is usually 8 or 10 inches in diameter, and must be provided with a handhole for examination and cleaning. The rear drum is made from 10 inches diameter, the Chicago practice, to 20 inches diameter, corresponding to St. Louis practice. For a drum as large as 20 inches a manhole can be inserted in the head, and this has been found useful in examining the condition of the water-grate pipes connecting the two drums, as by attaching a candle to a stick and passing it across the intersection of the openings in the small drum, a man in the large drum can ascertain the exact condition of each tube. A constructional

Fig. 267.

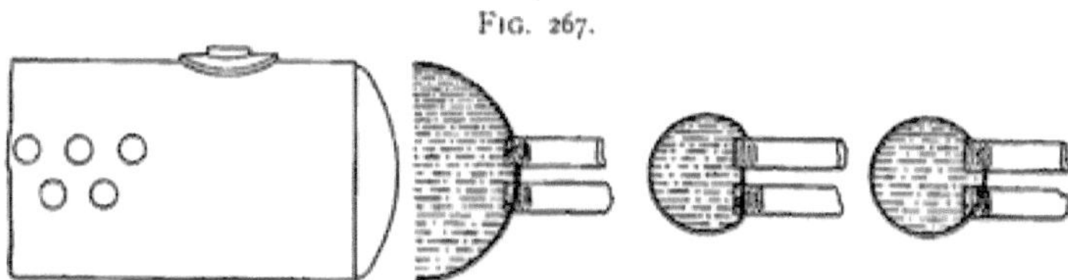

advantage which the large drum has over the small one is that when a zigzag row of tubes is desired no flattening of the drum is necessary to get the proper amount of thread for screwing in the water-grate pipes, which is necessary in the case of small drums. See Fig. 267.

The pipes forming the water-grates are commonly 2 inches in diameter. In the earlier forms they were set level, but it was found that by placing them on an incline of 2½ to 3 inches per foot an improved circulation was had, and the probability of burning off the tubes was greatly reduced.

The lower-grate surface is formed of common grate-bars, placed 18 inches below the lowest line of the upper grates; it is set level, or nearly so, one end resting on the boiler-front, the rear end on the bridge-wall.

The ash-pit is located under the lower-grate bars, and quite frequently at a level slightly below the fire-room floor. This arrangement is made necessary by the use of a fire-front three doors high, as the firing is at the upper door only. This necessarily raised the fire-door some eighteen inches higher than that to which firemen are accustomed, and to save the fireman the extra labor thus occasioned, the ash-pit floor was placed a few inches below the fire room level, as shown in the engraving. This depression in front of the boiler is about three feet wide, which does not inconvenience the fireman as much as lifting the coal to an unaccustomed height. The ash-pit slopes downward towards the bridge-wall, which facilitates the removal of ashes.

The fire-front is made three doors high. The upper one receives all the fresh fuel, the middle one opens into the lower grate, the lower one into the ash-pit. The middle door should always be kept closed, except when it is necessary to clean the lower fire,—perhaps three or four times a day.

The furnace combustion proceeds in two stages: first, the raw fuel is fed into the upper furnace, the air supply coming in through the upper-furnace door directly over the fresh charge of fuel. The direction of the current of gases is directly opposite that of an ordinary furnace, being downward instead of upward. The gaseous products of the upper furnace pass down into the lower furnace. These gases are combustible because the combustion in the upper furnace is incomplete. It is, in fact, a chamber for the distillation of gas from coal, rather than a furnace in the true sense of the word. The water-grates being set wide apart, the coked fuel of the upper furnace falls through the upper grates into the lower furnace, and is the source from which it receives its fuel supply. This brings us to the second stage, in which the heat from the incandescent body of coke burning on the lower grate raises the incoming flow of cooler gases from the upper furnace to that temperature necessary for their ignition, there being enough oxygen supplied by the excess of air mixed with the gases to complete the combustion begun in the upper furnace.

No raw fuel should be used in the lower furnace; if the upper furnace is properly managed it will supply all the coke necessary for the lower one. As about 90 per cent. of the entire work is done by the upper grate, the lower grates do not need a large fuel supply ; by slicing the upper fire enough half-burnt fuel will drop down in the lower furnace to supply its needs. Additional air supply for the lower furnace should be had through the ash-pit. An even fire 6 to 8 inches deep should be kept on the upper grates, the thickness depending on the size of the coal (which should be about the size of one's fist) and the intensity of the draught, a fine coal with moderate or poor draught requiring the thinnest fire.

The chimney draught must be ample to maintain a quick fire, otherwise the action of the furnace will be sluggish, producing smoke, and of less efficiency generally.

The standard form of construction is similar to that shown in the engraving, in which the furnace is under the front end of the boiler. Some 50 to 60 square feet of shell heating surface is thus made non-effective for evaporation, which has suggested the construction of an external furnace with its attendant advantages in capacity and efficiency.

Evaporative tests made by W. H. Bryan indicate that this furnace adds to the efficiency of improved water-tube forms of boilers, although the percentage of increase is not so great as with ordinary horizontal tubular boilers. The claims for this furnace are that it prevents at least 95 per cent. of smoke, saves 10 to 40 per cent. of fuel, and increases boiler capacity 25 to 50 per cent.

Reynolds Furnace.—A horizontal return tubular boiler with a furnace externally fired is shown in Fig. 268. It is well known that a much higher furnace temperature can be had if its walls are made of refractory and non-absorbing material than if the top or sides be made of an absorbing body like the shell or fire-box of a steam boiler. Combustion is more easily perfected in a furnace at a high temperature than at a low one. Smoke abounds most in furnaces of low temperature. This furnace has for its object better combustion and a reduction of smoke where the fuel is bituminous coal ; this end is sought to be attained by the highest temperature which can be had by the combustion of bituminous coal in a separate chamber.

This furnace as applied to a steam boiler, shown in Figs. 268 and 269, consists of three parts,—the furnace proper where the coal is burnt upon the grates, a combustion-chamber at the rear of the furnace, and a diffusion-chamber underneath the boiler. The products of combustion on leaving the external brick furnace pass through a contracted opening or throat, where they mingle with a supply of air which is admitted through the brick walls. The entrance of this air not only enhances the combustion, but carries back into the furnace some of the heat which would otherwise escape through radiation from the brickwork, and it thus serves a twofold purpose. The products of combustion then come in contact with an overhanging brick arch, by means of which their direction of motion is changed and they emerge into a combustion-

16

chamber, from which they escape through a checker-work arch into the space beneath the boiler-shell.

Experiments by W. H. Bryan with this furnace and setting applied to a 66-inch x 18-foot horizontal tubular boiler with 56 4-inch tubes, a total heating surface of 1276 square feet, the grate surface 25 square feet, a ratio of heating to grate surface of 51 to 1, resulted as follows: The coal used was Cherokee slack, having a calorific value of 11,335 thermal units per pound, with 15 per cent. of ash, burning 30.6 pounds per square foot of grate per hour. The equivalent evaporation per pound

Fig. 268.

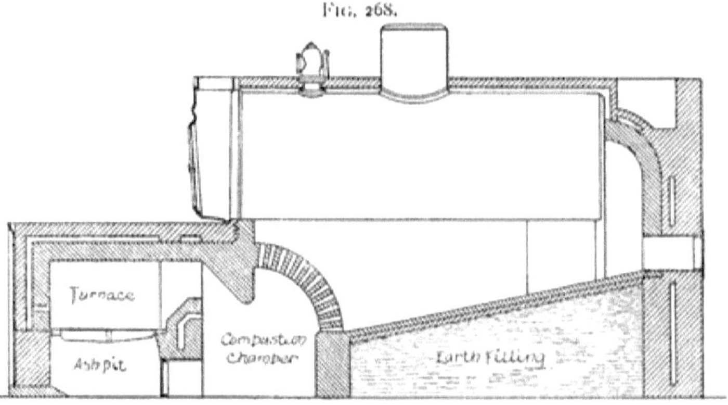

Fig. 269.

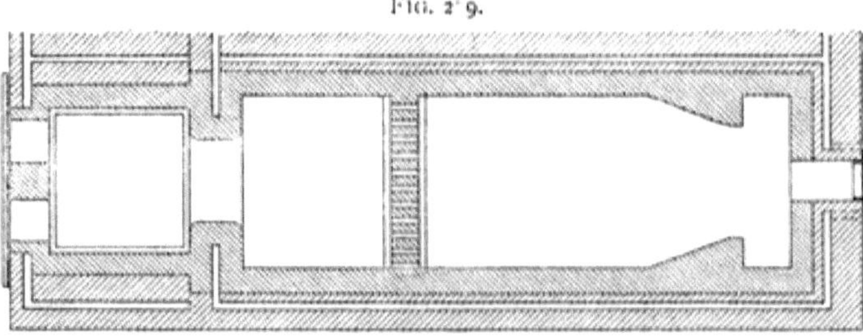

of combustible from and at 212° Fahr. was 7.83 pounds; the temperature of the escaping gases was 671° Fahr.; the efficiency or percentage of total calorific power utilized was 56.6 per cent. This is not a bad showing considering the kind of fuel used and the high temperature of the escaping gases.

As proof of the excellent working of the furnace, the smoke escaping from the chimney in a daily run varied, according to the St. Louis standard, from 0.6 of 1 per cent. to 2.8 per cent., according to the amount of work done.

Admission of Air over the Fire.—It has long been thought essential to the proper and economical combustion of bituminous coal that air should be admitted over the fire, and within certain limitations this is true and is recommended; but an excess of air is wasteful, because it lowers the temperature of the furnace at a time when the air is not needed to assist in the work of combustion. An opinion based upon practical results has been gaining, that if an excess of air is to be permitted in the furnace at all, it is better for it to pass up through the fire than to be admitted over it; and the reason is quite obvious, for whatever oxygen is needed for the combustion of the incandescent fuel on the grate is taken up by this fuel during the passage of air through it, and any excess of air simply passes through the fuel, its temperature being raised to that of the other products of combustion. Should this heated air come in contact with combustible gases not yet combined, the conditions for ignition and combustion are more favorable than if the air were admitted over the fire instead of through it. It is for this reason principally that the admission of air through the sides of a furnace, through the bridge-wall, or in the combustion-chamber back of the bridge-wall has so often failed of its purpose, such currents of air acting as a cooling medium rather than uniting with the combustible gases under the conditions necessary for making the combustion complete.

Distance between Under Side of Boiler and Top of Grate.—Neglecting all other fuels than anthracite and bituminous coals, it may be stated generally that anthracite coal requires the least distance, and for horizontal tubular or flue boilers this approximates 24 inches when the larger sizes are used for steam purposes, such as Nos. 1 or 2 chestnut; but coal of this size is now seldom used in steam-boiler furnaces on account of its cost; pea, buckwheat, and rice sizes, being much cheaper, are used instead; such coals may have the grates placed within 20 inches of the boiler-shell. For bituminous coal the grates may be say 30 inches for non-caking coals like Indiana block and kindred varieties, to 36 or 48 inches for fatty or gaseous coals; 36 inches is a good distance for average bituminous coals; this gives ordinarily all the cubic space needed for a thorough mixing of the gases, and such furnaces, if not forced too hard, are commonly smokeless. This latter quality ought not to be ignored, even though no direct saving in fuel occurs when the distance exceeds 36 inches. Anthracite and bituminous coals cannot be economically burnt in the same furnace, therefore furnaces should be adapted for either one or the other variety of coal.

The height of the level of the grate surface above the fire-room floor is commonly 24 inches, occasionally 26 inches; this latter distance is rarely exceeded.

Ash-Pit.—The vertical depth of the ash-pit is fixed by the height of the grate surface above the fire-room floor; this distance will, therefore, approximate 20 inches clear space if the ash-pit floor be level, which is ordinarily the case. Sometimes the ash-pit floor inclines to the rear, making it deeper at the bridge-wall than at the furnace front. This makes an easier angle for the removal of ashes, especially if the grate-bars are 6 feet or so in length. The bottom of the ash-pit should have a tight brick or cement floor.

Bridge-Wall.—This wall must be strong enough to take the thrust of the implements used in cleaning the fire, and must be thick enough that the joints do not loosen by this action. The top of the bridge-wall should be not less than 13 inches, or 1½ bricks, in thickness, and never less than 18 inches thick where it receives the weight and thrust of the grate-bars. The bridge-wall should be faced next the fire and capped with fire-bricks laid in fire-clay. The top of the bridge-wall is sometimes curved to follow that of the boiler, so as to present an equal area for the passage of the gases. This is altogether a matter of fancy, and no additional valuable results are had over carrying the bridge-wall straight across the furnace. This latter has long been, and is now, the common practice in building bridge-walls. It is customary to make a sloping surface from the top of the grates to the top of the bridge-wall, as shown in Fig. 293. So far as affecting combustion, it makes no difference whether this wall is straight or inclined. Hollow spaces are frequently provided in bridge-walls by which air is introduced into the furnace for the better combustion of bituminous coal. Whilst this device has in many cases been productive of good results, it is not in itself sufficient to make good the loss occasioned by having the furnace itself too small.

Fire-Brick Lining.—Steam-boiler furnaces should be lined with fire-brick at least as far back as the rear end of the bridge-wall. This lining should be capable of repair or complete removal without disturbing the side walls. In making provision for relining a furnace with fire-brick, one or two courses of fire-brick "headers" should be used in finishing the top courses of the lining, as shown in Fig. 270. This will permit the removal of the lower without disturbing the upper courses in the wall. One thickness, 4 or 4½ inches, depending on the width of the brick, will suffice; these bricks should be set in fire-clay mortar. The fire-brick lining should begin at the bottom of the grate-bars, and should include the stepped courses where the top of the furnace is closed in upon the side of the boiler, as in Fig. 270, which represents a wall extending vertically downward from that distance allowed at the top of the furnace for the hot gases to come in contact with the shell of the boiler near the water-line. When the furnace is narrowed to the width of the boiler, or sometimes less, the fire-brick should not be trimmed to fit the angle of the furnace sides, but should be used their full width and backed up by the red brick walls, as in Fig. 271.

The fire-brick lining between two boilers when set singly is carried out in the same manner as that just described for the side walls.

FIG. 270.

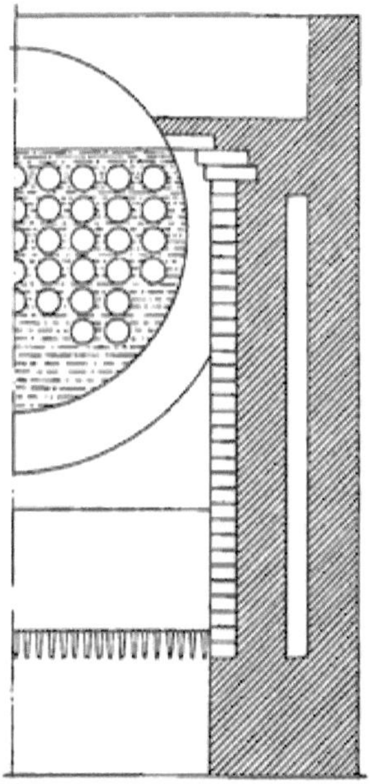

FIG. 271.

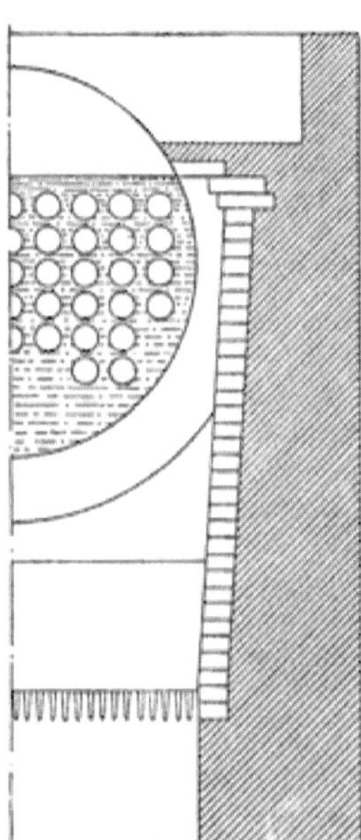

The distance from the side of the boiler at its centre line to the side wall of furnace, as at *A*, Fig. 272, varies in practice from 3 inches for 36-inch boilers to 6 inches for 72-inch boilers. This latter distance is

FIG. 272.

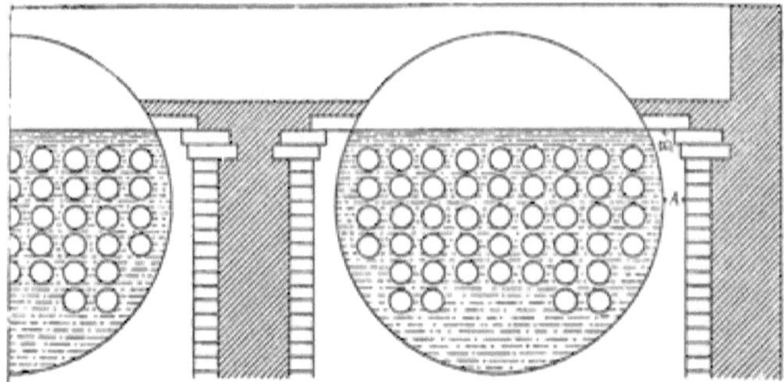

greater than is necessary ; it need not, for all practical purposes, be more than 4 inches. The under side of the top line of fire-bricks where they join the boiler, as at *B*, should not be much higher than the top of the upper line of tubes ; if this distance were extended up to the water-level no harm could occur, but if above the water-level there would be danger of burning the boiler along that line when forcing the fire before steam was raised in the boiler.

Thickness of Walls.—For single walls the thickness should not be less than 2 bricks, varying from 17 to 18 inches, depending upon the locality, and this must be exclusive of any fire-brick lining subject to renewal. The thickness of walls between two boilers ought not in any case to be less than 1½ bricks, say 13 inches, to which is to be added the two fire-brick linings of 4 or 4½ inches each, adding 8 or 9 inches more to the thickness, as shown in Fig. 272.

Double Walls.—A considerable economy can be effected by building the furnace-walls double, with an air-space between the outer and inner walls, as shown in Fig. 270. Walls thus constructed prevent radiation of heat and are, therefore, recommended. When constructing hollow walls headers should extend occasionally, say every 2 feet or so, to give support to the walls and prevent the air-space closing up. These headers should merely touch, and must in no case enter or be fastened to the opposite wall.

The width of the air-space may be 3 inches for boilers up to 46 inches, and 4 inches for boilers 48 inches and larger in diameter.

When two or more boilers are set singly in a battery the thickness of the division walls should be such that an air-space is had of the same width as that of the outer walls.

The walls when made hollow are commonly thinner than when they are single. Any cracking of the interior wall is not likely to cause much leakage into the furnace, because the outer wall may still be intact ; if not, the cracks are in plain sight and can be easily remedied. If the walls carry straight down, as in Fig. 270, each wall and the air-space will be parallel from top to bottom, but if the grate is diminished to the diameter of the boiler, as in Fig. 273, the air-space becomes less in width at the top ; this applies also in the case of the dividing walls, as shown in the same engraving. The rear wall of the boiler setting should have an air-space as well as the side walls. See Fig. 207.

Air-spaces in furnace-walls have been objected to by some engineers, who never include them in furnace designs, the objection being that two thin walls are weaker than a single wall equalling the combined thickness of the two, the thin wall is more liable to crack than a thicker one, and that the saving in radiant heat is not worth the extra cost of the brickwork ; but these objections are not generally held by engineers : when single walls are used, it is usually because the construction costs less money.

Brickwork.—Nearly all furnaces for steam boilers are constructed of red bricks. These should be quite hard, so as to produce a ringing sound when struck ; they must be flat, square to each other when laid together, and of uniform size and color. The sizes vary for different localities, but will average not far from 8⅜ x 4⅛ x 2¼ inches each. The harder the brick the less water it will absorb and the greater will be its ability for carrying a heavy load. The safe working load on a fair quality of hard-burnt red brickwork well bedded in domestic cement mortar is approximately 10 tons per square foot; but furnace-walls are not usually thus laid, the mortar being ordinarily of lime and sand, with very little cement, if any be used at all. For light-colored red bricks in common mortar the safe working load would not be more than half the above. Buff or salmon bricks when resulting from deficient burning should not be used.

Fig. 273.

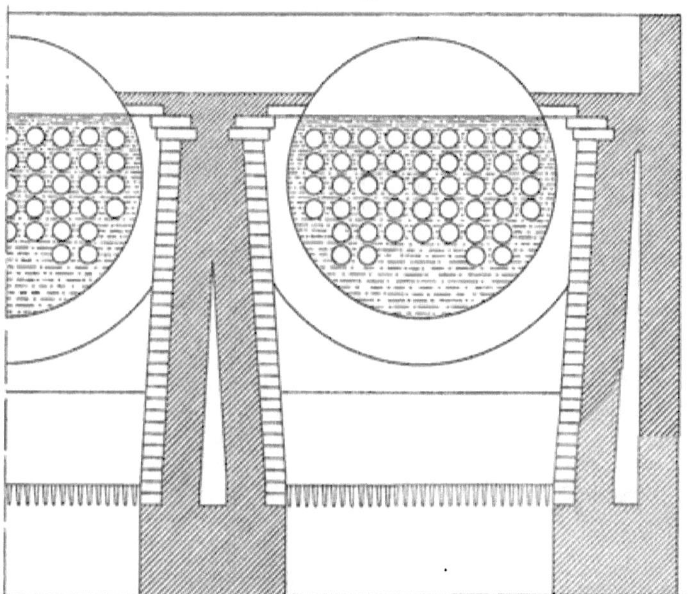

Mortar Joints.—The thickness of mortar joints will approximate ¼ inch; that is, the average height of four courses of common brick work is 10 inches.

The mortar used for boiler settings has commonly been of lime only, but a stronger wall is had if a small proportion of cement be added to it. A very good mortar is made by mixing three parts of good lime mortar with one part of hydraulic cement mortar. The cement may be of domestic manufacture. The sand should be clean and sharp.

Fire-bricks should be laid only in fire-clay. Arch bricks should be carefully fitted dry, and when the arch is completed the bricks should then be set with a thin fire-clay paste, from the same clay as that of the bricks if practicable. The sizes of fire-bricks vary according to locality, but they are commonly larger in every way than red bricks.

Buck-Staves.—The cracking of any brick wall, hot on one side and cold on the other, cannot be entirely prevented. Thick walls offer a greater resistance to cracking, but it is only a question of time when the continued action of the furnace-heat will cause any furnace-wall to crack, regardless of its thickness. When furnace-walls begin to open, the joints should be filled with a thin fire-clay grout, poured into the opening, which will prevent the admission of cold air through the walls into the furnace. A buck-stave is shown in Fig. 274. Stay-rods extend through from side to side of the furnace at top and bottom, as shown. Buck-staves are commonly made of cast iron. Their use is to prevent the spreading of the furnace-walls; to best secure this end, the web should be of considerable depth, from 4 to 6 inches, depending on the diameter of the boiler. These are usually placed 4 to 5 feet apart on the side walls.

FIG. 274.

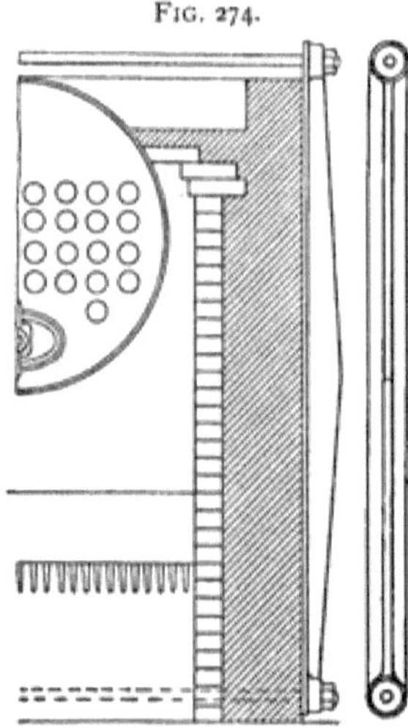

Combustion-Chamber.—The space back of the bridge-wall is popularly known as a combustion-chamber, but it is doubtful if much, or if any, combustion takes place after the gases pass the bridge-wall, or beyond the immediate vicinity of the bridge-wall, in case hot air is admitted at that point. This space is thought by some to be useful as a reservoir for the heated gases, breaking up the rapid current which would otherwise flow in a line parallel to the bottom of the boiler and thence through the tubes to the chimney, not giving out as much heat as would be the case if this volume of hot gases could be interrupted in its passage and proceed with a slower movement, but this argument is far-fetched and probably not true. Boiler tests made with a combustion-chamber, as shown in Fig. 275, have given practically the same results as when filled in, as shown in Fig. 217, showing that it is immaterial whether a combustion-chamber is provided or not. The flow of furnace gases over the top of the bridge-wall will not work downward to fill this large space with heated gases at the same temperature as the furnace, but will follow along the boiler-shell to the rear end, where they turn upward and flow through the tubes to the chimney. As no combustion takes place in this chamber, and the temperature is much higher than that of the atmosphere, the side walls must of necessity radiate heat. For this reason it has become a common practice to fill in the combustion-chamber with earth, as shown in Fig. 295, leaving room enough at the rear end for a cleaning door for taking out the ashes which accumulate there.

Back Connections.—These are constructed usually in two ways, the practice being quite evenly divided between the plate, as shown in Fig. 277, and the arched connections, as shown in Fig. 278. The arch consists of a cast-iron skeleton filled in with fire-brick and afterwards with earth to make it air-tight. The back plate affords easy access to the end of the boiler and plenty of light at the same time. Either of the two are good, but the covering of the joints with earth should not be neglected, or cold draughts of air at that point will lower the temperature of the gases passing into the tubes.

The arched connection at the rear of the boiler is made up of cast-iron segments, as shown in Fig. 279. These rest

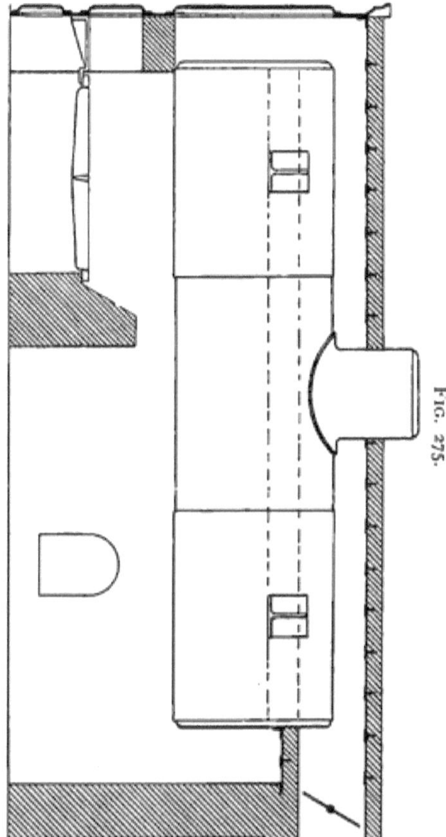

FIG. 275.

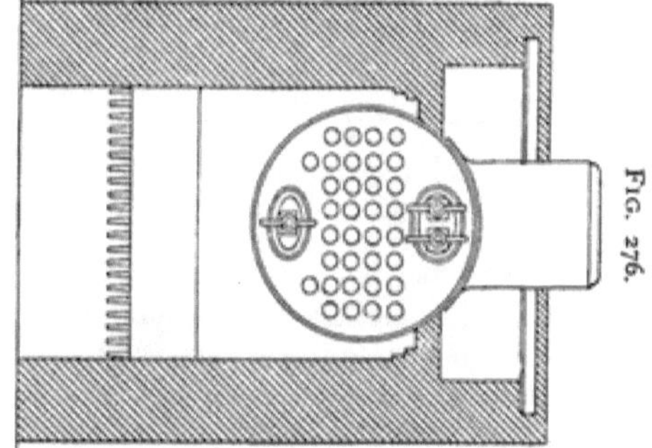

FIG. 276.

Fig. 277. Fig. 278.

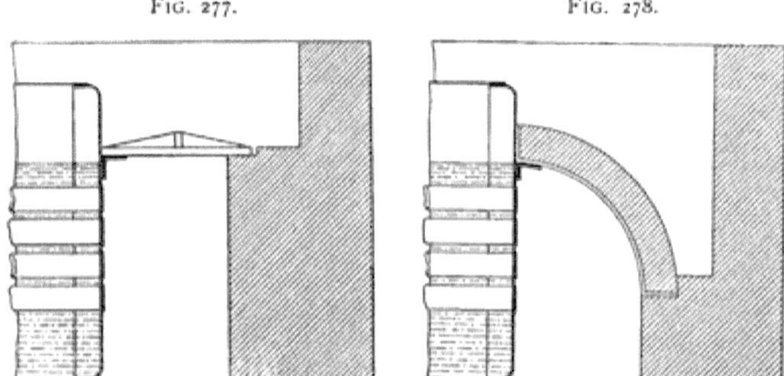

upon the rear wall and against the boiler, an angle-iron about 2 x 2 inches being attached to the boiler-head to form a suitable support. Another form of arched connection is shown in Fig. 280, in which an arch springs from the two side-walls ; this is also shown in Fig. 299.

Fig. 279.

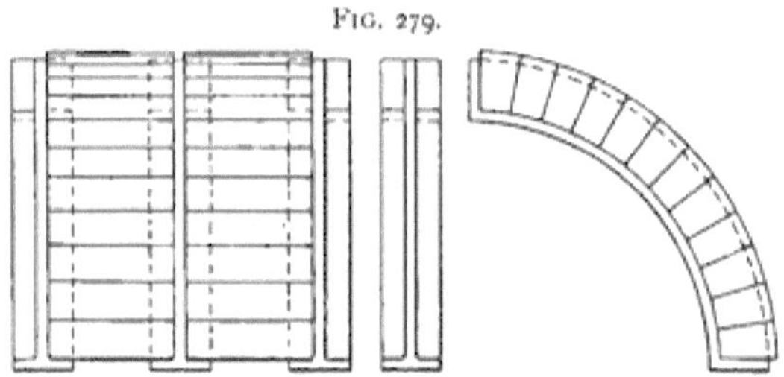

Boiler Covering.—Any non-conducting covering will answer, provided it contains nothing that will act injuriously on the shell of the boiler in case of a steam leak. Boilers ought not to be covered until

Fig. 280.

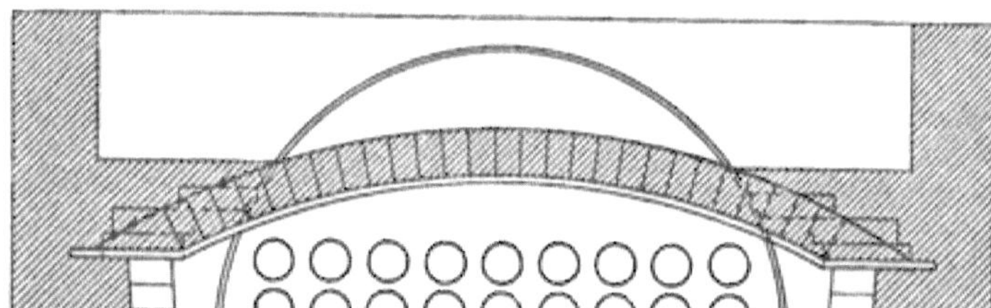

they have steamed sufficiently to be sure that no leaks occur in the joints. A good covering consists in laying narrow strips, say 2 or 3 inches wide, of pine boards 1 inch thick over the top of the boiler from the brick ledges on either side, and then covering all with a 4-inch brick arch resting on these narrow strips, as shown in Fig. 281. The only objection to this covering is, that in case of a leak almost the whole top must come off to get at a joint. Another method is, to cover the whole surface of the top of the boiler with asbestos board about ⅛-inch thick, and on top of this narrow strips of pine, say 2 inches wide and 1 inch thick, and covering again with clay to a depth of 4 to 6 inches.

FIG. 281.

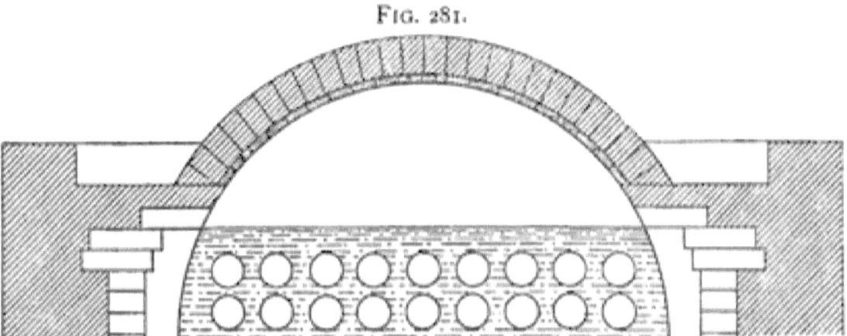

This clay can be easily shovelled off when not wanted, and the pine-strips removed, the asbestos board preventing the loose clay coming in contact with the boiler. Asbestos board with additional hair-felt clothing makes a good covering.

Carrying the products of combustion over the top of a boiler, as shown in Fig. 275, is quite generally practised in some localities. This, of course, prevents any radiation from the boiler, because the temperature of the escaping gases is always higher than that of the steam.

A jacket of hot gas over the top of the boiler, as shown in Fig. 282, is frequently employed, and affords complete protection against heat radiation from the boiler.

Both of these styles of boiler settings have been seriously opposed because of the supposed liability of the hot gases passing over the top of a boiler-shell to cause overheating, especially in the interval between starting a fire under a cold boiler and filling the steam-space with steam. Referring to Fig. 275, the products of combustion from the grate pass along the under side of the boiler to the rear, then return through the tubes to the front, and thence back again, along and over the top sheets of the boiler-shell to the rear, where they are finally conducted to the chimney. The overheating of the top sheets, so far as the writer is aware, has never occurred, nor ought it to be expected in any properly designed boiler setting not using a powerful fan-blast. No injury could accrue to the shell at a temperature below red heat (900° Fahr.), and it is scarcely possible that any such temperature, or more than the half of it, ever reaches the top portion of the shell of a boiler not under steam. In any boiler setting in which there is a ratio of 30 square feet of heating to 1 square foot of grate surface, employing a natural draft, there is probably no danger whatever that the top of the sheets will ever become overheated.

FIG. 282.

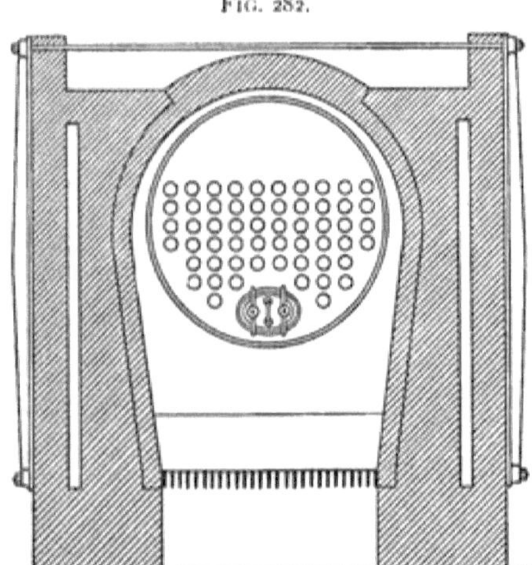

Smoke-Connections.—Flue boilers are almost invariably and tubular boilers quite frequently made with flush ends, without smoke-box extension. The smoke-box is therefore made to bolt directly to the end of the boiler, and for a single boiler is similar to that shown in Fig. 283, in which case an iron chimney may start directly from the smoke-box, or a pipe may convey the products of combustion into a chimney alongside.

FIG. 283.

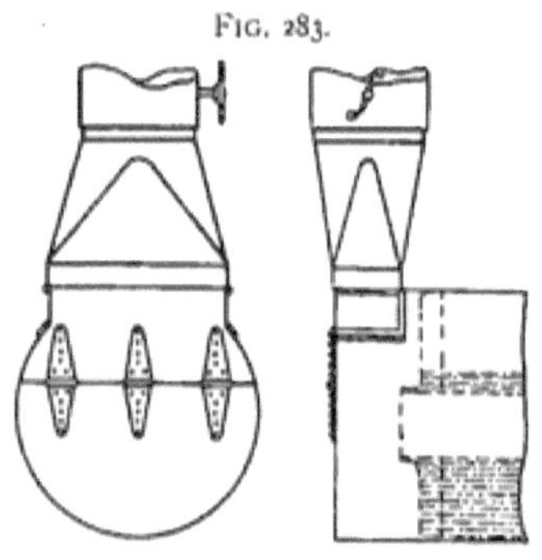

A design for a smoke-connection for two or more boilers is shown in Fig. 284. Each boiler should be provided with its own damper when they are set separately. These smoke-connections empty into a common breeching, from which a central chimney of wrought iron is intended to be fitted, but if a brick or other chimney be located at the side of the boiler, a combined smoke-box and breeching, as shown in Fig. 285, is commonly fitted.

Half-Arch Front.—For tubular boilers the very general practice is to extend the front sheet sufficiently to make a smoke-box by simply

Fig. 284.

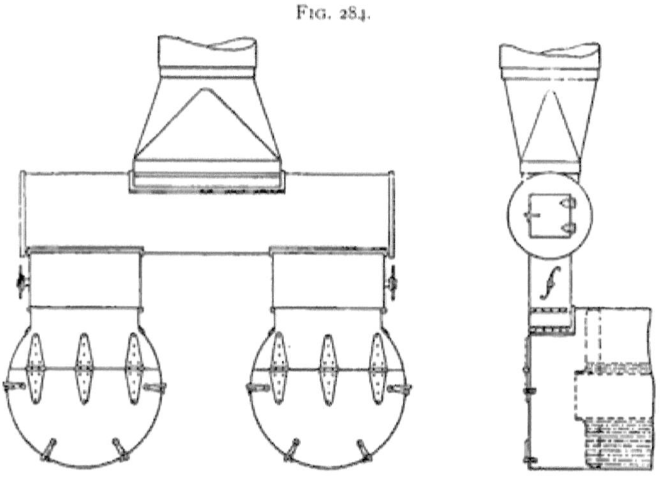

Fig. 285.

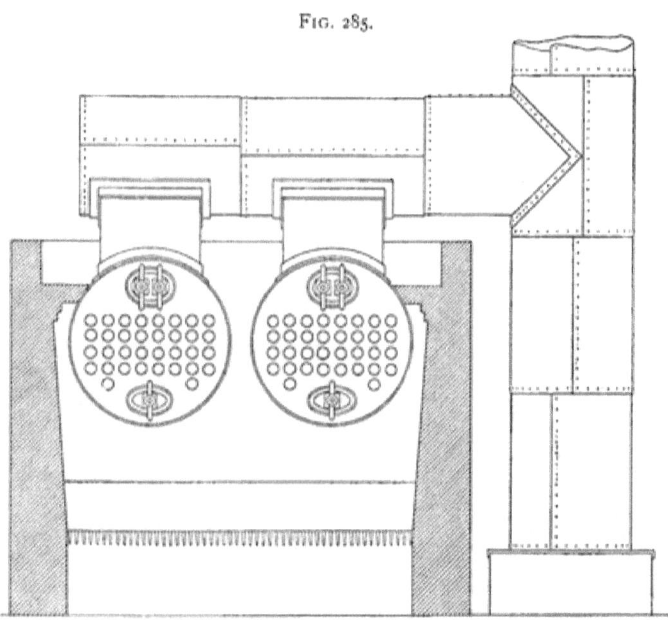

closing the end with a cast-iron front with a hinged door, as shown in Fig. 286. In this case, whatever distance the smoke-box projects beyond, the head extends into the fire-room, and is of more or less annoyance to the fireman. The extension of the shell makes a stronger base

FIG. 286.

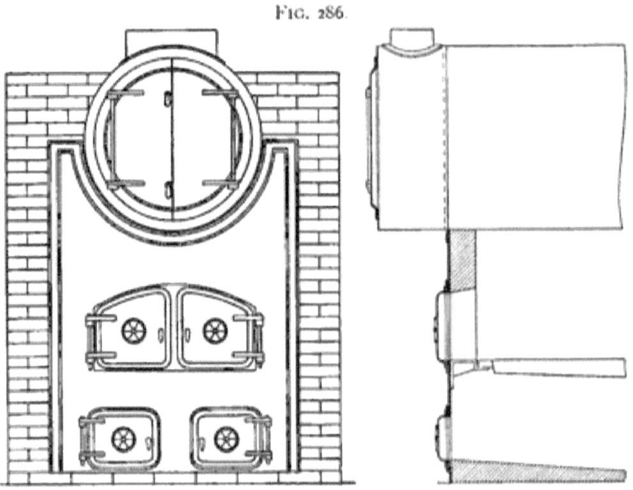

for the wrought-iron chimney to rest upon than when this smoke-box is bolted to the front head; when a brick chimney is employed a wrought-iron pipe connects it with the smoke-box.

Full Square Front.—Another method is to bring the whole front out far enough to include the depth of the smoke-box between the outer line of the fire-front and the inner face of the fire-brick lining. Such a construction is shown in Fig. 287. One objection to this method of boiler-setting is that the fire-brick lining of the front must equal the depth of the smoke-box, and this necessitates fire-door liners from 13 to 18 inches deep, making it somewhat more difficult to attend and clean the fires than is the case with the ordinary depth of linings. A cast-iron plate, Fig. 288, serves as a base for a wrought-iron chimney or to receive a breeching-pipe connecting with a brick chimney at one side. This plate is anchored in the brickwork immediately back of the fire-front. This style of front presents a neat appearance and usually receives more or less of decoration, according to the fancy of the designer.

Naylor's fire-front, shown in Fig. 289, overcomes the objection to the deep fire-brick lining, referred to above, by recessing the cast-iron front, bringing the fire-door to within the narrow thickness of a half-arch front, and filling in the space between the arch over the fire-door and the bottom of the boiler with fire-brick,—a detail clearly shown in the two sectional drawings which accompany the front elevation.

Fire-Doors.—For boilers of the locomotive type and internally fired boilers generally these are commonly fitted to cast-iron frames bolted to the shell of the boiler. Such a door is shown in Fig. 290.

FIG. 287.

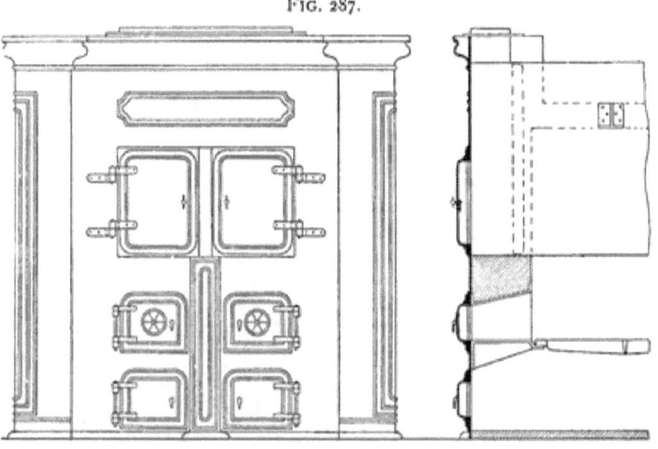

The door and frame should have planed surfaces to make a tight joint; the hinges should be of unusual strength to withstand the rough usage of a fire-room. The perforated lining shown may be of cast iron; the perforations are simply round holes about ½ inch in diameter, and may be drilled in the pattern, so that the rough casting will have all the holes included in it. A bolt passing through the lining, distance-piece, and the door make a simple and substantial fastening. A butterfly register opening is usually included in the door-casting.

FIG. 288.

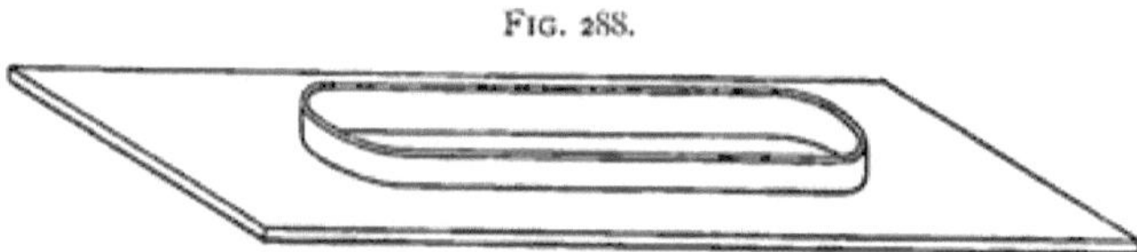

Butman's fire-door is shown in Fig. 291. The frame of the door is bolted against the fire-front, or against the boiler, as the case may be. Unlike the ordinary fire-door, this door is hinged to open upward, the counter-weight being slightly more than sufficient to overbalance the door and raise it when the latch underneath the door is carried down a sufficient distance to release it. The weight has a segment of a gear

Fig. 289.

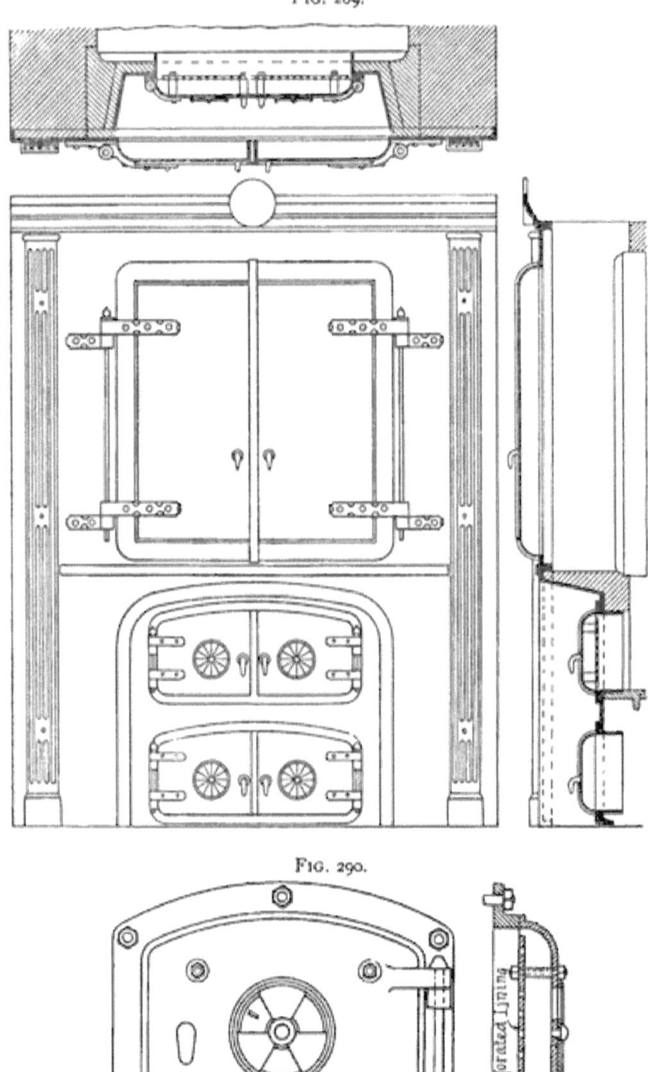

Fig. 290.

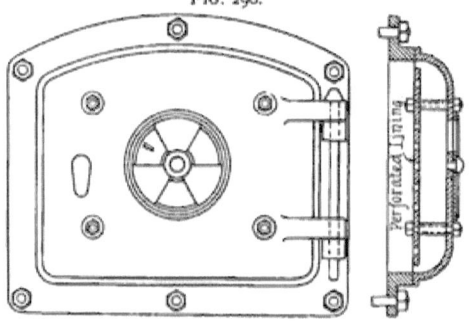

cast inside, as shown in the engravings, into which a similar toothed segment, fitted to the door, is geared, and thus the movements of the door and weight are controlled by each other. On the same central shaft, around which the weight oscillates, is also secured a deflecting-plate. When the door is closed, as in Fig. 291, the deflecting-plate is wholly within the housing. A butterfly register is attached to the door, through which a greater or less quantity of air may be admitted. When the deflecting-plate is down, as shown in Fig. 291, the air passes underneath its lower edges, and is thus brought into close surface contact with the burning fuel. When the furnace door is opened to supply fresh fuel the deflecting-plate is thrown out horizontally, as shown in

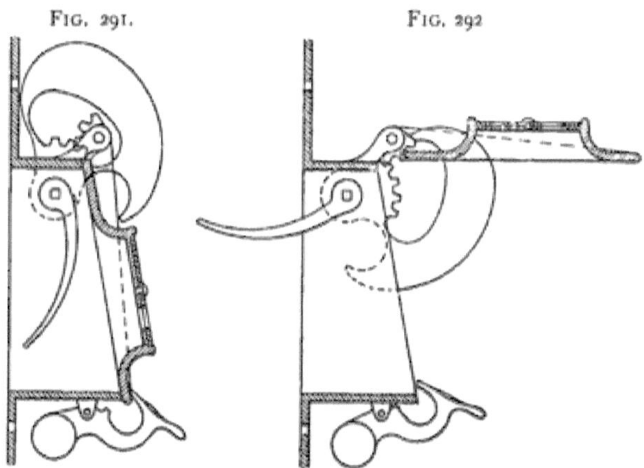
FIG. 291. FIG. 292

Fig. 292. The object of this deflecting-plate is to prevent the cold air from impinging directly against the bottom of the boiler, but to so direct its course that it shall mingle with the heated gases immediately over the fire.

Examples of Furnace Construction.—In Chapter VI., Fig. 209, is given an illustration of the ordinary method of setting cylinder boilers; Fig. 217, that of flue boilers, except that the earth filling is not always included; indeed, for flue boilers this filling is the exception rather than the rule.

Horizontal tubular boilers are set with a considerable variety of detail. Taking the country at large, the ordinary setting shown at Fig. 293 is, perhaps, more largely employed than any other; it represents a boiler suspended on side wings and with a half front, as shown in Fig. 286. The distance from the underside of the boiler to the top of the grates may be 24 inches for anthracite coal, 26 to 30 inches for semi-bituminous coals, and 36 to 48 inches for rich bituminous coals.

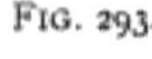

Fig. 293.

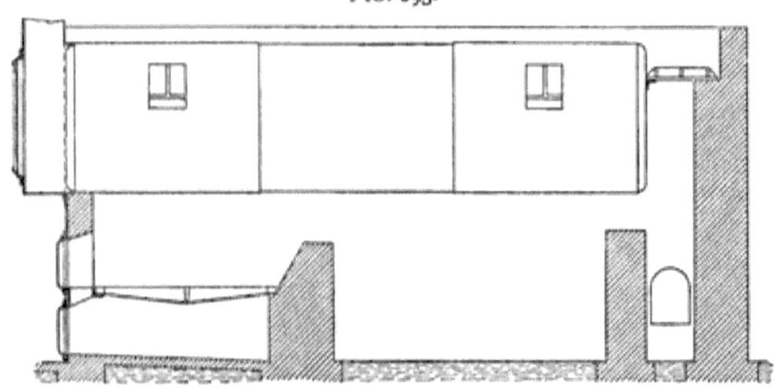

Fig. 294.

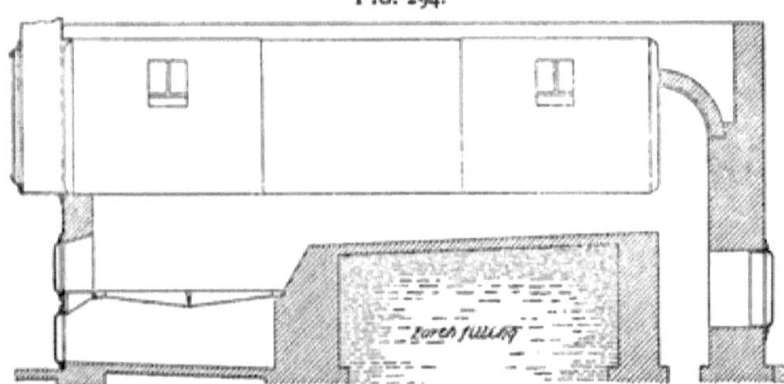

Fig. 295.

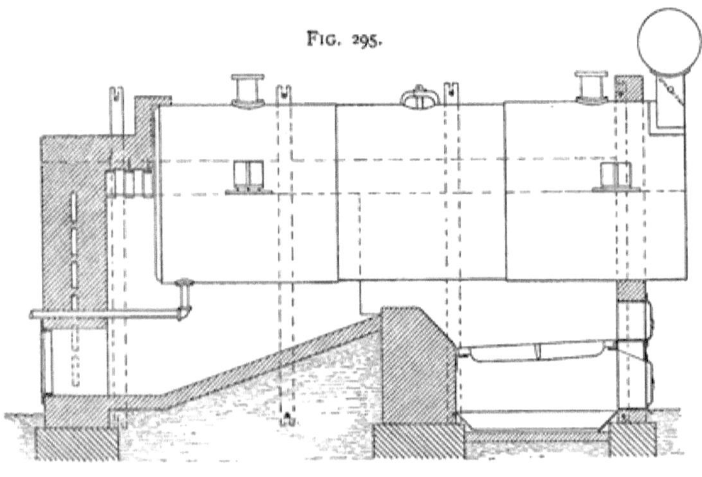

The distance from boiler to top of bridge-wall may be 16 inches for all furnaces, regardless of diameter of boiler or kind of fuel used. The distance from bottom of boiler to top of rear division wall may be 12 inches for all sizes of boilers. It has long been the practice to curve the top of this wall, as shown in the curved line in Fig. 273, but no special advantage is had by so doing.

It is becoming the usual practice to fill in the space between the bridge-wall and the rear division wall with earth, as shown in Fig. 294, covering the top of it with a single layer of red brick; in this case the top of the wall is not curved, but extends straight across the furnace, the same as the bridge-wall.

A design of furnace for horizontal tubular boilers by the Hartford Steam-Boiler Inspection and Insurance Company is shown in sectional elevation in Fig. 295. In this design the rear division wall is omitted; the space back of the bridge-wall is filled in at the inclination shown in the drawing and afterwards covered with brick. A cleaning-door is shown at the rear end of the boiler-wall. A cross-section through the furnace showing the width of grate and the details of the fire-brick lining is given in Fig. 296. A plan of the boiler showing the air-space between the walls is given in Fig. 297; so also the points of suspension of the boiler, showing the details of brickwork upon which the side wings rest. A large number of boilers are set in this manner and yield excellent results.

FIG. 296.

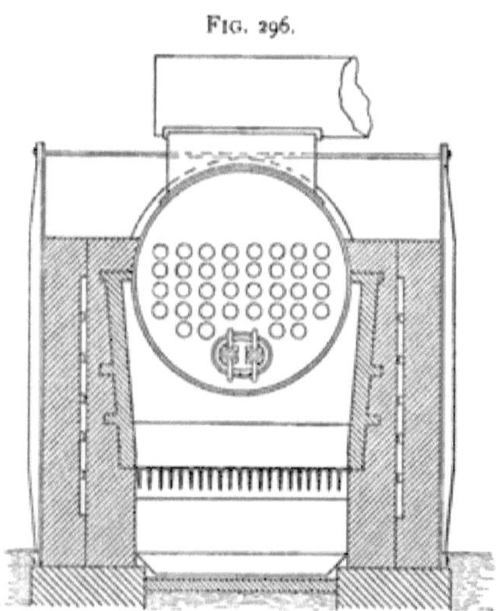

A furnace design for carrying the products of combustion over the top of the boiler is shown in Fig. 275. In this drawing there is no rear division wall and no filling in back of the bridge-wall. This furnace was designed for burning bituminous coal, and it was thought that this rear combustion-chamber might be advantageous in affording a better admixture of gases and better combustion,—results which were probably not realized.

A furnace designed for two boilers, set singly, by Charles Edgerton, is shown in longitudinal elevation in Fig. 298. In this case the combustion-chamber is filled up to the level of the grates. It will be noticed that the bridge-wall is of unusual depth. The products of combustion pass to the rear of the boiler, thence through the tubes to the front, and over the top of the boiler to the chimney. A damper is placed near the exit of the gases for controlling the draft.

FIG. 297.

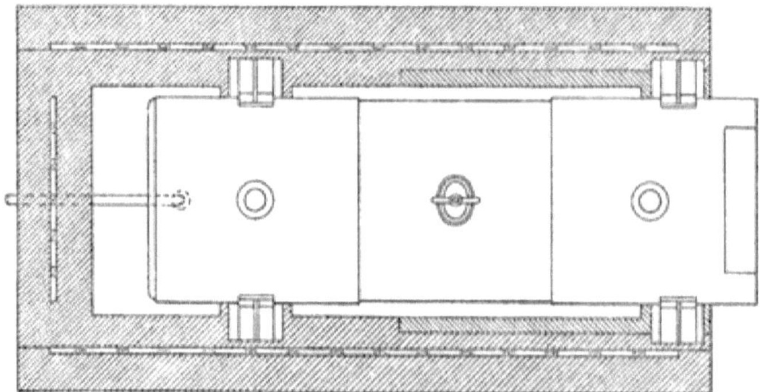

Fig. 299 gives two cross-sectional elevations: one through the furnace showing details of the fire-brick lining, as well as a flue over the top of the boiler for conducting the furnace gases to the chimney; the other half of this engraving illustrates a detail of fire-brick lining to the cast-iron fronts, forming also the smoke-chamber around the sides and at the front end of the boiler. A plan of the flues on top of the boilers is shown in the sectional drawing, Fig. 300. This drawing also shows the location of the chimney, which in this case was made of metal and attached to a cast-iron base-plate resting upon one corner of the boiler setting, as shown.

The horizontal tubular boiler setting shown in sectional elevation, Fig. 301, is by W. Barnet LeVan. The boiler has a superheating drum attached by a wrought-metal connection, also shown in the illustration. This detail was prepared in recognition of the fact that steam cannot ordinarily be superheated when in contact with the water from which it was generated. The drum is, therefore, isolated and attached to the main shell by a single connection, located at the front end of the boiler. The delivery of steam is from the rear end of the superheating drum. The latter is of sufficient size to allow ample time for the highly heated gases which outwardly surround and also pass through the tubes in the drum to evaporate any entrained water in the steam-drum before it passes off to the engine.

It will be noted that in the furnace details a bridge-wall of somewhat unusual construction is provided. The combustion-chamber beyond is

Fig. 298.

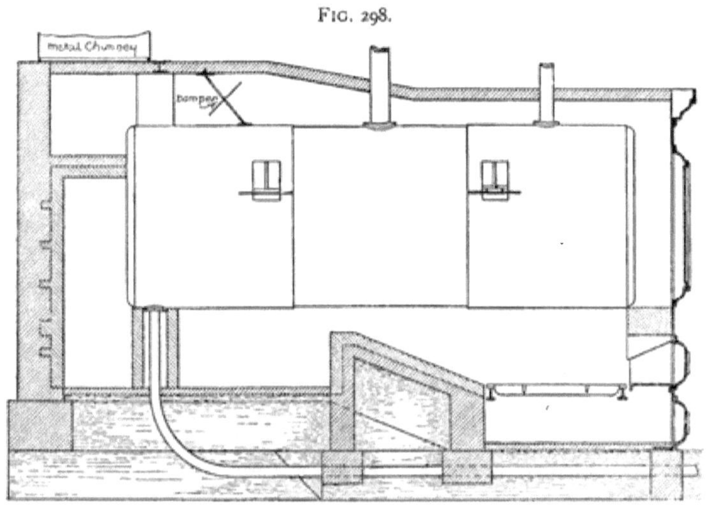

Fig. 299.

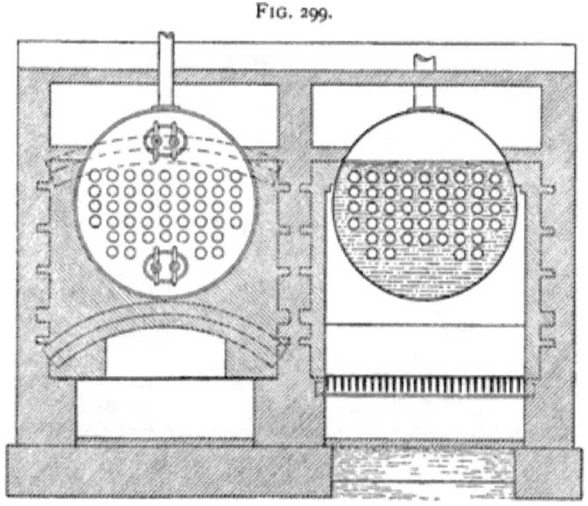

FIG. 300.

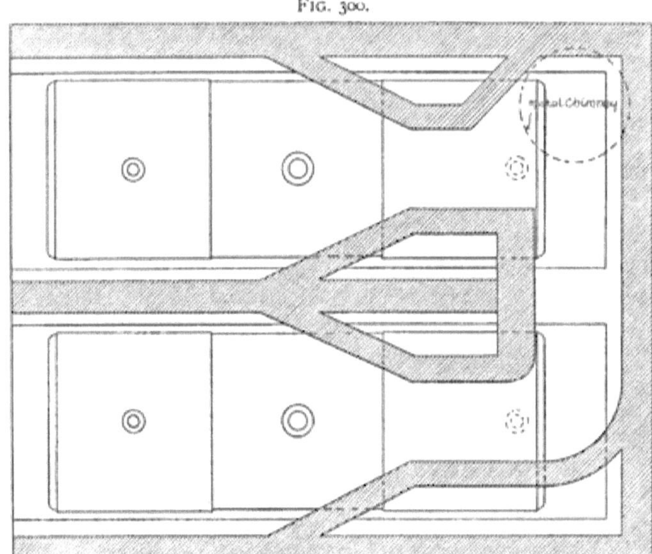

FIG. 301.

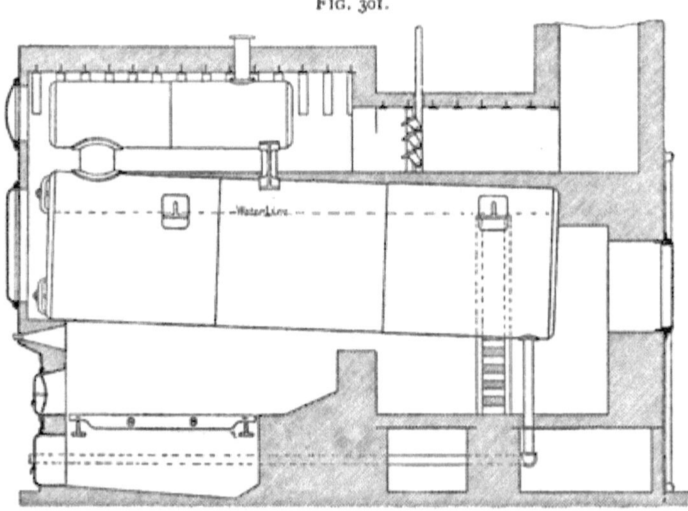

on the same level with the grates. In this combustion-chamber is a rear division wall, with checkered openings, as shown in the cross-sectional elevation, Fig. 302. This wall becoming highly heated is intended to act as a regenerator and to promote combustion of the uncombined gases on their way to the rear of the boiler. The gases are baffled in their flow, and in passing through these openings are thoroughly mingled and receive an additional supply of heated air conducted into this chamber through register openings not shown in the engraving. After the gases pass through this division wall, they are then in the rear chamber; from thence they pass through the tubes of the boiler to the front end. They are then conducted upward into another chamber, in which is located the superheating drum. The gases not only surround the drum, but pass through the tubes with which this superheating drum is fitted. They then pass through a damper of somewhat unusual construction, and from thence to the chimney.

Fig. 302.

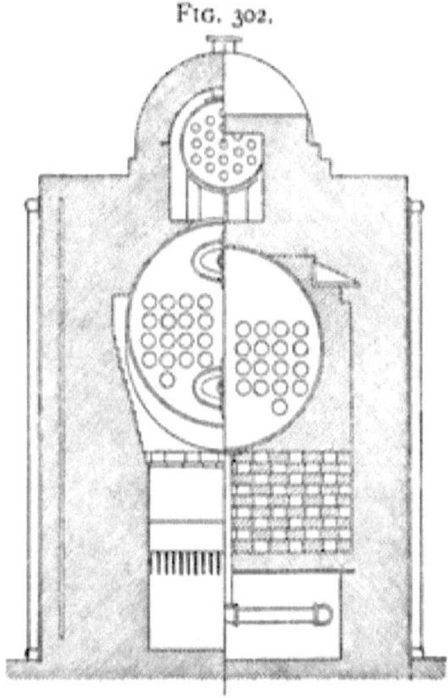

CHAPTER VIII.

INTERNALLY FIRED BOILERS.

INTERNALLY fired boilers are those in which the furnace is included within the structure of the boiler itself. The commonest varieties of internally fired boilers are :

The ordinary vertical tubular boiler, a cylindrical shell in which is enclosed a cylindrical fire-box, with numerous tubes leading directly from the furnace to the upper head of the shell. Sometimes vertical boilers are made with a single flue, especially in localities where the water is bad.

The Cornish boiler, a cylindrical horizontal shell fitted with a single horizontal flue, one end of which is made the furnace. This boiler is rarely met with in this country.

The Lancashire boiler, a cylindrical horizontal shell with two horizontal flues and furnaces.

There are a variety of designs for marine and land boilers, consisting in the main of a shell of large diameter in which are one or more large flues fitted with grate-bars ; the furnace being constructed within the flue, the products of combustion are returned from the combustion-chamber at the rear end of the boiler forward to the fire-room, and from thence to the chimney.

The rectangular enclosed fire-box combined with an outer shell, to which is attached a cylindrical shell containing numerous horizontal tubes leading directly from the furnace to the smoke-box. It is a type of boiler well known to engineers as the locomotive boiler, though its use is not restricted to such service. Such boilers are constructed in great variety, both as to size and detail.

Vertical Tubular Boilers.—The ordinary construction of a vertical tubular boiler consists of an outer shell containing a cylindrical fire-box and vertical tubes, as shown in Fig. 303. Boilers of this type are made in considerable numbers for small powers in combination with steam engines, pumps, etc. These boilers are ordinarily fitted with tubes 2 inches in diameter, and the largest diameter of tubes even for very large boilers seldom exceeds 2½ inches. The upper end of the tubes are located in the steam-space, and result in giving the steam a slight superheating.

For ordinary vertical boilers as many tubes are placed in the top of the furnace or crown-sheet as can be conveniently arranged ; the diameter of furnace is made as large as the outer shell will permit. The aggregate tube area will approximate one-fourth that of the grate area, a higher ratio than obtains in horizontal boiler practice.

The fire-box affords an excellent evaporating surface, and for that reason it should be as high as possible and maintain a good circulation between it and the outer shell. The circulation will be improved if the fire-box be made slightly conical. The furnace height will be governed somewhat by the fuel to be used: if bituminous coal, the height may be from 30 to 48 inches, according to the size of the boiler; if anthracite coal, the height may be from 24 to 36 inches. In general, furnaces for soft coal should be high enough to form an ample combustion-chamber and prevent the formation of smoke, the usual accompaniment of low furnaces and small combustion-chambers.

Fig. 303.

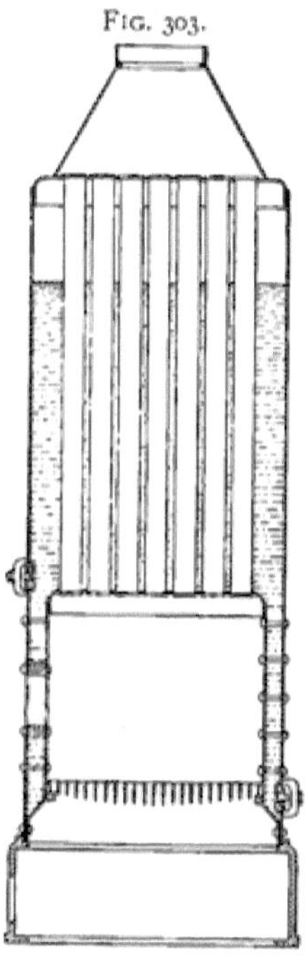

The length of tubes in vertical boilers will vary according to the purpose for which they are designed. These boilers are largely used by contractors, who move them around from place to place, according to the necessities of their business. As economy of fuel is not so generally practised in contracting as in a fixed business, an increase in diameter with its larger fire-box is usually preferred to that of height, which gives merely additional length of tube. If, however, it is to be used as a stationary boiler, the height should be as great as is permissible for the location. Short tubes under heavy firing are apt to burn out at the upper tube-sheet, and this is one argument for increasing their length, that more absorbing surface may intervene, with the consequent reduction in temperature of the gases escaping at the chimney. Notwithstanding the increased length of the tube, there is danger at all times in overheating the tops of the tubes when raising steam from cold water, and for this reason the fire should not be urged until after steaming has begun. To obviate this fault, boilers are sometimes made with submerged tubes, as in Figs. 304 and 305, which show two methods of constructing the upper chamber. The conical one is to be preferred, as giving more steam-room than would be the case with a chamber having vertical sides and a flanged head. The cost of manufacture would also be less, but as no protection is afforded the plate surface, which is thus made to take the place of the tube surface, the change is at best one of doubtful utility.

TABLE XLV.

PRINCIPAL DIMENSIONS OF VERTICAL BOILERS WITH FULL-LENGTH TUBES AS FURNISHED BY THE TRADE, FIG. 303.

Commercial Rating.	Shell.			Furnace.			Flanged Heads.	Tubes 2 Inches in Diameter.		Heating Surface.	Diameter of Stack.
	Diameter.	Height.	Thickness.	Diameter.	Height.	Thickness.	Thickness.	Length.	Number.		
H.-P.	Ins.	Ft.	In.	Ins.	Ins.	In.	In.	Ins.		Sq. Ft.	Ins.
4	24	4	1/4	20	24	1/4	3/8	24	31	44	12
5	24	5	1/4	20	24	1/4	3/8	36	31	60	12
6	24	6	1/4	20	24	1/4	3/8	48	31	75	12
8	30	5	1/4	25	27	1/4	3/8	33	55	92	14
10	30	6	1/4	25	27	1/4	3/8	45	55	121	14
12	30	7	1/4	25	27	1/4	3/8	57	55	150	14
15	36	6 1/2	1/4	31	27	1/4	3/8	51	77	189	15
18	36	7	1/4	31	27	1/4	3/8	57	77	210	15
20	36	8	1/4	31	27	1/4	3/8	69	77	250	15
25	42	7 1/4	9/32	37	27	1/4	3/8	60	109	307	18
30	42	8 1/4	9/32	37	27	1/4	3/8	72	109	364	18
35	42	9 1/4	9/32	37	27	1/4	3/8	84	109	422	18
40	48	8 1/2	5/16	43	30	1/4	3/8	72	149	496	20
45	48	9	5/16	43	30	1/4	3/8	78	149	535	20
50	48	10	5/16	43	30	1/4	3/8	90	149	613	20
60	54	9	5/16	48	30	1/4	3/8	78	201	716	24

TABLE XLVI.

PRINCIPAL DIMENSIONS OF VERTICAL BOILERS WITH SUBMERGED TUBES, AS FURNISHED BY THE TRADE, FIG. 304.

Commercial Rating.	Shell.			Furnace.			Flanged Heads.	Tubes 2 Inches in Diameter.		Height of Chamber.	Heating Surface.	Diameter of Stack.
	Diameter.	Height.	Thickness.	Diameter.	Height.	Thickness.	Thickness.	Length.	Number.			
H.-P.	Ins.	Ft.	In.	Ins.	Ins.	In.	In.	Ins.			Sq. Ft.	In.
4	24	5 1/2	1/4	20	24	1/4	3/8	24	31	18	44	12
5	24	6	1/4	20	24	1/4	3/8	30	31	18	52	12
6	24	6 1/2	1/4	20	24	1/4	3/8	36	31	18	60	12
8	30	6	1/4	25	27	1/4	3/8	27	55	18	83	14
10	30	6 1/2	1/4	25	27	1/4	3/8	33	55	18	98	14
12	36	6 1/2	1/4	31	27	1/4	3/8	33	77	18	133	15
15	36	7	1/4	31	27	1/4	3/8	39	77	18	155	15
18	36	8	1/4	31	27	1/4	3/8	51	77	18	196	15
20	42	7 1/2	9/32	37	27	1/4	3/8	39	109	24	215	18
25	42	8	9/32	37	27	1/4	3/8	45	109	24	244	18
30	42	9	9/32	37	27	1/4	3/8	57	109	24	301	18
35	48	9	5/16	43	30	1/4	3/8	51	149	27	370	20
40	48	9 1/2	5/16	43	30	1/4	3/8	57	149	27	409	20
45	48	10	5/16	43	30	1/4	3/8	63	149	27	448	20
50	54	9 1/2	5/16	48	30	1/4	3/8	54	201	30	518	24
60	54	10 1/2	5/16	48	30	1/4	3/8	66	201	30	623	24

A vertical flue boiler, as shown in Fig. 306, was formerly much used in the South and West for small powers in localities where the water was bad. The absence of tubes made it a comparatively easy matter to rid the boiler of any accumulated scale on the crown-sheet. The fire-box in boilers of this kind was commonly higher than was the case with tubular boilers; the average height was approximately one-half the vertical height of the boiler. In some cases an internal sheet-metal lining was inserted in the central flue as far down as the water-line, the object of which was to prevent overheating the flue when getting up steam from cold water. This kind of boiler yields fairly good results, but is not as much in use as formerly.

FIG. 304. FIG. 305. FIG. 306.

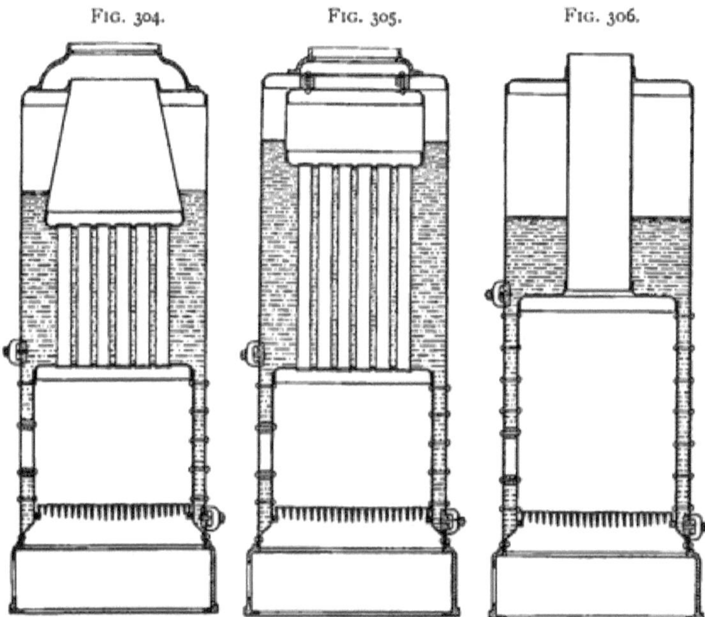

The expansion of the interior portion of a vertical tubular boiler over that of the outer shell, when arranged as in Fig. 303, is considerable, caused by the upward movement of the fire-box as well as the lengthening of the tubes, these being of higher temperature than the outer shell, to which they are both attached. This expansion generally manifests itself by the continual leakage of tubes in the crown-sheet.

Economy of floor space combined with moderate first cost has contributed much towards making vertical boilers popular when the conditions are favorable for their installation, the water actually evaporated per pound of coal equalling that of any other type. The principal objection to vertical boilers of large power is the height, a 100-horse-power boiler approximating 24 feet. On the other hand, vertical boilers secure an economy of floor space not equalled by any other type. The boiler above referred to would require not more than 7 feet square for its foundation; a horizontal tubular for the same power would require a space of about 10 x 20 feet.

Cleaning a vertical boiler and removing any deposits on the crown-sheet is at all times a difficult and uncertain operation, the causes for which are easily to be seen by an inspection of the arrangement of tubes shown in Fig. 307, which fairly represents the common method of tube spacing and construction. A better arrangement of tubes is

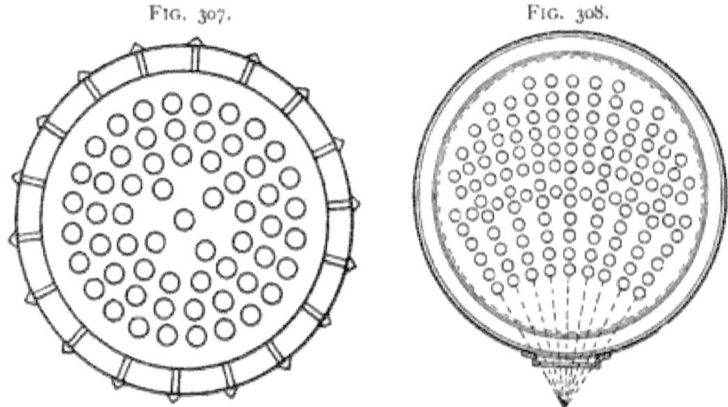
FIG. 307. FIG. 308.

that designed by Reynolds for vertical tubular boilers, and shown in Fig. 308. Handhole-plates must be inserted at the crown-sheet level, as well as at the bottom of the water-leg, for cleaning purposes, not less than three at each line.

Priming is a common fault in vertical tubular boilers, due mostly to the insertion of an unnecessarily large number of fire-tubes in the fire-box head.

The commercial rating of vertical tubular boilers is 12 square feet of heating surface per horse-power, but this is a misleading factor, because one-quarter to one-half of the tube surface is above the water-level, and in no wise assists evaporation. The tube surface not being as effective as the fire-box surface for evaporation, the area through the tubes need not, therefore, be more than that necessary for the proper escape of the gases. The fire-box heating surface is highly effective and does the greater part of the work, but, making allowance for this, no less than 12 square feet of heating surface should be allowed per horse-power. Retarders or radiators may be used to advantage in vertical-boiler tubes, descriptions of which are given on pages 197 and 199.

The Reynolds Boiler.—The tubes in this design are set in rows radiating from a large manhole located over the fire-door and bottom tube-sheet, as shown in Fig. 308. The tubes and crown-sheet over the furnace can be inspected and cleaned when the manhole cover is removed. Handholes are located opposite the manhole for admitting light for inspecting and inserting a hose-nozzle for washing the tubes and crown-sheet. Handholes are placed at intervals around the base, whereby any sediment collected in the water-legs may be removed. The feed-water is pumped into the internal reservoir through the feed-pipe, shown in the sectional elevation, Fig. 309. This reservoir being closed at the bottom, the discharge into the boiler is over the top, and, it being so much larger than the feed-pipe, the current upward is very slow; consequently the feed-water gains the same temperature as the water in the boiler before it is discharged into the boiler. This action is effective in precipitating nearly all of the heavy impurities carried in with the feed-water, which can be blown out of the reservoir by a blow-off arranged for the purpose. By carrying the water in the boiler slightly above the top of the reservoir, it can be utilized as a surface blow-off to free the boiler of scum or light impurities collected on the surface of the water.

The smoke-hood on top of the boiler is furnished with a revolving top having a movable cover. For the purpose of cleaning the flues this cover is removed, and only a small portion of the total number of flues is exposed at one time. This arrangement enables the fireman to clean the flues while the boilers are in operation.

Vertical boilers usually furnish dry steam by reason of the tube surface above the water-level. Frequent tests of this boiler show from 10° to 40° Fahr. superheating.

The Manning Boiler.—This boiler is represented in sectional elevation in Fig. 310, accompanied by three cross-sectional drawings representing sections through the fire-box, the tubes, and the smoke-box. It will be observed that the fire-box is much larger than is common in vertical tubular boiler construction. In all cases it is larger than the diameter of the waist, or upper cylindrical shell. This is made possible by the double-flanged connection between the waist and the outer fire-box shell. This connection serves another purpose, in providing for the expansion and contraction between the tubes and the outer shell. The ability to increase the diameter of the fire-box to the exact point where the proportion of grate area to heating surface is such as to give the best possible results with the most economical firing is a valuable one.

The tubes have their length so proportioned to their diameter that the temperature of the escaping gases is not higher than that necessary to produce a good draught, about 500° Fahr., the proportion in the case of a 100 horse-power boiler being 15 feet in length for a 2½-inch

Fig. 309.

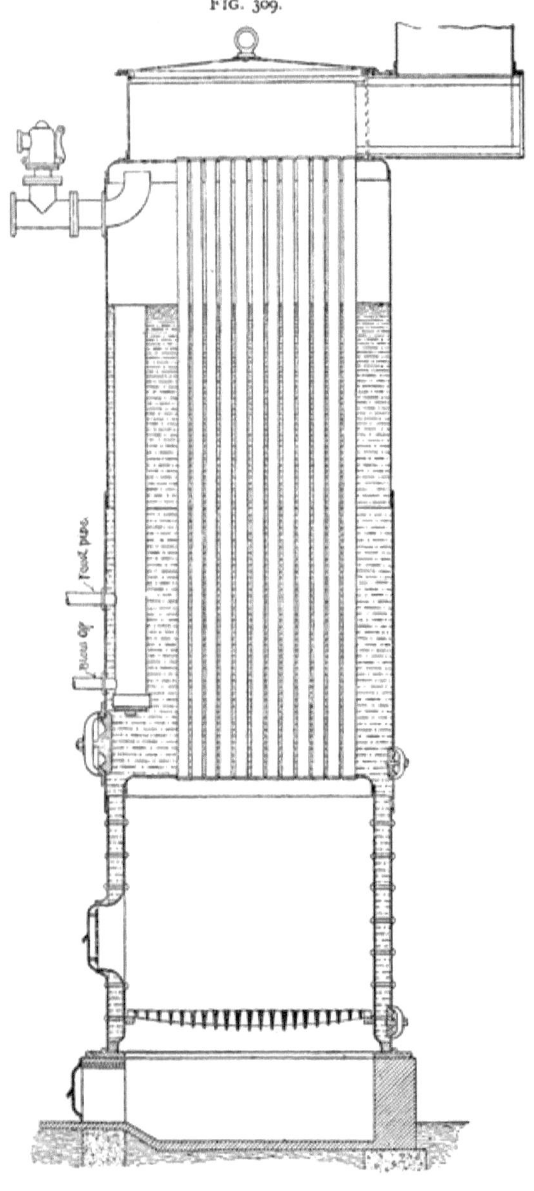

Fig. 310.

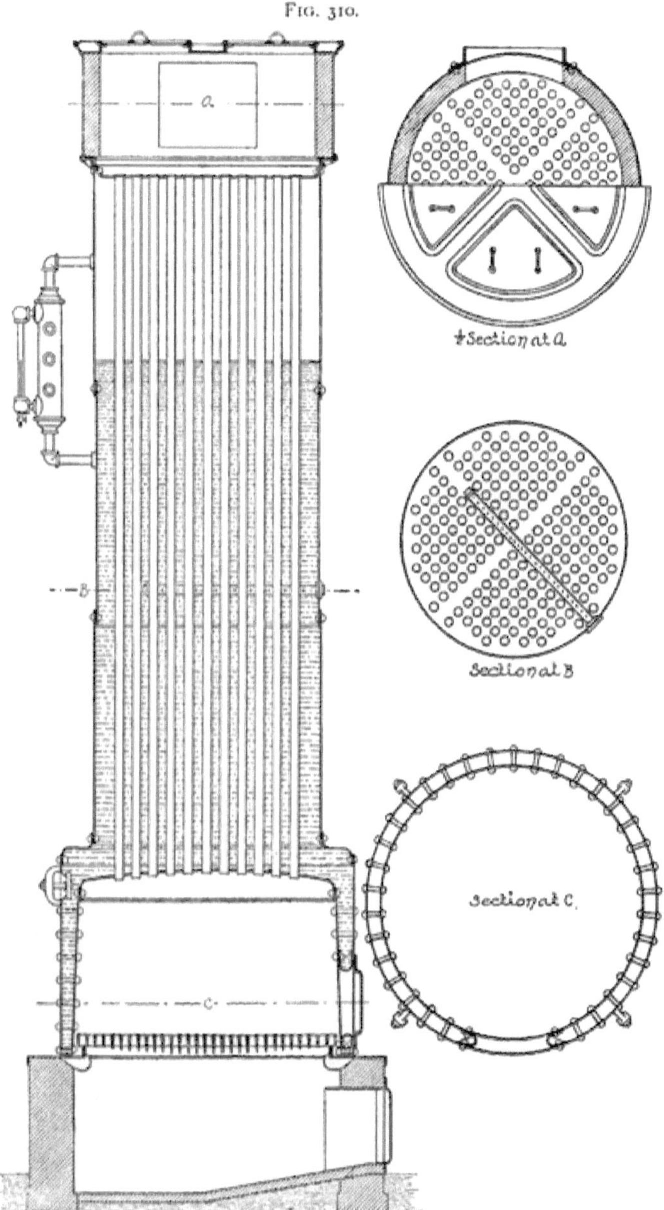

tube, or 72 to 1. This proportion insures ample draught and prevents injury to the tubes in the top head that are unprotected by water. Too much draught in a vertical boiler is worse than too little, as the fuel is wasted and, what is even worse, the tubes are soon burnt out and destroyed. With an insufficient draught the boiler will, of course, fail in efficiency.

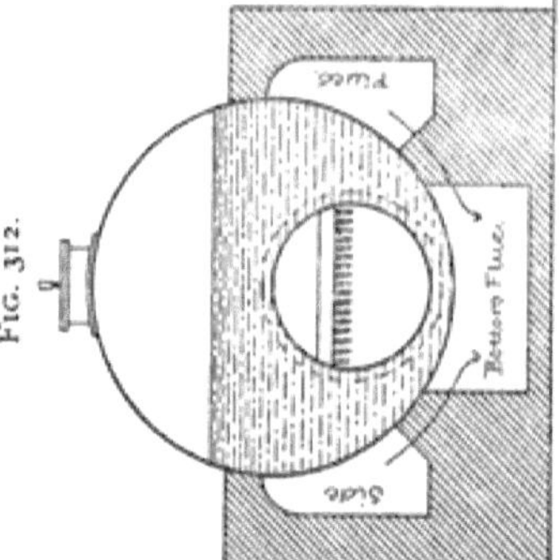

Fig. 312.

The outer fire-box shell is carried well above the head, and handholes are placed exactly on a line with the crown-sheet. The tubes are placed in straight rows, and at right angles to one another extend two cleaning-channels of ample size, shown in sections *A* and *B*. A bent tube connected with a hose can be inserted through the handholes and between the rows of tubes, whereby the crown-sheet can be thoroughly washed and cleaned.

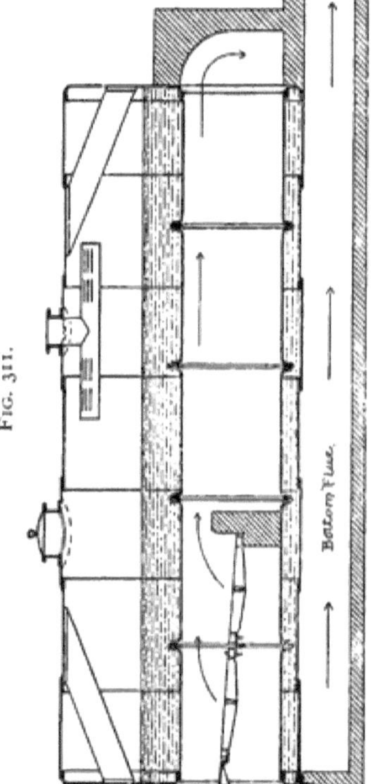

Fig. 311.

The outer-shell plates, which bear the greatest strain and which it is impossible to brace or strengthen by stay-bolting, receive no heat from the fire and can therefore be made of any required thickness. In the fire-box, on the other hand, where the sheets come in direct contact with the fire, they can be made thin enough to prevent their burning, and the requisite strength is gained by stay-bolts. The tubes, being of standard dimensions, have a thickness far in excess of that required to enable them to bear any collapsing strain to which they may be subjected.

In the water-leg are placed a number of handholes and a cleaning-chain, by means of which any sediment that may have accumulated there can be stirred up and removed.

The Cornish Boiler, Figs. 311 and 312, derives its name from the circumstance that boilers of this type were first used in the Cornish mines; they are still used in England for small and medium powers. It consists of an outer shell, within which is a flue of sufficient size to permit its being fitted with a furnace for the combustion of the fuel. The products of combustion pass to the rear end of the boiler, where they divide and return along the sides of the boiler to the front, where they are again united and pass into a flue underneath the shell to the rear end of the boiler and from thence to the chimney. The external flues alongside the boiler are built of ordinary red bricks lined with fire-bricks.

By reason of the large diameter of the flue and its liability to collapse under a high pressure, the latter was formerly restricted to 45 pounds per square inch, but with improved construction these boilers are now made for any ordinary pressure, though commonly not more than 100 pounds.

The principal dimensions of the ordinary sizes used in England are:

Diameter of shell, 3 feet 6 inches, 4 feet 3 inches, 5 feet, 5 feet 6 inches, 6 feet.
Length of shell, 8 feet, 12 feet, 15 feet, 18 feet, 22 feet.
Diameter of flue, 2 feet 2 inches, 2 feet 4 inches, 2 feet 9 inches, 3 feet 3 inches, 3 feet 6 inches.

The efficiency of the Cornish boiler was tested incidentally in a series of fuel experiments made in England, in which it appeared that in a mean of thirty-seven experiments, using Welsh coal, 9 pounds of water at 212° Fahr. were converted into steam at any working pressure per pound of fuel; a mean of eighteen experiments, using Newcastle coals, gave 8.37 pounds of water evaporated as above; and a mean of twenty-eight experiments, using Lancashire coals, gave 7.94 pounds evaporation under the same conditions.

A Cornish boiler 6 feet in diameter by 28 feet long, having a single flue 3 feet 6 inches in diameter, fitted with grates 4 feet 6 inches long, yielding 15 square feet of net grate surface, burned 7.24 pounds of coal per square foot of grate, or 0.19 pound per square foot of heating surface per hour, evaporating 9.9 pounds of water from 51° Fahr. into steam at 60 pounds pressure by gauge, an evaporation of 11.86 pounds of water from and at 212° Fahr. per pound of bituminous coal. Taking into account the calorimetric value of the coal, the efficiency of the boiler was 0.77.

Lancashire Boilers.—When the outer shell of a boiler on the Cornish plan exceeds 5½ or 6 feet in diameter, a flue of excessive diameter would be required to get a proper width of grate. It is well known that flues are less able to resist a collapse as they increase in diameter. Inasmuch as a proper width of grate can be secured by the use of two smaller flues without the risks attending the use of one large flue, it is a better construction if boilers are thus fitted, and this particular construction of the Cornish boiler is called a Lancashire boiler.

The principal dimensions of the three leading sizes used in England are here given :

Diameter of shell, 6 feet, 6 feet 6 inches, 7 feet.
Length of shell, 20 feet to 28 feet, 20 feet to 30 feet, 24 feet to 30 feet.
Diameter of each flue, 2 feet 3 inches, 2 feet 6 inches, 2 feet 9 inches.

In practical working it is customary to fire the furnaces alternately, so that while the one is giving off smoke and unburnt hydrocarbon gases, the other is burning briskly and with the greatest heating effect. By this arrangement, when the gases from the two furnaces mix in the external flues, the unburnt gases given off by the freshly charged fire are burned by the excess of air which has passed through the other furnace, being raised to the point of ignition by the great heat of the gases from the bright fire.

The Galloway Boiler.—This boiler, as manufactured by the Edgemoor Iron Company, is a modification of the Lancashire boiler. A sectional elevation of a boiler 7 feet in diameter by 28 feet in length is given in Fig. 313. It has two furnaces 2 feet 10 inches in diameter by 7 feet 9 inches in length ; the metal is ⅜ inch thick, welded seams, flanged at each end to form three lengths riveted together with a central stiffening ring similar to the detail shown in Fig. 161. These two furnaces merge into a combustion-chamber of segmental cross-section, shown in Figs. 314 and 315. This chamber is fitted with tapered water-tubes for the purpose of increasing the effective heating surface of the boiler and to promote a better circulation of water ; they also act as stays, largely increasing the strength of the flue to which they are fitted. The diameters of the flanges and necks of these tubes are such that an opening which will allow the smaller flange to pass through will be of proper diameter for the inside of the larger flange. They are riveted in place as shown.

A sectional-plan view of the boiler is given in Fig. 316, in which the arrangement of the two furnaces and the conical water-tubes is clearly shown. This view also shows six corrugations along each side of the combustion-chamber, adding to the stiffness of the latter and assisting in the deflecting and diffusion of gases back of the furnace.

The shell of the boiler is ½ inch thick, heads $\frac{9}{16}$ inch thick, the fire-box and conical tubes are ⅜ inch thick ; all the other plates are ½ inch thick except the gusset-plates, which are $\frac{7}{16}$ inch. The gusset-plates are shown in Figs. 317 and 318, which represent front- and rear-end views of this boiler.

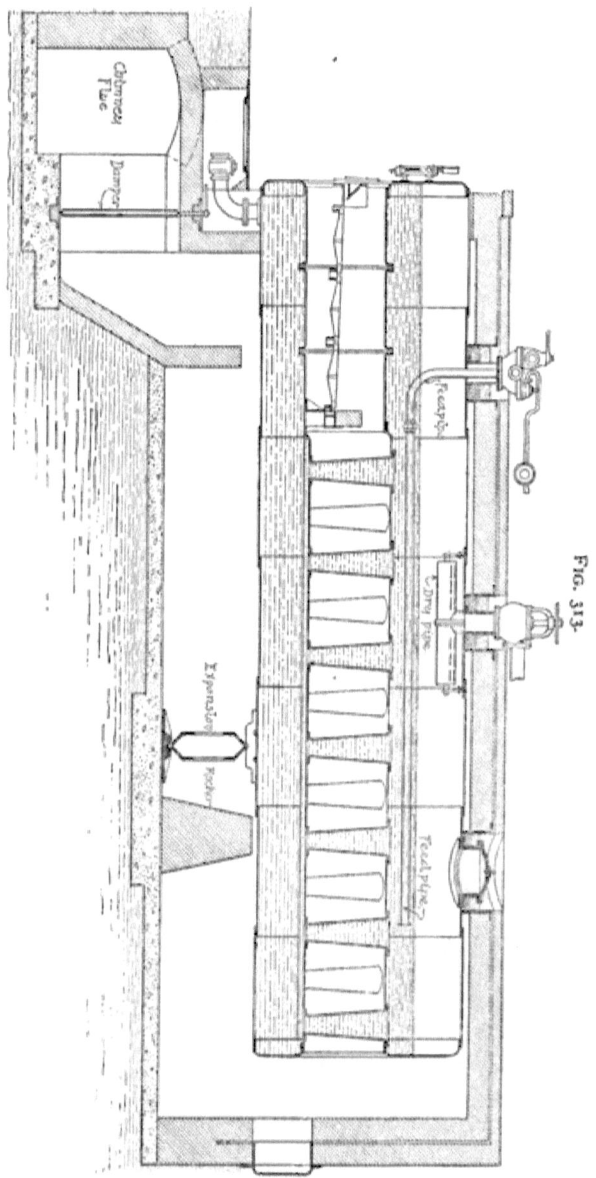

FIG. 313.

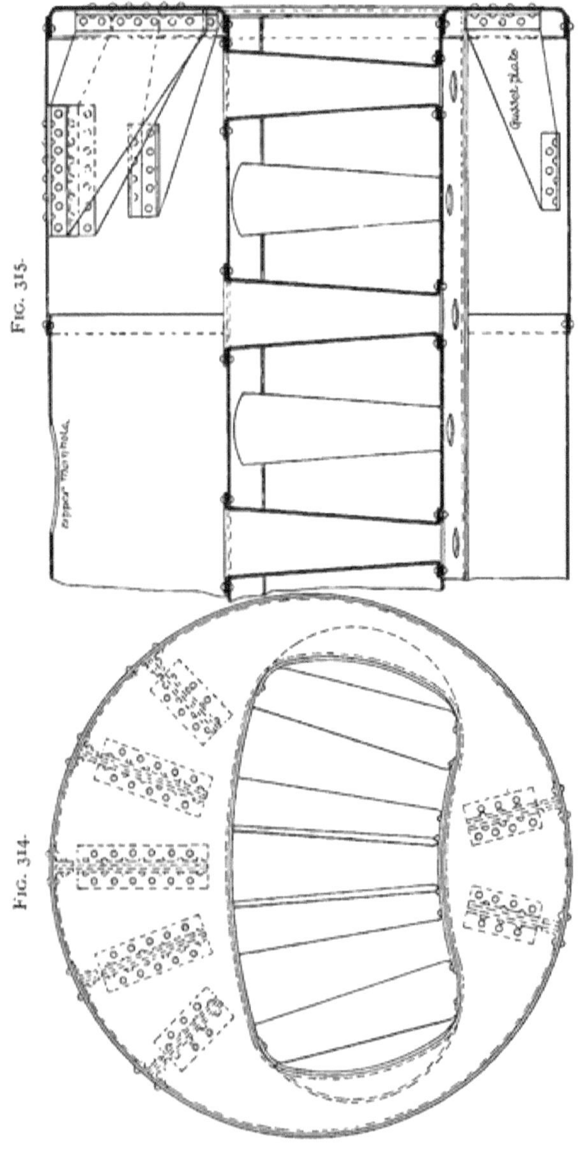

Fig. 315.

Fig. 314.

Brick Setting.—The boiler is supported at the front by means of a brick pier, which is built tight to the boiler for about one-third of its circumference. See Figs. 313 and 319. The remainder of the brickwork at the front end of the boiler is built free from the shell, allowing a space of about one-half inch to be filled with asbestos-fibre or other non conducting and elastic substance. After the inner side and end walls and the first 9-inch arch-covering is laid, they are covered with a layer of mortar, and a wrought-iron case (No. 22 black iron) is then put on and bedded into the layer of mortar. The iron sheets forming the case are united with roofers' standing joints, which are laid down flat with the rest of the casing. This casing covers the ends, sides, and top arch, and after it is put on the external courses of brick are laid. See Fig. 320. This iron casing prevents air being drawn into the boiler chamber and insures the temperature of the gases surrounding the boiler to be higher than that of the contained steam.

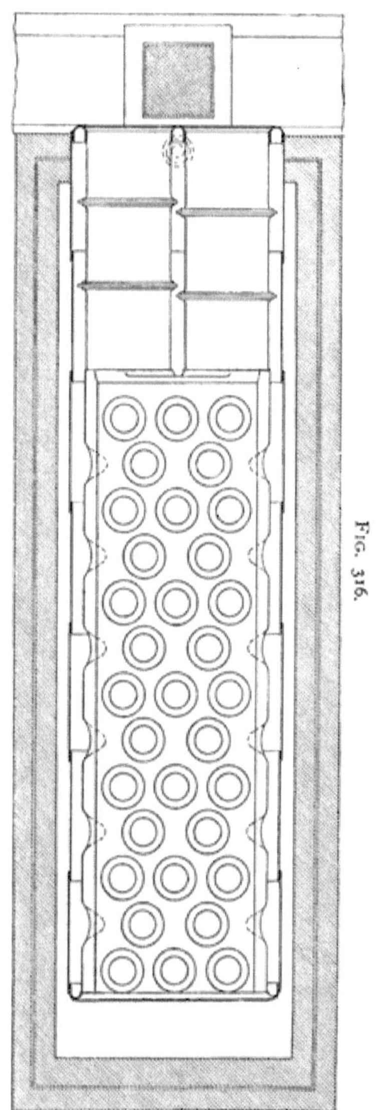

Fig. 316.

The rear end of the boiler is supported on a cast-iron expansion-rocker, shown in Fig. 313, behind which is built a small brick pier, which does not touch the boiler by about three quarters of an inch. This pier is built as a precautionary measure against accident to the rocker-casting. Without this pier, a failure of the rocker support would allow the boiler to fall with disaster to itself and its setting. The boiler set as above described

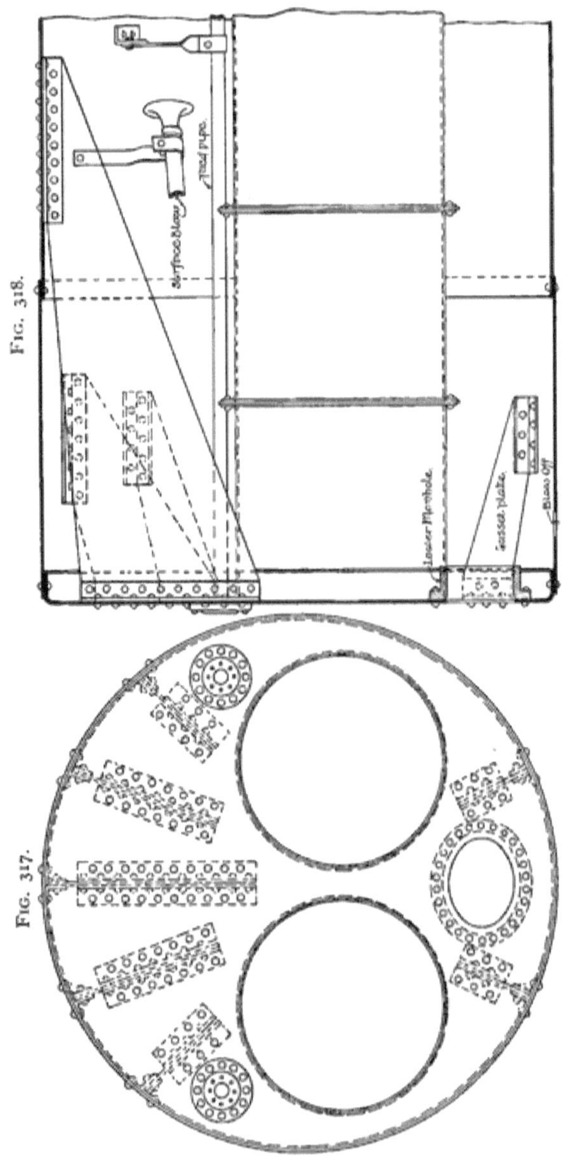

Fig. 318.

Fig. 317.

is free to expand and contract without touching the brickwork of the setting, except at the front end, where it is built fast and is stationary. The brickwork, being exposed to but low temperatures, will not be affected by the expansion and contraction of the boiler. It can neither be cracked by the heat nor racked by the motion of the boiler, as must occur with the ordinary brick setting.

Fig. 319. Fig. 320.

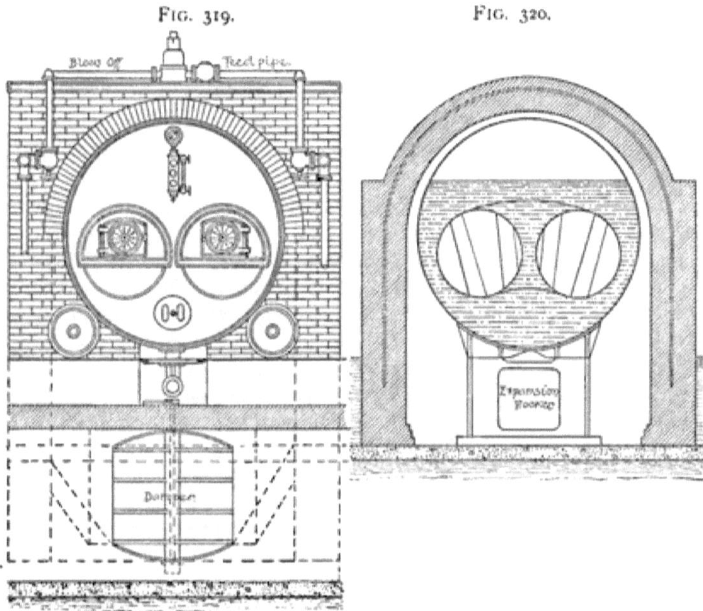

The nozzles and manhole project through the covering arch in the brick setting. A space is left in the brickwork surrounding the nozzle to be filled with asbestos-fibre or other non-conducting and elastic substance impervious to the passage of air.

The damper-flue leading to chimney is shown relatively to the boiler in Figs. 313 and 319. So, also, the location of the bottom-blow under the grating at the boiler front; the location of the surface-blow is shown in Fig. 318. The feed-pipe enters either from the top, as shown in Fig. 313, or from the front head, as in Fig. 318. A fusible plug, which will melt out when the water-line is about three inches deep over the fire-boxes, is screwed into each fire-box. The Philadelphia Rules and Regulations require two safety-valves to be set for service on each steam boiler. Fig. 313 shows one of these to be of the lever and weight variety, the outer a spring safety-valve similar to Fig. 418.

The steam- and water-spaces are very large,—larger, perhaps, than any other type of boiler of similar capacity. Large steam-liberating surface, if the circulation underneath is good, is usually productive of dry steam, tending to coal economy and higher efficiency in the engine.

Scotch Boiler.—An internally fired boiler by the Continental Iron-Works is shown in sectional elevation, together with two cross-sections, front and rear views, in Figs. 321–325. It will be observed that the boiler consists of a horizontal cylindrical shell having an internal furnace-flue, in one end of which the fire-grate is located, the other end terminating in a fire-brick-lined back connection or combustion-chamber, which is contained in the casing forming an extension of the shell proper. The products of combustion are conducted therefrom through horizontal tubes and delivered into the sheet-iron breeching attached to the front head of the boiler. This breeching may be connected directly with the smoke-stack, or, where a group of boilers is employed, it may be attached to an uptake, leading to a common chimney.

Several modifications of this design may be made. The boiler, if large enough in diameter, may be arranged to contain two or more furnaces, or it may be provided with a comparatively short furnace, terminating in a combustion-chamber, from which the tubes lead direct to a smoke-connection at the rear of the boiler without returning forward over the furnace, forming what is known as a gun-boat boiler, numerous examples of which are in use in the United States navy and by several water-works corporations, Fig. 326 being an illustration of recent installations by the Philadelphia Water Department.

The furnace is the important part of an internally fired boiler. Morison's suspension furnace is the one illustrated here. It is necessary that the furnace be able to resist the collapsing pressure to which it is subjected. It must be longitudinally elastic, and of such external form as to admit of its being readily freed from accumulations of scale, grease, or other deposit. The furnace shown in Fig. 321 is corrugated, the details of which are better illustrated in Fig. 160. It meets successfully all the above somewhat exacting conditions, and, in addition, the undulations in the furnace surfaces act as baffle-plates to the passage of the furnace gases, causing them to mechanically mix, thus producing a better combustion than is practicable with smooth furnaces. The additional heating surface in the most active portion of the boiler, due to the corrugations, adds to its evaporative power.

Gun-Boat Boiler.—The city of Philadelphia recently installed 24 boilers, from designs by J. E. Codman, in the Queen-Lane Water-Works Pumping Station, of which a longitudinal sectional elevation is shown in Fig. 326, and a cross-sectional elevation in Fig. 327, the latter being on a larger scale than the former. This design will be recognized as one much used for marine purposes. The boiler illustrated is 8 feet 6 inches in diameter by 20 feet long. The shell-plates are ⅞ inch thick, with double-welt butt-joints on the longitudinal seams and double-riveted in

FIG. 321.

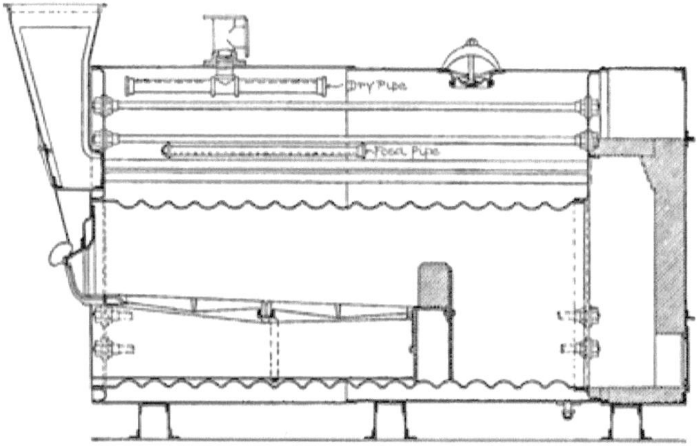

FIG. 322.

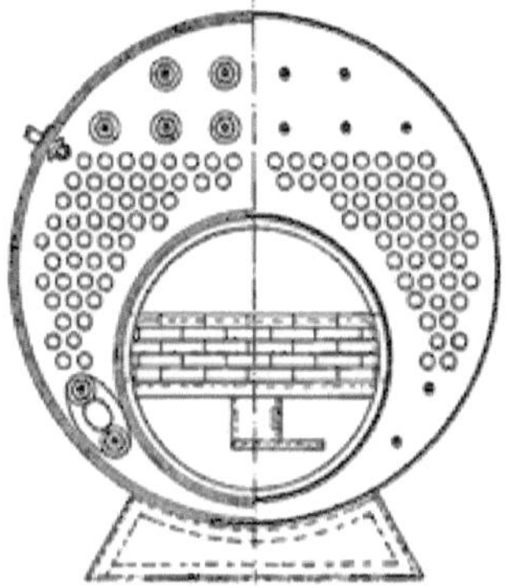

FIG. 323.

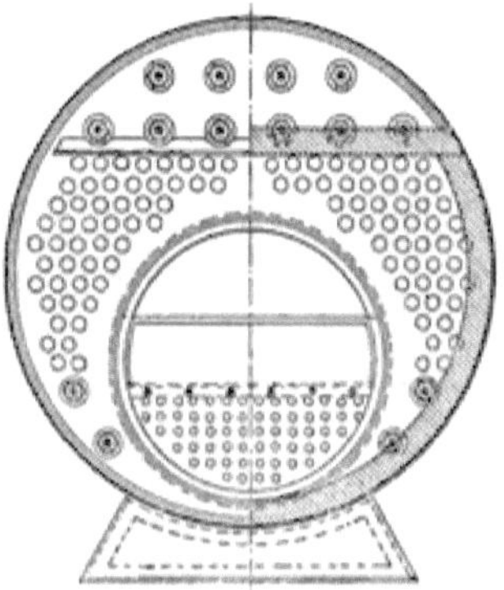

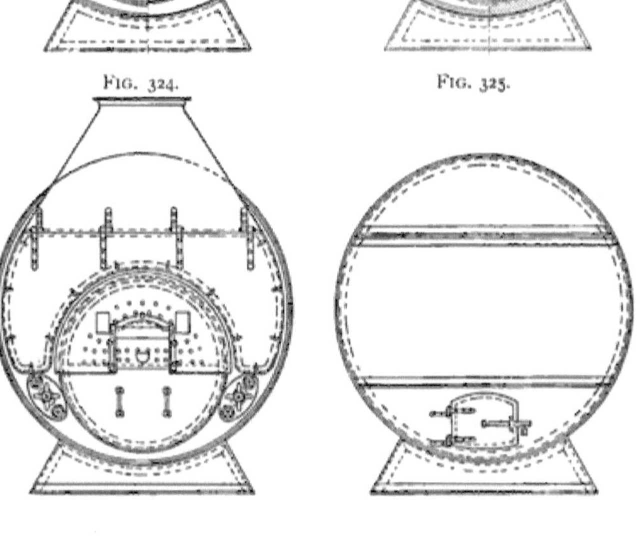

FIG. 324.

FIG. 325.

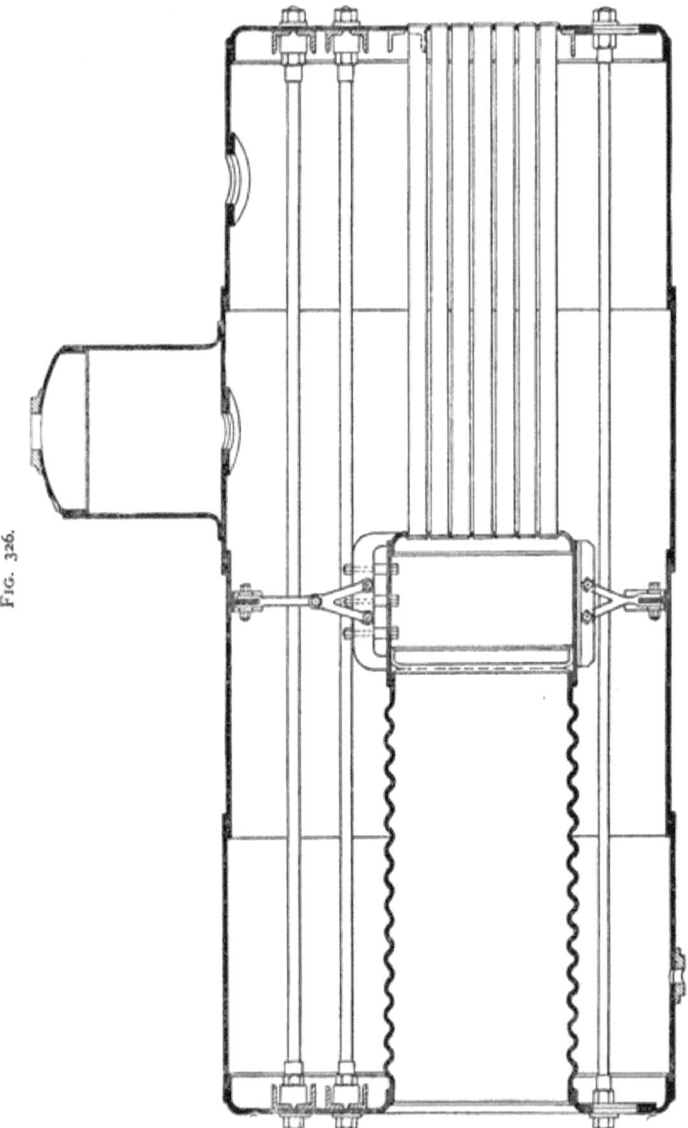

Fig. 326.

FIG. 327.

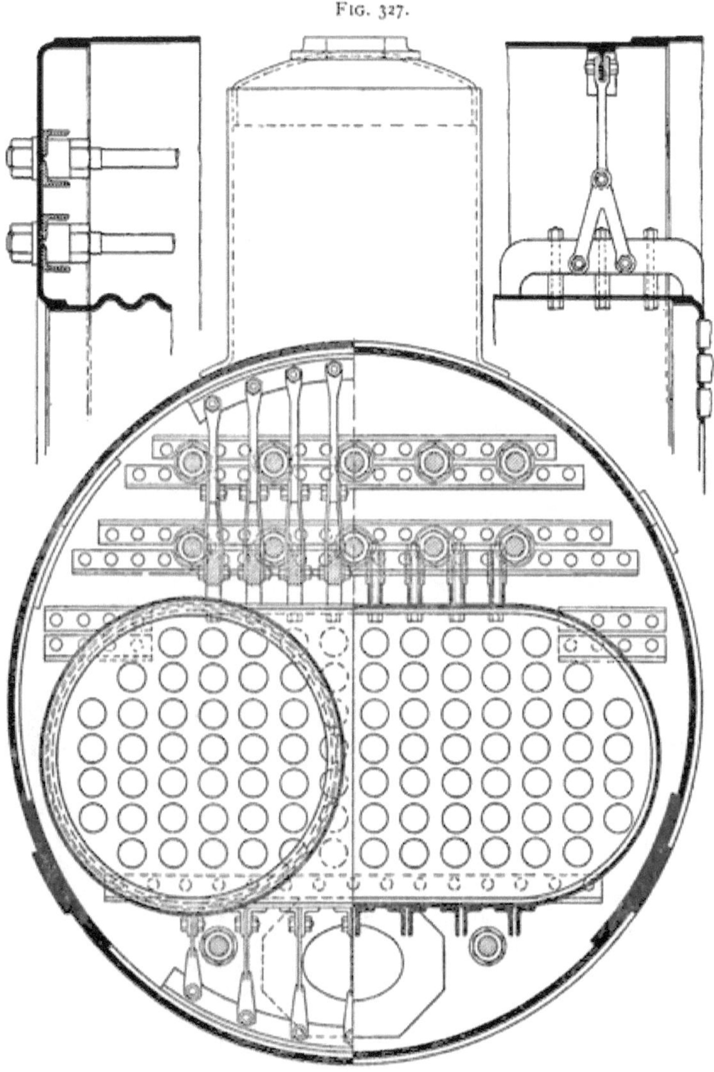

the circumferential seams; a detail of this joint is given in Fig. 328. The heads are ⅝ inch thick. There are two Fox corrugated furnaces 3 feet 6 inches in diameter by 8 feet long, the longitudinal seams of which are welded. The metal is $\frac{17}{32}$ inch thick; the corrugations are on 6-inch centres.

Fig. 328.

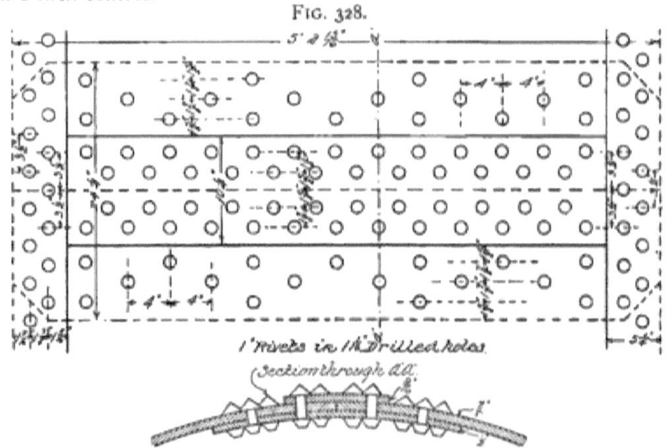

The combustion-chamber is situated midway in the boiler; its form is clearly shown in the two elevations. The front head is flanged to receive the rear ends of the two furnaces; the back head is fitted with 90 tubes 4 inches in diameter. The front and rear heads, as well as the plates joining them, are ⅝ inch thick. The combustion-chamber is stiffened longitudinally by crown-bars 5⅛ x 1½ inches placed on 6-inch centres, shown in Fig. 329. Each crown-bar carries 3 stay-bolts 1 inch in diameter on 7-inch centres, also shown in place in Fig. 326, but enlarged in separate detail in the upper right-hand corner, Fig. 327. Double angle-irons 4 x 3½ x ½ inch riveted to the bottom shell of this chamber on 8-inch centres perform the same office. Connecting stays provide for the transfer of collapsing strains, to which these plates are subjected, to the outer shell, the latter being fitted with T-sections 4 x 4 x ½ inch to make a substantial link connection.

Fig. 329.

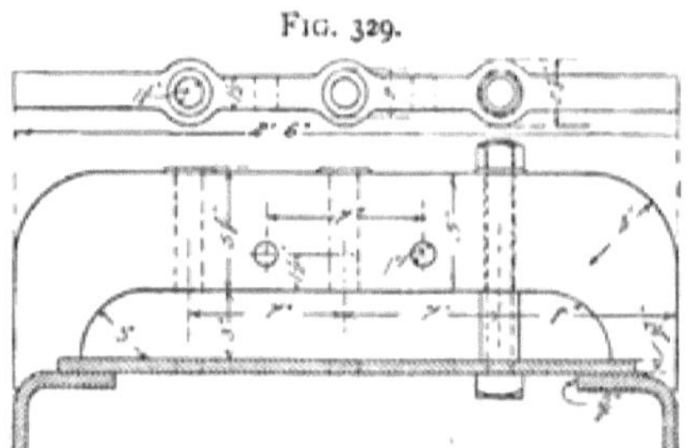

The 90 tubes above mentioned are 9 feet 4½ inches to outside of heads, spaced on 6-inch horizontal and 5-inch vertical centres.

The bracing of the heads of this boiler is clearly shown in both sectional engravings. Four 3½ x 4 x ½ inch angles extend across the front and back heads in pairs, as shown in both sectional drawings; these are detailed on a larger scale in Fig. 330. The angle-irons are riveted to the heads by ¾-inch rivets on 4-inch centres. Between each pair of angles are 5 through-going stay-rods 2½ inches in diameter with upset ends to 2¾ inches in diameter. Each end is fitted with nuts, a collar being furnished at the inside to clear the angles. Two similar bolts are placed under the furnace on either side of the under manhole.

FIG. 330.

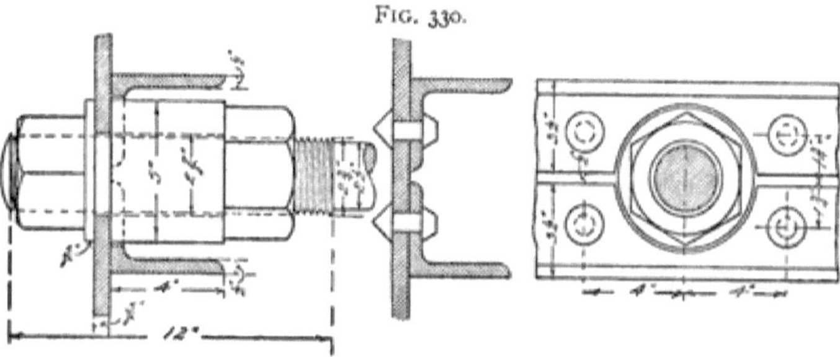

Three 11 x 15 inch manholes are provided, one at the top and one near the bottom of each head. For the bottom openings a $\frac{9}{16}$-inch plate 24 x 27 inches, with the corners clipped, as shown in Fig. 327, is secured to the head by 32 rivets, to restore in part the weakness occasioned by the opening. Over the rear manhole is a 3½ x 6 x ½ inch angle riveted to the head by ¾-inch rivets on 5-inch centres. This angle extends across the head to the beginning of the flange-curve. The manhole on the top of the boiler has a ring 4½ inches wide by ¾ inch thick, double-riveted to the shell.

The steam-dome is 30 inches in diameter by 3 feet 6 inches high over all; the shell is $\frac{7}{16}$ inch thick, double-riveted to the boiler; the dome-head is also $\frac{7}{16}$ inch thick, bumped to a radius of 3 feet. An 8-inch diameter opening is made in both the boiler and dome-head; the latter is fitted with a cast ring riveted to the dome-head, and to which the steam fittings are bolted. The opening into the boiler is reinforced by a ring 4 inches wide by ¾ inch thick, double-riveted to the boiler-shell.

The details of the riveting are so clearly shown on the dimensioned drawing, Fig. 328, as to require no further explanation.

Locomotive Boiler.—This name is applied to a certain design in boilers much used in stationary-engine practice because of its universal adoption as a type in locomotive construction. Fig. 331 represents in sectional elevation a boiler designed for stationary purposes by the Pennsylvania Railroad. This boiler is used for testing and for other purposes. The fire-box is well adapted for burning bituminous coal.

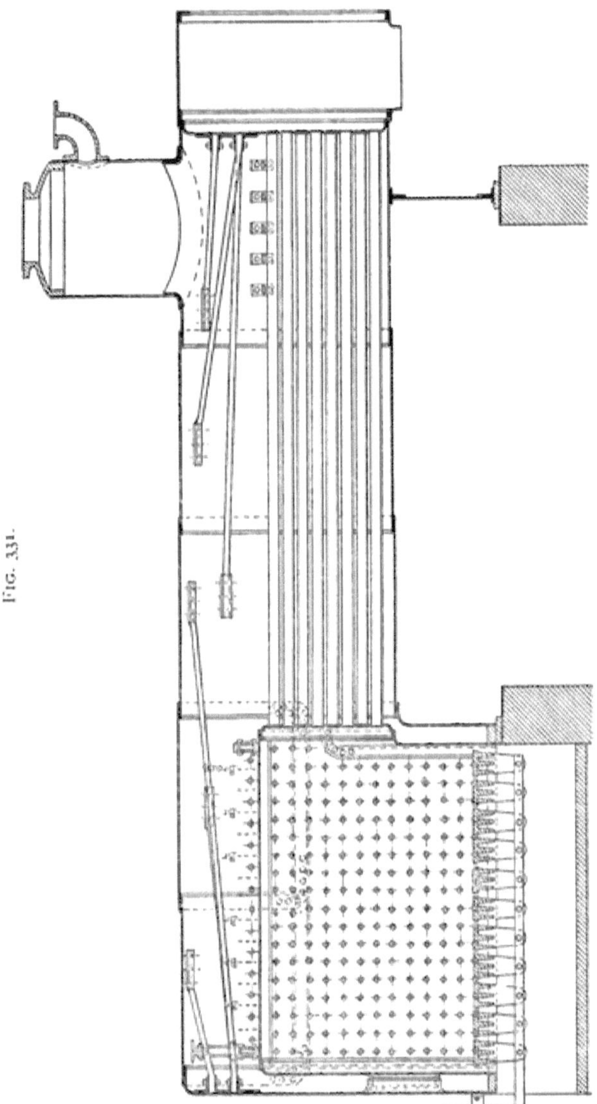

Fig. 331.

The dimensions are: 49⅜ inches in width at the bottom; the top measures 42 inches in width at the front and 38 inches at the rear; the taper at the sides is well illustrated in Fig. 332; the inside length of the furnace is 72 inches; the height from the top of the grates to the underside of crown-sheet is 50 inches; the crown-sheet is ⅜ inch; the other inside sheets are $\frac{5}{16}$ inch thick; the water-space around the bottom of the fire-box is 4 inches; the fire-door opening is 16 inches in diameter and is flanged similar to Fig. 192; the bottom ring is 2 inches deep.

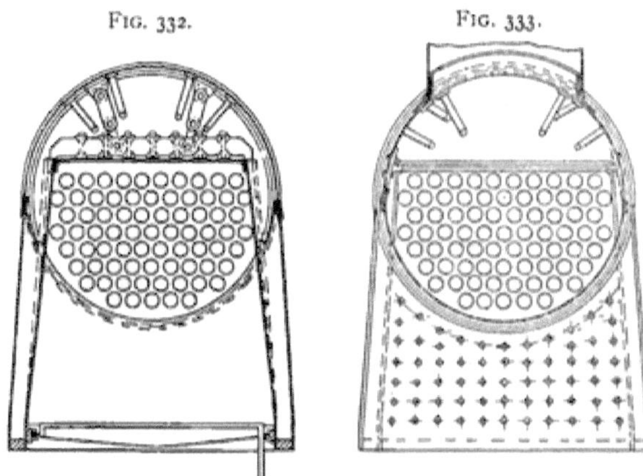
FIG. 332. FIG. 333.

The barrel of the boiler is 50 inches in diameter, straight on top, and provides a distance of 18 inches from crown-sheet to shell. The laps of the sheets are such that the drainage will be into the water-space around the fire-box without pocketing. The barrel-sheets are ⅜ inch thick, so also the sheets forming the outside of the fire-box. There are 74 tubes 3 inches in diameter by 11 feet in length, making the total length of the boiler 20 feet, including 26 inches depth of smoke-box. The fire-box tube-sheet is ½ inch thick and constructed somewhat differently from the ordinary, in being a separate head flanged around the edges and riveted to a corresponding flanged opening in the fire-box sheet, clearly shown in Fig. 331.

The steam-dome is 30 inches in diameter by 30 inches in height, plates ⅜ inch thick. The circumferential sheet to which it is riveted is flanged to receive the dome, as shown in Fig. 196. Underneath the opening for the dome are five cross-stays 2 x ⅝ inches of cross-section extending across the barrel and riveted at both ends, as shown in Fig. 333.

The stays for both front and rear heads are secured at one end by bolts passing through angle-irons riveted to the head, at the other end by riveting to the circumferential shell, both of which details are shown in Fig. 334. These stay-rods are 1⅛ inches diameter. The fire-box stay-bolts are ⅞ inch diameter, placed on approximately 4-inch centres.

Fig. 334.

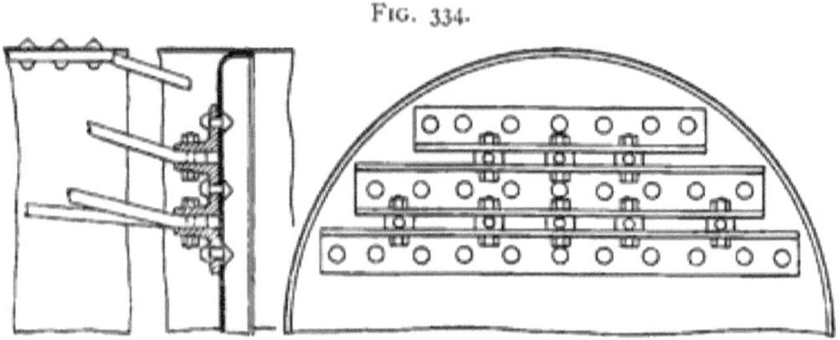

The crown-bars are double, each 4 inches deep by 2½ inches in width over all, carrying 7 stay-bolts of ⅞ inch diameter. The arrangement of head, thimble, washer, and nut for one of the stay-bolts is shown in Fig. 335, which also shows the proportionate dimensions of crown-bars. There are 14 horizontal through-going stays, one between each crown-bar. These are ⅞ inch diameter, with one end enlarged to 1 inch for passing the ⅞-inch stay through the larger tapped hole. These pass directly over the crown-sheet, as shown in Fig. 331. There are 15 crown-bars, every other one being fitted with two links, one on either side of the centre, connecting with the outer shell, as shown in Fig. 332 and in enlarged view in Fig. 335.

Fig. 335.

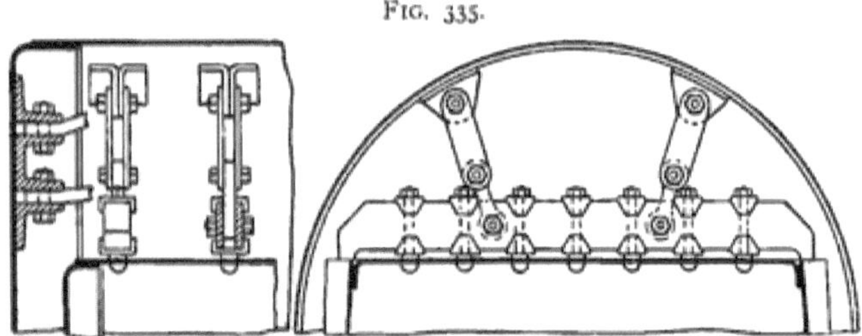

The sectional area of the tubes is 3.123 square feet. The grate area is 22.2 square feet, the ratio of tube to grate area being 22.2 ÷ 3.123 = 7.1 to 1. The total heating surface is 734 square feet, of which 96 is in the fire-box and 638 square feet in tube heating surface, the ratio of total heating surface to grate surface being 734 ÷ 22.2 = 33 to 1.

A shaking-grate is provided, the details of which are shown in Fig. 331. Three bars are cast together with one inch air-space between, the whole width of one grate being 5⅞ inches. A lever projects downward from each grate for making connection with a link-bar, to which all the grates are attached. This link-bar is operated by a lever pivoted to the outside of the fire-box as shown. On each end of each grate-bar is a trunnion 1¾ inches diameter resting in a groove included in the side-bar castings, fastened to the sides of the fire-box.

The smoke-box extends 26 inches beyond the tube ends; it is fitted with smoke-doors and a saddle for making connection with the chimney. In this case the smoke-flue is downward, an underground flue leading to the chimney.

TABLE XLVII.

PRINCIPAL DIMENSIONS OF PORTABLE BOILERS WITH OPEN BOTTOM AND WATER-FRONT.

Commercial Rating.	Diameter of Barrel.	Furnace.			Tubes 3 Inches in Diameter.		Thickness.			Dome.		Diameter of Stack.	Length of Boiler over all.
		Length.	Width.	Height.	Length.	Number.	Shell.	Furnace.	Tube-Sheets.	Diameter.	Height.		
H.-P.	Ins.	Ins.	Ins.	Ins.	Feet.		In.	In.	In.	Ins.	Ins.	Ins.	Feet.
25	40	48	34	33	8	34	9/32	5/16	⅜	22	24	18	13¼
30	42	50	36	34	8	40	9/32	5/16	⅜	22	24	20	14
35	44	50	38	36	8½	44	9/32	5/16	⅜	26	28	20	14½
40	44	50	38	36	10	44	9/32	5/16	⅜	26	28	20	16
50	48	54	42	40	10½	54	5/16	5/16	⅜	26	28	22	16¾
60	54	60	48	44	11	60	5/16	5/16	⅜	30	34	24	18
70	56	60	50	44	12	66	5/16	5/16	⅜	30	34	26	19
80	58	60	52	48	12	76	5/16	5/16	⅜	32	36	26	19
90	58	60	52	48	14	76	5/16	5/16	⅜	32	36	26	21
100	62	60	56	50	14	90	⅜	5/16	⅜	32	40	30	21¼
110	64	60	58	52	14	100	⅜	5/16	⅜	36	40	30	21¼
125	66	60	60	58	15	108	⅜	5/16	7/16	36	40	32	22½

The Belpaire Fire-Box is unlike the ordinary fire-box of the locomotive type in many of its details of construction, but especially in the method of bracing, the former having no crown-bars and no radial stay-bolts made to accommodate curved surfaces seldom or never parallel to each other. The crown-sheet in the Belpaire fire-box, as applied to locomotive boilers, is flat, and the roof-sheet lies parallel to it, as shown in Fig. 336. Stay-bolts extend in straight lines and at right angles from the crown-sheet to the roof-sheet. In locomotive practice, for 160 pounds steam pressure, these stay-bolts are ⅞ inch diameter in the body, with screw-ends enlarged to 1 inch, 12 threads, and placed on 4½- to 4¾-inch centres. Horizontal tie-rods extend across the fire box to stay the flat sides; in locomotive boilers these are usually three in number, 1 inch diameter in the body with screw-ends 1⅛ inches. They are spaced longitudinally to the same centres as the stay-bolts between which they are located; vertically the distance varies according to the height over the crown-sheet, but the approximate centre to centre is about 5½ inches, with the centre of the lower tie-rod about two inches above the crown-sheet. This method of staying is advantageous, in having all the stays and tie-rods in tension in straight lines, as compared with radial stays fitted to curved sheets, in which it frequently happens that not more than one or two full threads are had in the outer sheet, whereas by placing the sheets parallel full threads are had in both.

FIG. 336.

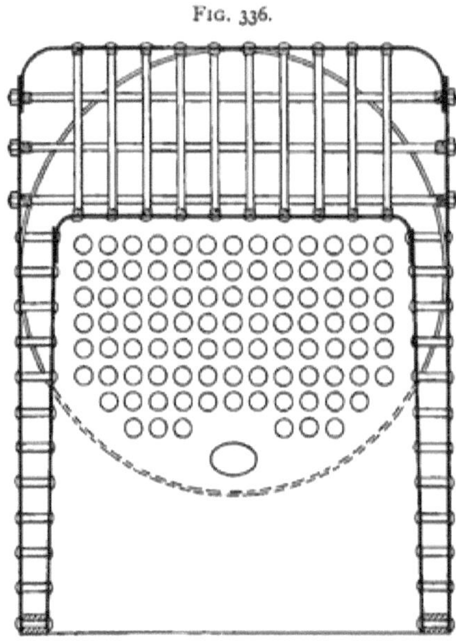

A Belpaire Fire-Box Boiler, differing in many respects from the preceding, and shown in the longitudinal sketch, Fig. 337, is from designs by E. D. Leavitt, Jr., who has long used this type of boiler with that success which characterizes his engineering designs. The boiler here illustrated is 7 feet in diameter of shell by nearly 35 feet in length from out to out, and is intended to work under a continuous steam pressure of 135 to 140 pounds per square inch.

Two fire-boxes are included in this design, with a central water-leg between, as shown in Fig. 338. Each furnace is connected with a large central combustion-chamber, in which a thorough admixture of the gases and final combustion occurs before these gaseous products enter the tubes. The crown-sheets are flat, but the roof-sheet of the fire-box end is curved to the same radius as the barrel of the boiler ; this necessitates a somewhat unusual scheme of bracing, that the vertical stay-bolts shall pass through no curved surfaces in the roof-sheet. This is accomplished by the introduction of forged frames placed on about 5-inch centres. One of these frames is shown in place in Fig. 338 ; the two ends of it rest upon and are riveted to the two crown-sheets ; nearly vertically above these riveted joints the frame is riveted to the roof-sheet. Horizontal offsets are then made for a stay-bolt on each side of the boiler, and a vertical offset for a riveted stay passing through a distance-piece on each side of the boiler, after which the frame extends horizontally across the fire-boxes. Six vertical stay-bolts, with distance-pieces between, fix the horizontal portion of this forged frame and transmit any strains received to the curved roof-sheet above. This arrangement permits the insertion of 8 vertical stay-bolts, 4 on each crown-sheet, with a central space through which a man may crawl for inspection of the interior; it also permits the removal and replacing of any of the vertical stays. The front, central, and outside water legs have ⅞-inch stay-bolts on 4½-inch horizontal by 4⅜-inch vertical centres. The outer and inner sheets of the front head are flanged for two fire-door openings, each 21 inches wide by 24 inches high, the bottom of each opening being about 12 inches above the bottom of the boiler. The front head above the crown-sheets is stayed by means of two pairs of angles and one T-iron, all riveted to the head, as shown in Fig. 339. Eight tie-rods extend from end to end of the boiler; these are 1⅞ inches diameter with 2¼-inch ends. The arrangement of nuts and distance pieces between the angle-irons is also shown in the above illustration. The relative positions of these rods from centre to centre, as arranged for carrying the strains, is shown in Fig. 340 and Fig. 338. The sides of the fire-box swell from 7 feet diameter of barrel to 9 feet spread at the bottom of the fire-box. This introduces some difficulties not usually experienced in boiler design. The ring of plates corresponding to the gusset-sheet has to be greatly strengthened. The three

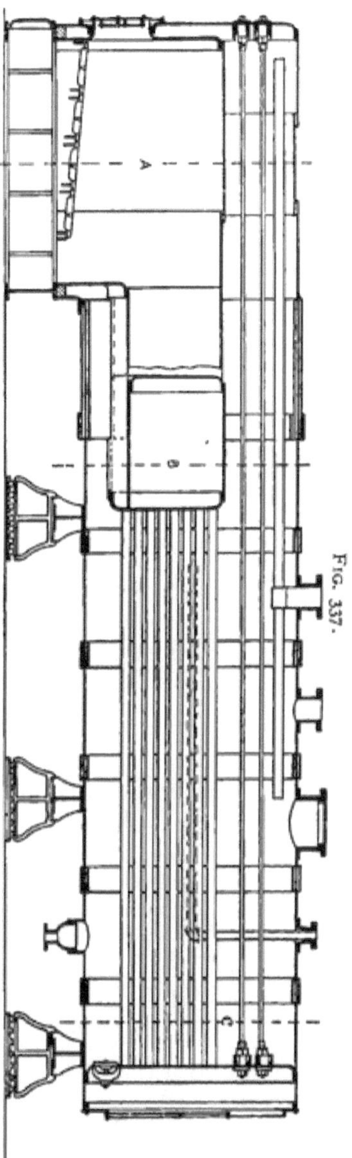

FIG. 337.

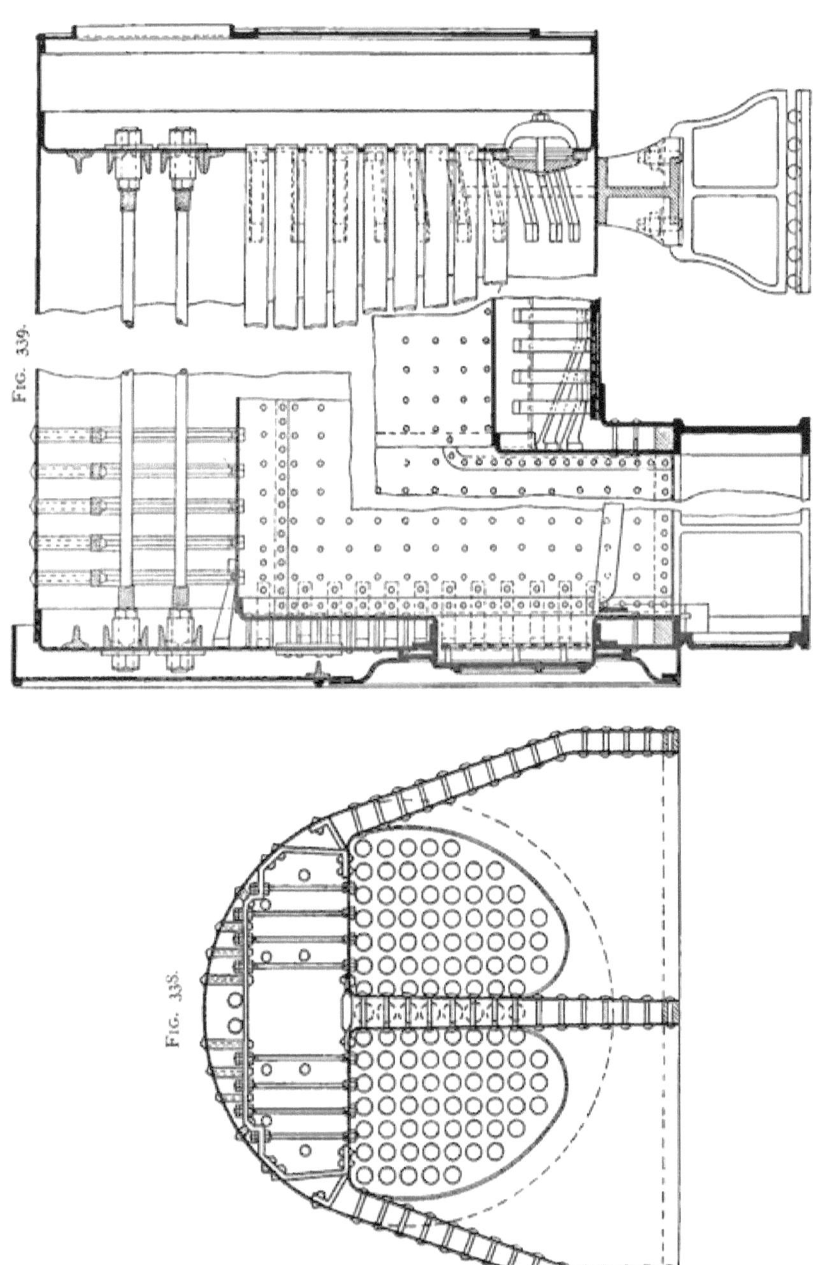

Fig. 339.

Fig. 338.

thicknesses and throat-flange are shown in longitudinal section in Fig. 339, as are also the brackets for stiffening the fire-box sheet around the throat.

FIG. 340.

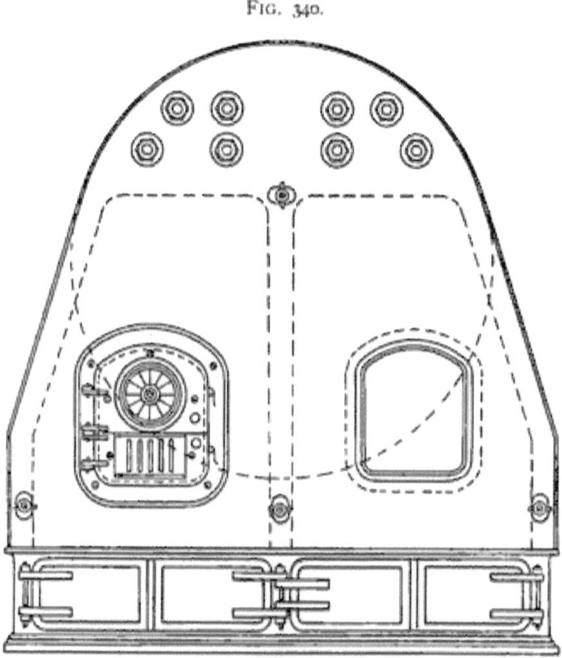

The combustion-chamber is shown in cross-sectional elevation in Fig. 341. Its crown-sheet is on the same level as that of the furnace. Inasmuch as there is no central water-space at this point, additional stay-bolts are inserted at the centre of this crown-sheet. The opening provided over and between the fire-box crown-sheets is here continued, whereby a man may crawl through the centre of the bracing over the combustion-chamber by employing a frame similar to that already described for the fire-box crown-sheets and suspending another frame below it ; this provides for five short vertical stays, shown in the drawing. An end view is given of the tube-sheet which closes the rear end of the combustion-chamber ; it provides for 118 tubes 3½ inches in diameter. These tubes are 18 feet long inside the heads. Underneath the combustion-chamber is a space large enough to permit a man to examine or clean that portion of the boiler. Forged stays, fitting the outer and inner curves of the plates to be joined, and which do not admit of radial stays, are also shown. The details of the rear head are shown in Fig. 342, and further supplemented by the longitudinal section, Fig. 339. Angle- and T-irons are riveted to the head above the tubes, with provision for the eight longitudinal tie-rods. The feed-pipe enters through a nozzle at the top of the boiler, branches to either side, and continues forward between the shell of the boiler and the tubes to within a couple of feet of the combustion-chamber, shown by dotted line in Fig. 337. These pipes are of brass with perforations along the submerged horizontal branches for diffusing the feed-water throughout the entire length. The fire-tubes are sufficiently above the bottom of the barrel of the boiler to admit a man through its whole length. A series of brackets on about 6-inch centres are placed radially around that portion of the head not sufficiently stayed by the fire-tubes. These brackets are riveted to both head and shell, and are shown in Figs. 339 and 342. Brackets for similarly supporting the front head are shown in the left-hand portion of Fig. 339. A rear elevation is shown in Fig. 343. A cast-iron plate closes the entire end, in which is a suitable opening for attaching the plate-metal connection leading to the chimney, underneath which are two hinged doors giving access to the tubes.

Fig. 341.

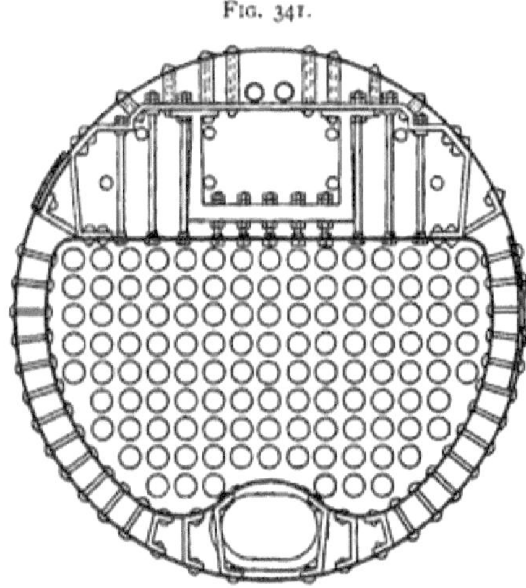

Fig. 342.

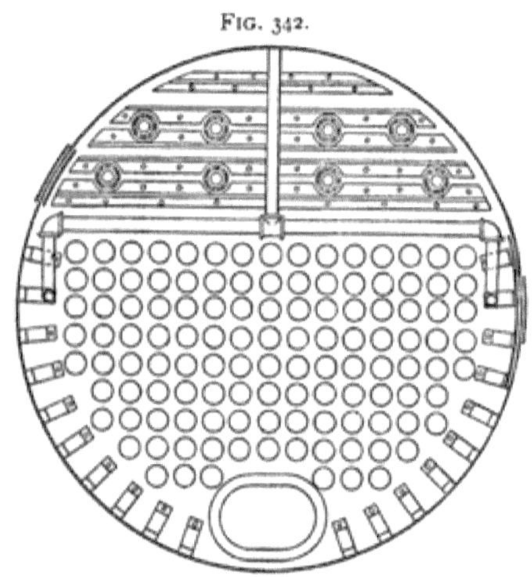

The fittings and attachments include a manhole-opening under the tubes; the usual handhole-openings in the water-legs and at the surface of the crown-sheet; a blow-off connection under the barrel of the boiler near the rear end; an 18-inch ring manhole-opening on the top of the barrel of the boiler; a safety-valve nozzle; a steam-pipe nozzle. Two dry pipes, perforated along their upper sides, extend fore and aft from the latter nozzle, shown in Fig. 337. A cast-iron front, shown in Fig. 339, covers that end of the boiler, making an attractive finish and protecting the non-conducting covering at a point always liable to injury.

Fig. 343.

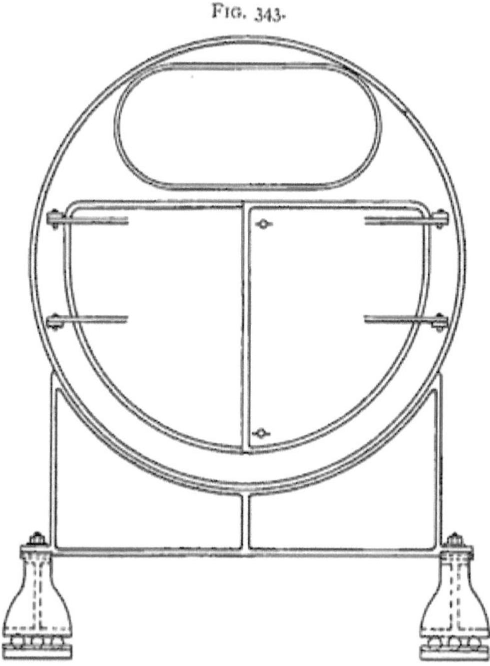

The covering consists of a coating of plaster-of-Paris and sawdust 2½ inches thick, over which is placed a layer of hair-felt 1 inch thick and a painted canvas cover. The fire-box doors are cast iron and are divided horizontally, as shown in Fig. 340. They have perforated baffle-plates and register openings in each half.

The plates of the boiler are $\frac{9}{16}$ inch for the main shell and $\frac{7}{16}$ inch for the fire-boxes, all of mild steel of 60,000 pounds tensile strength, with an elastic limit of 40,000 pounds. The joints of the shell are double butt-joints, the horizontal seams being triple-riveted, the vertical ones double-riveted.

The expansion of this boiler is towards the rear, the front end resting upon a cast-iron ash-pit provided with cleaning-doors, shown in Figs. 337 and 340, no part of which is below the floor level. Three additional supports are provided under the barrel of the boiler, the latter resting upon cast-iron cradles, shown in Fig. 343. These cradles have a broad base at each end, supported upon balls which permit a free movement in any direction, and are provided with adjusting screws, shown in Fig. 339, by means of which the weight can be equally distributed over the entire surface of the balls.

FIG. 344.

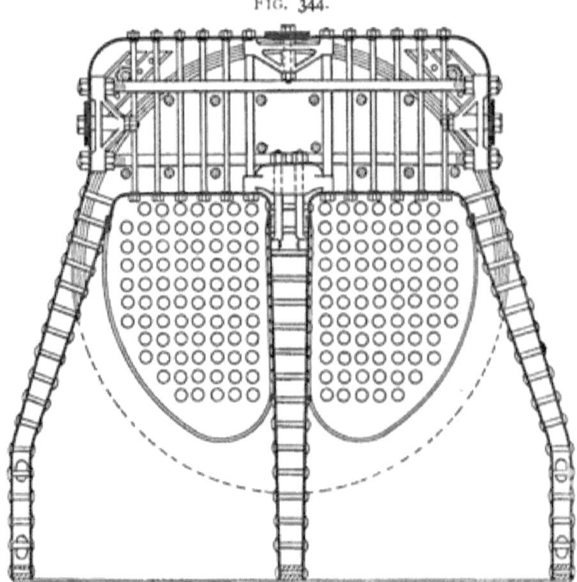

A design for a Belpaire fire-box, by F. W. Dean, is shown in cross-section through the fire-box in Fig. 344, the details being those in sight when looking towards the rear end of the boiler. This design applies to those fire-boxes which are approximately rectangular in form over the crown-sheet, as indicated in the engraving. The usual longitudinal tie-rods extending from front to rear heads are shown above the crown-sheet in section. Additional to these are transverse tie-rods extending across the steam-space. Vertical stay-bolts connect the roof-sheet of the boiler with the crown-sheet below, as shown. It is almost imperative in large boilers that the fire-box bracing be accessible for examination or repairs. This demands a central space through which a man may crawl into and examine the interior of the boiler. In order to accomplish this it is necessary to do away with the central transverse tie-rod commonly used and also two or more of the central vertical stay-bolts. The removal of these tie-rods and stay-bolts is a source of weakness; therefore, to make good the loss of strength occasioned by the omission of these parts, trussed connecting-bars are here employed, and these are provided with a central hub through which a bolt passes and by which the outer shell of the fire-box is securely bolted to the trussed connecting-bars on the inside. In this manner the fire-box shell is as well strengthened as if a through-going tie-rod was used. The horizontal tie-rods have outside and inside nuts, by means of which the trussed connecting-bars are firmly secured to the outer shell. The loss of strength occasioned by the removal of any two of the central vertical tie-rods is made good by increasing the diameters of the two inner tie-rods and in providing a trussed connecting-bar similar to the one above referred to, except that the inner nuts for clamping it to the outer shell are omitted. A series of these trussed connecting-bars are located centrally at the top of the boiler. Connecting-bars of somewhat different shape, but answering the same purpose, span the central water-space; vertical bolts riveted to the inner side of the central water-leg pass up through these connecting-bars, each bolt being supplied with nuts, as shown.

Fig. 345.

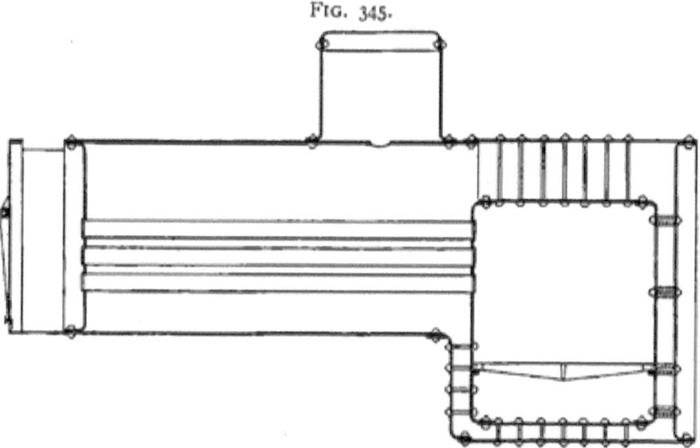

Portable Engine Boilers.—These are usually of the locomotive pattern, and are very generally provided with a water-bottom, as shown in Fig. 345. The object of the water-bottom is to prevent live coals falling upon the ground, thereby lessening the fire risk in agricultural districts, where portable engines of small and medium powers are principally used.

This type of boiler permits of furnace length and height suited to any kind of fuel. The width of furnace is governed by the diameter of the barrel and the water-space at the sides and bottom of the furnace.

The sides of the fire-box are somewhat deficient in circulation, and as a result are liable to fill with deposits if hard water is used; for this reason handhole plates should be provided at the corners that any sediment may be easily removed.

A water-front is shown in the sectional illustration, Fig. 345, which also shows a water-bottom. The outer fire-box head is reversed so as to afford the best facility for riveting the head and shell together. The head next to the smoke-box has its flange turned outward, as at the fire-box end. A handhole should always be provided for cleaning the barrel of the boiler.

The fire-door opening is commonly made with a wrought-iron ring riveted through, as shown in Fig. 187. Some of the cheaper boilers in the market have cast-iron rings, but such rings are not always safe, because liable to crack without warning.

Another method of constructing a furnace is shown in Fig. 346, in which the furnace-sheet extends through and is riveted to the outside sheet at the fire-box end, the latter being flanged to receive it. This opening is covered by a cast-iron front, in which is included the fire-door opening, a bearing-bar for the grates, and an ash-pit opening. As this cast plate is subjected to an intense heat, its interior must be covered with fire-brick tiles. The fire-door is fitted with a perforated plate, as in stationary-boiler practice.

Portable engine boilers when mounted on wheels are commonly from 8 to 15 horse-power, seldom more than 20. The diameter of the waist or barrel for such boilers ranges from 26 to 32 inches, consequently the width of the fire-box varies from 21 to 28 inches. This easily permits the use of an arched crown-sheet supported by radial stays instead of crown-bars necessary to the proper supporting of flat crown-sheets. Radial stays are much to be preferred to crown-bars for portable boilers because of the little room occupied by them in the boiler as well as the lesser weight. Such stays provide a satisfactory transfer of the collapsing strains upon the crown-sheet to the outer shell. Radial stays obstruct the circulation much less than crown-bars; they also permit a greater thoroughness of inspection and cleaning. The stay-bolts joining the outer and inner plates around the fire-box are commonly $\frac{3}{4}$ inch in diameter for $\frac{1}{4}$-inch plates and $\frac{7}{8}$ inch in diameter for $\frac{5}{16}$- to $\frac{3}{8}$-inch plates. The radial stays and those used in the side and bottom of the fire-box are commonly of the same diameter. For long stay-bolts the outer plate should be tapped enough larger than the inner one that the main body of stay shall pass through without being obliged to screw the stay its whole length through the outside hole. This necessitates the use of two taps, the upper one of which is a hob secured to a shank, forming the smaller tap. Both threads must be of the same pitch and so placed that the threads shall start at the same point in a revolution; this necessitates the same provision being made in threading the stay-bolt. The front and rear heads are braced similarly to those for horizontal tubular boilers; that is to say, the stays are riveted to the head, extend back, and are then riveted to the shell of the boiler.

The tubes are commonly 3 inches in diameter for all portable and semi-portable boilers, whatever may be their size.

FIG. 346.

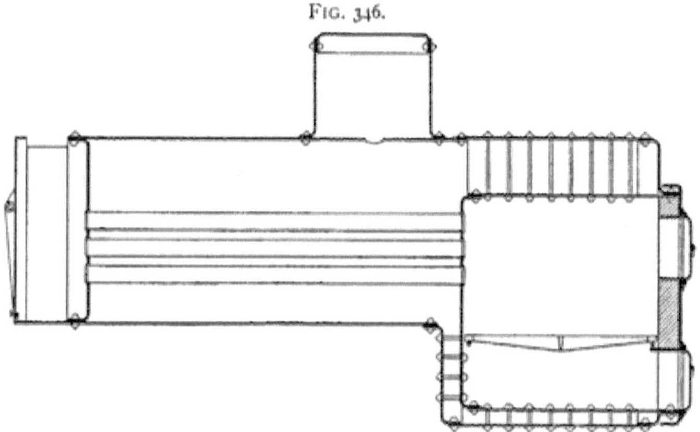

A steam-dome is commonly furnished portable boilers, the opening into which is not usually any larger than the diameter of the pipe leading to the engine. Small holes should always be provided at the bottom where the dome and barrel intersect for the drainage of any water of condensation back into the boiler. See Fig. 195.

The smoke-box is commonly an extension of the barrel of the boiler and fitted with a cast-iron frame and door, giving easy access to the tubes.

A saddle for the smoke-stack attaches to the opening provided at the top of the smoke-box. In the case of portable boilers it is further provided with a hinged ring, permitting the stack to be laid down upon the boiler when transporting it from place to place.

Handhole plates must be inserted in the water-leg for the removal of scale or mud which may accumulate there.

In some tests made for the writer, a very inferior bituminous coal being used for fuel, the feed-water entering the boiler at 65° Fahr., an eight horse-power boiler evaporated 12 cubic feet of water per hour, a ten horse-power boiler evaporated 15 cubic feet in the same time. The evaporation was under a pressure of 80 pounds per square inch. The boilers were in the condition in which they are usually delivered to the trade, and the firing in the test was as near as possible the same as it would have been in the hands of the purchaser, except that it was conducted with a view to ascertain the actual evaporative capacity of the boiler instead of being an economy trial.

TABLE XLVIII.

PRINCIPAL DIMENSIONS OF PORTABLE BOILERS WITH WATER-BOTTOM AND CAST-IRON FRONT, FIG. 346.

Commercial Rating.	Diameter of Barrel.	Furnace.			Tubes 3 Inches in Diameter.		Thickness.			Dome.		Diameter of Stack.	Length of Boiler over all.
		Length.	Width.	Height.	Length.	Number.	Shell.	Furnace.	Tube-Sheets.	Diameter.	Height.		
H.-P.	Ins.	Ins.	Ins.	Ins.	Feet.		In.	In.	In.	Ins.	Ins.	Ins.	Feet.
6	26	34	21	29	5	15	1/4	9/32	3/8	12	15	12	9¼
8	28	36	22	33	6	18	1/4	9/32	3/8	14	16	14	11
10	30	38	24	35	7	20	1/4	9/32	3/8	16	18	14	12
12	32	38	26	35	6½	24	1/4	9/32	3/8	18	20	16	11¼
15	32	44	26	35	7	24	1/4	9/32	3/8	18	20	16	13
20	34	52	28	37	8	28	1/4	9/32	3/8	20	22	16	14½
25	36	52	30	40	8½	32	1/4	9/32	3/8	20	22	18	15
30	40	60	34	43	8	38	9/32	9/32	3/8	22	24	20	15
35	40	60	34	43	9	40	9/32	9/32	3/8	22	24	20	16
40	40	60	34	43	10½	40	9/32	9/32	3/8	22	24	20	17½
50	44	64	38	50	11½	46	5/16	5/16	3/8	22	24	22	19¼
60	48	64	42	52	12	52	5/16	5/16	3/8	26	28	22	20

The "Electric" return tubular portable boiler, shown in Fig. 347, is not unlike the locomotive fire-box pattern of portable boiler, except that the crown-sheet is placed lower in order that return tubes may be

Fig. 347.

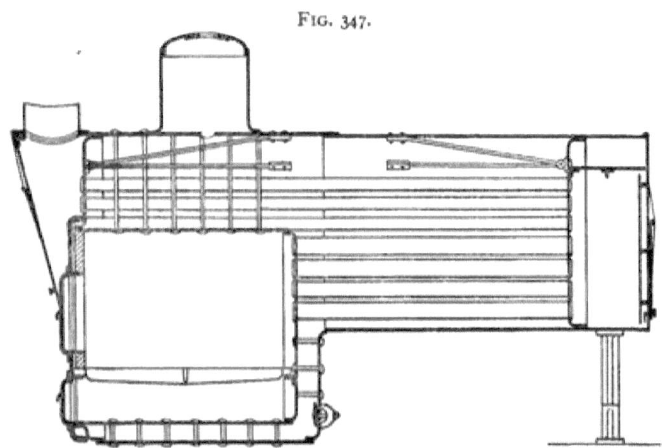

located above it. The fire-box does not differ from the ordinary; it is provided with a water-bottom and a cast-iron front lined with fire-brick tiles. The tubes are of two diameters,—the lower ones being 4 inches, the upper ones 3 inches,—the combined areas for each diameter approximately equalling those of the other. The products of combustion pass through the shorter and larger tubes to the back connection and from thence return through the smaller and longer tubes to the front connection leading into the smoke-stack. The crown-sheet, which is usually the highest part of the locomotive type of boiler, has the least water to protect it and is the first part exposed to low water. The protection afforded the crown-sheet in this design is all that can be desired; in general, it is a compact boiler for the power developed and has good steaming qualities.

FIG. 348.

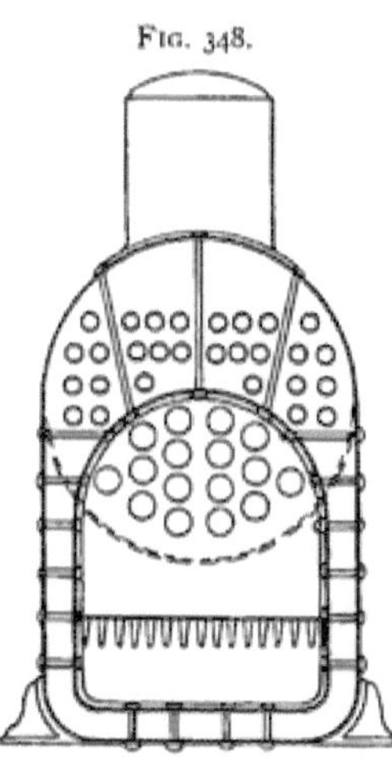

The "Economic" return tubular boiler, shown in Fig. 349, is of somewhat unusual construction. The front end is cylindrical in form, the rear end is oval, the length of each portion being about one-half that of the entire length of the boiler. A furnace is attached to and underneath the front end. The lower portion of the rear end extends down far enough

FIG. 349.

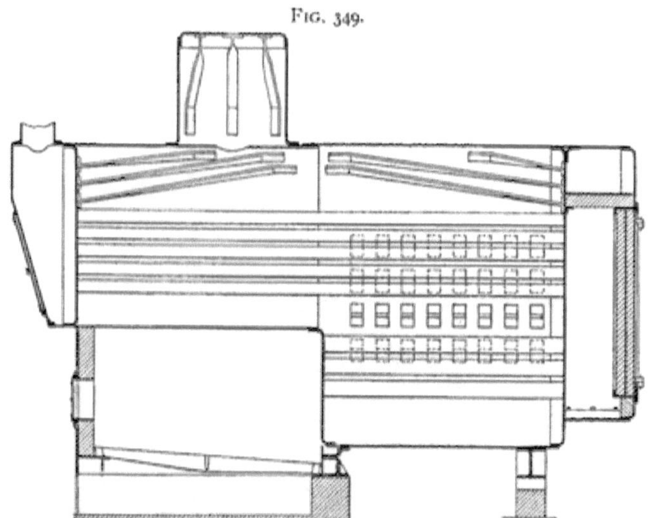

to hold the short tubes leading from the furnace to the back connection. The lower half of the front end of the boiler, being over the fire, forms the crown-sheet: it will be noticed that it occupies a position relatively to the fire quite the reverse of ordinary portable boiler practice, the latter commonly having the crown-sheet arched over the fire and not recessed towards it; this design insures that the crown-sheet is always well supplied with water.

Fig. 350.

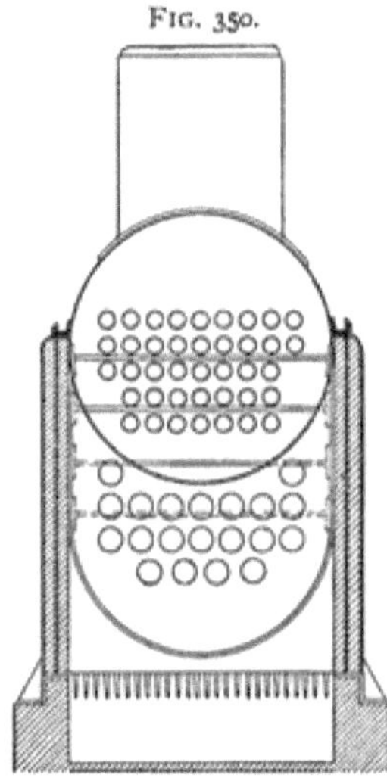

The furnace consists of iron plates attached to the cylindrical shell of the boiler, which plates extend to any depth suitable for the location of the grates adapted to fuel to be burned. The side walls and front of the furnace are lined with fire-brick, and these are held in place by iron rods protected from the fire. Both rods and fire-brick can be removed and replaced whenever necessary. Fig. 350.

The advantages claimed for this boiler over the ordinary internally fired portable boiler of the locomotive type is that it occupies no more space, is an equally rapid steamer, there are no water-legs in which to deposit sediment, and that it combines with the safety of the stationary return tubular boiler the convenience and portability of those of the locomotive type.

An internally fired boiler of about 40 horse-power, by Tonkin, shown in Fig. 351, is of somewhat unusual construction for a semi-portable

Fig. 351.

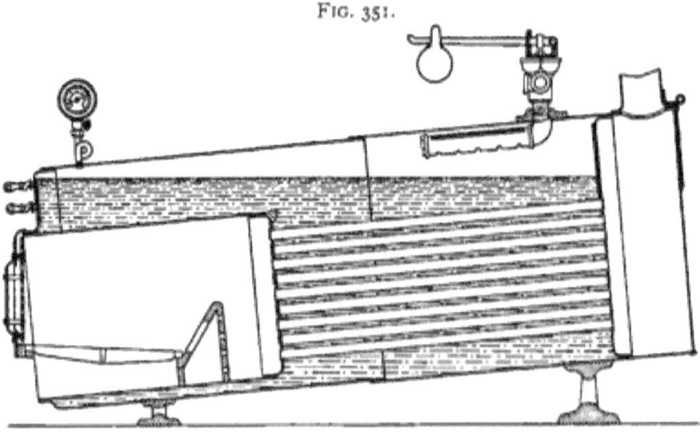

boiler. This boiler is 54 inches in diameter and about 15½ feet long. The furnace is 36 inches in diameter and 54 inches long, approximating

13.5 feet of surface. It is closed by a cast-iron front, which includes the fire-door opening, grate bearing-bar, and an ash-pit opening. The bridge-wall, or what corresponds to a bridge-wall in ordinary boilers, is made of cast iron and perforated for the admission of air through slots provided in the casting. Those included in the inclined portion of the casting are located under the fire, and practically answer the purpose of an extension of the grate-bars so long as they are kept covered with fuel. The openings at the rear admit air into the combustion-chamber between the end of the bridge-wall and the tube-sheet.

There are sixty tubes 3 inches in diameter by 9 feet in length. The ratio of tube area to grate surface is $13.5 \div 2.5 = 5.4$ to 1.

The total heating surface is 474 square feet, the ratio to grate surface being $474 \div 13.5 = 35.1$ to 1.

This boiler is set at an inclination to the horizontal for the purpose of increasing the depth of water at the fire-box end.

CHAPTER IX.

SECTIONAL AND WATER-TUBE BOILERS.

SECTIONAL boilers are those made up of a number of units similar in size and shape and so constructed that when suitably assembled they will admit of performing all the functions of a steam boiler. To the late Joseph Harrison, Jr., is due the credit for the invention of what may be considered an ideal sectional boiler, consisting, as it does, of a series of simple units, spherical in form, convenient in size, and capable of aggregation to any extent required in practical operation.

Mr. Harrison's invention dates back some thirty years. At that time boiler pressures did not often exceed 75 pounds per square inch. Mild steel had not yet been introduced, but the quality of cast iron was in all respects satisfactory for his purpose, because such castings were not only of ample strength, but were found to better resist the corrosive action of acids in the feed-water than wrought iron, the only other material then in use for land boilers. Castings were also less affected by oxidation in the furnace and were free from blisters.

This boiler possessed at that time the unique distinction of being constructed wholly of cast iron, a material hitherto considered unsafe in boiler construction. Its use was adopted, and since continued, because it contributes better than any other material to the manufacture of spherical units with their connecting necks for grouping into "slabs," as these aggregations of units are called. Harrison's selection of this material was the result of certain guiding principles in design, which led him to the adoption of the spherical form of unit because its strength is in no respect dependent upon any system of stays or braces; further, these spherical units need not be of any great size or weight, allowing the easy displacement and replacement or interchange of one or more of these units without disturbing the remaining portion of the structure.

The spherical form was adopted because it combined greatest strength with greatest heating surface, the hollow sphere being twice as strong as a tube of equal diameter. These spheres as now made are 8 inches in external diameter, with curved necks 4 inches in external diameter, the thickness varying from ½ inch in the necks to $\frac{3}{16}$ inch in the central line of the sphere. These units are subjected to a hydrostatic pressure of 350 pounds per square inch before they are passed for the erecting-shop. Two, three, or four spheres are included in a single casting, joined together by suitable necks, an illustration of which is given in

Fig. 352. These units must of necessity be uniform and duplicate. No present means are known whereby either wrought iron or mild steel can be economically wrought into such shape. The units have milled faces made by special machinery, and are combined in vertical sections by means of bolts passing through their centres which are entirely surrounded by water. The joints so made are perfectly steam- and water-tight, iron to iron, without the aid of any packing whatever. In case of repair, it is accomplished by the insertion of a new unit of exactly the same size and shape as the defective one removed, which does not detract from the original construction. Cast iron easily permits all parts to be tooled to a uniform standard gauge, and such parts will go together without forcing.

FIG. 352.

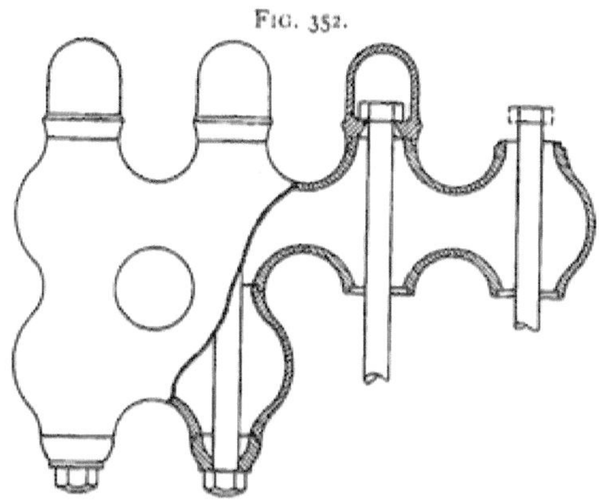

A slab, as already explained, is an aggregation of units held together by wrought-iron bolts, the heads of which are protected from the fire by round caps, shown in Fig. 352, while their upper ends extend through the common caps and terminate with a screw-thread and close-ended nut. The slabs to the required number, and of varying sizes according to the size of the boiler, are suspended side by side from a suitable iron framework, and are also connected at top and bottom to form a proper steam and water coupling. The slabs are hung one inch apart, and are slightly inclined upward from front to back, so as to secure a uniform area of steam-liberating surface, while the height of the water-line may vary.

Safety was the first consideration in Harrison's mind when designing this boiler; his one object was to make it absolutely secure from destructive explosions, even when carelessly used. From the method of joining, a rupture cannot extend beyond the unit in which it originates. The effect of the rupture is, therefore, localized, so that the adjacent parts of the boiler are not affected except in so far as they may be injured or displaced by the rupture of a defective unit. A boiler of this construction was tested by a committee of the Franklin Institute, during which the pressure was increased to 875 pounds per square inch, when a sudden discharge of steam took place, after which the pressure fell to 450 pounds, at which it stood when the fire was drawn.

The Wharton-Harrison Boiler, a modified form of the original Harrison boiler, is shown in Fig. 353, which, it is claimed, possesses certain advantages over the earlier forms. For example, the changed relative positions of the units to the radiant heat and impinging hot gases is an improvement, and with this there is an increased depth of water over the fire which admits of a gauge-glass one-half as long again

FIG. 353.

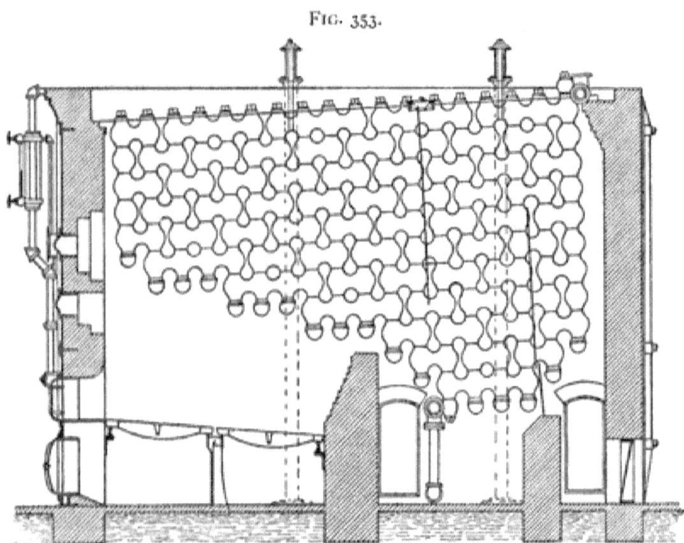

in this form of slab as was possible in the earlier forms, relieving the attendant from anxiety about the position of the water-level. Injury to the boiler from low water, whether due to neglect or accident, and which would have caused leaky joints in the old style, is practically obviated in the present design.

A longer travel for the steam after its liberation from the water impinging against the highly heated surfaces in its flow outward from the boiler results in greater dryness than was the case formerly.

The cost of repairs is reduced to a small percentage of what it once was, because they are more easily made.

Longer travel for the products of combustion between the slabs and around the units before escaping to the chimney-flue results in the better absorption of heat by the boiler, and the consequent reduction in temperature of the waste gases; there is thus secured an increase in the amount of power for a given space.

The present design affords increased facility for thorough interior cleaning, as, when the bolts are removed, any soft sediment or loose scale can be washed down through the vertical openings.

The brickwork for this type of boiler, when compared with other boilers, is, by reason of its nearly cubical form, quite economical, the cost being much less. As the entire weight of the boiler proper is borne by the metal columns and beams, the brickwork is reduced to mere enclosing walls carrying no load: they can be set, repaired, or removed without disturbing the boiler. The covering over the boiler is formed by special tiles.

The construction of these boilers is favorable for introduction into cellars and other inconvenient places, or for mule-back transportation over mountainous localities, because the boiler can be shipped in sections, no one of which exceeds 200 pounds in weight.

Water-Tube Boilers.—This type of boiler has a history extending back for more than a century; but the water-tube boiler of to-day can hardly be said to extend further back than the invention of Wilcox in 1856, which invention, changed in form but without loss of identity, appeared again in 1867 in the Babcock & Wilcox design, one widely known and probably more generally copied than any other.

A water-tube boiler consists usually of an assemblage of tubes filled with water, located in a furnace, and connected with two receivers,—the lower one of small diameter called the mud-drum, and an upper and usually larger one called the steam-drum. The mud-drum is placed in the coolest part of the furnace, sometimes protected from the heat of the furnace by a non-conducting covering, and not infrequently it is placed outside of the furnace altogether. The upper receiver, or steam-drum, is placed in the hottest part of the furnace, and performs the double function of a storage for water needed for circulation, as well as providing a large disengaging surface for the liberation of steam. The water-surface being usually at the centre of the drum, the upper half is steam-room, from which the supply of steam is taken.

The tubes are inclined to the horizontal, though in some cases the inclination is from the vertical, and so arranged that they receive water through the lowest headers, sometimes from the mud-drum, discharging the water through the upper headers into the steam-drum, the tubes being so located in the furnace that the flame and hot gases pass over and between them, usually at right angles to their axis. The tubes are both straight and curved, depending on the design of the boiler. The Babcock & Wilcox, Heine, Cahall, and variations of these designs have straight tubes; the Hogan, Stirling, Morrin, and boilers of that class have bent tubes.

The manufacturers of boilers with bent tubes claim that the bending is advantageous, because such tubes adapt themselves readily to the expansion of the boiler and prevent leaky joints. The manufacturers of boilers with straight water-tubes claim that such tubes can be more easily cleaned, examined, and renewed than is the case with bent tubes.

The present tendency in boiler design for steam pressures higher than 100 pounds per square inch is in the direction of water-tube boilers, of which there is now a great variety of designs in active competition. This type of boiler, when properly designed and built, is well calculated to carry high steam pressures. The tubes are seldom more than four inches in diameter; they are not only quite thin, but possess a large factor of safety. The evaporative efficiency is fully equal to any type of boiler yet introduced. It permits also a large power-rating in moderate space.

Circulation.—Water-tube boilers require a free and abundant supply or circulation of water through them to prevent overheating. This circulation adds directly to the efficiency of the boiler; it also contributes to both durability and safety. The cause of circulation in water-tube boilers has been attributed to a difference in density caused by a difference in temperature of the water within the boiler by the action of the fire, such that one portion of the boiler becomes more highly heated than another. The circulating movement is very gentle at first, because all the water in the boiler is at approximately the same temperature; but as the temperature rises and the tubes become filled with foam or steam-bubbles, it becomes more intense, partly because of the entraining action of the bubbles, until a powerful circulation is had throughout the whole interior, which, when once secured, may be indefinitely prolonged by the continued application of heat.

Wet steam is a common fault in water-tube boilers. To obviate this the steam generated in the tubes should have a free and direct passage through suitably designed and properly located headers leading directly into the steam-drum. Attempts have been made to counteract this fault by the employment of superheating surfaces which were not altogether satisfactory, though some designs permit superheating with less risk than others.

This lack of sufficient steam- and water-room is a serious defect of many water-tube boilers. A small water-space may be attended by a capacity for quickly getting up steam to a high pressure when all escapes for steam are shut off, but unless there is a sufficiently large volume of heated water in reserve ready to convert itself into steam as rapidly as the draught of steam is made from the steam-drum, the pressure will fall quicker than it was raised in spite of a fairly good fire, unless the boiler is of more liberal capacity than usually furnished for a given power. Consequently a closer attention will be required on the part of the fireman than with ordinary flue or tubular boilers.

Draught Area.—This receives considerable attention in horizontal tubular-boiler practice, but apparently less in the case of water-tube boilers. The cross-sectional area in fire-tube boilers varies from $\frac{1}{7}$ to $\frac{1}{8}$

that of the grate surface, an average of 1/8 being considered good practice. Too large a draught opening in any boiler has a tendency to reduce its efficiency. An editorial in vol. xxxi. of the *Engineering Record* calls attention to this defect in furnace proportion, and makes reference to water-tube boilers known to have been erected with a draught opening as large as 1/4 of the grate surface, resulting in a wasteful temperature of chimney gases. The draught area for water-tube boilers need not be greater than has been found necessary for return tubular and internally fired boilers. In the water-tube boiler most largely in use the direction of the current of gases is subjected to less abrupt change than in return tubular boilers, for in the latter the current is forced to change to an exactly opposite direction as the gases enter the tubes, while in the former the motion is continuously forward, though not in a continuous straight line. Consequently a less force of draught and a smaller total area should suffice in the water-tube boiler than in the return tubular.

The heating surfaces in water-tube boilers are of the most effective kind, because the heat acts upon the large aggregate tube surface and a comparatively small quantity of water for any given cross-section through the tubes; if the circulation is not interfered with, such heating surfaces permit steam to be quickly and continuously made.

The facilities for inspection, cleaning, and repairing are of the utmost importance, because these have a direct bearing upon the efficiency and durability of the boiler. Water-tube boilers are complicated structures at best, and if to this be added inaccessibility for cleaning, the efficiency will be quickly lowered and result in permanent injury. One of the advantages possessed by nearly all the later designs of water-tube boilers is the facility afforded for the renewal of parts by a system of duplication which obtains in all modern and properly equipped manufactories.

Freedom from disastrous effects of explosion is one of the merits particularly emphasized by makers of water-tube boilers. It is true that no overwhelming disaster and wreckage of property, such as often accompanied the explosion of Cornish, Lancashire, horizontal tubular and cylinder boilers has been chargeable to water-tube boilers, nor is it ever likely to be, but the casualty list resulting through the breakage of headers and other constructive details is by no means small.

The transportation and setting up in place of water-tube boilers is in most cases both easy and economical; there are, however, excellent water-tube boilers which cannot be taken apart for easy transportation, nor more easily erected in places difficult of access than is the case with horizontal tubular boilers. For ordinary supply this is a matter of little consequence, but if a boiler is to be erected in a city basement or in an inconvenient mountain district, it must be selected with special reference to the facts governing the case.

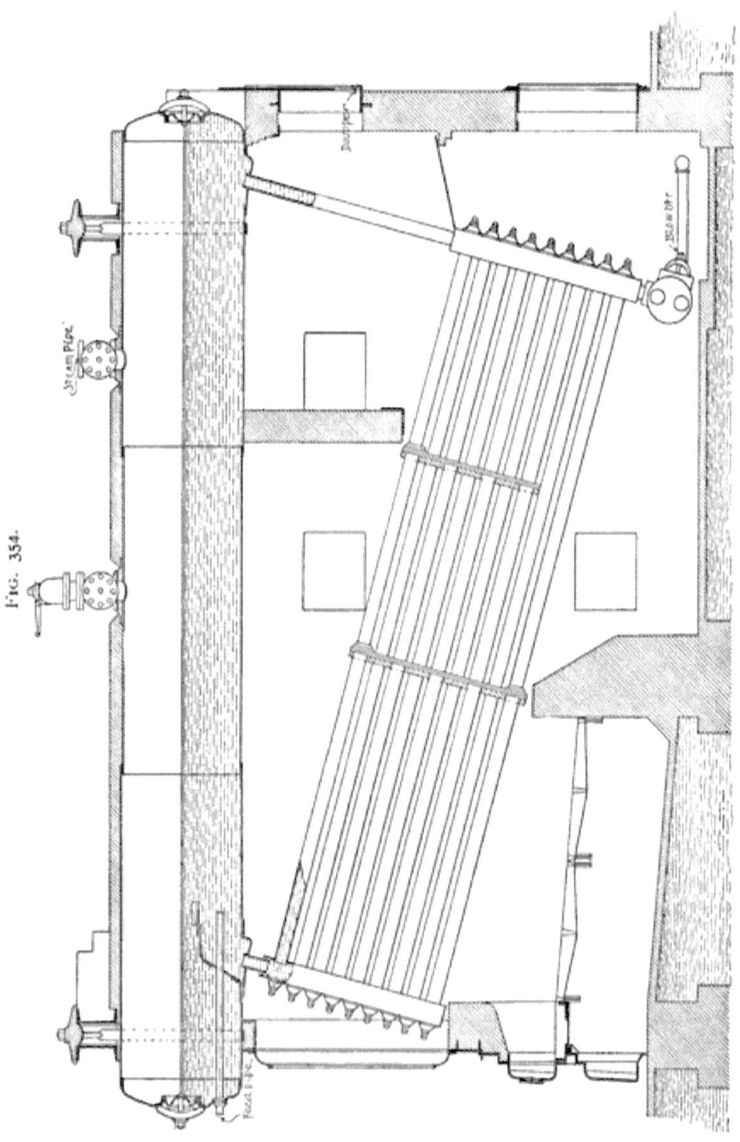

Fig. 354.

The Babcock & Wilcox Water-Tube Boiler.—The general design of this boiler is shown in Fig. 354. It is composed of an assemblage of lap-welded wrought-iron tubes, an overhead horizontal drum extending the whole length of the tubes, and a mud-drum below extending across the rear of the boiler at the bottom of the combustion-chamber, all of which are connected together in one complete circulatory system. The particular example here illustrated is 9 tubes high and 7 tubes wide by 20 feet in length, the tubes being placed at an inclination of 15° to the horizontal, as shown.

Fig. 355.

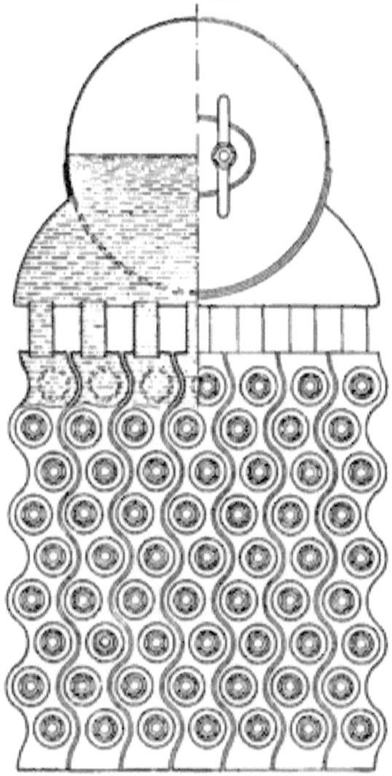

These tubes are 4 inches in diameter, arranged in series by means of a vertical connection called a header. A number of these headers can be placed side by side to make any requisite width, as shown in Fig. 355. In this boiler the headers are of such form that the tubes are staggered, or placed zigzag, so that the flow of gases upward among the tubes is interrupted and made to impinge against the sides and bottom of each tube instead of passing through a straight opening. The holes are bored and reamed tapering in the headers, into which the tubes are expanded so as to make a tight joint under pressure: a detail of this joint is shown in Fig. 356. Opposite each of these openings is a handhole, detailed in Fig. 357. The handhole plates and headers are faced by milling the surfaces to accurate metal contact, making a tight joint without the use of packing of any kind. The front header is attached to the water- and steam-drum in two ways, one underneath the drum by a cross-box, as in Figs. 355 and 356, the other in the head of the drum, as in Fig. 358.

A drum cross-box is riveted to the underside and to each end of the upper drum. After having been bored and reamed, these receive the nipples which connect the headers, as shown in Fig. 355. In case connection is made directly into the head of the drum, as shown in Fig. 358, a cast-iron head is employed.

The headers are made of cast iron and of forged steel. Both kinds

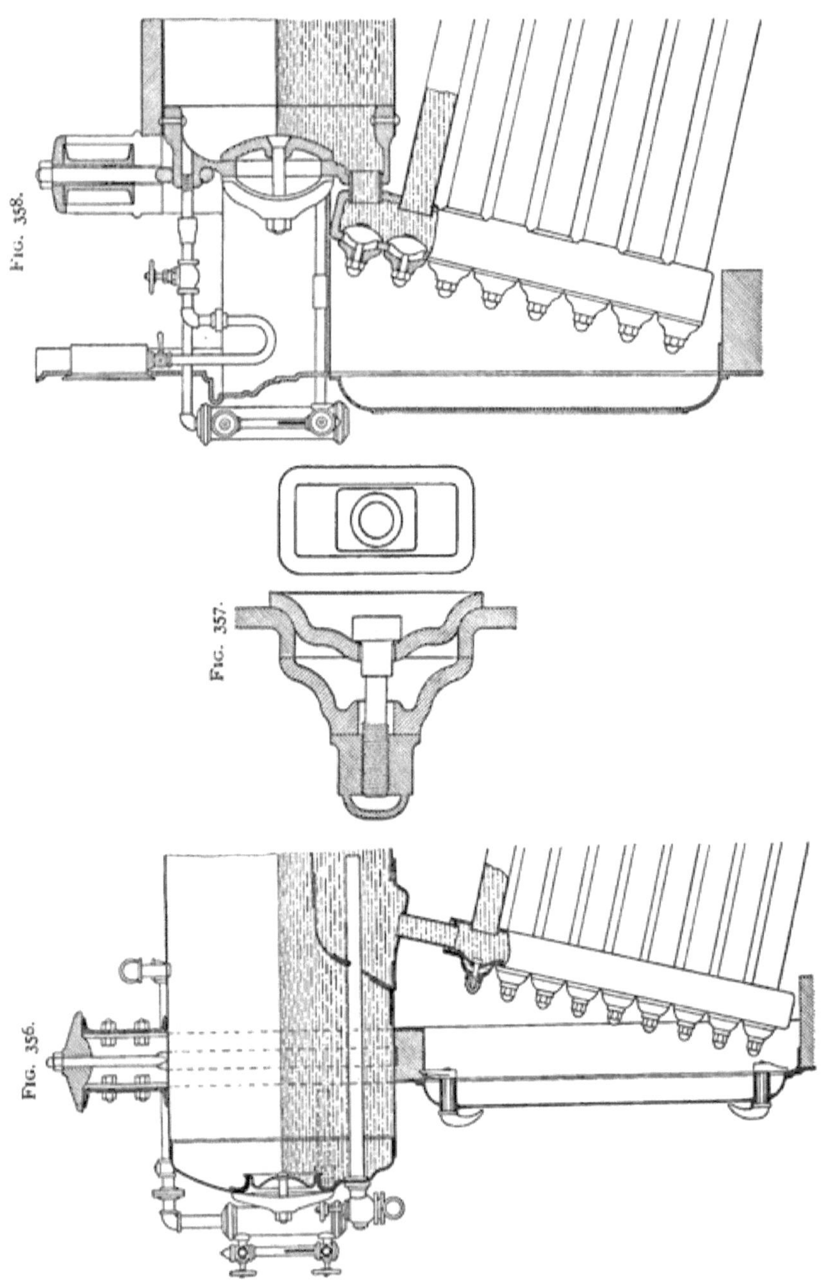

Fig. 358.

Fig. 357.

Fig. 356.

are now in use,—the former for pressures approximating 100 pounds, the latter for higher pressures. The connection of each of the rear headers to the drum is in all respects similar to that at the front just described, having a pipe instead of a nipple connecting the rear cross-drum to the headers arranged below. From the bottom of each of the rear headers is a nipple, forming a connection with the mud-drum below, which is located at the lowest part of the combustion-chamber back of the bridge-wall.

The drum is of a diameter and length suited to the size of the boiler. It is usually made up of three sheets, as shown in Fig. 354, the longitudinal seams being butt-strapped inside and out, with two rows of rivets passing through both straps and the shell, and two rows through the shell and inner strap only, as shown in Fig. 359. Manhole-openings are provided in the heads, as shown in Fig. 356, for such as are made of plate metal, and in Fig. 358 for those made of cast iron. A deflector is placed in the front end of the drum when connection is made from below, as shown in Fig. 356, the object of which is to prevent a violent agitation of the water at the front end, caused by the high velocity of the upward current from the headers in the body of the water in the drum, the deflector changing the direction of the upward current and directing it towards the rear. The steam- and water-gauge fittings are attached to the front of the drum in the ordinary manner. The feed-pipe enters the front head of the drum, as shown in Fig. 354, extending through the deflector, so that the end of the feed-pipe is under it and to the rear of the openings in the cross-box above the line of header connections.

FIG. 359.

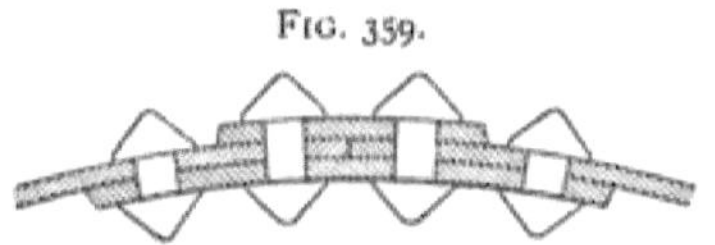

The completed boiler is suspended in the furnace by a strap at the front and rear end of the drum, as shown in Fig. 354. The ends of these straps are fitted with screw-threads passing through cast-iron washers resting upon channel-beams, which extend across from side to side of the furnace and rest upon iron columns, which carry the weight of the boiler on suitable foundations below the ground level. By this means the entire boiler may be, and is, suspended in place before the brickwork is begun. The furnace-walls may, therefore, be constructed independently of the boiler; so, also, any portion of the furnace-walls may be repaired, taken down, or rebuilt without affecting the boiler itself.

The furnace front has large doors, by which complete access can be had to all of the handhole plates in the front headers. The ordinary fire- and ash-pit doors are underneath, as is common to stationary boilers.

The mud-drum is made of cast iron, commonly 12 inches in diameter, bored and reamed for expanding the wrought-iron nipples, as is done with the headers. It is tapped for blow-off connections and furnished with handhole for cleaning.

Operation: The fire is made under the front and higher end of the tubes. The products of combustion pass up between and around the tubes into a combustion-chamber above and underneath the steam- and water-drum. From thence the gases pass down between the tubes, thence once more up through the spaces between the tubes, and off to the chimney. The water inside the tubes, absorbing heat from the gases, is heated and tends to rise towards the higher end; if it receives heat enough, the water is converted into steam. The mingled column of steam and water, being of less specific gravity than the solid water at the back of the boiler, rises through the headers into the drum above, where the steam completely separates from the water, the latter flowing back to the rear of the drum and down through the connecting tubes into the lower headers and inclined tubes in a continuous circulation. If the passages are all large and free, this circulation will be very rapid, sweeping away the steam as fast as formed and supplying its place with water; absorbing the heat of the fire to good advantage; causing a thorough commingling of the water throughout the boiler and a consequent equal temperature; preventing also to a degree the formation of deposits or incrustations upon the heating surfaces, these being deposited in the general course of circulation in the mud-drum, whence they are blown out.

FIG. 360.

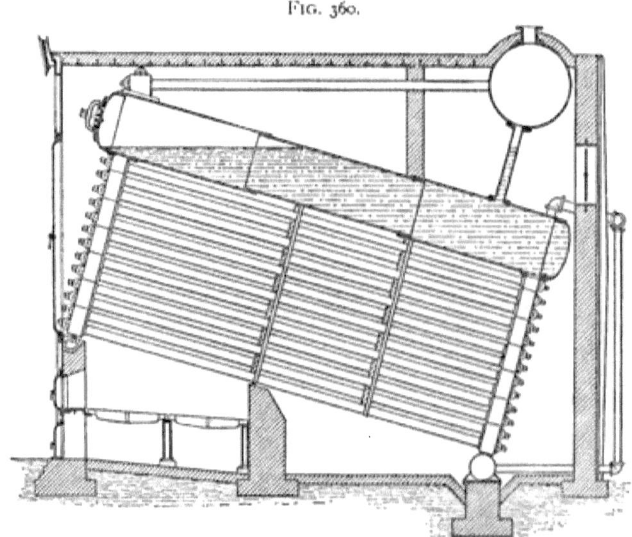

Zell Water-Tube Boiler.—This boiler is shown in sectional elevation in Fig. 360. It consists of an assemblage of tubes and a water-drum placed at an angle of about 15° to the horizontal. These tubes are equally spaced both vertically and horizontally, and arranged in zigzag vertical rows which are connected at the rear by suitable header

Fig. 361.

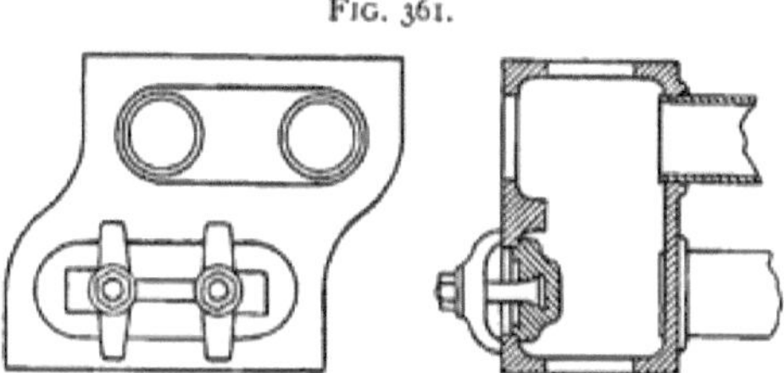

connections, which also attach to a mud-drum located at the bottom of the combustion-chamber. At the front end of the boiler these tubes connect with similar headers, which are joined together by suitable connections with the water-drum.

Fig. 362.

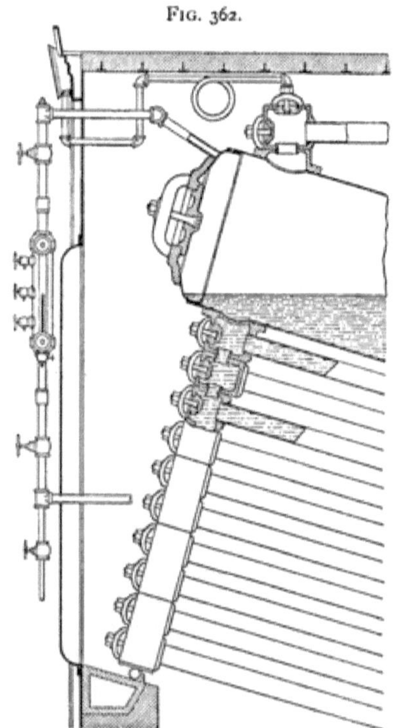

The tubes are standard 4-inch lap-welded wrought-iron boiler tubes arranged in sections or headers of four tubes each. An elevation and section of one of these headers is shown in Fig. 361. These headers are made of cast iron bored and reamed to receive the tubes which are expanded in the holes thus prepared; each header is fitted with two internal handhole plates, each of which covers an opening across two tubes. The handhole plates are fitted to the inside of each header, as shown, the plates being held in position by T-bolts, the heads of which fit into slots on the outer faces of the plates. The pressure of steam holds the plates up to the joints without the aid of the bolts, a much safer method than that of putting these plates or caps on the outside and securing them by bolts from the inside, for if the threads slip or if the bolts break under steam pressure, the plate will fly off and steam and boiling water would rush out, to the danger of life and property. The headers are connected in series vertically by means of nipples expanded into bored holes, as shown in Fig. 362. The number of these headers to be placed in vertical and horizontal rows will depend upon the size of the boiler or the power required.

A horizontal mud-drum 12 inches in diameter is placed immediately below the bottom of the rear headers and below the line of circulation; it is connected with the several headers extending across the furnace by means of short expanded wrought-iron nipples. Since cast iron has been proved to be the best material to resist the action of corrosion, these mud-drums are made of that material. They are tested up to 500 pounds hydraulic pressure, and have the necessary handholes and plates for cleaning. The plates closing these handholes have the same internal surface-bearing and general arrangement as those closing the handholes in the headers. Suitable blow-off and feed openings for flushing, with proper fittings, are placed at opposite ends of the drum.

The inclined water-drum at the top of the tubes is attached to the headers by wrought-iron nipples expanded into bored and reamed holes, as shown in Fig. 362. A cast-iron saddle is riveted to the front and back ends of the water-drum, and each saddle has sufficient area of opening for the tubes below. These are connected with the top headers by means of expanded wrought-iron nipples in the same manner that the headers are connected together. This drum is made of mild steel plate with convex heads and manhole. A saddle is riveted to the top of the highest part of the water-drum, to which is attached by an expanded nipple connection a manifold having a number of openings for pipes which convey the steam to the steam-drum at the rear of the boiler, shown in Fig. 362 and on enlarged scale in Fig. 363.

Fig. 363.

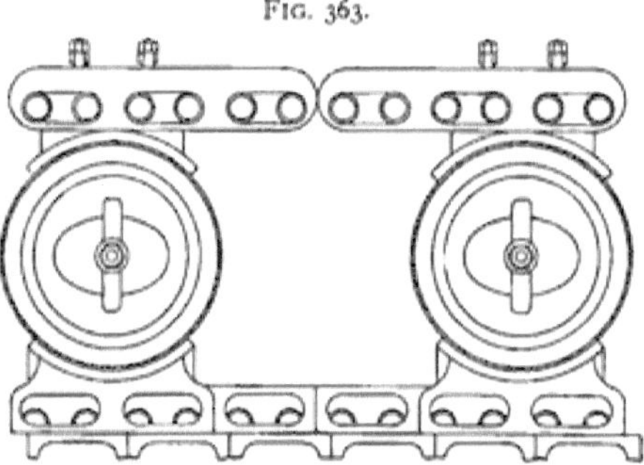

The steam-drum is located to the rear and at the top of the boiler and furnace. It is made of mild steel. The heads are convex in shape and will withstand high pressure without bracing. There is a manhole at one end of the drum, the cover-plate for which has its bearing on the inside of the drum in a similar manner to the headers. This drum is located in a horizontal position at the rear end of the water-drums, and its length is somewhat greater than the combined width of the headers. It acts as a superheater and reservoir for the storage of steam and contains no water, thereby differing from the usual water- and steam-drums which accompany other forms of water-tube boilers. A connection is made from it to the water-drum below by means of wrought-iron tubes expanded at each end into the drums. These connecting tubes are incased in cast-iron columns, not shown in the engraving; they are flanged at the ends, giving a uniform bearing on each drum. These columns are the support of the steam-drum and relieve the wrought-iron tubes from all strains of that nature. Being made in half sections and bolted together, they can readily be removed to inspect the enclosed tubes. The proper openings and fittings are provided for safety-valves and steam piping.

Superheating pipes connect the manifold, which extends across the whole width of the front of the boiler, to the steam-drum at the rear; these are lap-welded wrought iron, placed in a single row and expanded in both the manifold and the steam-drum.

The boiler is supported at the rear end by resting upon two or more cast-iron saddles placed under the mud-drum; these castings are set upon a substantial brick wall. The front end of the boiler rests upon a roller, which is placed upon the top of the arch-box, shown in Fig. 362, thus allowing freely for expansion and contraction. This arch-box, through which there is a constant circulation of air, is covered on the side nearest the furnace with large and especially moulded fire-brick, which protects it from injury by the heat. It rests upon cast-iron stands, one on each side of the furnace, which are embedded in brickwork and anchored to the foundations. The entire weight of the boiler is carried by the foundation supports mentioned above; that is, it is not suspended, but supported directly from underneath. The manufacturers of this boiler claim that all water-tube boilers that are suspended from above must have the combined weight of the boiler and its contained water carried by the wrought-iron nipples in the expanded joints which tie the structure together, subjecting such joints to strains for which they were never designed or intended, which strains are mainly eliminated when supporting the boiler from below.

The flame-plates, shown in Fig. 360, consist of sections made of cast iron and are faced with fire-brick; they are usually two in number, and extend from side to side of boiler and from top to bottom of tubes.

The feed-water pipe connects at the rear end of the water-drum. The water descends into the rear headers and is then distributed through the tubes. By this means of feeding it is expected that more or less of the impurities in the feed-water will be precipitated on contact with the hot water in the boiler and, gravitating downward, will collect in the mud-drum below.

The operation is such that from the fire in the furnace the products of combustion pass up between the staggered tubes and water-drums, then down and around the same between the flame-plates to the bottom of the combustion-chamber, thence up again, making three runs across the water-tubes, and out through the space between the steam- and water-drums to the flue, and thence to the chimney. The water in the inclined tubes next to the furnace is first heated and raised to a rapid state of ebullition. This mingled body of steam and water, being lighter than the solid water in the rear of the drum, is forced rapidly up through the front headers into the water-drum, where the steam and water separate; the steam passes up through the manifold into the superheating tubes, which, being surrounded with hot gases, absorb additional heat, evaporating any entrained water in its passage to the steam-drum; this latter being jacketed by the waste gases in their passage to the chimney and having a temperature of a few degrees higher than the steam within the drum, more heat is absorbed by the steam and it becomes further superheated.

Fig. 364.

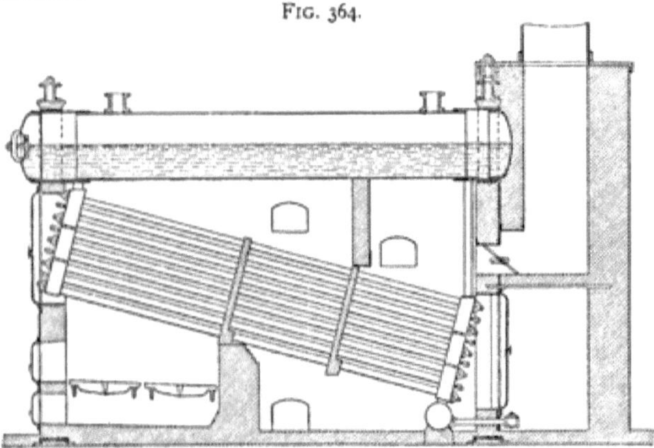

The Gill Boiler in its general features is not unlike the Babcock & Wilcox boiler, and is shown in Fig. 364. The water-tubes are 4 inches in diameter and spaced about 3 inches apart; they incline at an angle of about 15° to the horizontal. The tubes are grouped in sets of four or more, which are expanded into bored holes in a cast-iron box or header at each end in such a way that the tubes are staggered instead of being placed one above the other, shown in Fig. 365. By making these boxes short, and by connecting them by slightly flexible tubes, the danger of breakage is practically eliminated. In the engraving it will be seen that the column of headers at the left hand shows a 5- and 6-hole header with the caps removed, together with the curved tube-connection with the steam-drum; the column at the right shows similar headers, with their inside handhole, caps, bolts, and dogs in place. The middle column shows a 4- and 6-hole header in section at the middle of the headers, showing the manner of connecting the headers with each other and with the steam-drum. The small illustration at the bottom shows a section through the three headers and four tubes at the line A. Handholes are provided in the headers at the front and rear of all the tubes through which a scraper can be used to remove any sediment or scale that may have lodged in the tubes or in the headers. The tubes may also be removed through these handholes when necessary and new ones inserted. These openings are closed with caps placed inside of the headers and packed with thin rubber gaskets. The assemblage of tubes is divided into two or three sections by one or two flame-walls made of fire-brick, secured by rods running through them and into the side-walls. The flame and heated gases are thereby compelled to traverse the tubes two or three times before finding their exit at the rear end of the furnace.

FIG. 365.

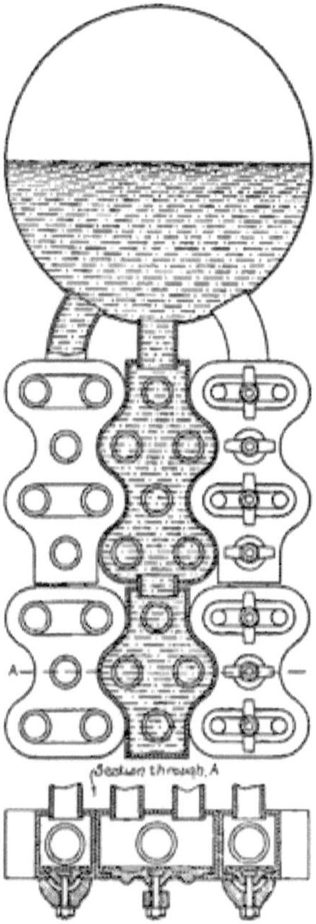

The steam-drum is made of mild steel, in diameters varying from 30 to 50 inches, or so proportioned as to allow a cubic-foot capacity for steam-space and a cubic foot for water storage in the drums for each horse-power to be developed. Thirty-inch drums have been found to be the smallest diameter that will allow of the production of dry steam under conditions of rapid driving. The water-tubes located over the fire receive the most intense part of the heat, and absorb a considerable portion of it before it reaches the drum; the latter is not, therefore, subjected to the same destructive energies that would otherwise occur, and thus large drums may be employed with comparative safety in water-tube boilers.

A mud-drum, usually 18 inches in diameter, is placed under the rear row of headers, the connection being made of short nipples. This drum serves to collect the sediment floating in the water and hold it until it is blown off.

A structural framework composed of wrought-iron I-beam columns and channel-bar cross-beams is erected, within which the drums, tubes, and headers are suspended, thus allowing the working parts to expand and contract independently of the structural parts.

The Caldwell Water-Tube Boiler, Fig. 366, differs principally from those already described in the construction of the header and in

Fig. 366.

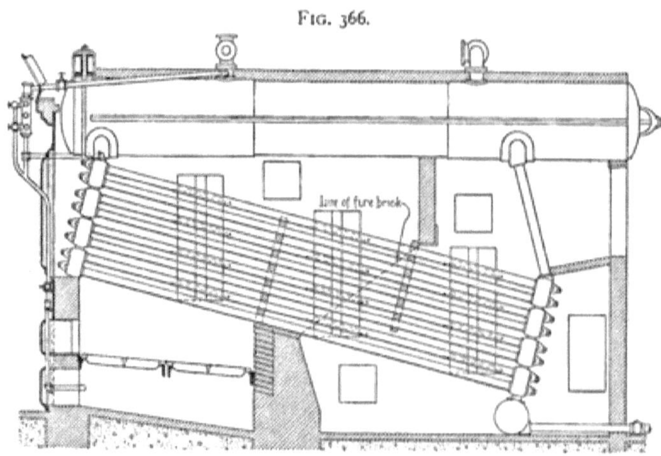

the distribution of the current of gases in the furnace, the combination of the horizontal water- and steam-drum, the inclined assemblage of tubes, and the mud-drum underneath being substantially that of the Babcock & Wilcox design.

Fig. 367.

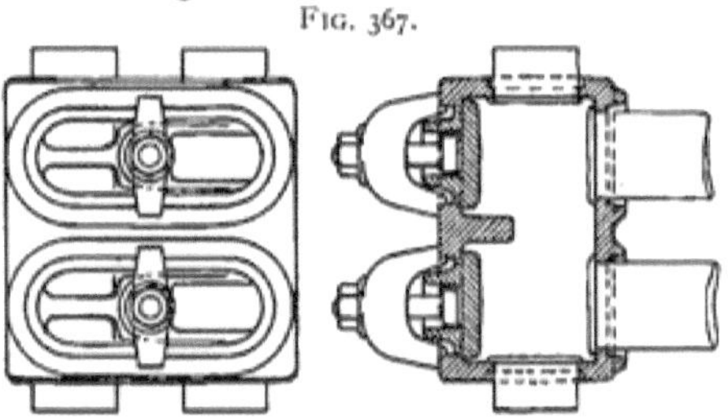

The tubes are of wrought iron, 4 inches in diameter, built up in sections of 4 tubes, each with a cast header at each end of the tubes, into which they are secured by expanding. These sections are set over each other three, four, and five high, and nippled together by means of short pieces of 4-inch extra heavy tubing, expanded into bored holes of uniform diameter, making as a whole a series of flexible members yielding to the inequality of expansion among the tubes.

The header castings are quadrangular in shape, and fully detailed in Fig. 367. The openings and handholes in front, through which access is had to the tubes, are covered with plates which close on an inner lip, whereby a tight joint is secured, the pressure assisting.

The baffle-bricks used in giving direction to the current of gases in the furnace are shown in Fig. 368. The uses of these bricks are to absorb the heat of the furnace while the furnace doors are shut, to give it out again when they are opened and the cold air is rushing in during the firing, thus saving the tubes from some of the evil effects of sudden changes in temperature to which such boilers are commonly subjected. It is claimed they also offer a more positive impingement of the gases on the tubes than is gained by merely staggering the tubes. It is further claimed that the gases passing upward between the bricks and the tubes tend to keep the tops of the tubes clean,

Fig. 368.

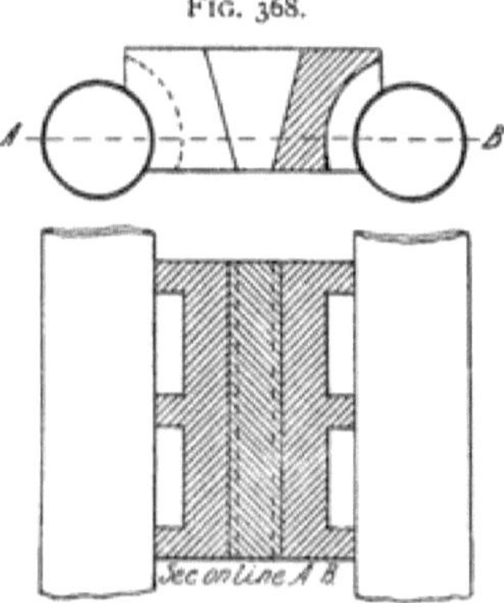

whereas in water-tube boilers not thus equipped there is always a ridge of soot which gathers on the upper side of each tube. The baffle-bricks are only used in the first and last passes of the tubes, these being upward passes, and not in the middle, which is a downward pass. Baffle-bricks are not used on the three or four rows of tubes immediately over the fire, these being within the influence of the radiant heat from the fire, and require no baffling to hold the gases back. The baffling is provided for on the top rows only. These tubes being away from the furnace and radiant heat, the circulation of gases between the upper and lower tubes is thus equalized and their efficiency increased. The baffle-bricks lie on and partly embrace the top portion of the tubes, as shown in Fig. 368; there is, therefore, a constant absorption of heat by the bricks, to be afterwards given out by them to the upper portion of the tubes. Inasmuch as flame and heat in ascending form the cone-shaped flame of a candle, so also the last pass of the gases at the upward and outward corners of the assemblage of tubes do not get as much heat as the middle tubes, unless the gases are directed towards these places by means similar to that just described. It was found when operating at rated capacity that the temperature of flue gases was lowered from 480° to 410° Fahr. by a proper distribution of the baffle-bricks in the last pass of the gases. They also serve to reduce the draught between the tubes, which would ordinarily be too great to attain the best results in water-tube boilers; they also furnish a ready means for cutting down the draught area in the last compartment in the furnace to that required for a proper distribution of the heated gases over and around the tube surfaces, a necessity for realizing the highest economy in water-tube boilers.

The Root Boiler.—The general design of this boiler is shown in Fig. 369. It may be briefly described as consisting of an assemblage of inclined water-tubes connected at their front and rear ends with a series of horizontal overhead water- and steam-drums; the fire being applied under the front and raised ends of the tubes, its action is to cause an upward and forward flow in the tubes, the water thus forced from the tubes into the front end of the horizontal drums being constantly replaced by a downward flow from the rear end of the same drums. This continuous circulation is common to other water-tube boilers, and is an essential requisite to successful operation.

FIG. 369.

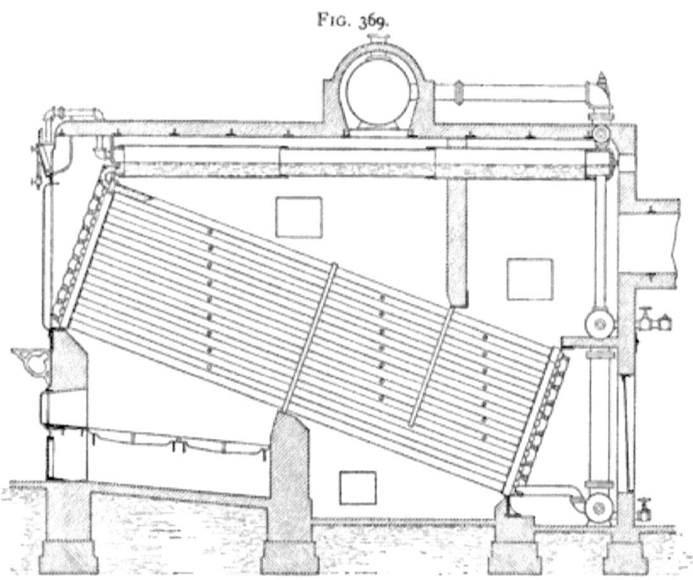

The tubes are expanded in pairs at both ends into bored holes in cast-iron headers, the pair of tubes with their headers forming what the manufacturers call a package. See Fig. 371. The two tubes in one package are on the same horizontal level when set up, and are in identical conditions as regards exposure to flame and temperature of contents; therefore, the expansion and contraction of both must be identical. The headers are stacked alongside of and over each other in such a manner as to form straight horizontal rows of tubes, while the rows of tubes up and down are staggered as shown. There is no connection between any one header and another lying either to the right or left of it, but each header is connected with the one above it by means of a return bend, shown in Fig. 371. The uppermost header of each vertical row, front and back, is connected respectively with the front and back of an overhead horizontal drum, of which latter there are as many in a boiler as there are vertical rows of headers.

A horizontal drum at the rear with its underlying row of headers and the tubes therein form a separate section of the boiler. The rows of headers at the front and back of each section are all connected with the ends of the overhead drums to insure a complete circulation.

FIG. 370.

The joints between headers and bends are made in the following manner: in line with the tubes and on the face of the header are milled recesses, in which fit rings of gun-metal which have a bearing at the bottom of the recesses. These rings are bored on a taper, and into this taper is engaged the end of the return bend, turned to the same taper. Lugs are cast on the bend, whereby two T-headed bolts have their heads pocketed in recesses cast on the outer face of the headers, and these serve to draw the bend home into the gun-metal ring. The other end of the bend is likewise secured to the next header (above or below), so that each bend is held in place by four bolts against two separate headers. No other packing than this gun-metal ring is used in making the joint. Much importance is attached to the elasticity of this joint, for if one package of tubes expand more than the one above it, the gun-metal rings in the two joints on the connecting-band yield enough to adjust themselves to the difference in expansion without leakage. The return bends, with similar metallic joints, are also used to connect the uppermost front header of each section with the front end of the overhead drum.

The expansion and contraction of the tubes and drums take place at the front end of this boiler. All the rear headers are locked together by means of plates or wedges, engaging under suitable lugs cast on the outer face of the headers. The front headers, being held by the bends with flexible joints, can yield slightly without leakage, permitting every pair of tubes in the boiler and each one of the overhead drums to expand and contract independently.

A cross steam-pipe lies over the top and at the rear of the overhead drums, being connected with each by means of a nipple expanded into bored holes. From this common steam-pipe two outlets connect with an overhead cross steam-drum located over the centre of the boiler, to which are attached the main stop- and safety-valves. This drum is provided with a drip-pipe leading below the water-line, so that any condensation may drain back into the boiler.

Fig. 371.

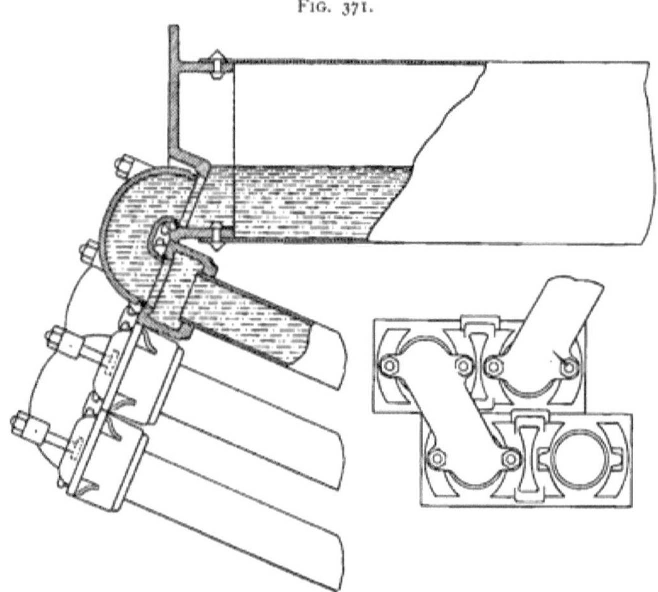

The steam-drum is enclosed in brickwork and surrounded by an air-space connecting with the combustion-chamber, but far enough removed from the direct path of the gases as to prevent injury to the metal in the drum by overheating.

From the rear end of each one of the overhead drums is a connection with the rear headers of its corresponding section. In the smaller boilers, having tubes 10, 12, or 15 feet long, the connection is made direct from the drum to the uppermost header, and the undermost rear headers of all the sections are connected with a common mud-drum located at the lowest point of the boiler. In the larger boilers, having tubes 18 feet long, the downtakes from the several overhead drums all lead into one horizontal drum, to which the feed-pipe is attached, and which is connected with the mud-drum set at the lowest part of the boiler by means of two large standpipes; the mud-drum, as in the case of smaller boilers, is connected with the undermost rear headers of all the sections. The path of the downward circulation in the large boiler is, therefore, from each overhead drum through its connected downtake into the common feed-drum, from this through the standpipes to the mud-drum, and from the latter the water is distributed to the lower headers of all the sections.

The connections of the mud-drum are all made at the top, so that the contents are not stirred up by the circulation. The object of this arrangement of feed-drum and standpipes is to heat the entering feed-water by mingling with the descending currents from the overhead drums, both in the feed-drum and in the standpipes, the cross-section of which latter in proportion to downtakes is made sufficiently large to retard the velocity of flow and give the feed-water time enough to become heated to the point where it will precipitate most of its lime in the mud-drum before entering the boiler.

FIG. 372.

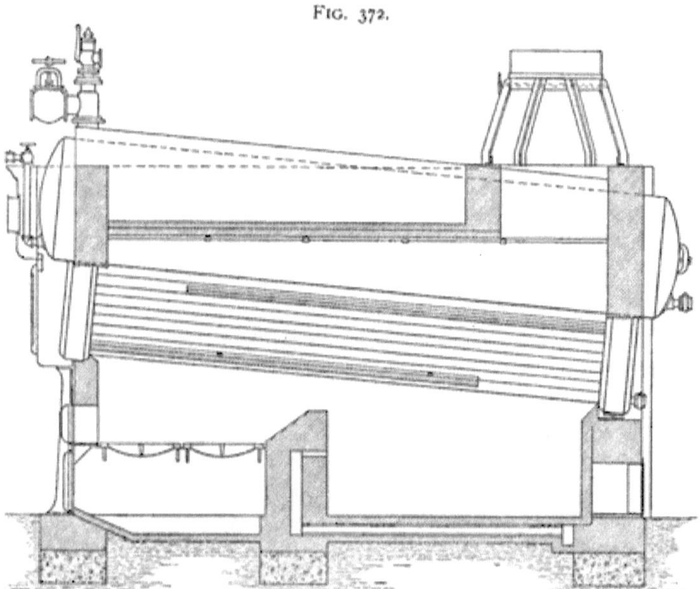

The Heine Boiler belongs to the water-tube variety, as indicated by the illustration, Fig. 372, consisting, as it does, of an assemblage of inclined water-tubes over the furnace, these tubes connecting at front and rear with an overhead water- and steam-drum, the connections corresponding to headers in ordinary water-tube boilers being water compartments, commonly known as water-legs, extending across the furnace and into which the tubes are expanded, thus making a continuous water circulation similar to that of the water-tube boilers previously described.

The water-legs are of approximately rectangular shape, drawn in at top to fit the curvature of the shells, as shown in Fig. 373. The area of throat or opening at the top of each water-leg is approximately equal to the combined area of tubes for that leg. Each is composed of a head-plate and a tube-sheet, flanged all around and joined at bottom and sides by a butt-strap of the same material strongly riveted to both, as shown in Fig. 374. The water-legs are further stayed by hollow stay-bolts of hydraulic tubing, of large diameter, so placed that two stay-bolts support each tube and handhole, also shown in Fig. 374. The water-legs are joined to the shell by flanged and riveted joints and the drum is cut away at these two points to make connection with the inside of water-leg, the opening thus made being strengthened by bridges and special stays, so as to preserve the original strength, the details of which are shown in Fig. 375. The tubes extend through the tube-sheets, into which they are expanded with roller expanders. Opposite the end of each, in the head-plates, is placed a handhole of slightly larger diameter than the tube and through which it can be withdrawn. These handholes are closed by small cast-iron plates shown in Fig. 374.

FIG. 373.

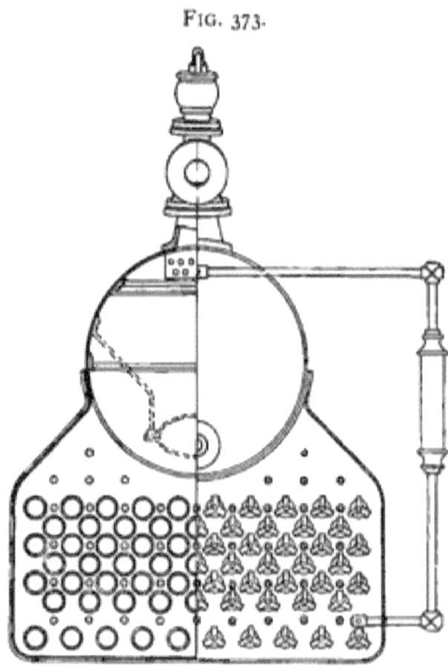

The water- and steam-drum is a cylinder with heads dished to form part of a true sphere and requiring no stays. Both the cylinder and its spherical heads are free to follow their natural lines of expansion when put under pressure. To the bottom of the front head a flange is riveted, into which the feed-pipe is screwed. This pipe is shown in Fig. 376 with an angle-valve and check-valve attached.

On top of the drum, near the front end, is riveted a steam-nozzle, to which is bolted a T-fitting; this fitting carries the steam-valve on its horizontal branch, the safety-valve being placed on top. Just under

Fig. 374.

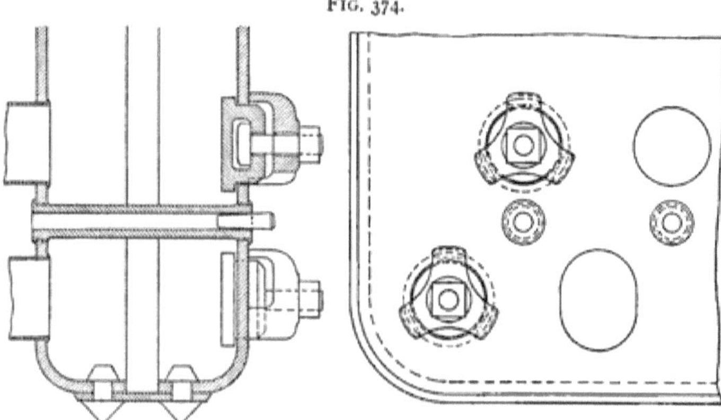

the steam-nozzle is placed a dry-pipe, and underneath that an inclined deflecting-plate which extends from the front head of the drum to some distance beyond the mouth or throat of the front water-leg. The rear

Fig. 375.

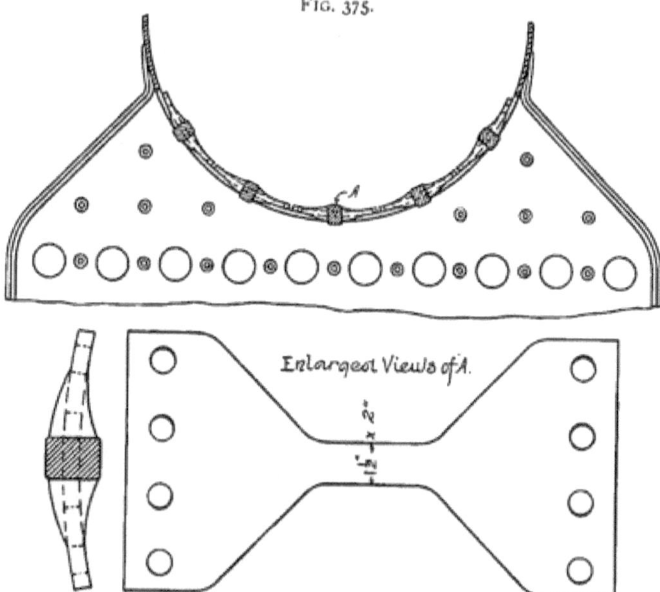

head carries a blow-off-flange of about the same size as the feed-flange and a manhead curved to fit the head, the manhole supported by a strengthening ring on the outside. On each side of the drum a tile-bar

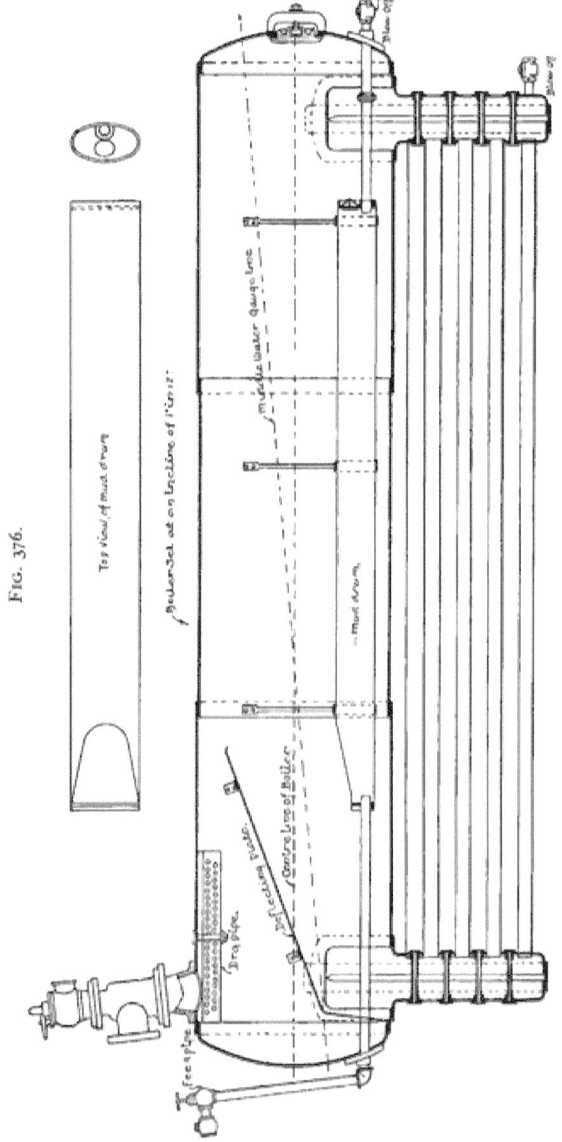
Top view of mud drum
Boiler set at an Incline of 1" in 12"
Middle Water Gauge line
Mud drum
Centre Line of Boiler
Deflecting Plate
Dry pipe
Feed pipe
Blow off
Blow off

FIG. 376.

rests loosely in flat hooks riveted to the drum. This bar supports the side tiles, whose other ends rest on the side walls, thus closing in the furnace on top. The top of the tile-bar is 2 inches below low-water line. The bars rise from front to rear at the rate of 1 inch in 12. When the boiler is set the tile-bars must be exactly level, the whole boiler being then on an incline,—*i.e.*, with a fall of 1 inch in 12 from front to rear; this makes the height of the steam-space in front about two-thirds the diameter of the drum, while at the rear the water occupies two-thirds of the drum, the whole contents of the drum being equally divided between steam and water.

The mud-drum is located inside of the water- and steam-drum; it is placed well below the water-line, usually parallel to and 3 inches above the bottom of the shell. See Fig. 376. It is thus completely immersed in the hottest water in the boiler. It is of oval section, as shown in the engraving at the top of Fig. 376, and slightly smaller than the manhole; it is made of strong sheet iron with cast-iron heads; it is entirely closed except about 18 inches of its upper portion at the forward end, which is cut away nearly parallel to the water-line. The feed-pipe enters the mud-drum through a loose joint in front; the blow-off-pipe is screwed tightly into its rear head and passes by a steam-tight joint through the rear end of the main drum.

In setting this boiler the front leg is placed firmly on a set of cast-iron columns bolted and braced together by the door-frames, dead-plates, etc., and forming the fire-front. This is the fixed end. The rear water-leg rests on rollers which are free to move on cast-iron plates firmly set in the masonry of the low and solid rear wall. Wherever the brickwork closes in to the boiler broad joints are left which are filled in with tow or waste saturated with fire-clay or other refractory but pliable material. Thus the boiler and its walls are each free to move separately during expansion or contraction without loosening the joints in the masonry.

FIG. 377.

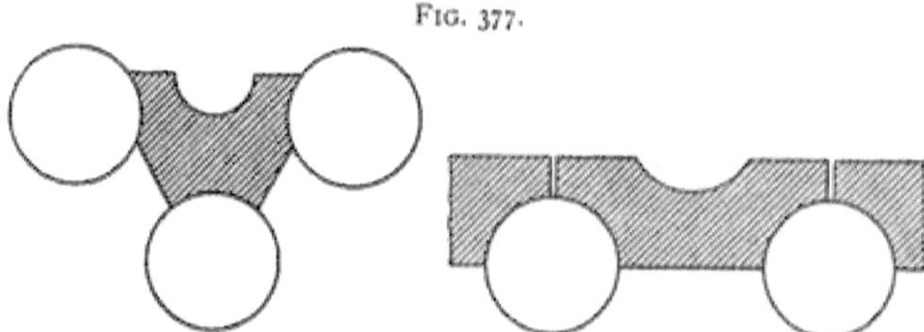

Light fire-brick tiles, shown in Fig. 377, are placed between the upper tubes. The lower tier extends from the front water-leg to within a few feet of the rear one, leaving there an upward passage across the rear ends of the tubes for the flame, etc. The upper tier closes in to the rear water-leg and extends forward to within a few feet of the front one, thus leaving the opening for the gases in front, as shown in Fig. 372. The side tiles extend from side of walls to tile-bars on the drum and close up to the front water-leg and front wall, and leave open the final uptake for the waste gases over the back part of the shell, also shown in Fig. 372. The rear wall of the setting and one parallel to it are arched over the shell a few feet forward to form the uptakes; on these and the rear portion of the side walls is placed a light sheet-iron hood, from which the breeching leads to the chimney. When an iron stack is used this hood is stiffened by L- and T-irons, so that it becomes a truss, carrying the weight of such stack and distributing it to the side walls, as shown in Fig. 372.

Fig. 378.

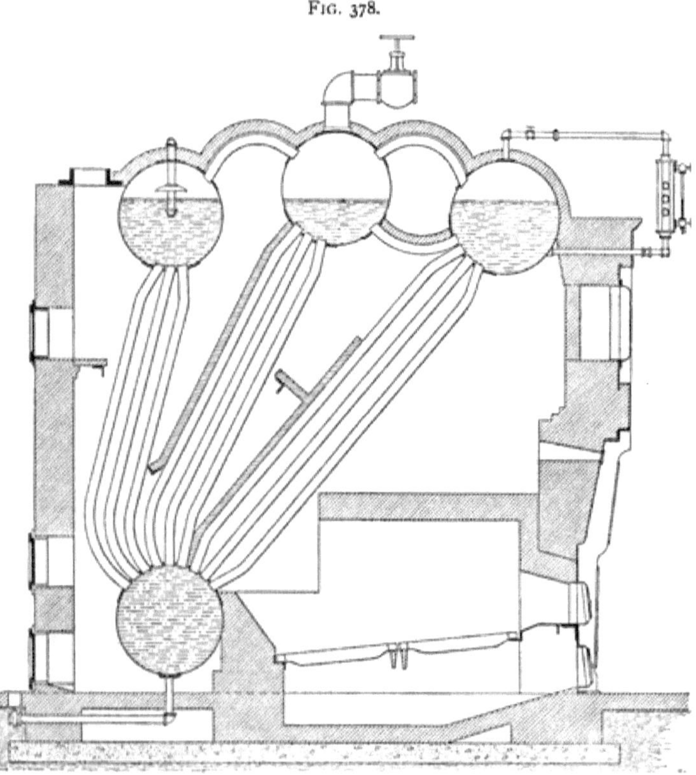

The Stirling Boiler as now made consists of three upper or steam-drums and one lower or mud-drum, all connected together by means of tubes, which are bent slightly so as to allow them to enter the drums radially, as shown in Fig. 378, and in sectional and half-front elevation in Fig. 379. All of the upper or steam-drums are connected by steam-circulating tubes, but the front and middle drums only are connected by water-circulating tubes. The tubes used are 3¼-inch lap-welded mild steel; the drums are also made of mild steel. These drums and tubes form the boiler proper, no cast metal entering into its construction.

Fig. 379.

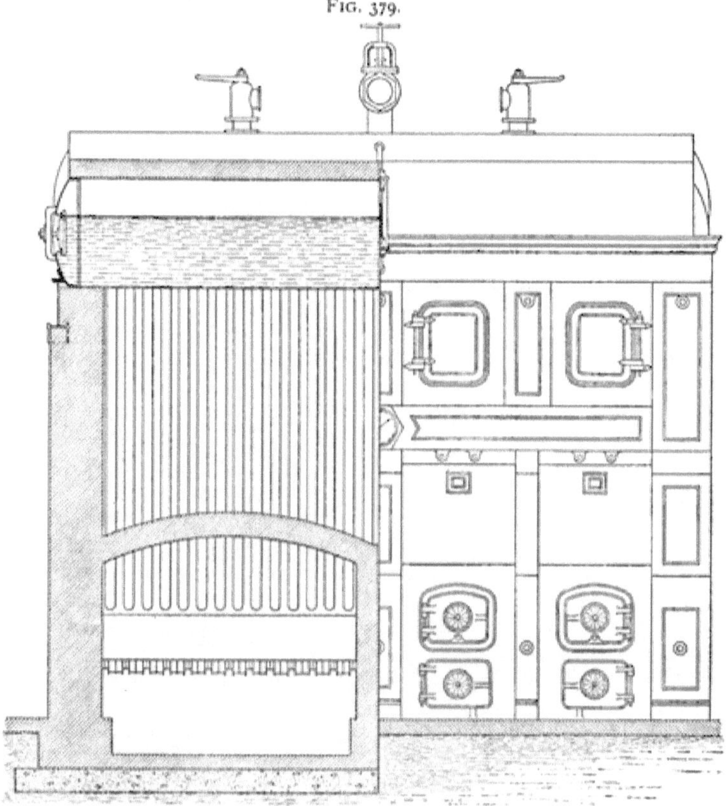

The front and middle bank of tubes, through which rapid circulation takes place, receive their supply of water from the mud-drum. Fig. 380 is an ideal representation of the ascending and descending currents induced by the action of the furnace. The tubes, *D*, receiving heat sufficient to vaporize the contained water, cause an upward movement of the water into the drum, *B*, having water communication with the adjoining drum, *A*, and it by descending pipes with the mud-drum, *C*, at the rear of the bridge-wall. A circulation once established is maintained so long as fire is continued in the furnace.

The feed-water enters at the top of the rear upper drum. This being the coolest part of the boiler, the temperature of the feed-water is gradually brought to the steaming-point in its descent through the rear bank of tubes to the mud-drum below, these being surrounded by the hot gases escaping to the chimney.

The mud-drum, protected from the fierce heat of the furnace by an ample bridge-wall, acts as a settling-chamber, the circulation in that portion of the boiler being comparatively slight. Reaching the mud-drum, the impurities in the feed-water descend to the bottom in the form of sludge or mud, which can be blown out as often as may be found necessary. Whatever solid impurities or precipitate adhere to the interior surfaces of the rear bank of tubes must be washed off with a hose or removed with a scraper.

FIG. 380.

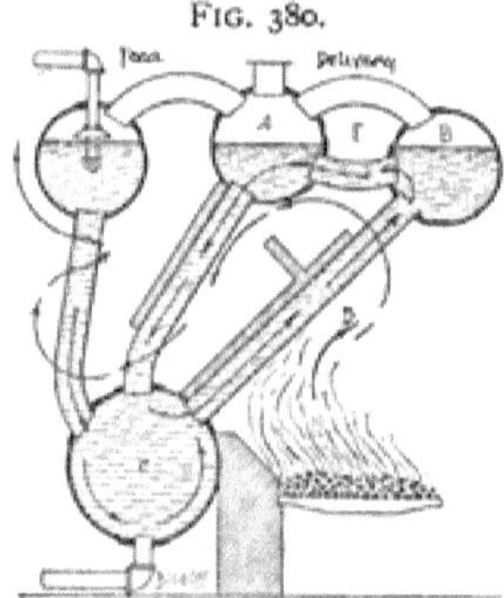

The interior surfaces of this boiler are rendered accessible by the removal of four manhole plates, which exposes to view the two ends of every tube in the boiler and, for all practical purposes, the entire area of the interior heating surface. The boiler attendant can then enter the drums and remove the mud and other deposits, scraping the tubes from end to end if need be with a chain or jointed scraper adapted for that purpose.

This boiler is erected entirely independent of the brickwork, so that the latter may be removed or replaced without disturbing the boiler or its connections. The three upper or steam-drums are supported by wrought-iron beams resting on wrought-iron columns, with cast-iron bases properly secured, whilst the mud-drum is suspended and left free to allow for contraction and expansion.

The Hogan Boiler, shown in Fig. 381, is made up of a water-and-steam-drum at the top, a distributing-drum at the bottom, and a series of bent tubes connecting these two drums, the shape of the tubes being such that the heat will not injuriously strain the expanded joints or distort the form of the tubes. There are no headers or stayed tube-plates to prevent the free expansion of all tubes.

The circulation begins in the upward movement of the water over the fire and into the water-and-steam-drum at the top of the furnace, the water returning through the rear tubes, marked "water-heating tubes" and "circulating-tubes," both series of which are connected with the upper and lower drums as shown.

It will be observed that the steaming-tubes, which are 2 inches in diameter, deliver the steam formed in them above the water-level of the upper drum. A mechanical extractor relieves the steam from any en-

FIG. 381.

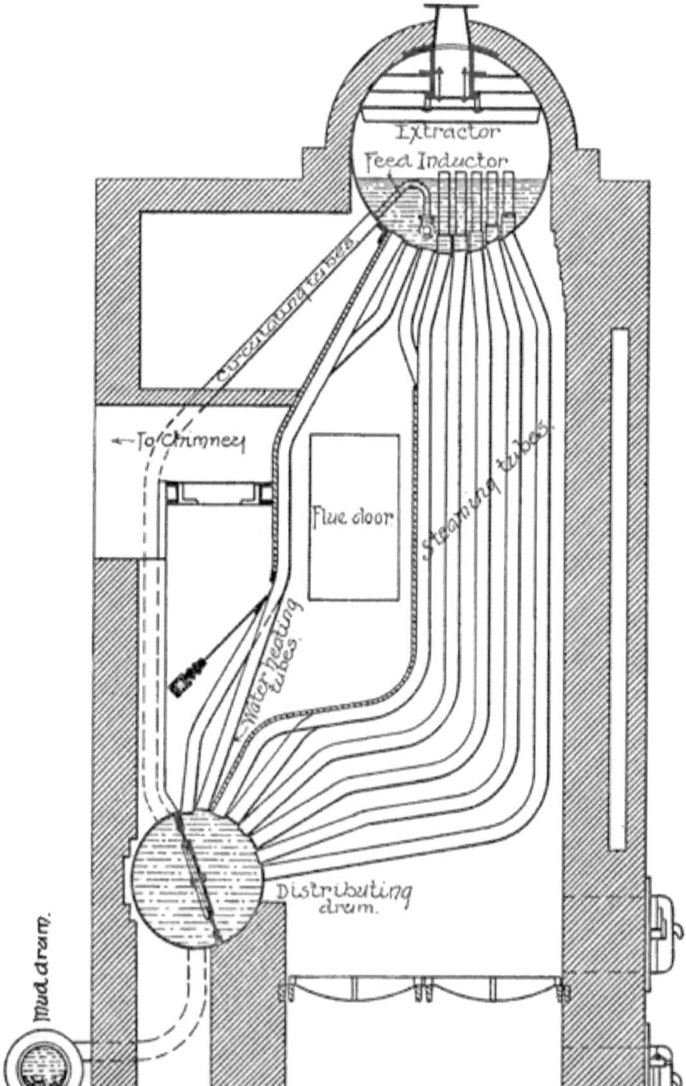

Extractor
Feed Inductor
Circulating tubes
To Chimney
Flue door
Steaming tubes.
Water heating tubes.
Distributing drum.
Mud drum.

trained water, which falls back into the drum and passes down the 3-inch circulating-tubes to the distributing-drum below. The extent of water surface from which steam has to escape as it is produced in ordinary boilers is the same under all conditions of firing. The effect of this limitation of water surface is great agitation of the water, which takes place under moderate or forced conditions of firing, as indicated by the motion of water in the gauge-glass and by the wet condition of the steam.

The feed-water enters the inductors in the upper drum and passes thence to the circulating-tubes, where precipitation takes place below the water-level and at a temperature within a few degrees of that of the steam. The circulating-tubes are not exposed to the heat of the gases. If the foreign substances and sediments held in suspension in water are not allowed to precipitate or come in contact with the surfaces in steam boilers which are exposed to the fire and the gases no scale will form on these surfaces. It is claimed for this boiler that the precipitation occurring in the distributing-drum, which is not heated, passes to the mud-drum, which is an external vessel and not exposed to any heat, where it may be blown out at intervals.

The heating-tubes in this boiler are not claimed to be self-cleaning; on the contrary, the makers state that no water-tube in a steam boiler, be it straight or bent, can be self-cleaning; but what is claimed for it is, that the precipitation of the scale-forming substances is so nearly accomplished by the methods here shown that deposits of scale do not occur. A soft deposit has been found in the distributing- and mud-drums, but no hard scale has been brought into existence.

The Hazelton Boiler, shown in Fig. 382, has a central vertical cylinder or standpipe which varies in diameter and height according to the power required; it rests upon a circular cast-iron foundation-plate, placed upon a supplementary foundation of brick, raised one course above level of foundation, so as to prevent any water in the ash-pit from coming in contact with the boiler; nor is it fastened to the foundation, but left free to expand and contract according to variations of temperature. That portion of the stand-pipe below the grate-bars forms the mud-drum, into which a manhole is placed for entering the boiler and which permits access to every portion of its interior surface.

Radial tubes extend outwardly from the standpipe at regular intervals, as shown in Fig. 383; the diameter, length, and number depend upon the size of the boiler. The outside end of each tube is closed upon itself by welding, forming a hemispherical end; in the process of closing this end is slightly thickened. The open end of the tube is expanded into the standpipe. The tube, extending outward horizontally and being secured at one end only, can expand and contract without strain.

The steam-pipe and steam-drying tubes are located at the top of the boiler, as shown in Fig. 382. A wrought-iron flange is riveted to the under side of the top head of standpipe, and abundant threadhold secured by cutting a thread through both head and flange. A heavy nipple is screwed in from the outside, the lower end with a long thread extending below the head two or three inches. At the upper end of this nipple is a spherical tee, from one outlet of which is the steam-pipe extending horizontally to the outer line of brickwork to receive steam-valve; similarly, another pipe extends in the opposite direction, which connects with the pop safety-valve. At the lower end of this central nipple and inside of the standpipe, connected with it by a pair of flanges bolted together, is a vertical pipe open at the upper and closed at the lower end; into this vertical pipe are screwed a series of wrought-iron pipes of small diameter open at both ends; these extend radially almost to the outer end of the uppermost steam-drying tubes in the standpipe, as shown in Fig. 382. The steam as it becomes disengaged from the surface of the water must enter the steam-drying tubes and pass to their outer

FIG. 382.

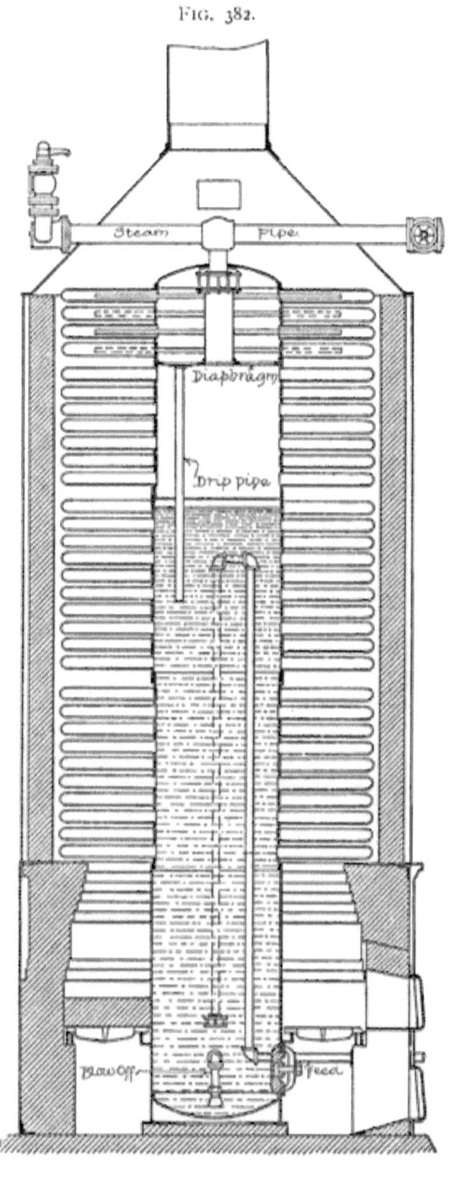

ends before entering these small pipes, which convey the dry steam to the steam-pipe.

The feed-water pipe is led beneath the grate-bars and screws into the standpipe, passing through which it is then carried upward to a short distance below the water-line, thence downward to about the grate level, where it discharges into the lower part of the standpipe, that portion having little or no circulation and where precipitation is most likely to occur, hence called the mud-drum.

Fig. 383.

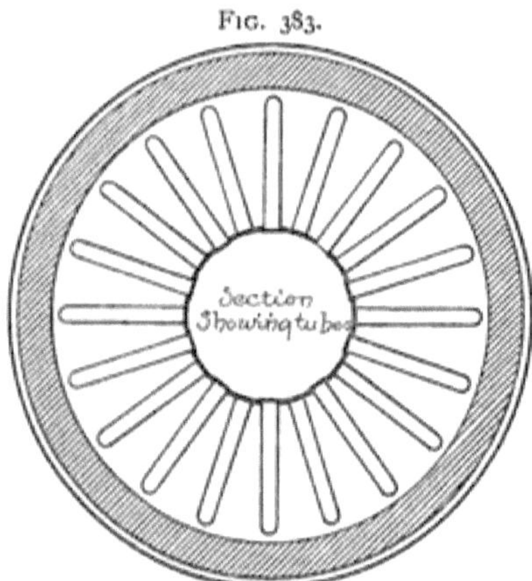

The blow-off-pipe is similarly connected to the standpipe, but the extension in this case is downward to within a short distance of the bottom head.

A plan of the furnace and grate-bars is shown in Fig. 384, the grate area showing the herring-bone pattern, this being the manufacturer's preferred style of grate. The shaded portion back of the standpipe represents a bridge-wall, and is not counted in the grate area, but any portion of this could be made available for grates if necessary. That portion of the brickwork lining of the square furnace casing, from the floor level of the ash-pit to a point a little above the top of the grates, is built of uniform thickness on the four sides, and by means of corbelling, shown in Fig. 382, the brickwork is brought in so as to form a circular deflecting furnace. The furnace casing is reinforced by angle-irons and braces, and at its top supports a circular metal shelf, upon which rests the upper brick lining inside of the circular upper jacket of the boiler. A space of

Fig. 384.

several inches is left between the top of the deflecting furnace and this metal shelf for expansion of furnace and to remove the weight of the superstructure from the furnace.

Fig. 385.

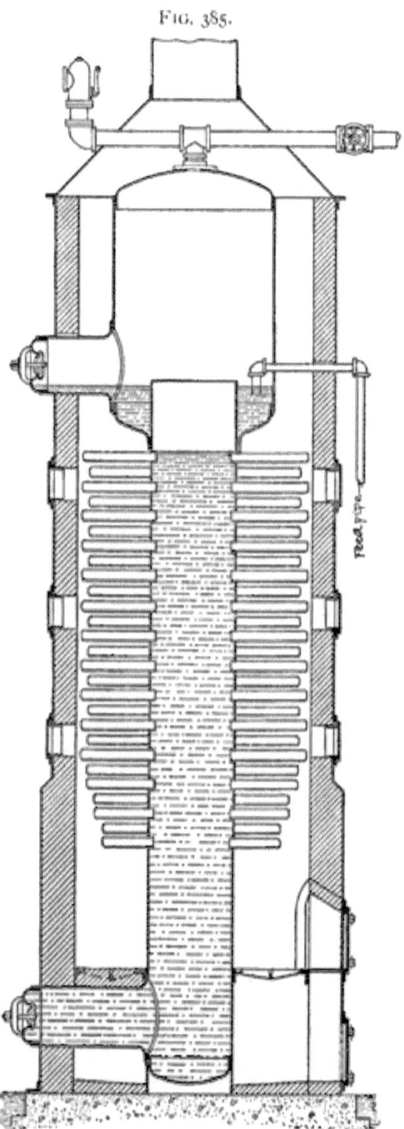

The smoke-hood has a door placed in it for ready access to exterior of upper part of standpipe. The steam and pop safety-valve pipes extend outward through openings in this hood, and are covered with movable slides, permitting the standpipe to expand and contract without straining the joints.

The Adams Boiler, shown in Fig. 385, has some points of similarity to the boiler previously described, yet is widely divergent in its details. Water-tube boilers having radially projecting tubes closed at their outer ends have many stanch advocates, who assert that such boilers will evaporate more water per pound of coal and raise steam more rapidly than boilers of any other type. The two principal objections raised against boilers of this type have been: first, that on account of the small settling-chamber, where water containing impurities in large degree is used, the tubes are liable to fill up, requiring constant cleaning; second, when crowded beyond their rated capacity they are very liable to give wet steam. This boiler has been designed to overcome these difficulties not only, but to embody such other necessary requisites as shall make it a durable and economical steam generator.

A feed-water reservoir is formed by extending the centre tube up into the steam-dome, as shown. This is practically a live-steam heater and purifier. As the feed-water falls into the reservoir it is raised to the same temperature as the steam, and the impurities settle in the form of sediment to the bottom of the reservoir, whence they can easily be removed. From the reservoir the hot water passes slowly through a large down-pipe or outside connection not shown in the engraving from the extension of the upper dome to the extension from the bottom of the centre tube. Should any of the impurities held in suspension flow over from the reservoir, they will be retained and settle in the bottom extension formed at the bottom of the centre tube, thus leaving the inside of the steam-tubes clean and free from scale. To provide against the possibility of imperfect separation of the steam from the water, a large dome is placed on top of the centre tube above the water-line; this secures ample room for separation and a large storage for steam, insuring a steam-supply free from saturation and to some extent superheated. Manholes at the outer ends of the two extensions allow access to both top and bottom of the boiler. Fig. 386 shows a plan of the furnace.

Fig. 386.

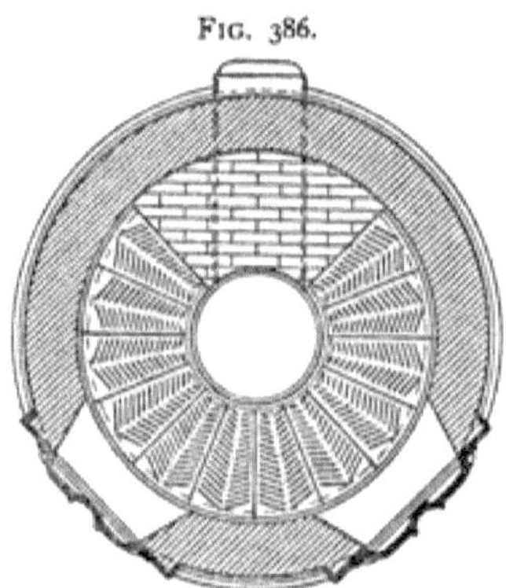

The Morin Boiler, shown in Fig. 387, has a vertical central standpipe with bumped heads. The lower end of the bottom sheet extends so far below the head that the latter carries no weight and is above contact with the foundation. The weight of the boiler and its contents is borne by a heavy cast-iron ring riveted to the projecting sheet below the bottom head. This ring has a broad base resting upon the brick foundation underneath.

The lower part of the vertical shell, having little or no circulation, acts as a mud-drum, in which foreign substances, held in solution and precipitated by the action of the heat, accumulate and may be blown out as required.

The construction of the cylindrical shell into which the tubes are inserted is the same as that of any boiler-shell, with the exception that it is welded longitudinally instead of riveted, which allows a uniform spacing of the holes for the tubes.

The tubes are from 1½ to 3 inches in diameter, depending upon the

Fig. 387.

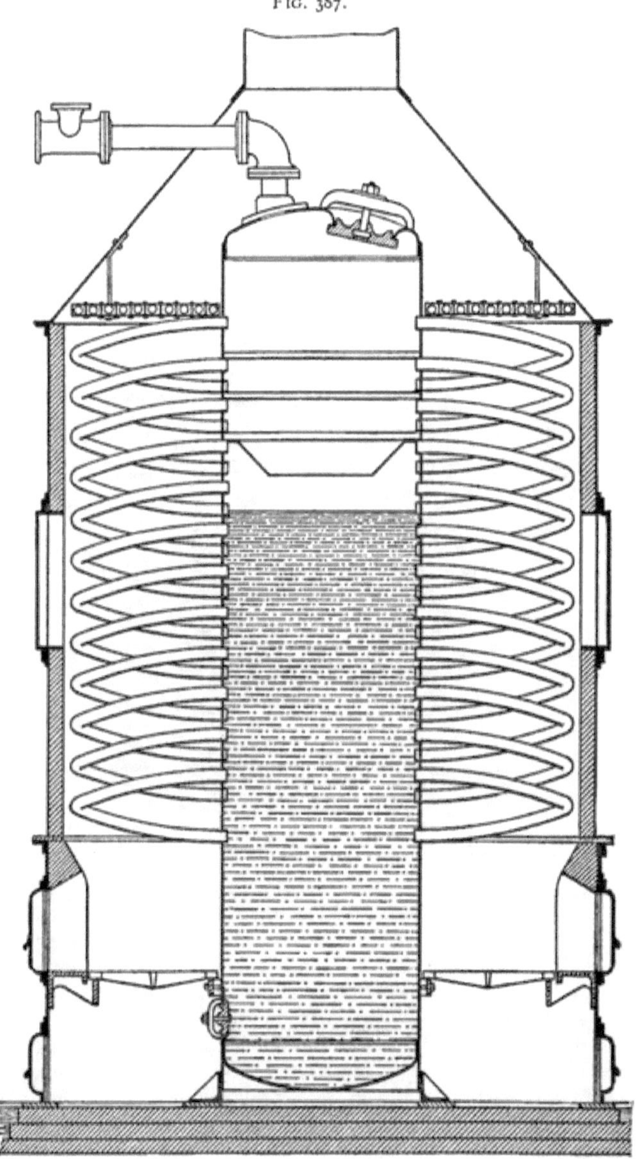

size of the boiler. They are bent to the shape indicated in Fig. 388, in addition to which they also have a vertical twist, so that the bent tube re-enters the cylinder at about 18 inches above the lower entrance. The general appearance of a series of tubes when put together in the boiler is that of a number of spirals forming annular rings in series one above the other to such a height as may be required for a given boiler. The tubes must be identical in their shape to fit into place in the series to which each belongs. To insure exact duplication they are bent over formers, giving each tube the precise shape needed for the boiler for which it is intended.

FIG. 388.

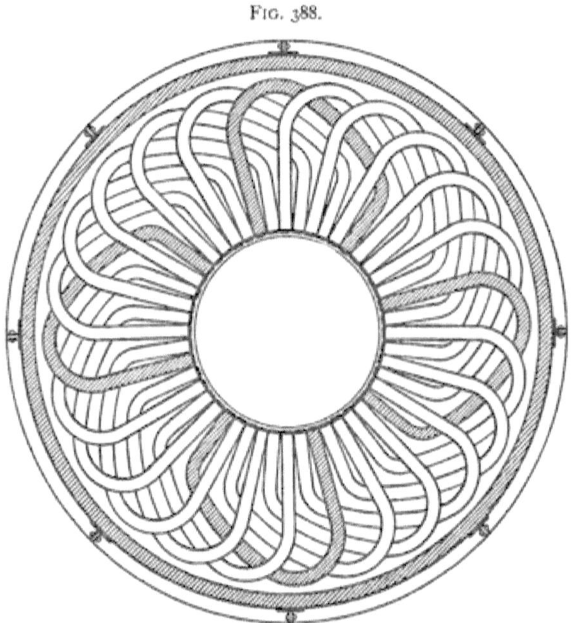

The tubes, being expanded in the main shell, allow for unequal expansion and contraction, while the bends provide for unequal expansion in the individual tubes. The expanded joints are well protected from the action of the fire, being for the most part under water; consequently they are not only well protected, but any tube can be easily replaced, and at small expense. The lower tubes and the central shell are filled with water, as shown in Fig. 387. The heat circulates among the tubes, around the central shell, and thence to the chimney-flue. The upper tubes and the central shell above the water-line act as superheating surfaces.

A manhole is placed in the upper head, by which access is had to the interior of the boiler for examination or repairs. A handhole-plate is located in the mud-drum for convenience in cleaning.

A deflector-plate is inserted in the central shell a short distance above the water-level, which tends to throw back any water that may be carried up by the steam; and a series of diaphragms divide the upper portion of this cylinder, forming a series of superheating-chambers, through which the steam is successively compelled to circulate by the connecting bent tubes, drying the steam and slightly superheating it. The steam then passes into a reservoir provided for it above the top row of tubes.

A water-heating coil or economizer of 1½- to 3-inch pipe is located above the central shell, resting on the top row of tubes; the length may be from 100 to 300 feet, according to the size of the boiler. This coil is welded and without screw-joints. The feed-water enters and flows through this coil of pipe before it enters the boiler, absorbing considerable heat from the waste gases, and then passes to the central shell, where it is taken up by the curved tubes, in which a rapid circulation is maintained by reason of their upward course and the intense heat by which they are surrounded.

The fire-box surrounds the central vertical cylinder, and is, therefore, annular in form. It is enclosed in a casing of iron, bolted together in sections, and lined with fire-brick. Ordinarily three or four fire-doors are provided, according to the size of the boiler.

The outside casing of the boiler is sectional and bolted together; it can be easily removed, wholly or in part, should it be necessary to remove a defective or substitute a new tube. Each section of the outside casing of the boiler is provided with a cleaning-door, through which the ashes and soot may be blown off the tubes by means of a steam-hose.

In erecting this boiler the central shell is stood upon the cast-iron base-plate secured to the foundation, after which the work of placing the tubes in the shell is commenced. After all the tubes have been expanded, the whole is subjected to a hydrostatic test of 300 pounds per square inch, replacing any defective tubes; this completed, the castings, which have been lined with fire-brick, are then placed in position and securely bolted. The smoke-hood is then placed and the usual boiler attachments made, until every part is in proper position.

The Cahall Boiler, shown in sectional elevation in Fig. 389, consists of two steel drums arranged one above the other, and connected with 4-inch lap-welded tubes. These tubes are vertical, are perfectly straight throughout their entire length, and are expanded into the drums at each end. The upper or steam-drum has an opening through its centre for the exit of waste gases. The water-line in the upper drum is about a foot above the bottom of the drum, the drum itself being about

FIG. 389.

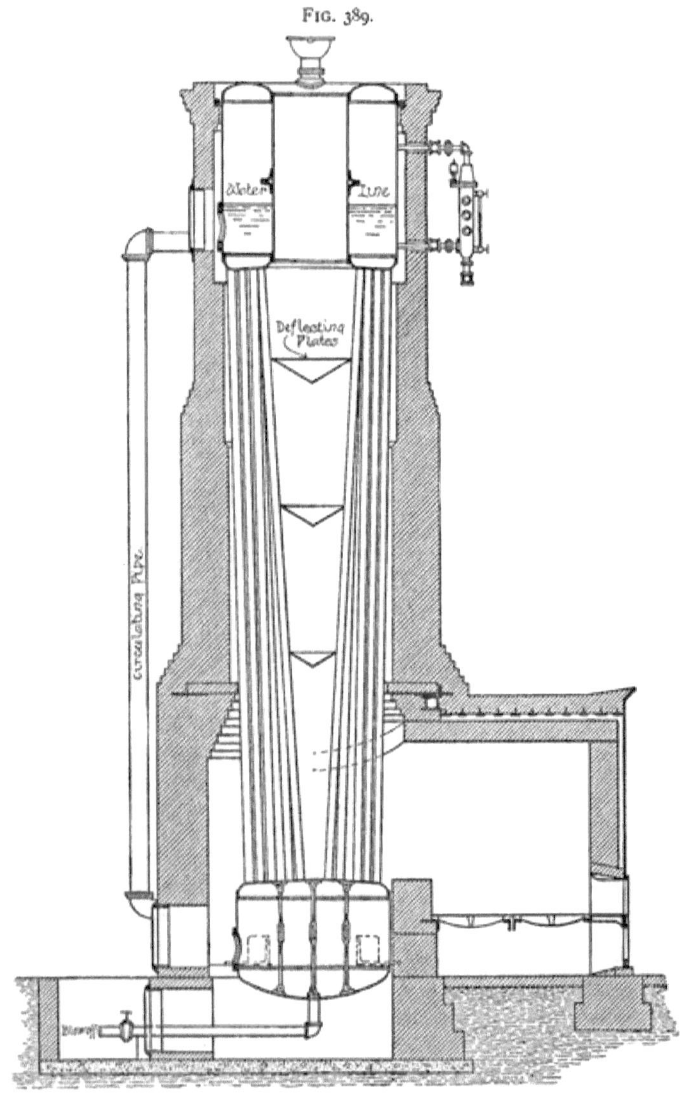

6 feet high, leaving a space of 5 feet between the surface of water and the point at which the steam is drawn off from the boilers. A circulating-pipe is shown at one side of the engraving; this is attached to the upper or steam-drum, just below the water-level, and is carried downward, outside the brickwork, to a point just below the tube-sheet of the lower drum, where it enters that drum. Through some oversight these connections were not shown in the engraving. There is no steam in this external circulating-pipe and no possibility of making any, but the tubes connecting the two drums are steam making tubes, consequently the water in the external pipe has a greater specific gravity than the mixture of steam and water in the steaming-tubes, and a rapid and positive circulation is thus set up. The water in the tubes connecting the drums, ascending to the steam-drum, delivers the mixture of water and steam there; the steam separating at once from the water, the latter enters the circulating-pipe and is carried down to the mud-drum, and again rises with its mixture of steam. As this mixture of steam and water coming from the upper end of the tube is about half steam and half water in bulk, and as steam at 100 pounds pressure will occupy about 218 times the space occupied by the water itself, the water in the boiler will circulate through the boiler 218 times before finally becoming steam. This insures not only a rapid and steady circulation, but also insures a uniform temperature of water in all the tubes, as fresh feed-water, being thus circulated 218 times before evaporation, must necessarily mingle in such minute parts with the water already present in the boiler that the water in one ascending tube cannot be different in temperature to that in the others. The boiler is thus relieved from any possibility of destructive strains from unequal expansion.

A central opening is provided in the upper drum through which the gases escape to the chimney; the upper tube-sheet has, therefore, a circular opening in its centre, leaving a central open space between the tubes, which gradually narrows to the bottom tube-sheet. Advantage is taken of this space, which is in the form of an inverted cone, to introduce deflecting-plates, which cause the gases to be alternately thrown out and in throughout the whole heating surface, giving them a sweep at nearly right angles to the tubes, thereby extracting from these gases their heat until they come to very nearly the temperature of the water contained in the boiler.

Manholes are provided in both the upper and lower drums, to which are fitted the swinging manheads shown in Fig. 170. By simply taking off the nuts from the manheads and swinging them open, the attendant can place a light in the lower drum of the boiler and get into the upper drum and examine the condition of the tubes and that of the interior of the boiler. In case scale is discovered in any of the tubes, he can run a scraper through them, for which purpose the one in use is made in sections a trifle less than 6 feet long. Four of these sections are used, and the man who is cleaning the boiler takes them into the upper drum and pushes the first section down as far as it will go, then simply hooks the second section to that, and continues doing this until the scraper has gone entirely through the tube, forcing any scale that may have been deposited on the sides of the tube straight through to the bottom drum, where it can be removed through the bottom manhole.

The boiler rests upon four iron brackets riveted to the lower or mud-drum, supported upon four piers of the foundation, the entire structure standing without contact with the brickwork, thus allowing the boiler every freedom for expansion without in any way straining the brick setting. In all places where pipe connections are made to the boilers through the walls they are encased in expansion-boxes.

Defective tubes may be removed from the boiler in the following manner: In the upper or top head of the steam-drum there are placed six handholes, not shown in the engraving, which are closed by means of plate, yoke, and bolt in the usual manner; there are in addition two other holes,—one for the steam-pipe connection, the other for the pop safety-valve connections. By means of these eight openings any of the tubes needing removal, after having been cut loose from the tube-sheets, can be pushed up through the tube-hole from which it has just been cut and through the most convenient of these openings in the top head and removed from the boiler. The new tube to replace the defective one is passed into the boiler through the same openings.

FIG. 390.

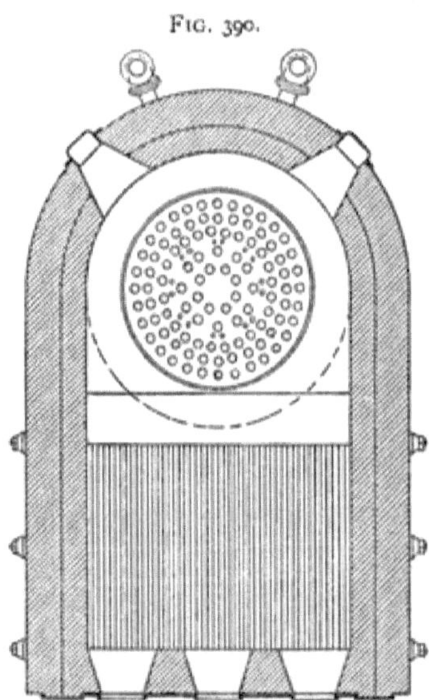

The furnace is external, as shown in Fig. 389. The combustion-chamber is roofed with a heavy fire-brick arch, which becomes incandescent shortly after the boiler is fired and radiates its contained heat directly on top of the green coal; furthermore, owing to the direct upward passage of all gases and full free openings, a comparatively short stack will furnish a draught pressure not usually had with most other boilers; for instance, it is claimed that in tests with a stack only 50 feet high a draught pressure in the furnace of over ½ inch was attained, but the temperature of the escaping gases is not furnished. This is a result which could hardly be expected from any other water-tube boiler with a stack 100 feet high. The heavy draught causes a rapid combustion of fuel per square foot of grate, with the consequent high initial temperature of gases, both of

FIG. 391.

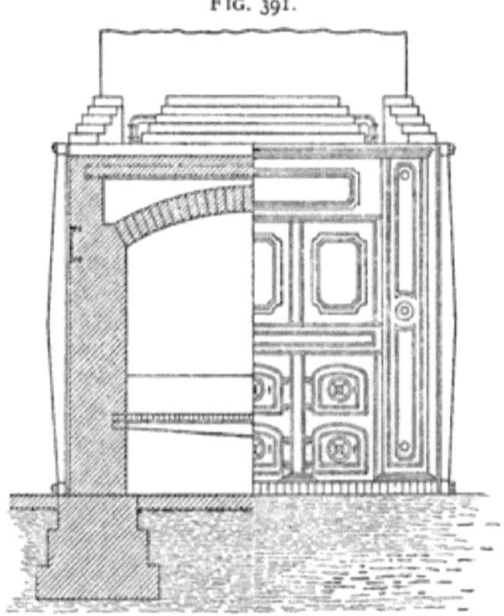

which are requisite to either efficiency or economy in boiler practice. A plan of the furnace is shown in Fig. 390; a half-front and half-sectional elevation through the arch of the furnace is shown in Fig. 391.

BOILER PERFORMANCE.

ADDITIONAL TO CHAPTERS VI, VIII, IX.

IT was no part of the original scheme to enter upon questions relating to boiler performance, whether for economy, capacity, or durability, but rather to confine the subject-matter of this book to the details of construction and arrangement of parts. As the work progressed, however, it became more and more apparent that a table of comparative evaporative tests would prove interesting even though it had no special value. Table XLIX. is not, strictly speaking, comparative, but is simply an aggregation of boiler performances arranged in the order in which the several types of boilers are considered in the preceding chapters, all of which are in use in this country. The tests from which this table was compiled varied from eight to twenty-four hours in duration; but to form a better basis of comparison, all the tests were reduced to a common basis of ten hours.

As nearly all of the tabular numbers have been changed from the final sheets prepared by the experts who conducted the tests, it did not seem fair to use their names in connection with the tests without presenting the figures as originally prepared. We will say, however, that the experts thus quoted are all well known and recognized as being thoroughly competent for conducting such expert trials.

From the mass of material at our disposal, only such boilers were selected as seemed to meet the ordinary conditions of service. Extraordinary results, above or below what is recognized as good performance, have not been used, except in the case of one or two boilers which were overfired, as indicated by the high temperature of the escaping gases. One object in the preparation of this table was to present as nearly as possible the conditions under which the several steam boilers are operated in actual service.

A. *Cylinder Boilers.*—Number of shells, three; each shell 36 inches in diameter by 30 feet in length; area of water-heating surface, 518 square feet; area of grate surface, 47.5 square feet; ratio of water-heating surface to grate surface, 10.9 to 1. Arrangement of furnace and setting similar to Figs. 209 to 212. Fuel, anthracite pea coal; moisture, 3.2 per cent.

B. *Two-Flue Boilers.*—Number of shells, three; each shell 48 inches in diameter by 36 feet in length; two 16-inch flues in each shell; area of water-heating surface, 1225 square feet; area of grate surface, 90 square feet; ratio of water-heating surface to grate surface, 13.6 to 1. Arrangement of furnace and setting similar to Fig. 217. Fuel, bituminous coal, Pittsburgh run of mine.

C. *Three-Inch Tubular Boiler.*—Number of boilers, one; shell, 48 inches in diameter by 16 feet in length, with 50 tubes 3 inches in diameter; area of water-heating surface, 763 square feet; area of grate surface, 25 square feet; ratio of water-heating surface to grate surface, 30.52 to 1. Arrangement of furnace and setting similar to Fig. 293. Fuel, anthracite pea coal; moisture, 3.5 per cent.

D. *Three-and-One-Half-Inch Tubular Boilers.*—Number of boilers, two; each shell 60 inches in diameter by 16 feet in length, with 62 tubes 3½ inches in diameter in each boiler, or 124 in all; area of water-heating surface, 2100 square feet; area of grate surface, 52 square feet; ratio of water-heating surface to grate surface, 40.38 to 1. Arrangement of furnace and setting similar to Fig. 293. Fuel, bituminous coal, Pittsburgh run of mine.

E. *Four-Inch Tubular Boilers.*—Number of boilers, two; each shell 72 inches in diameter by 20 feet in length, with 68 tubes 4 inches in diameter in each boiler, or 136 in all; area of water-heating surface, 3385 square feet; area of grate surface, 62.24 square feet; ratio of water-heating surface to grate surface, 54.39 to 1. Arrangement of furnace and setting similar to Fig. 293. Fuel, bituminous coal, Mount Olive lump.

F. *Six-Inch Tubular Boilers.*—Number of boilers, two; each shell 60 inches in diameter by 24 feet in length, with 18 lap-welded tubes 6 inches in diameter in each boiler, or 36 in all; area of water-heating surface, 1879 square feet; area of grate surface, 58 square feet; ratio of water-heating to grate surface, 32.4 to 1. Arrangement of furnace and setting similar to Fig. 294. Fuel, bituminous coal, small lump.

G. *Double-Deck Horizontal Tubular Boiler.*—Number of boilers, one, of a series of four; diameter of boiler-shells, 60 inches and 50 inches; length of boiler-shell, 15 feet; number of 4-inch tubes in each boiler, 60; area of water-heating surface, 1286 square feet each; area of steam-heating surface, 141 square feet each; area of grate surface, each boiler, 24 square feet; ratio of grate to water-heating surface, 53.6 to 1. Arrangement of furnace and setting similar to Fig. 234. Fuel, anthracite coal, rice, 3 per cent. moisture, Wilkinson stoker, force-blast.

H. *Triplex Boiler* (see Fig. 236).—The two lower shells are 58 inches in diameter by 16 feet 9 inches long, with 62 4-inch tubes in each shell. The upper drum is 48 inches in diameter by 16 feet in length. The connecting necks are 15 inches in diameter. In the test here given the Hawley down-draft grate was used. The original design was for a standard grate which would give data for the test as follows: area of the standard grate, 11 x 5 feet, 55 square feet; total area of heating surface, 2705 square feet; ratio of heating surface to grate area, 49.2 to 1; coal consumed per hour, 893.8 pounds; coal per hour per square foot of standard grate, 16.25 pounds; capacity of boiler, as stated in the contract, 250 horse-power; capacity of boiler averaged during test 277.1 horse-power (excess 11 per cent.); capacity of boiler in preliminary test, 315.2 horse-power (excess 26 per cent.).

The lump coal when weighed contained 1.6 per cent. of moisture.

The total heat of combustion by calorimeter per pound of dry coal in B.T.U. was 12,765.

Efficiency, 80.6.

I. *Fire-Box Tubular Boiler.*—Trial 48 hours; test records results of one boiler out of a battery of five boilers working together. Grate surface, 54½ square feet each; water-heating surface, 1562 square feet each; ratio of water-heating to grate surface, 28.7 to 1. Arrangement of boiler details similar to Fig. 331. Fuel, anthracite coal, bird's eye; 7.4 per cent. moisture.

K. *Belpaire Boiler, Double Furnace.*—Two boilers in use; record of one boiler. Diameter of shell, smallest inside, 82 inches; inside length of fire-box, 8 feet 1½ inches; inside width of each fire-box (2), 7 feet 6 inches; length of combustion-chamber, 7 feet; length of tubes, 16 feet; diameter of tubes, outside, 3 inches; number of tubes, 159; length of grate, 7 feet 1 inch; width of grate during trial bricked up to 5 feet 3 inches; heating surface, 2240 square feet; grate surface during trial, each boiler, 37.185 square feet; ratio of heating to grate surface during trial, 60.24 to 1. Pocahontas coal; average moisture, 2.60 per cent.; calorific value of 1 pound of coal by analysis, 14,924 B.T.U.; design of boiler approximates Fig. 337.

L. *Gun-Boat Boiler* (Fig. 326).—Results of one boiler in a battery of two boilers working together. Diameter of boiler, 8 feet 6 inches; length of boiler, 20 feet; furnaces in each boiler, 2; type of furnace, Fox corrugated; diameter of furnaces, 3 feet 6 inches; length of furnaces, 7 feet 6 inches; number of 4-inch tubes in each boiler, 90; length of tubes, 10 feet; heating surface, 1 boiler, 1119 square feet; heating surfaces, 2 furnaces, 136 square feet; heating surface, combustion-chamber, 47 square feet; heating surface of tubes, 936 square feet; area of cross-section of tubes, 1 boiler, 8 square feet; ratio of tube cross-section to grate area, 5.25 to 1; grate surface, 1 boiler, 42 square feet; ratio of water-heating to grate area, 26.67 to 1. Semi-anthracite pea coal; 9 per cent. moisture. Heavy rain during test; coal was brought in from yard perfectly saturated.

M. *Vertical Tubular Boiler.*—One boiler; shell 48 inches in diameter by 9 feet in height; 148 tubes, 2 inches in diameter, 72 inches long; fire-box, 42 inches in diameter by 36 inches in height. Heating surface: fire-box, 35; water-tubes, 349; superheating, 116; head, 7; making a total of 505 square feet. Upper end of tubes pass through steam-space for 18 inches. Area of grate surface, 9.62 square feet; ratio water-heating to grate surface, 40.44 to 1. Arrangement of boiler details similar to Fig. 303. Fuel, anthracite stove coal.

N. *Manning Vertical Boiler.*—Results of one boiler in a battery of six boilers. Diameter of shell, 5 feet; diameter of fire-box, 6 feet; height of fire-box, 3 feet 6 inches; number of tubes, 180; diameter, 2½ inches; length of tubes, 15 feet; grate area, 28.7 square feet; water-heating surface, 1383 square feet; superheating surface, 471 square feet; ratio of water-heating surface to grate, 48.2 to 1; ratio of super-heating surface to grate, 16.4 to 1. Arrangement of boiler details similar to Fig. 310. Fuel, bituminous coal, Pocahontas; 4.75 per cent. moisture.

O. *Galloway Boiler.*—Diameter, 85 inches, by 28 feet in length; number of conical tubes, 33; diameter of conical tubes, 5½ and 10 inches; length of conical tubes, 2 feet 11 inches; diameter of furnace. 2 feet 10 inches; grate, 6 feet 7 inches long by 2 feet 9 inches wide total heating surface, 1058 square feet. Arrangement of boiler and setting similar to Fig. 313. Fuel, bituminous coal, Cumberland.

P. *Wharton-Harrison Boiler.*—The details of this boiler are in all respects as described on page 298 and following. Figs. 352 and 353 illustrate the details common to all boilers of this manufacture.

Q. *Babcock & Wilcox Boiler.*—Boiler drum, 36 inches in diameter by 16 feet 6 inches in length; grate, 60 inches wide by 72 inches long, 30 square feet; 64 tubes 4 inches in diameter by 16 feet in length; total water-heating surface, 1253 square feet; ratio of water-heating surface to grate surface, 41.8 to 1. Arrangement of furnace and setting similar to Fig. 354. Fuel, anthracite coal.

R. *Caldwell Water-Tube Boiler.*—Fig. 366 shows the general arrangement of this boiler. It is distinguished from the Babcock & Wilcox boiler in the design and grouping of the headers, a detail of one of which is shown in Fig. 367; also a novel design of baffle bricks used, shown in Fig. 368. On a capacity test by G. H. Barrus, additional to the one given in the table, a 206 horse-power boiler developed 65.7 per cent. more than its rated power, and this was done with a draught of 0.47 inch in the flue, which corresponds to that of an ordinary good chimney. The actual draught at command was more than double this force, and in this particular instance the excess of capacity could, if desired, have been increased much beyond the amount realized.

S. *Heine Safety Boiler* (Fig. 372).—Number of boilers, one; 1 shell 48 inches in diameter by 19 feet 9 inches in length; 92 tubes 3½ inches in diameter by 16 feet in length; grate, 81 inches wide by 48 inches long, 27 square feet; water-heating surface, 1406 square feet; ratio of water-heating surface to grate surface, 52 to 1. Fuel, bituminous coal.

T. *Stirling Boiler.*—Number of boilers, one; number of water-drums, 1; number of steam- and water-drums, 3; diameter of water-tubes, 3¼ inches; grate surface, 52 square feet; water-heating surface, 2300 square feet; ratio of water-heating surface to grate surface, 44.1 to 1. Arrangement of furnace and setting similar to Fig. 378. Fuel, Youghiogheny bituminous coal.

U. *Hogan Boiler* (Fig. 381).—Number of boilers in test, one; diameter of steam-drum, 42 inches; diameter of distributing-drum, 24 inches; diameter of mud-drum, 10 inches; number of steaming- and heating-tubes, 384; number of circulating-tubes, 36; grate surface, 80 square feet; water-heating surface, 4100 square feet; ratio of water-heating surface to grate surface, 51.25 to 1.

V. *Hazelton Boiler.*—Shell, 42 inches diameter by 34 feet in height; 1070 tubes 4 inches in diameter by 3 feet long; grate, 129 inches outside by 44¼ inches inside diameter, 80 square feet; water-heating surface, 2371 square feet; superheating surface, 1049 square feet. Arrangement of furnace and setting similar to Fig. 382. Fuel, bituminous coal.

W. *Morin Climax Boiler.*—Total heating surface, 10,000 square feet; grate surface, 113.6 square feet; ratio of total heating to grate surface, 88 to 1. Arrangement of furnace and setting similar to Fig. 387. Fuel, Youghiogheny bituminous coal.

X. *Cahall Water-Tube Boiler.*—Number of boilers in test, one; diameter of upper drum, 6 feet 8 inches; diameter of internal flue, 34 inches; 108 tubes 4 inches in diameter by 18 feet long; area of grate surface, 40 square feet; area of water-heating surface, 2064 square feet; area superheating surface, 50 square feet; ratio of water-heating to grate surface, 51.6 to 1; draught area through internal flue, 6.3 square feet, or a trifle less than one-sixth of the grate. Arrangement of furnace and setting similar to Fig. 389. Fuel, Pocahontas bituminous coal; 3.3 per cent. moisture.

	R.	S.	T.	U.	V.	W.	X.
	Caldwell Boiler.	Heine Boiler.	Stirling Boiler.	Hogan Boiler.	Hazelton Boiler.	Morin Boiler.	Cahall Boiler.
Dimensions and Proportions.							
Grate surface, area in square feet	51.3	27	52	80	80	113.6	40
Water-heating surface, square feet . . .	2300	1407	2300	4100	2371	Total Surfaces.	2064
Superheating surface, square feet . . .	. . . .	. . . .	. . . .	None.	1049	10,000	50
Ratio of water-heating surface to grate .	44.8 to 1	52 to 1	44.1 to 1	51.25 to 1	29.63 to 1	88 to 1	51.6 to 1
Average Pressures.							
Steam pressure by gauge	77.1	123.3	129	92	90	68.8	110.7
Chimney draught in inches of water . .	.45	.65	.28	.143	. . . .	1.23	.14
Average Temperatures.							
External air, degrees Fahr.	. . . .	90.5	50	90	82.5	37	. . . .
Escaping gases from boiler, degrees F.	453	644	480	530	498	501	619
Feed-water entering boiler, degrees F. .	39	84	168	83	91	35.3	71.1
Fuel.							
Kind and size	Pocahontas.	Bituminous.	Youghiogheny.	. . . .	Bituminous.	Bituminous.	Bituminous.
Coal fed to furnace, deducting moisture, pounds	6644	8550	6929	18,115.41	12,801	36,252	6507
Percentage of ash	7.1	8	5.5	32.14	4.21	3.75	6
Total combustible, pounds	6170	7866	6549	12,293.12	12,281	33,123	. . . .
Coal burned per square foot of grate surface per hour	12.9	31.7	13.32	22.52	16	31.9	16.3
Coal burned per square foot of water-heating surface per hour	.29	.608	.332	.439	.541	.363	.315
Quality of Steam.							
Moisture in steam, per cent.	.11	.75	.91	.58	. . . .	6.81	. . . .
Superheated, degrees Fahr.	. . . .	. . . .	. . . .	. . . .	1.2	. . . .	4.4
Water.							
Weight of water fed to boiler, pounds .	57,528	77,995	71,784	103,285.14	109,980	31,433	64,772
Equivalent water evaporated into dry steam from and at 212° Fahr., pounds	69,839	91,878	78,574	120,245.45	127,906	38,034	77,108
Evaporation.							
Water evaporated per pound of coal under actual conditions	8.658	9.12	10.36	5.71	8.58	8.34	9.97
Equivalent evaporation from and at 212° Fahr. per pound of coal	10.511	10.74	11.34	6.64	9.99	10.07	11.84
Equivalent evaporation from and at 212° Fahr. per pound of combustible .	11.314	11.60	12.00	9.79	10.28	11.02	12.60
Equivalent evaporation from and at 212° Fahr. per square foot of water-heating surface per hour, pounds . .	3.03	6.53	3.41	2.93	3.74*	3.80*	3.2
Commercial Horse-Power.							
On a basis of 34½ pounds of water evaporated per hour from and at 212° Fahr.	202.4	280	229	348.53	370.8	1058.8	223.4
Square feet of water-heating surface per H.-P.	11.4	5.03	10.04	11.76	6.39	9.44	9.28

TABLE XLIX.

EVAPORATIVE TESTS OF STEAM BOILERS.

	A	B	C	D	E	F	G	H	I	K	L	M	N	O	P	Q	R	S	T	U	V	W	X
	[illegible]	[illegible]	[illegible]	[illegible]	[illegible]	[illegible]	[illegible]	[illegible]	[illegible]	[illegible]	[illegible]	[illegible]	[illegible]	[illegible]	[illegible]	[illegible]	[illegible]	[illegible]	[illegible]	[illegible]	[illegible]	[illegible]	[illegible]
Dimensions and Proportions.																							
Grate surface, area in square feet	[illegible]	[illegible]	[illegible]	[illegible]	[illegible]	[illegible]	[illegible]	[illegible]	[illegible]	[illegible]	[illegible]	[illegible]	[illegible]	[illegible]	[illegible]	[illegible]	[illegible]	[illegible]	[illegible]	[illegible]	[illegible]	[illegible]	[illegible]
Water-heating surface, square feet	[illegible]	[illegible]	[illegible]	[illegible]	[illegible]	[illegible]	[illegible]	[illegible]	[illegible]	[illegible]	[illegible]	[illegible]	[illegible]	[illegible]	[illegible]	[illegible]	[illegible]	[illegible]	[illegible]	[illegible]	[illegible]	[illegible]	[illegible]
Superheating surface, square feet	[illegible]	[illegible]	[illegible]	[illegible]	[illegible]	[illegible]	[illegible]	[illegible]	[illegible]	[illegible]	[illegible]	[illegible]	[illegible]	[illegible]	[illegible]	[illegible]	[illegible]	[illegible]	[illegible]	[illegible]	[illegible]	[illegible]	[illegible]
Ratio of water-heating surface to grate	[illegible]	[illegible]	[illegible]	[illegible]	[illegible]	[illegible]	[illegible]	[illegible]	[illegible]	[illegible]	[illegible]	[illegible]	[illegible]	[illegible]	[illegible]	[illegible]	[illegible]	[illegible]	[illegible]	[illegible]	[illegible]	[illegible]	[illegible]
Average Pressures.																							
Steam pressure by gauge	[illegible]	[illegible]	[illegible]	[illegible]	[illegible]	[illegible]	[illegible]	[illegible]	[illegible]	[illegible]	[illegible]	[illegible]	[illegible]	[illegible]	[illegible]	[illegible]	[illegible]	[illegible]	[illegible]	[illegible]	[illegible]	[illegible]	[illegible]
Chimney draught in inches of water	[illegible]	[illegible]	[illegible]	[illegible]	[illegible]	[illegible]	[illegible]	[illegible]	[illegible]	[illegible]	[illegible]	[illegible]	[illegible]	[illegible]	[illegible]	[illegible]	[illegible]	[illegible]	[illegible]	[illegible]	[illegible]	[illegible]	[illegible]
Average Temperatures.																							
External air, degrees Fahr.	[illegible]	[illegible]	[illegible]	[illegible]	[illegible]	[illegible]	[illegible]	[illegible]	[illegible]	[illegible]	[illegible]	[illegible]	[illegible]	[illegible]	[illegible]	[illegible]	[illegible]	[illegible]	[illegible]	[illegible]	[illegible]	[illegible]	[illegible]
Escaping gases from boiler, degrees F.	[illegible]	[illegible]	[illegible]	[illegible]	[illegible]	[illegible]	[illegible]	[illegible]	[illegible]	[illegible]	[illegible]	[illegible]	[illegible]	[illegible]	[illegible]	[illegible]	[illegible]	[illegible]	[illegible]	[illegible]	[illegible]	[illegible]	[illegible]
Feed-water entering boiler, degrees F.	[illegible]	[illegible]	[illegible]	[illegible]	[illegible]	[illegible]	[illegible]	[illegible]	[illegible]	[illegible]	[illegible]	[illegible]	[illegible]	[illegible]	[illegible]	[illegible]	[illegible]	[illegible]	[illegible]	[illegible]	[illegible]	[illegible]	[illegible]
Fuel.																							
Kind and size	[illegible]	[illegible]	[illegible]	[illegible]	[illegible]	[illegible]	[illegible]	[illegible]	[illegible]	[illegible]	[illegible]	[illegible]	[illegible]	[illegible]	[illegible]	[illegible]	[illegible]	[illegible]	[illegible]	[illegible]	[illegible]	[illegible]	[illegible]
Coal fed to furnace, deducting moisture, pounds	[illegible]	[illegible]	[illegible]	[illegible]	[illegible]	[illegible]	[illegible]	[illegible]	[illegible]	[illegible]	[illegible]	[illegible]	[illegible]	[illegible]	[illegible]	[illegible]	[illegible]	[illegible]	[illegible]	[illegible]	[illegible]	[illegible]	[illegible]
Percentage of ash	[illegible]	[illegible]	[illegible]	[illegible]	[illegible]	[illegible]	[illegible]	[illegible]	[illegible]	[illegible]	[illegible]	[illegible]	[illegible]	[illegible]	[illegible]	[illegible]	[illegible]	[illegible]	[illegible]	[illegible]	[illegible]	[illegible]	[illegible]
Total combustible, pounds	[illegible]	[illegible]	[illegible]	[illegible]	[illegible]	[illegible]	[illegible]	[illegible]	[illegible]	[illegible]	[illegible]	[illegible]	[illegible]	[illegible]	[illegible]	[illegible]	[illegible]	[illegible]	[illegible]	[illegible]	[illegible]	[illegible]	[illegible]
Coal burned per square foot of grate surface per hour	[illegible]	[illegible]	[illegible]	[illegible]	[illegible]	[illegible]	[illegible]	[illegible]	[illegible]	[illegible]	[illegible]	[illegible]	[illegible]	[illegible]	[illegible]	[illegible]	[illegible]	[illegible]	[illegible]	[illegible]	[illegible]	[illegible]	[illegible]
Coal burned per square foot of water-heating surface per hour	[illegible]	[illegible]	[illegible]	[illegible]	[illegible]	[illegible]	[illegible]	[illegible]	[illegible]	[illegible]	[illegible]	[illegible]	[illegible]	[illegible]	[illegible]	[illegible]	[illegible]	[illegible]	[illegible]	[illegible]	[illegible]	[illegible]	[illegible]
Quality of Steam.																							
Moisture in steam, per cent	[illegible]	[illegible]	[illegible]	[illegible]	[illegible]	[illegible]	[illegible]	[illegible]	[illegible]	[illegible]	[illegible]	[illegible]	[illegible]	[illegible]	[illegible]	[illegible]	[illegible]	[illegible]	[illegible]	[illegible]	[illegible]	[illegible]	[illegible]
Superheated, degrees Fahr.	[illegible]	[illegible]	[illegible]	[illegible]	[illegible]	[illegible]	[illegible]	[illegible]	[illegible]	[illegible]	[illegible]	[illegible]	[illegible]	[illegible]	[illegible]	[illegible]	[illegible]	[illegible]	[illegible]	[illegible]	[illegible]	[illegible]	[illegible]
Water.																							
Weight of water fed to boiler, pounds	[illegible]	[illegible]	[illegible]	[illegible]	[illegible]	[illegible]	[illegible]	[illegible]	[illegible]	[illegible]	[illegible]	[illegible]	[illegible]	[illegible]	[illegible]	[illegible]	[illegible]	[illegible]	[illegible]	[illegible]	[illegible]	[illegible]	[illegible]
Equivalent water evaporated into dry steam from and at 212° Fahr., pounds	[illegible]	[illegible]	[illegible]	[illegible]	[illegible]	[illegible]	[illegible]	[illegible]	[illegible]	[illegible]	[illegible]	[illegible]	[illegible]	[illegible]	[illegible]	[illegible]	[illegible]	[illegible]	[illegible]	[illegible]	[illegible]	[illegible]	[illegible]
Evaporation.																							
Water evaporated per pound of coal under actual conditions	[illegible]	[illegible]	[illegible]	[illegible]	[illegible]	[illegible]	[illegible]	[illegible]	[illegible]	[illegible]	[illegible]	[illegible]	[illegible]	[illegible]	[illegible]	[illegible]	[illegible]	[illegible]	[illegible]	[illegible]	[illegible]	[illegible]	[illegible]
Equivalent evaporation from and at 212° Fahr. per pound of coal	[illegible]	[illegible]	[illegible]	[illegible]	[illegible]	[illegible]	[illegible]	[illegible]	[illegible]	[illegible]	[illegible]	[illegible]	[illegible]	[illegible]	[illegible]	[illegible]	[illegible]	[illegible]	[illegible]	[illegible]	[illegible]	[illegible]	[illegible]
Equivalent evaporation from and at 212° Fahr. per pound of combustible	[illegible]	[illegible]	[illegible]	[illegible]	[illegible]	[illegible]	[illegible]	[illegible]	[illegible]	[illegible]	[illegible]	[illegible]	[illegible]	[illegible]	[illegible]	[illegible]	[illegible]	[illegible]	[illegible]	[illegible]	[illegible]	[illegible]	[illegible]
Equivalent evaporation from and at 212° Fahr. per square foot of water-heating surface per hour, pounds	[illegible]	[illegible]	[illegible]	[illegible]	[illegible]	[illegible]	[illegible]	[illegible]	[illegible]	[illegible]	[illegible]	[illegible]	[illegible]	[illegible]	[illegible]	[illegible]	[illegible]	[illegible]	[illegible]	[illegible]	[illegible]	[illegible]	[illegible]
Commercial Horse-Power.																							
On a basis of 34½ pounds of water evaporated per hour from and at 212° Fahr.	[illegible]	[illegible]	[illegible]	[illegible]	[illegible]	[illegible]	[illegible]	[illegible]	[illegible]	[illegible]	[illegible]	[illegible]	[illegible]	[illegible]	[illegible]	[illegible]	[illegible]	[illegible]	[illegible]	[illegible]	[illegible]	[illegible]	[illegible]
Square feet of water-heating surface per H. P.	[illegible]	[illegible]	[illegible]	[illegible]	[illegible]	[illegible]	[illegible]	[illegible]	[illegible]	[illegible]	[illegible]	[illegible]	[illegible]	[illegible]	[illegible]	[illegible]	[illegible]	[illegible]	[illegible]	[illegible]	[illegible]	[illegible]	[illegible]

* [illegible]

CHAPTER X.

BOILER MOUNTINGS AND SAFETY APPARATUS.

THE necessary attachments to a steam boiler include the feed- and blow-pipes, safety-valve, gauge-cocks and water-gauge, pressure-gauge, steam-delivery pipe, and the connections for operating the damper.

The Feed-Pipe may enter the boiler at any convenient place, but the valves and other feed-controlling devices should, if possible, be placed in the fire-room rather than elsewhere, because the water-level requires constant attention on the part of the fireman. There is no uniformity regarding the location for the admission of feed-water, but almost any place is better than directly over the furnace or crown-sheet. The coolest part of the boiler is generally thought to be at the bottom of the rear end, probably because of the greater amount of sediment which collects there, indicating that the circulation is least in that portion of the boiler. This location has long been a favorite one for the admission of feed-water. When a mud-drum is used the feed-water sometimes enters there, but this is not now considered good practice. One might easily infer from certain articles in technical journals on this subject that the normal condition of the water in a steam boiler was in layers of unequal temperature, which is probably not true, except as to the underlying fact that it requires a difference in temperature to begin and continue circulation. The important thing is to prevent local injury to plates or tubes by not admitting cold water.

FIG. 392.

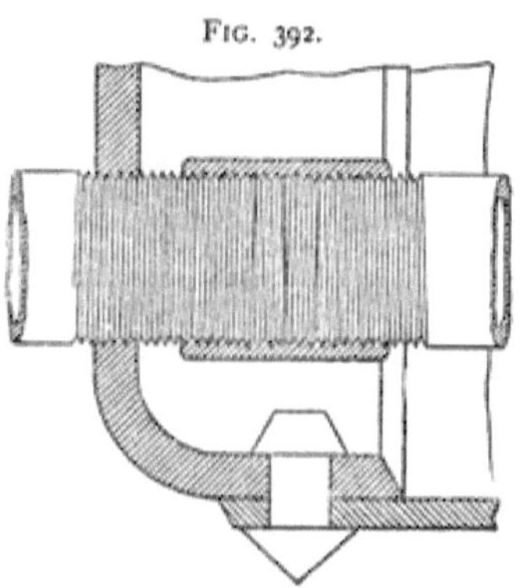

An internal feed-pipe is recommended. The opening into the boiler may be at any convenient place—the front or back head will answer as well as any. After tapping the head put in a nipple with a long thread and screw a coupling on the inside of the boiler, as shown in Fig. 392, to which a pipe may be attached leading to within a few inches of the other end of the boiler; an elbow will then permit the pipe to be extended upward between the tubes (if the latter are not placed zigzag), with another elbow leading towards the opposite end of the boiler. The far end of this last pipe may be bent downward so as to direct the feed-water down among the tubes, where it will be taken up by the general circulation. Feed-water entering the boiler in this manner will have acquired the same temperature as that of the water within the boiler before it is discharged into the general circulation.

The diameter of the feed-pipe should be quite liberal. The larger the pipe within the boiler the slower the movement of the water within, and the more heat it will have absorbed before it is discharged into the boiler. The least diameter recommended for the internal feed- or circulating-pipe is 1¼ inches for horizontal tubular boilers up to 48 inches diameter, and 1½ inches for larger diameters.

FIG. 393.

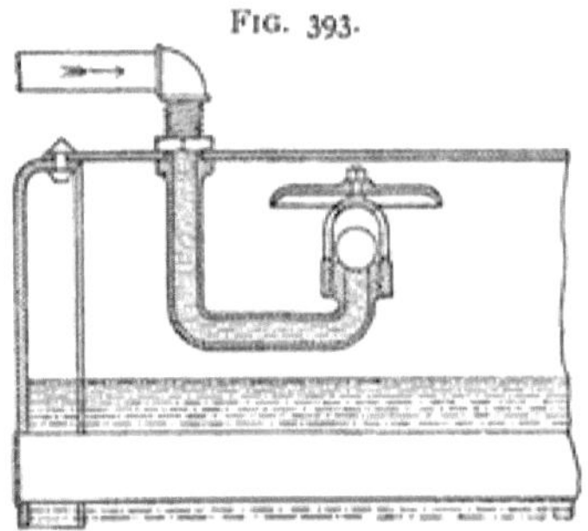

Feeding into the steam-room of a boiler is now quite common. Moore's device, shown in Fig. 393, has been in use some twenty years. The water is admitted, as shown by the overhead pipe and direction of the arrow, through the shell of the boiler into a return fitting, discharging upward at the ball check-valve. A guard over the top of the check-valve prevents any spray touching the overhead shell of the boiler. The feed-water, falling upon the water underneath, is immediately taken up by the circulation without coming in contact with any portion of the boiler.

A suspended pan in the steam-room, with the feed-pipe leading directly into it, as shown in Fig. 394, has been used in a number of boilers by the writer during the past twenty years with good results. The pan may be 3 or 4 feet long, 12 to 18 inches wide, and 3 to 5 inches deep, according to the size of the boiler. The top edge is serrated like a saw, so that the overflow will occur by drops and not in a stream, the drops being taken up by the circulation proceeding along the surface of the water.

FIG. 394.

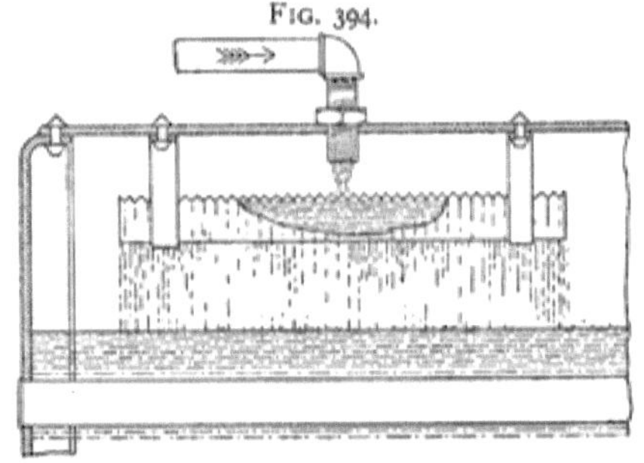

Feeding into an inverted frustum of a cone which overflows into a shallow pan and from the latter into the boiler, as developed in Ford's device, is shown in Fig. 395. This device, wherever used, has given satisfaction.

A purifying-chamber in the rear end of the water- and steam-drums of the water-tube boilers, made by the Standard Boiler Company, is unique in providing an apartment over which the pans are suspended in the steam-space, the overflow of which is not taken up directly by the circulation, but falls into a chamber partitioned off, so as to be uninfluenced by the circulation, as shown in Fig. 396. This purifying-chamber permits a precipitation of the scale-bearing matter into a separate compartment, where it can be discharged through a bottom-blow, also shown in the engraving. By this means the purifying-chamber can be completely emptied, if so desired, without disturbing the water-level in the boiler, a great advantage in any locality where the feed-water contains salts of lime and magnesia.

FIG. 395.

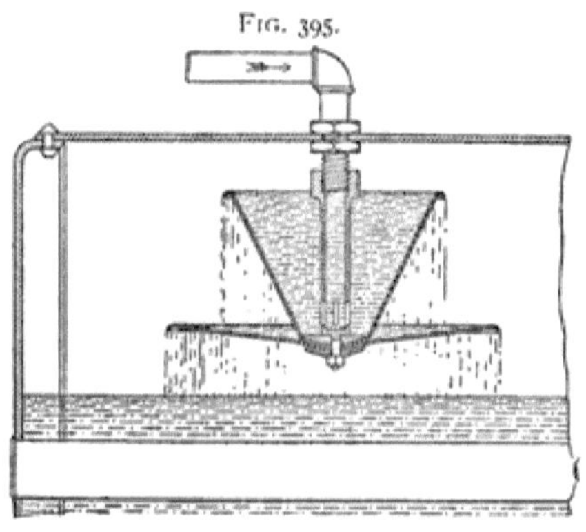

FIG. 396.

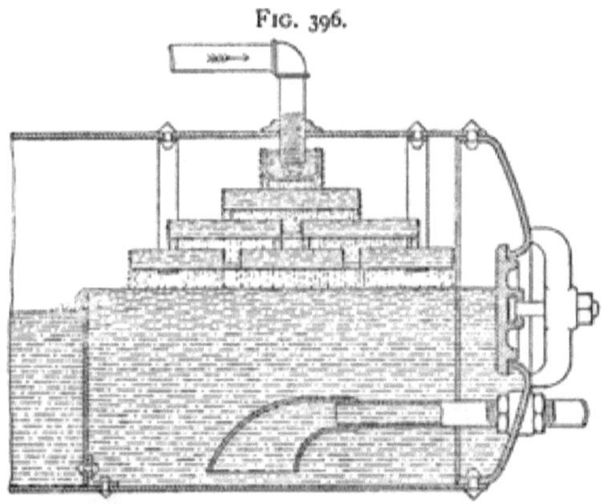

Check-Valves permit a flow of water into the boiler and prevent its escape through the feed-pipe upon withdrawal of pressure. They are made in considerable variety as to exterior form, but the principle of operation is always the same. For small powers the globe check-valve, shown in Fig. 397, is probably used more than any other. The globe, cap, and valve are all made of hard gun-metal; the ends of the globe are tapped for standard wrought-iron pipe-threads. The valve and seat are flat in this engraving, but they are also made with a bevel, as shown in the section of an angle valve, Fig. 398. Whether a globe

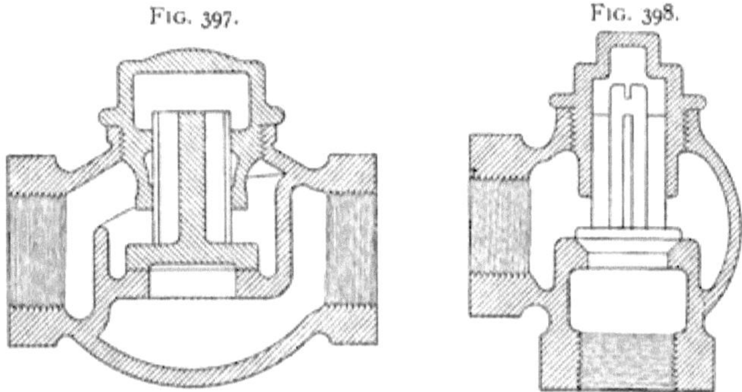
FIG. 397. FIG. 398.

or an angle check-valve shall be used will depend upon the details of piping for any given boiler.

A ball check-valve is shown in Fig. 399. This valve is not in as common use as the one preceding, and is not generally regarded with the favor accorded valves having flat or conical faces, which may be easily reground after distortion occasioned by wear.

The swinging check-valve, shown in Fig. 400, is now much used and has given general satisfaction. It gives a straight way for the passage of the feed-water. The valve is free to revolve on the face, having a pivoted connection in the swinging arm, as shown. These valves are wholly made of hard gun-metal.

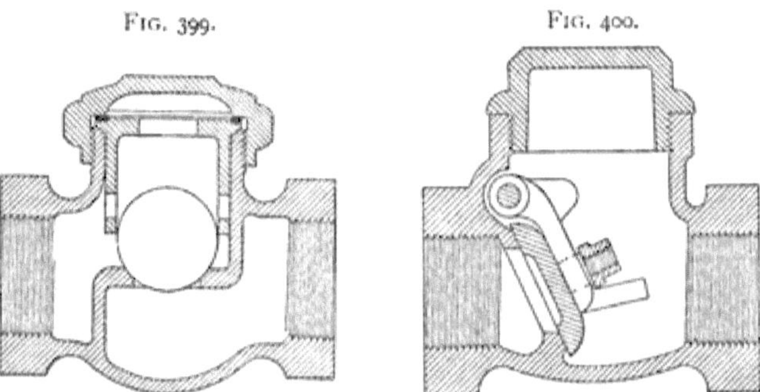
FIG. 399. FIG. 400.

An angle check-valve, much used with injectors, is shown in Fig. 401. It is provided with a tapered nipple for screwing directly into the shell of a boiler; the lower connection is that of a standard union common to wrought-iron pipe-work. The valve is winged and fitted to a bevelled seat. All the parts of this valve are made of hard gun-metal.

An angle valve, shown in Fig. 402, is much used in large installations where several boilers are connected with the same feed-pipe. The main shell and cap are of cast iron; the valve and bushing forming the seat are of hard gun-metal. The shell is bored the depth of the bushing with a recess, say of ¼ inch, into which the bushing, having a corresponding projection, is forced. The cap is provided with a yoke tapped to receive either a brass or an iron spindle, which, passing through a stuffing-box, also included in the cap, can be screwed down upon the check-valve so as to prevent it rising except at such times as feed-water may be required in the boiler to which it is attached. This valve is flanged, and must be fitted to corresponding flanges included in the system of piping, the whole being secured by through-going bolts and nuts.

FIG. 402.

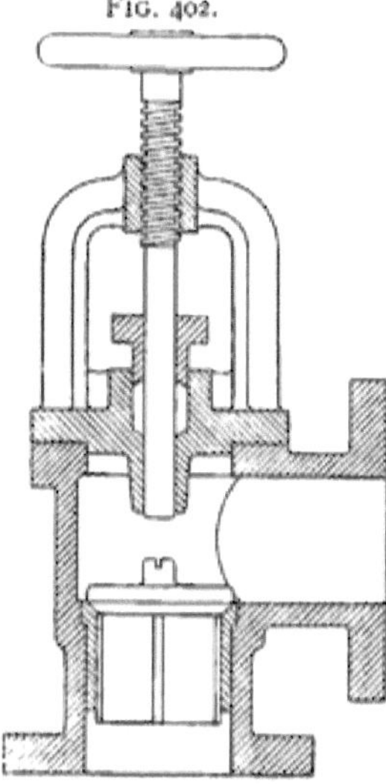

FIG. 401.

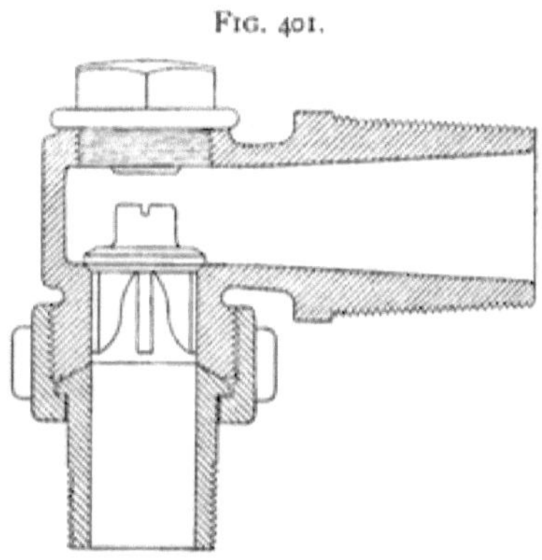

FIG. 403.

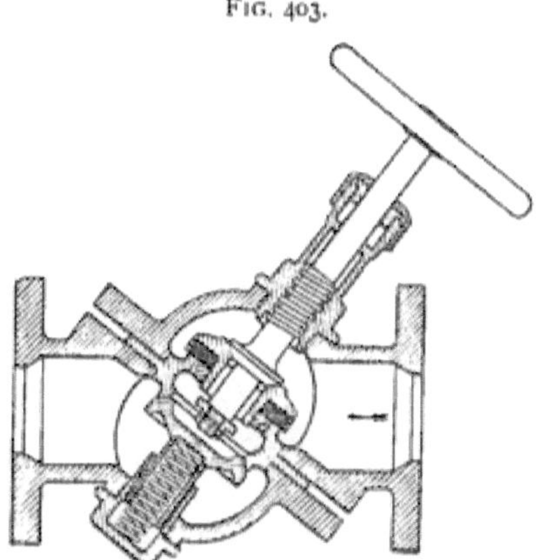

Culver's stop- and check-valve combined is shown in Fig. 403. The valve body is made of iron and in two parts, the gun-metal seat being located between the two, and all bolted together through flanges outside of the seat. This design is to all intents a straightway valve. An inspection of the drawing will make clear that it is first of all a stop-valve consisting of two parts, a small auxiliary valve having its seat on the back of the larger valve, and this larger valve having its seat on the central gun-metal ring lying between the two halves of the main body. The small valve is attached to the spindle operated by the hand-wheel the same as any globe valve. Raising the screwed spindle, however slightly, opens the small valve and water enters the space between the main valve and the check-valve below, thus balancing the upper disk. Upon a further raising of the spindle the nut under the upper valve is brought in contact with the latter and easily raised from its seat, because the pressure is the same on both sides. A square lug on the bottom of the check-valve passes through an opening in the lower half of the body, which allows it to work up and down freely and guides it to its seat, this guide keeping the check in perfect alignment with its seat. By unscrewing the bottom valve-cap and rotating the valve on its seat the check-valve can be reground in case of leakage. The seats are renewable by removing the bolts and springing the valve body far enough apart to remove the brass seat-ring and slip another in its stead; this can be done in less time than would be required to grind the valve if the leakage should be very great.

Gate Valves are much used for water-valves because they give a straightway passage of full diameter of the connecting-pipes. A sectional elevation of a Chapman valve is given in Fig. 404. It consists of a plug or gate in one piece, guided closely in the shell by means of ribs or splines, the latter taking all the strains, relieving the faces of the seats until the central plug is seated, the splines insuring a true and easy vertical movement of plug. The seats are hard gun-metal, and when placed in their proper positions in the body of the shell are held to their exact line by means of a screw-gland inserted through the threaded ends. These glands can be worked forward and back by means of a spanner fitting the splines in the inside of these screw-glands.

The plug is double-faced and equally tight on either face, and either end of the valve may be used for inlet or outlet. This valve is shown with a stationary spindle; that is, the valve rises and falls on the spindle; in other forms the spindle rises and falls through the stuffing-box corresponding to the position of the valve.

A Jenkins gate valve with travelling spindle is shown in Fig. 405. The details of the spindle are not shown, but they are similar to those of Fig. 450. The body has one vertical and one inclined side, both of which are faced true. The valve or plug has of necessity a vertical and inclined side to fit the seat. The back of the valve fits, metal to metal, against the body; the face is furnished with a vulcanized disk, shown black in the engraving; this has a metal guard outside of it, as shown; these disks are renewable by unscrewing the cap and removing the valve from the body. The spindle is independent of the disks, therefore not liable to stick in opening or closing the valve.

A Bottom Blow is used for emptying the boiler ; it is placed, therefore, at the lowest part of the boiler. If the latter is supplied with a mud-drum the bottom blow discharges from that. Horizontal boilers are usually set with an inclination of two to three inches to the rear. The

FIG. 404. FIG. 405.

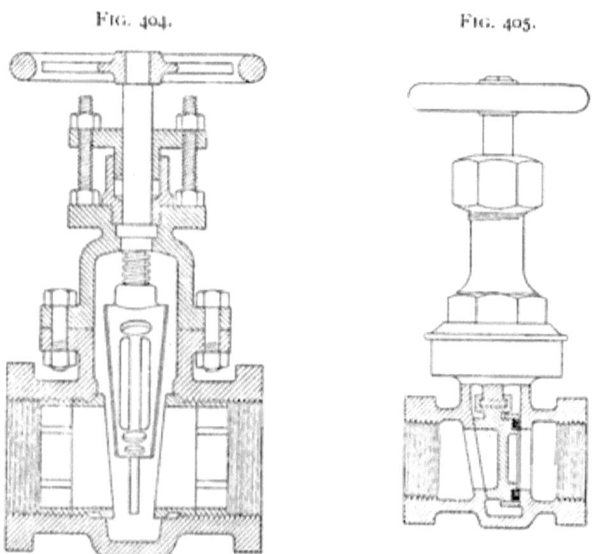

blow-off should be attached by preference at the bottom of the boiler rather than through the rear head, because, for practical reasons, the drilling and tapping into a boiler-head must occur above the curve of the flange, which leaves a couple of inches of water underneath the pipe that cannot be drained. The blow-off opening into the boiler in the case of pipes larger than two inches should be reinforced by a plate riveted to the shell, the pipe tap screwing through both plates, or a pipe-flange should be riveted to the boiler instead.

It is sometimes necessary to insert an elbow immediately under the boiler, as shown in Fig. 295, that the blow-off pipe may pass through the furnace-walls at a higher level than the bottom of the combustion-chamber. Some manufacturers furnish a cast-iron pipe, as shown in Fig. 406. In either case the pipe and fittings are wholly enveloped by hot gases. But a much better method is to carry the pipe vertically downward to the bottom of the combustion-chamber, where the temperature is much lower, and connect the elbow near or below the floor-level. An example of such an arrangement is shown in Fig. 298. A brick protecting wall is built around the blow-off pipe to the bottom of the combustion-chamber, and a long curve joins this pipe to the front of the boiler.

Another method is to build a pier around the pipe, as shown in Fig.

FIG. 406.

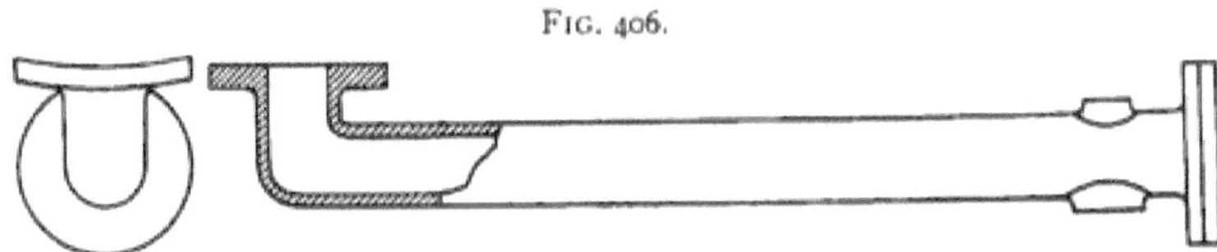

407. As this requires only the length of one brick it does not obstruct the flow of gases. Still another method is to wrap the blow-off pipe with coils of plaited asbestos packing ; the latter, being non-combustible, affords considerable protection to the blow-off pipe.

The diameter of the bottom blow-off pipe should be 1½ inches for boilers up to 42 inches in diameter, 2 inches for 44- to 60-inch boilers, and 2½-inch pipe for larger diameters. The blow-off pipe must be laid so as to completely drain at the discharging end, or it may in time fill with scale or mud. When using the bottom blow the velocity of flow should be rapid, otherwise scale is liable to lodge in the turns or other pipe-fittings and eventually clog the pipe so as to prevent full flow ; for this reason, when blowing, the valve or plug-cock should be wide open or nearly so.

FIG. 407.

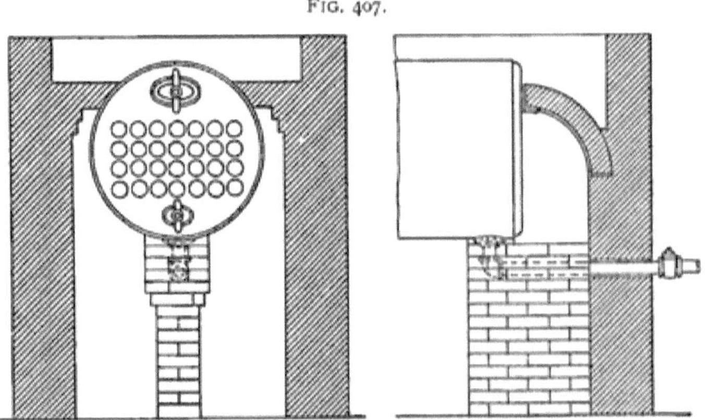

A Surface Blow is needed for boilers in which the feed-water contains impurities liable to separate during the process of ebullition and form a thick scum on the surface of the water. This scum is greatest in waters in which the carbonates of magnesia predominate. The surface blow should be located at the water-level and in that part of the boiler in which the surface agitation of the water is least. It should have a receiving or collecting funnel with a wide surface across the boiler, the wider the better.

An arrangement of surface and bottom blows as applied to a vertical boiler, shown in Fig. 408, was illustrated in the technical journals a few years ago, which, as a scheme of piping, is here reproduced. When the valve at the top of the boiler under the horizontal branch-pipe and the lower valve outside of the bottom tee are opened the water will be blown from the bottom of the blow-off pipe, located at about an inch above the surface of the crown-sheet, the latter being swept by the current of escaping water as it passes across it and up the pipe. So, also, any oil or scum that may be floating upon the surface of the water, and which would otherwise be deposited upon the crown-sheet, is blown off in the same way by opening the surface blow. After all the water that can be reached by the surface blow is carried off in this manner, the top valve may be closed and the bottom valve next the boiler opened, when the remainder of the water in the boiler may be blown out of the bottom blow. After all the water has been blown out of the boiler, the top blow-off valve being closed and the uppermost valve shown in the engraving opened, a hose can be coupled to the upper pipe and the crown-sheet washed off, the water striking the centre of the crown-sheet and washing all sediment into the water-legs, where it can be removed through the handholes.

FIG. 408.

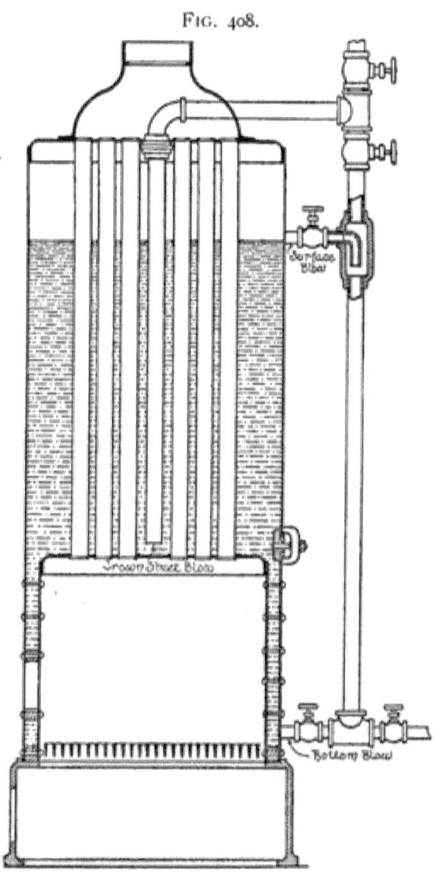

The Hotchkiss surface blow, illustrated in Fig. 409, has been in use in this country for the past twenty years. It consists of a funnel, an up-flow pipe, a reservoir, a return pipe, and a blow-off pipe. The funnel is made of iron, and as large as will pass through the manhole. The reservoir is a cast-iron spherical vessel, with a capacity of about eighteen gallons and of sufficient thickness to withstand the boiler pressure. Its action is as follows: in a boiler with the cleaner attached, the funnel is set near the surface, but partly submerged, and in such position that its opening will intercept the currents of hot water flowing towards it. By the action of gravity the hot surface water entering the funnel will flow into the reservoir through the up-flow pipe, displacing an equal quantity of cooler water therein, which latter returns to the boiler by the pipe shown between the flues. A constant circulation of water through the cleaner is thus maintained by the unbalanced columns of water so long as firing is kept up under the boiler. Any sediment once deposited in the reservoir remains there until removed through the blow-off pipe under control of the engineer.

Fig. 409.

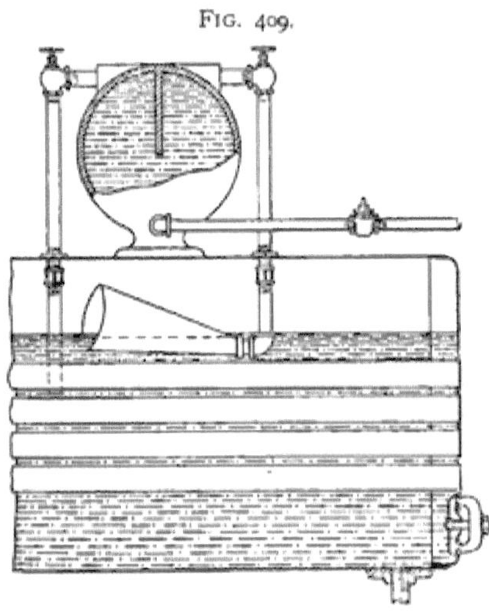

Plug-Cocks are generally used for the bottom blow of steam boilers. An objection to the use of ordinary plug-cocks has been the tendency to leak around the bottom of the plug, the loss of water not being of so much account as the drip upon the floor, which makes in some fire-rooms a disagreeable slop; also the tendency of the plug to stick to the shell if driven down slightly after closing it, making it difficult to afterwards move the plug when required. Both of these objections are overcome in the plug-cock shown in Fig. 410, which has a cast bottom through which no drips can occur. The sticking is less objectionable

Fig. 410.

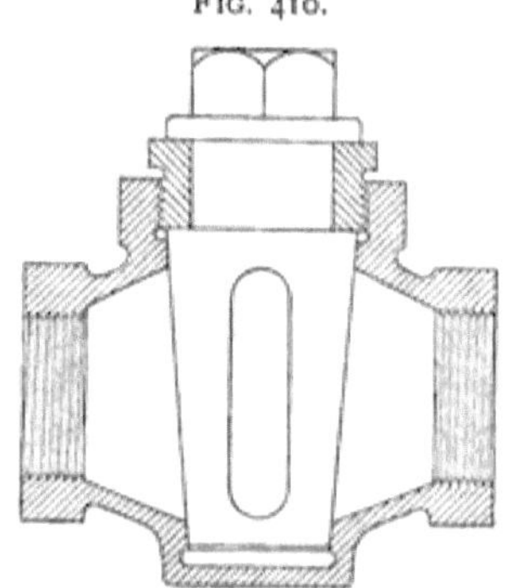

because the collar-nut at the top of the plug, which forces the latter down into its seat, will by a slight turn cause a vertical movement of the plug and thus loosen it sufficiently to move it with an ordinary wrench.

An asbestos-packed plug-cock is shown in horizontal and vertical sections in Fig. 411, the latter on a larger scale than the former. This cock has a closed bottom, the asbestos being driven solidly in the dove-tail grooves in the body of the cock and, being elastic, fits against the plug, making a tight joint with but little friction. Asbestos, being unaffected by heat and moisture, is quite durable. These cocks are in good repute wherever used.

FIG. 411.

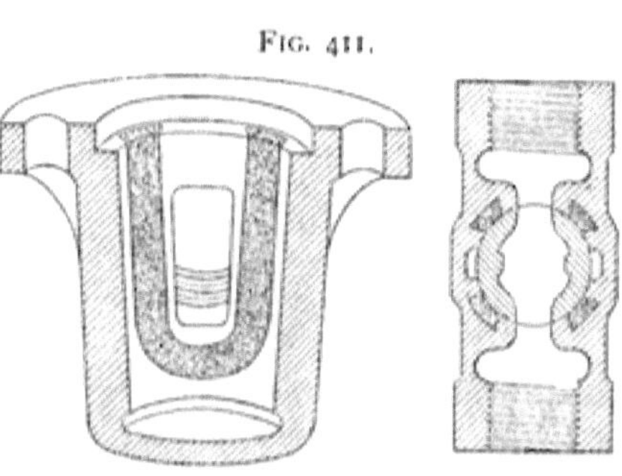

Ordinary globe valves should not be used as a bottom blow for steam boilers; a gate valve or one of the forms of angle valves in which no lodgment of scale can occur is much to be preferred. A Jenkins valve, similar to Fig. 412, or the Eastwood valve, shown in Fig. 413, have both been used with satisfactory results. Myers's blow-off valve as improved by Mowry is shown in Fig. 414. The seat of this valve is removable. Double wheels are so arranged that the large wheel opens and seats the valve. The small wheel rotates the valve on its seat without either raising or lowering the disk. After the valve has been opened and water blown off, the large wheel is used to close the valve upon its seat; then, reversing this wheel about an eighth of a turn, the smaller wheel is revolved to clean off both the valve-disk and the ring on which it is seated.

FIG. 412.

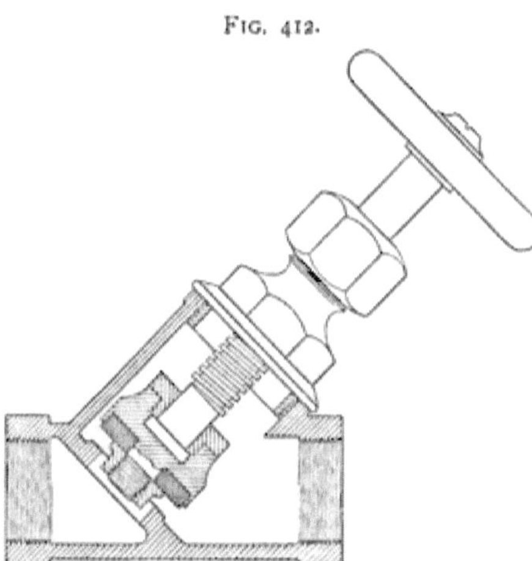

Safety-Valves.—The object of a safety-valve is to relieve the boiler and prevent accumulation of steam pressure above the limit at which the valve is set. The grate surface is now the commonly accepted unit by which to determine the size of a safety-valve. The rate of combustion varies with different boiler furnaces and the kind of coal used.

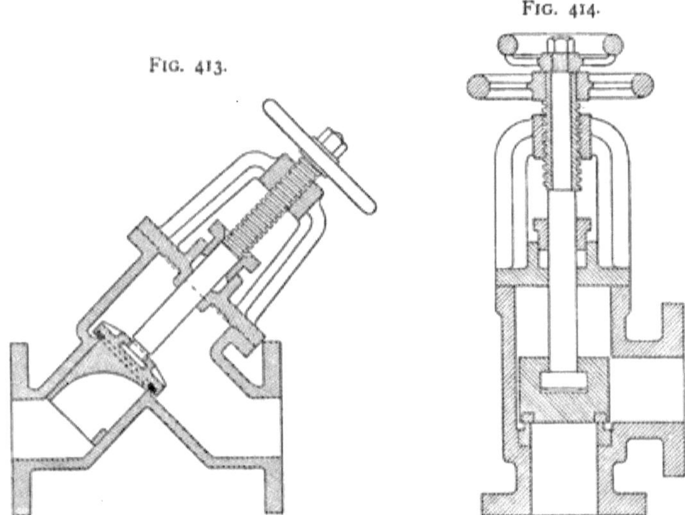

FIG. 413.

FIG. 414.

Records show from 12 to 40 pounds of coal per square foot of grate per hour, but furnaces with ordinary draught do not often burn more than 20 pounds. Good evaporation may be assumed to be 10 pounds of water per pound of coal.

Each boiler should have its own safety-valve. Some municipalities, Philadelphia, for example, require safety-valves to be in duplicate. Safety-valves should be attached directly to the shell or to the steam-dome of the boiler. There should never be a stop-valve intervening between the boiler and the safety-valve if, by closing the former, the latter is no longer in communication with the boiler pressure.

The United States Regulations for steam-vessels require that lever safety-valves shall have an area of not less than 1 square inch to 2 square feet of grate surface in the boiler, and the seats of all such safety-valves shall have an inclination of 45° to the centre line of their axes. These proportions obtain in good stationary-engine practice.

Fig. 415 shows a combined safety-valve and stop-valve, useful for situations in which several boilers deliver the steam into a common reservoir. Spring-loaded safety-valves, constructed so as to give an increased lift by the operation of steam, after the valve is raised from its seat, shall be required, according to the United States Regulations, to have an area of not less than 1 square inch to 3 square feet of grate surface of the boiler, and each spring-loaded valve shall be supplied with a lever that will raise the valve from its seat a distance of not less than one-eighth of the diameter of the valve opening. The seats of all such safety-valves shall have an angle of inclination to the centre line of their axes of 45 degrees. But in no case shall any spring-loaded safety-valve be used in lieu of the lever-weighted safety-valve without first having been approved by the Board of Supervising Inspectors.

FIG. 415.

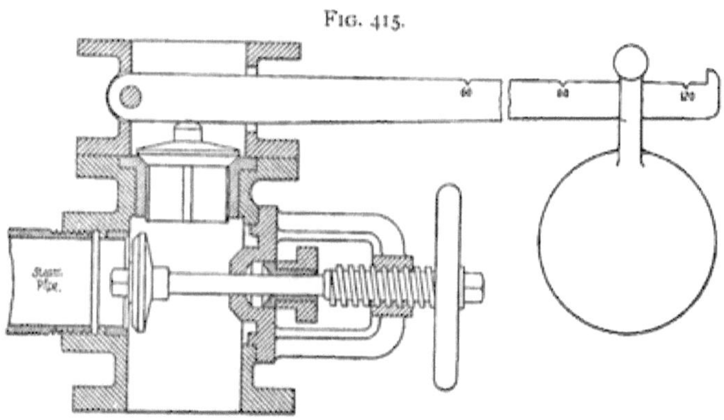

The diameter of a safety-valve is not a test of its efficiency. A valve is effective in direct proportion to its lift, other conditions being equal. Professor Burg, of Vienna, found by actual measurements that a lever safety-valve of 4 inches diameter rises from its seat according to the laws stated below:

With a boiler pressure of . .	12	20	35	45	50	60	70	80	90 pounds
The rise of a common valve is, in parts of an inch . . .	$\frac{1}{36}$	$\frac{1}{48}$	$\frac{1}{54}$	$\frac{1}{65}$	$\frac{1}{86}$	$\frac{1}{86}$	$\frac{1}{168}$	$\frac{1}{166}$	$\frac{1}{157}$

Other reliable authorities do not fix the rise of such a valve from its seat at more than $\frac{1}{23}$ of 1 inch when loaded at any pressure between 12 pounds and 90 pounds. And further experiments of Mr. Burg proved beyond a doubt that the higher the pressure the less will a common safety-valve rise; and in not rising it simply obeys the action of the forces exerted upon it.

According to the table of Professor Burg, the actual size of the venting capacity of a common lever and weight safety-valve of 3 inches diameter is but 0.56 of a square inch area at 70 pounds of steam. Pressure ought, therefore, to be taken into account when fixing upon the size of a safety-valve, especially for large powers, or in all cases where the flow of steam is intermittent and pressures are likely to accumulate nearly or quite to the danger limit.

The Philadelphia regulations for fixing the size of safety-valves take the pressure into account, and order that the least aggregate area of the two safety-valves required by law (being the least sectional area for the discharge of steam) to be placed upon all stationary boilers with natural or chimney draught may be expressed by the formula:

$$A = \frac{22.5\ G}{P + 8.62},$$

in which A is the area of combined safety-valves in inches; G is the area of grate in square feet; P is pressure of steam in pounds per square inch to be carried in the boiler above the atmosphere. The following table gives the results of the formula for one square foot of grate as applied to boilers used at different pressures:

Pressure per Square Inch.													
10	20	30	40	50	60	70	80	90	100	110	120	150	175
Valve Area in Square Inches, corresponding to 1 Square Foot of Grate.													
1.2	.79	.58	.46	.38	.33	.29	.25	.23	.21	.19	.17	.14	.12

Calculating the Load on a Lever Safety-Valve.—Two things are usually known in advance,—the size of the valve and the pressure at which it is to work.

Reference letters for lever safety-valves are given in Fig. 416. Let

A = area of valve in inches.
P = pressure at which valve is to lift.
S = short arm of lever in inches.
L = long arm of lever in inches.
W = weight of ball in pounds.
w = weight of lever in pounds.
g = fulcrum to centre of gravity in inches.
v = weight of valve and spindle in pounds, to which add:
T = actual work to be done, expressive of:
$A \times P - \left(v + \frac{w \times g}{S}\right)$, explained further on.
t = the effect of $v + \frac{w \times g}{S}$ in pounds.

These letters of reference are as far as practicable included in the accompanying sketches.

The ordinary formula,

$$W = \frac{A \times P \times S}{L},$$

is not complete, because it does not take into account the weight of the lever and valve.

The correct formula would be

$$W = P \times A - \left(v + \frac{w \times g}{S}\right) \times \frac{S}{L}.$$

This somewhat complicated-looking formula is made necessary because of the weight of the valve and the additional influence of the lever. The weight of the valve and spindle, v, may be had by the simple process of direct weighing. The influence of the lever upon the valve will be the weight of the lever, w, acting at its centre of gravity, g; this latter

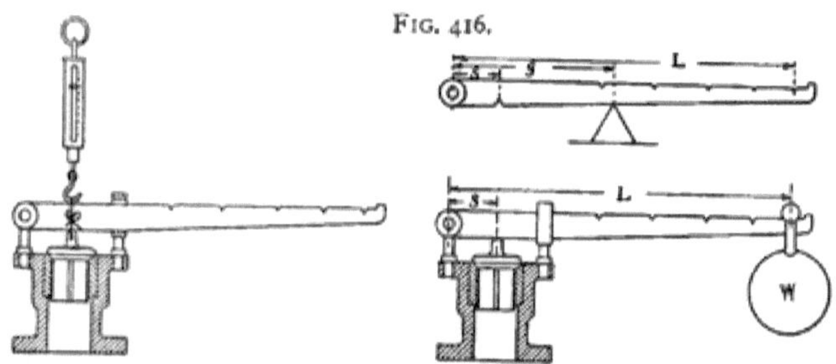

FIG. 416.

may be found by simply balancing it on a sharp edge, as in Fig. 416. The combined effect of the weight of the valve and that of the lever acting upon it as just indicated is that expressed in the formula as

$$v + \frac{w \times g}{S},$$

which is to be deducted from the work required of the weight, W.

However simple all this may appear to those who have it frequently to do, many practical men never feel wholly safe in such calculations, and much prefer to weigh the lever and valve, also shown in Fig. 416. Whatever the weight in pounds thus observed may be is simply deducted from the total pressure (A × P). This amount is that represented in the letters of reference as T.

To illustrate the foregoing, let us assume a 4-inch lever safety-valve of the following proportions:

Area of valve, A	12.57 sq. in.
Pressure, P.	100 pounds per sq. in.
Short lever, S	4 inches.
Long lever, L	36 inches.
Weight of lever, w	10 pounds.
Fulcrum to centre of gravity, g	16 inches.
Weight of valve and stem, v.	6 pounds.
Effect of $v + \frac{w \times g}{S}$ represented by t	46 pounds.

The value of T for this particular valve as calculated by the formulæ given in the letters of reference is

$$T = 12.57 \times 100 - \left(6 + \frac{10 \times 16}{4}\right) = 1211 \text{ pounds.}$$

Then

$$w = \frac{T \times S}{L}, \text{ or } w = \frac{1211 \times 4}{36} = 134.5+ \text{ pounds.}$$

Other calculations from the same data may be made thus:

$$T = \frac{W \times L}{S}, \text{ or } T = \frac{134.5 \times 36}{4} = 1211 \text{ pounds.}$$

The long arm of the lever thus:

$$L = \frac{T \times S}{w}, \text{ or } L = \frac{1211 \times 4}{134.5} = 36 \text{ inches.}$$

The short arm of the lever thus:

$$S = \frac{W \times L}{T}, \text{ or } S = \frac{134.5 \times 36}{1211} = 4 \text{ inches.}$$

To find the pressure, P, we may proceed thus:

$$P = \frac{\frac{W \times L}{S} + t}{A}, \text{ or } P = \frac{\frac{134.5 \times 36}{4} + 46}{12.57} = 100 \text{ pounds.}$$

Other sizes and other proportions may be similarly worked out.

Spring-Loaded Safety-Valves are those in which pressure is regulated by the tension of a spring instead of a weight and lever. Safety-valves of this kind do not ordinarily have plain bevelled valves, but are so constructed that when the pressure reaches the point at which it is to blow, the valve opens slightly at first and allows steam to escape. This escaping steam enters an annular chamber around the valve, and whilst the steam is not wholly prevented from escaping into the atmosphere, it is sufficiently retarded to accumulate pressure in this chamber, to which must also be added the work done by the escaping steam by impinging against the larger surface of the valve before it can make its escape downward, both of which effects, added to the valve already in balance, because of the pressure underneath, combine to force the valve suddenly upward to its full height of lift, producing a sound which has given this valve its characteristic name,—the pop safety-valve. The valve, being thus at its full height, quickly relieves the boiler of its pressure, and as the latter falls the valve is returned to its seat by the spiral spring above.

The American pop safety-valve is shown in sectional elevation in Fig. 417, which shows a flat valve on a flat seat, the kind usually furnished for stationary boilers, but for marine boilers the valves and seats are bevelled at an angle of 45°. The valve is originally set for any desired pressure by screwing down on top of the spring the sleeve-nut shown at the top of the case, afterwards securing it by the lock-nut, also shown at the top of the boss through which the sleeve-nut is adjusted. A combination of levers is shown by which the valve can be raised to a height equal to one-eighth of its diameter, to conform to the United States Regulations. To reset the valve, first loosen the lock-nut, then screw down the sleeve-nut to get increased pressure, or unscrew it to get a lower pressure. The blow-down is adjusted by a relief ring fitted with adjusting screws, shown on either side of the valve, which screws are adjustable in the outside flange; by raising or lowering this ring the distance between it and the outer lip of the valve can be adjusted to any desired area, and thus any desired height of lift of valve when blowing off.

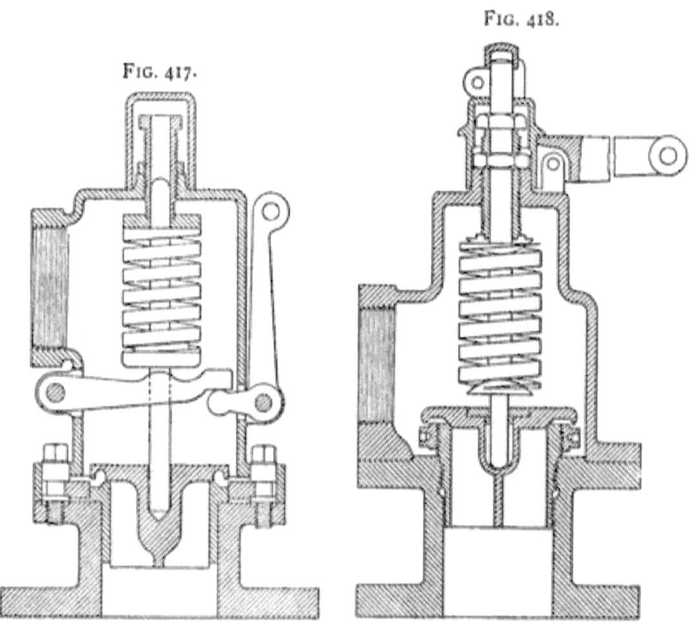
FIG. 417. FIG. 418.

The consolidated safety-valve is shown in Fig. 418. The flanged base is cast iron, into which is screwed a gun-metal seat; a winged valve is fitted into this seat, having a bevelled edge of 45°. The valve has a projecting annular lip, underneath which and screwed upon the outside of the valve-seat is an adjustable ring for regulating the lift of the valve when blowing off. The central stem by which the pressure of the spring is brought upon the valve has its lower end pivoted well below the valve-face, so as to insure a vertical movement at all times. This stem has two collars with curved faces, against which the spring is compressed, the thrust at the upper end being taken by an adjustable screw-bolt passing through the upper part of the cast-iron case, the screw being fixed at any determined point by means of a lock-nut, all of which are shown in the engraving. A lever is also provided by which the valve may be lifted off its seat one-eighth of its diameter.

Safety-valves should blow directly into the boiler-room, and not through a pipe into a chimney or other outlet. The objection to piping a safety-valve outlet is, that unless it passes directly upward through the roof there is always a possibility that water will accumulate in the valve chamber and be a direct cause of external corrosion, unless provided with suitable drips. Neglect to properly drain such pipes might end quite disastrously in the winter by freezing, if the pipes and valves are in an exposed position.

Low-Water Alarms.—These appendages to a steam boiler have long been in use, and are, therefore, of varying detail. They all have a common object, which is to sound an alarm in the event that the water in the boiler to which it is attached falls below its proper level. The alarm is commonly operated in one of three ways: by melting a fusible plug, by a differential expansion movement, as in the case of a wrought-iron and a copper pipe, or mechanically, by the use of floats; to which might be added, as a fourth, electrical devices. But these have never come into general use and are not now in favor.

The Ashcroft low-water detector is shown in Fig. 419, attached to the top of the shell of a horizontal tubular boiler. The ordinary water-level is shown; the alarm-level corresponds to that of the bottom of the pipe above the tubes; this distance is to be fixed for each particular case, and will probably vary from ½ to 1 inch, depending on the size of the boiler and the rate of evaporation. The plug-cock is always to be left open; this permits the water to flow into the upper chamber by the steam pressure acting upon the surface of the water below; there being no circulation in this chamber, the water cools to a temperature much below that of the boiler. A fusible plug which is shown, by double hatching closes the orifice leading into the water-chamber, and is held in place by a screw-plug having a central hole in it leading to the attached steam whistle. In the event of the water getting below the pipe, shown in the engraving, the upper chamber and its connecting-pipe is emptied, steam takes its place, the temperature of which is sufficiently high to melt the fusible alloy, and escaping through the whistle sounds an alarm. After a plug has melted the intermediate cock must be closed to stop the alarm. Water is then pumped up to its proper level. To fix a new fusible plug, unscrew the hollow plug by which it is to be held in place, and after carefully removing all fragments of the former fusible plug, insert a new one and replace the hollow plug, screwing it up sufficiently to make the joint water-tight; then turn on the water slowly to give it time to cool sufficiently so as not to melt the plug. When the pressure is on there must be no leak or drip, otherwise a circulation would be established, and the plug melted by the hot water flowing into the upper chamber.

Fusible plug-alarms have been objected to on the ground that they can never be tested, and new plugs have to be replaced at intervals of a few months at great inconvenience, the renewals being necessary on account of certain molecular changes which occur in the metal, which becomes non-fusible in use. They are also rendered inoperative by scale or sediment.

The Hardwick low-water alarm is shown in Fig. 420. This device is operated by the different expansions for the same temperature caused by steam entering two pipes, one of which is of wrought iron and the other of brass, copper, or other material in which the rate of expansion

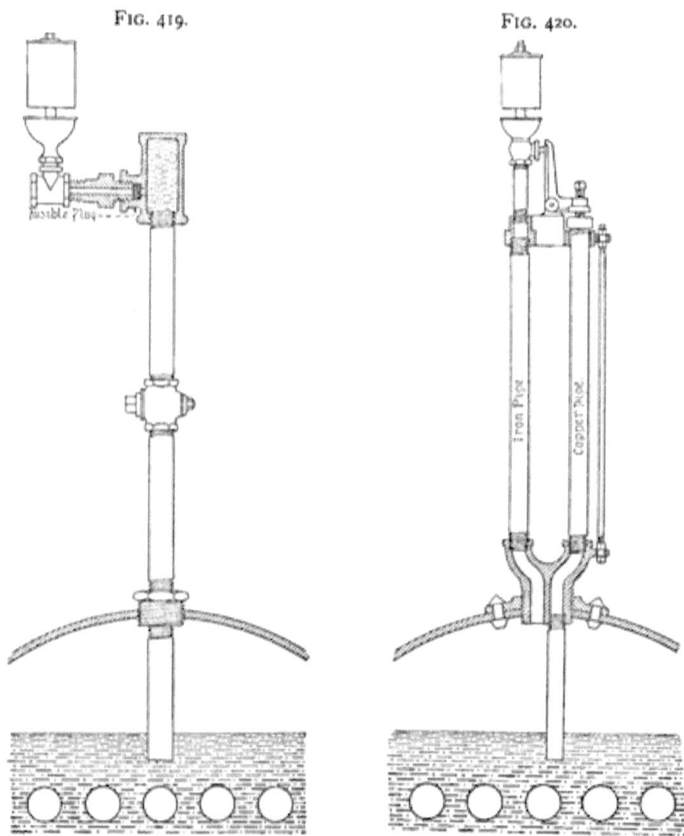

FIG. 419. FIG. 420.

is more than that of iron. The shell of the boiler is tapped to receive a casting having two pipe-openings, into one of which is screwed a wrought-iron pipe extending upward and terminating in a whistle; the other pipe, being made of brass or copper, also extends upward, and is fitted with a cap at the top. Near the top of this pipe and attached to the iron pipe is an arm carrying the fulcrum of a lever. This lever is in the form of a right angle, the upper end of which is intended to operate a steam-whistle; the lower end, or horizontal member of the lever, is fitted with an adjusting screw. Underneath the part which screws into the boiler is a short piece of pipe fitted to that opening which leads to the expansion-pipe. The length of this short piece of pipe is governed by the height to which the water is to be carried in the boiler, and which governs the sounding of the alarm.

When steam is raised on the boiler the water is forced up into the expansion-pipe and, as there is no circulation in this pipe, it is soon cooled, and will remain cool so long as the pressure is maintained in the boiler. In the event that the water should fall below the bottom of this pipe, the water contained in the expansion-pipe will fall by gravity and steam will take its place; the expansion of the brass or copper pipe, being greater than that of the iron pipe under the same steam pressure, will push upward the horizontal arm of the lever, which in turn will push the valve of the whistle inward and sound an alarm. It will be observed that there is a dry-steam connection from the top of the boiler directly to the whistle-valve, so that the operation of the whistle will not be interfered with by a combined mixture of water and steam; this might result if the steam were drawn from the water-line of the boiler, which is always in a more or less agitated state by reason of the disengagement of steam going on at that point. When the water-level is restored the brass pipe contracts in length and the whistle ceases blowing.

Expansion-tube devices have been objected to, notwithstanding the fact that nothing is more certain than the expansion of metal under heat, but the fact that the expansion of the metal in a low-water alarm is dependent upon steam taking the place of water—and the sounding of the whistle is dependent upon this condition and the proper adjustment and rigidity of the parts—makes, it is claimed, the greater expansion of one of the metals a matter of somewhat remote importance, also that the adjustment of these devices is commonly too fine for practical use in a boiler-room, inasmuch as the difference in temperature between hot water and steam is not sufficient to cause very great elongation of the tube. These strictures, while rather severe, are apparently supported by facts in the case of failure of expansion devices not wholly confined to low-water alarms. Particular care should be taken with such devices to see that they are at all times in good working order.

The Ashley low-water alarm combined with a water-column is shown in Fig. 421. The principle of its operation is based upon the difference in weight of a body suspended in air and immersed in water. The alarm has two connections with the boiler,—a steam- and a water-connection common to all water-columns. The top cover of the column is fastened in place by tap-bolts, is removable, and to this is attached all the mechanism of the device. Suspended from the under side of the cover is a valve working in combination with a double-ended lever, having its fulcrum between its two ends. From the ends of the lever two cylinders are suspended, the upper and smaller one of solid iron and the larger one hollow, with holes in its top, so that it is filled with water. When the water stands at the desired level in the boiler and column the solid cylinder is heavier than the larger immersed cylinder, and consequently keeps the valve closed. As the water in the column lowers, the hollow cylinder filled with water overbalances the weight of the solid cylinder

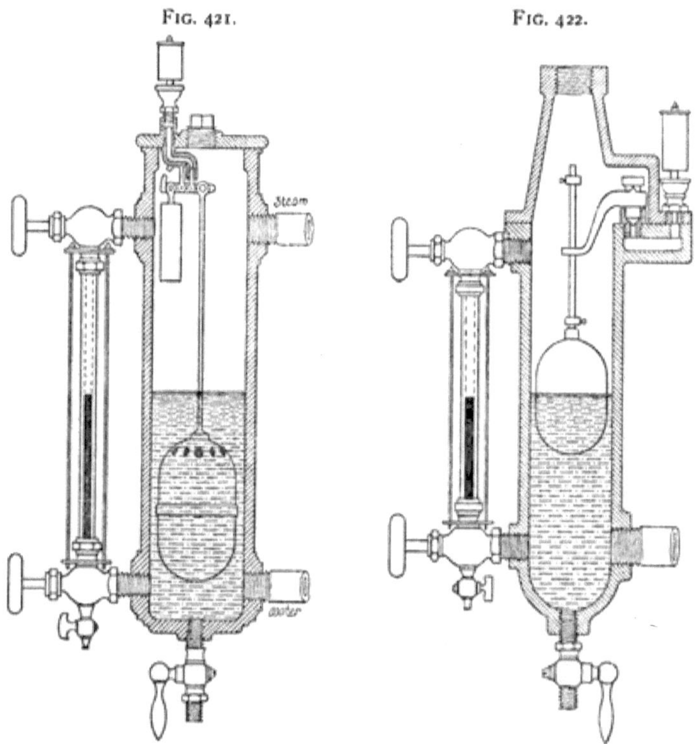

FIG. 421. FIG. 422.

when deprived of the buoyancy of the surrounding water and opens the valve, admitting steam to the alarm whistle. The solid cylinder regains its counterbalance on the admission of water to the boiler and column, closing the valve. Its working can be tested at any time by simply opening the valve at the bottom of the column, thus lowering the water in the column and sounding the whistle. By removing the bolts in the cover the whole of the mechanism can be lifted out for cleaning or inspection, and easily replaced without disturbing the boiler connections. The body of this alarm should always be attached in a vertical position with no valve between the boiler and the alarm. The position of the bottom of the gauge-glass should be at the line where the whistle is to blow, in which case the blow would occur when the water is about one inch from the bottom of the glass. The alarm can be tested by blowing the water out of the column, thus allowing the mechanism in the alarm to operate and blow the whistle.

The Pittsburgh high- and low-water alarm is operated by means of a copper float placed in a water-column, as shown in Fig. 422. As the water falls to the level of the bottom of the gauge-glass the float falls with it, the collar on the vertical rod resting upon and afterwards lowering the central end of the lever through which it passes; the other end, being fulcrumed above, opens the steam-valve by lifting it from its seat, sounding the alarm for low water. In the event that too much water is fed into the boiler, the float, rising in the water-column, carries with it a lower collar attached to the vertical rod near the float, which lifts the central end of the lever upward; the lever, centring upon the opposite fulcrum in the valve-chamber above, causes the alarm-valve to move upward with the float and blow the whistle. These collars can be adjusted to any desired variation in water-level.

Fig. 423.

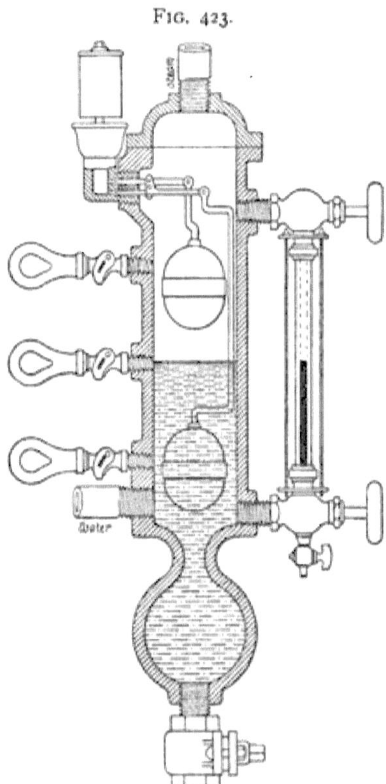

The Reliance high- and low-water alarm is shown in Fig. 423. There are two floats, one in the steam- and the other in the water-room of a vertical water-column, to which are also attached the gauge-cocks and glass water-gauge common to all combined water-gauges. A bell crank-lever and rod connects each float with a whistle-valve; when the water is at the proper height, the lower float, being submerged, presses upward, its steam-valve remaining closed; if from any cause the water gets low enough to rob the float of its support, it sinks of its own gravity, thus opening the valve and blowing the whistle.

The high-water alarm is simply the low-water alarm reversed. A bell crank-lever is turned over so that the weight of the float holds the valve closed until the water rises and carries the float up with it, thus opening the whistle-valve. The water cannot pass either the upper or lower limit without automatically blowing the whistle.

The spherical extension of the main water-column at the bottom is a sediment-chamber, into which all heavy particles fall, to be blown out into the ash-pit through a blow-off pipe. The sediment cannot get back into the column when the valve is opened, on account of the contracted neck connecting the sediment-chamber with the column proper.

Fusible Plugs.—The United States Regulations require that "cylinder boilers with flues shall have one plug inserted in one flue of each boiler; and also one plug inserted in the shell of each boiler from the inside, immediately before the fire-line, and not less than four feet from the forward end of the boiler. All fire-box boilers shall have one plug inserted in the crown of the back connection, or in the highest fire service of the boiler. All upright tubular boilers used for marine purposes shall have a fusible plug inserted in one of the tubes at a point at least two inches below the lower gauge-cock, and said plug may be placed in the upper head-sheet when deemed advisable by the local inspectors. All fusible plugs, unless otherwise provided, shall have an external diameter not less than that of a one-inch gas- or steam-pipe screw-tap, except when such plugs shall be used in the tubes of upright boilers. Plugs may be used with an external diameter of not less than that of a three-eighths of an inch gas- or steam-pipe screw-tap, said plugs to conform in construction with plugs now authorized to be used by this Board; and it shall be the duty of the Inspectors to see that these plugs are filled with Banca tin at each annual inspection."

A fusible plug as applied to land boilers is ordinarily a brass shell fitted with block-tin, Babbitt-metal, or other metal having its melting-point below the temperature of red-hot iron. These plugs are inserted in the crown-sheet, the upper part of a flue, or such other portion of a steam boiler as will be liable to dangerous overheating in case of low water. In order to fully protect a boiler, the fusible plug may extend upward a short distance inside of the boiler, as shown at Fig. 425. No escape will occur so long as the top of the plug is covered with water, but when the water gets below the upper surface the soft metal melts and runs out of the shell, followed by the steam, which either deadens or completely puts out the fire, or at least serves to warn the fireman that the water has reached the danger line in the boiler. Fusible plugs need to be carefully looked after, as a layer of scale or mud over the top will prevent the escape of steam, even though the metal underneath be melted and gone.

The insertion of a soft metal rivet, usually of lead, as at Fig. 424, in the crown-sheet or flue of a boiler was formerly much used for this purpose, but has been superseded by the better arrangement of a removable brass shell filled with soft metal, Fig. 425. Either of these will be effective if the top is not allowed to become covered with scale.

The Bailey plug, shown at Fig. 426, is an improvement over both the others. The body of the plug is permanently fixed. A screw-cap holds the fusible disk in place. The upper part of this disk is protected

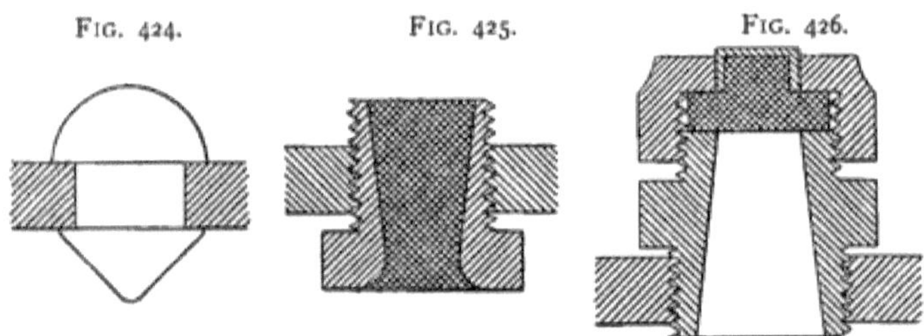
Fig. 424. Fig. 425. Fig. 426.

by a copper cap, shown in the drawing. This cap is intended to prevent the water coming in direct contact with the soft metal, thus maintaining its normal point of fusion.

Parry's safety-plug is shown in Fig. 427. The fusible metal is interposed between two brass shells, which protect it from the water and expose only a line of its surface to the furnace gases. The outer shell and the fusible metal fit against the sheeting in a hollow brass plug screwed into the crown of the furnace, and are held firmly in place by a cotter driven through slots in two lugs on the plug. The entire device can be cleaned both inside and out without any difficulty and can be taken to pieces without a wrench.

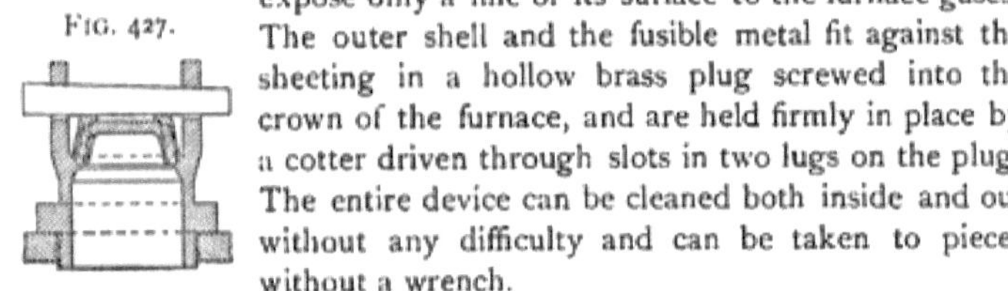
Fig. 427.

Gauge-Cocks are for the purpose of ascertaining the water-level in a steam boiler. They are usually three in number and, whenever practicable, should be placed directly in the boiler-shell, the centre one on the proposed water-line, the lower one about an inch above the top of the tubes, and the upper one at that point beyond which it is desirable that the water should not go.

The Mississippi gauge-cock, shown in Fig. 428, is largely used, and is of the simplest possible construction, being a hollow plug screwed

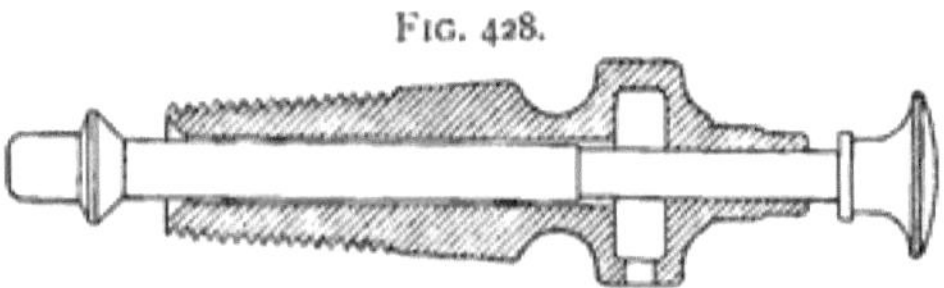
Fig. 428.

into the boiler, through which is a rod terminating in a valve having a bevelled face fitting into a corresponding seat in the end of the hollow plug, the pressure of the steam keeping the valve against its seat. When the valve is pressed off its seat the blow will indicate whether there is water or steam in the boiler at the level of that particular gauge-cock. These gauge-cocks are easily ground without removing them from the boiler.

The Williams rotating gauge-cock, shown in Fig. 429, is similar to the above, except that it has spiral wings attached to the valve-stem for

FIG. 429.

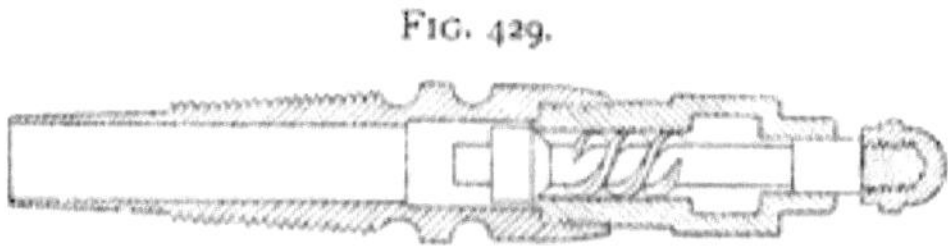

the purpose of making it self-grinding and self-cleaning. The water escaping from the boiler impinges against these wings, causing a rotary movement to the valve, which prevents it from seating twice in the same place, the rotary motion keeping the valve and seat both clean and tight.

The Bingham rotating gauge-cock is shown in Fig. 430. It is constructed on the same principle as the Mississippi gauge-cock, except that the wings are located back of the valve and next to the water in

FIG. 430.

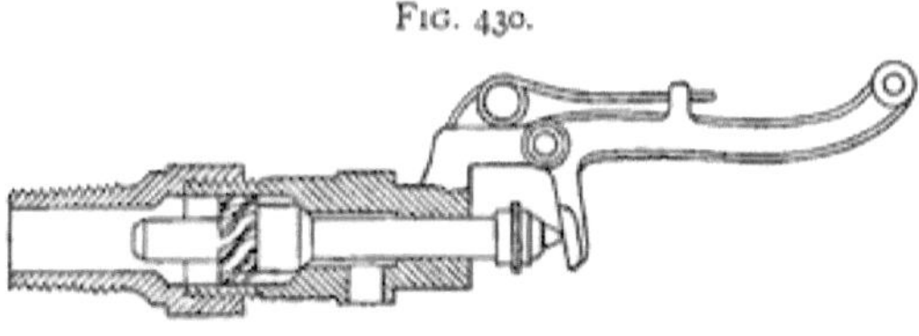

the boiler. It is provided with a lever for opening the valve by means of a cord or jack-chain. A rotary motion is given the valve when in use by the spiral wings making it self-grinding and self-cleaning.

The Register gauge-cock is shown in Fig. 431. It consists of a hollow plug screwed into the water- or steam-space, at the outer end of which is hung a weighted lever, having a strip of vulcanized rubber immediately over the central orifice of the plug ; the boiler is tested by simply lifting the weight, thus raising the rubber valve off its seat, causing the blow. For large boilers this gauge-cock is very convenient, as it can be lifted from its seat by a pole or rod from below, is an excellent form of gauge-cock, and not liable to get out of order, needing only a new strip of vulcanized rubber occasionally to make it good as new.

FIG. 431.

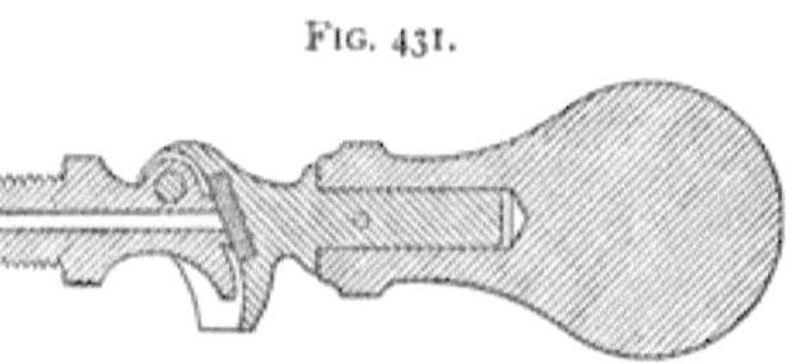

The Reliance gauge-cock, shown in Fig. 432, has its valve-stem passing through a chamber outside of the hollow plug leading to the boiler. One end of this chamber is fitted with a screw-cap. The valve-stem is provided with a collar, between which and the inside end of the

FIG. 432.

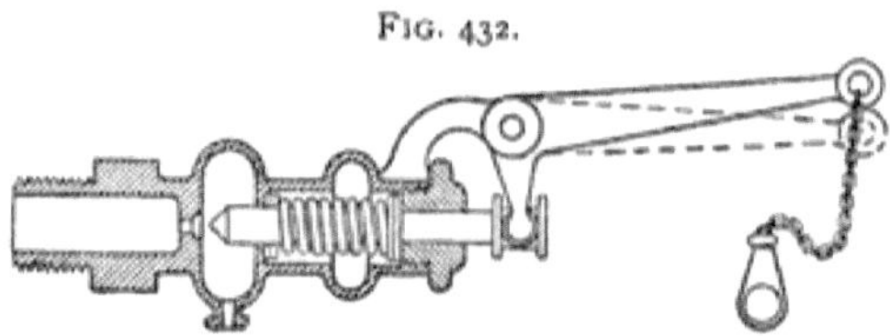

chamber is a spiral spring of sufficient tension to lift the chain and open the valve when no steam is in the boiler by simply lifting the weight. A right-angled lever with chain and weight keep the valve against the seat. The valve is not closed by the pressure behind it, but by means of the weight on the long end of the lever. By lifting the weight the combined action of the boiler pressure and that of the spring opens the valve, and the water-level is determined. Where these gauge-cocks are made up of sets of three, the levers are of graded lengths, the top one being longest, the bottom one shortest, so that the chains and handles hang clear of each other.

FIG. 433.

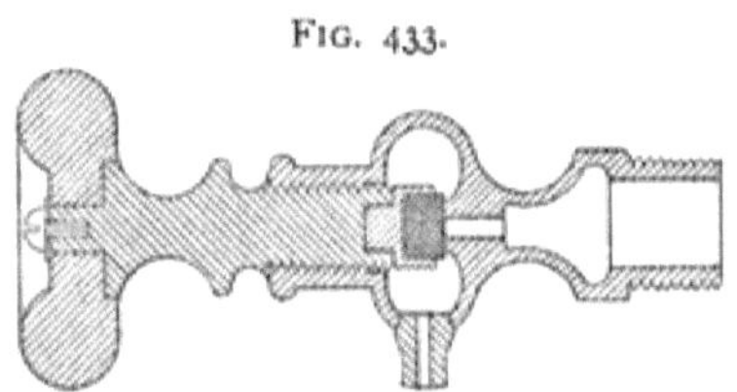

The compression gauge-cock, shown in Fig. 433, is fitted with a disk of soft metal or vulcanized rubber. The compression is had by turning the hand-wheel, the movement of the valve being effected by means of the screw attached to the hand-wheel. For small boilers where the gauge-cocks are within reach of the attendant, the compression-cock is the one commonly selected.

A Water-Gauge consists of an upper and lower angle valve, with a glass tube connecting the two, as shown in Fig. 434. When both valves are open, the water-level in the boiler is indicated by a corresponding level in the gauge-glass. It is important that the openings be kept free from obstructions of every kind, otherwise the true water-level will not be indicated. The small cock under the bottom of the valve is for the purpose of blowing through, to ascertain whether or not the openings in the boiler are free. If not free, the true water-level will not be indicated in the glass. If the bottom opening is completely obstructed, the steam will blow through, and the water will not rise to its proper level upon closing the pet-cock. In the event of a glass breaking, the bottom- or water-valve should be closed first, and then the steam-valve. The ends of the broken glass tubes may now be removed and a new glass inserted, with new packings in the stuffing-boxes.

FIG. 434.

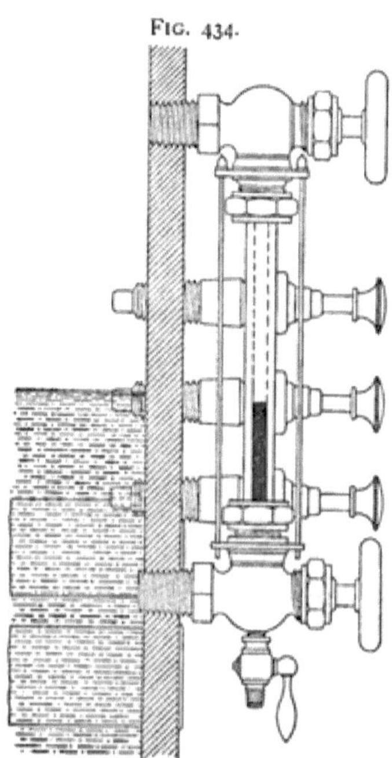

Glass tubes for water gauges must be made of clear, transparent glass and of considerable toughness. Those most in use, probably, are the imported Scotch tubes, easily known by a certain fibrous appearance in the glass lengthwise of the tube. Occasional breakage of glass tubes is to be expected, owing to the brittle nature of the material and to the unequal expansion incident to the action of the hot water and steam inside and the cold air on the outside. As glass tubes never give previous warning of failure, spare tubes should always be kept on hand.

The lower valve and fittings of a glass water-gauge are shown in Fig. 435; the pet-cock underneath is for the purpose of blowing out. A valve under control of the hand-wheel on the outside opens or closes communication with the boiler as desired. The glass water-tube need not project below the bottom of the stuffing-box farther than is shown in the drawing. This box is commonly filled with a soft packing, usually of plaited lamp wick.

FIG. 435.

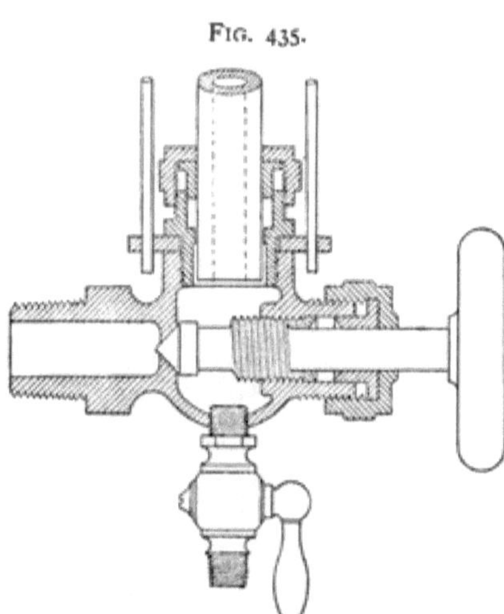

Upon the breakage of a water-glass, especially if the fittings are at a considerable height, it is not only difficult, but sometimes dangerous, to approach the gauge for the purpose of shutting off the water and steam blowing into the fire-room. Several devices have been brought out having an automatically closing valve which will remain open so long as the glass is whole, but which will immediately close upon the breakage of a tube. Such a fitting is shown in Fig. 436. The lower of the two valves shown is a metal ball resting against a pin to prevent its rolling into the boiler or into the barrel of the water-gauge fixture. So long as there is no circulation in the water-glass the ball-valve will remain in the position shown, but as soon as a current is set up, whether by the water or steam, the ball-valve will roll forward and close the orifice leading to the water-glass, remaining there as long as the pressure continues behind it. The upper one of the two valves shown is of a different type, being a mitred valve with wings arranged spirally about its body. The steam and water will pass through the spiral openings and indicate the true water-level in the glass. Should this glass break, the pressure within will force the valve up against the seat and prevent the flow of either steam or water into the fire-room. Once this valve is closed, it requires to be opened from without. A ready means for opening this valve is had by projecting the valve-rod through the seat and forcing the valve back against a pin shown in the drawing. It will be understood that in this illustration the valve controlled by the handle should be withdrawn further than appears in the drawing; if not, it would be impossible for the inside valve with the spiral wings to seat itself, and would not, therefore, prevent the flow of either steam or water into the fire-room.

Fig. 436.

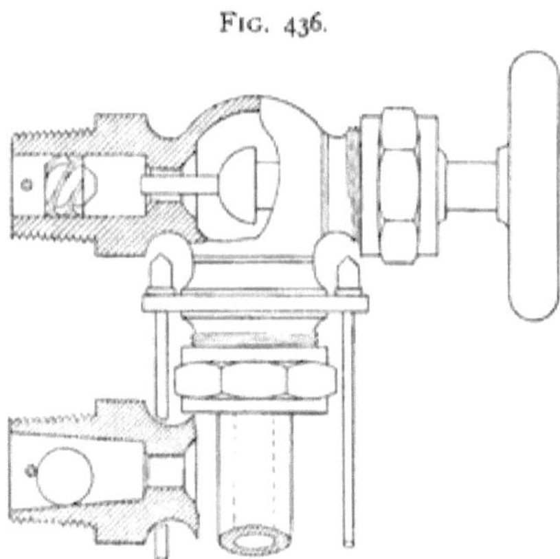

Glass tubes may be cut to any desired length by using a 6-inch half-round fine-cut Stubb's file. Place the glass tube on an even surface, such as a flat board; then place the sharp edge of the file at the point at which the tube is to be cut and bear on lightly at first, rolling the tube back and forth by pushing and pulling on the file, having the cut or mark meet around the tube; bear on a little harder as the cut grows deeper and roll until the tube flies apart. Tubes thus cut have square ends, and when properly managed need never result in failure, even for short lengths. Another method, and a good one, is to take a piece of steel wire; sharpen at one end and bend that end into a right-angle hook and harden same; run this hook into the inside of the glass and make a scratch around the interior at the point where the break is to be made. The glass can then be readily broken without cracking.

There is a want of uniformity in the matter of locating the glass water-gauge fixtures with reference to top of tubes, fire-box, or other heating surface liable to be exposed in case of low water. A glass water-gauge ought to show when the water is near or, perhaps, at the danger limit. In case of low water it is particularly important that the exact height of water above the line of heating surface be definitely known. A permanent marker should be attached to the lower glass gauge fixture to accurately indicate the top of the highest heating surface inside the boiler. The height of water above this line is the one important fact the fireman must have constantly before him.

A Combined Water-Gauge consists of a barrel to which are attached the gauge-cocks and the glass water-gauge. This barrel has two pipes connecting with the steam and water portions of the boiler; these pipes are of considerably larger diameter than would ever be used with any of the fittings taken singly. The lower pipe being removed from the path of circulation, there is less disturbance of water-level in the barrel of the combined water-gauge during foaming than occurs within the boiler itself. The illustration, Fig. 437, shows the glass tube and the gauge-cocks opposite each other; they are not usually thus placed, but at right angles to each other.

Fig. 437.

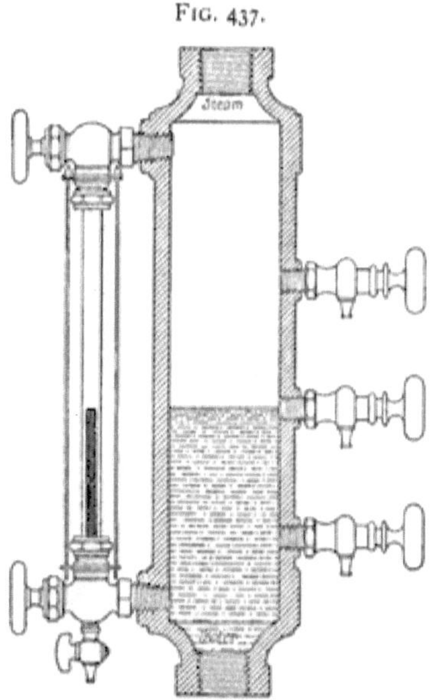

There are no fixed proportions for water-gauge barrels. They vary from 2½ to 5 inches in diameter, with glass gauges ranging from 12 to 20 inches in length. The smallest barrels are not often fitted with less than ¾-inch pipe connections to the boiler, the larger sizes being tapped for 1-inch and 1¼-inch wrought-iron pipe. Stationary-engine boilers more than 48 inches in diameter should have water-gauge barrels not less than 4 inches in diameter and should have pipe connections not less than 1¼ inches. The gauge-glass for such a barrel need not be more than 12 inches between stuffing-boxes.

The piping of a water-column must be properly done or the apparatus will fail to indicate the true water-level in the boiler. The steam-pipe should lead from a point above the highest gauge-cock and the water connection to a point below the lowest gauge-cock. The drainage of these pipes must be carefully attended to, and must be arranged to drain completely dry, either into the boiler or into the water-column. No water-pockets or trapping must be permitted. A blow-off and drain-pipe should lead from the bottom of the water-column into the ash-pit; this pipe will be useful when blowing out any sediment which might have been carried into the water-barrel. The bottom gauge-cock should be placed at what is to be considered the danger line. In Fig. 434 it is shown at the upper side of the top row of tubes. This is too low down; it should be at least 1 inch above for boilers 46 inches in diameter or less, and at least 1½ inches above for 48-inch and larger boilers. The middle gauge-cock should be placed on the water-line at which the boilers are intended to work; this may be from 2 to 4 inches above the lower gauge-cock, depending upon the diameter of the boiler. The upper gauge-cock is usually placed at the same distance above the central gauge-cock that it is above the bottom one.

Safety Water-Columns are those in which a signal is given automatically to warn the attendant when the water in the boiler is either more or less than the established water-line, and for which provision was made in the apparatus itself. The sectional elevations, Figs. 421, 422, 423, show the interior arrangements of such safety water-columns as are in common use. The barrel includes the gauge-cocks and glass water-gauge usual in all water-columns. In addition to these there are devices arranged for operating a small steam whistle.

Fig. 438.

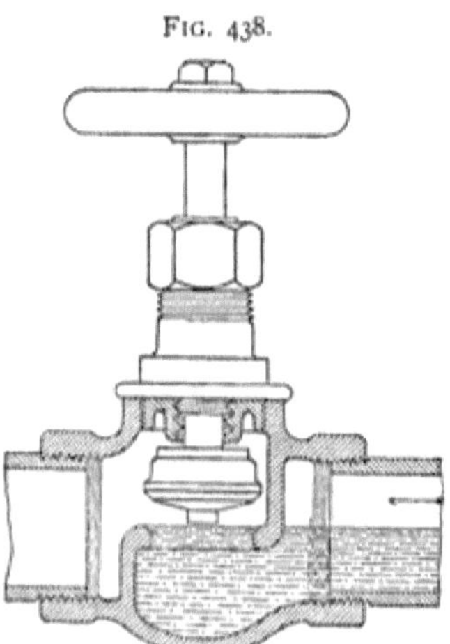

The Piping of Water-Columns, and, in fact, all the piping for a steam boiler where subject to an accumulation of sediment, should have crosses fitted with plugs at the turns rather than elbows. By the removal of a plug opposite the pipe likely to accumulate sediment, an iron bar or scraper can be forced

Fig. 439.

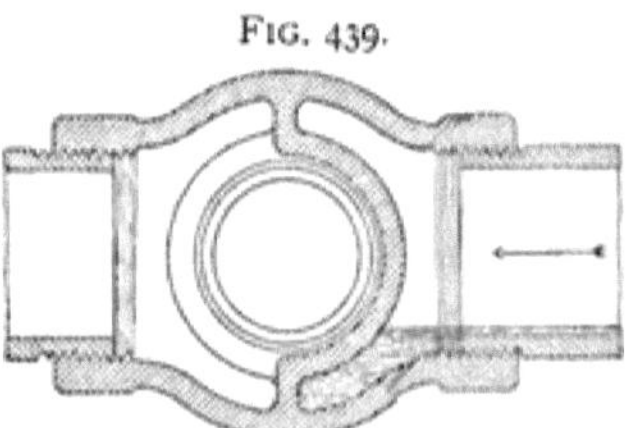

through and the pipe kept open. If globe valves are used, they should be placed with the spindle horizontal instead of vertical in all steam-pipes likely to trap the condensed water. The effect of this condensation and trapping is shown in Fig. 438, and the non-accumulation of water by placing the valve-spindle horizontal in Fig. 439.

A Steam-Pressure Gauge is an instrument for indicating the pounds pressure per square inch in a steam boiler above that of the atmosphere.

The Schæffer diaphragm pressure-gauge is shown in Fig. 440. This gauge is the pioneer of all spring gauges, being the first one to supersede the old-style mercury gauges. It consists of a corrugated steel diaphragm secured between two concaved flanges. The fluid pressure is admitted underneath this diaphragm, which will, by reason of the flexibility allowed by its corrugations, permit a central rise, the extent of which is determined by the amount of pressure and the stiffness of the diaphragm. A central rod connects this diaphragm with a sector attached to the cylindrical case above. This sector has teeth which engage a small pinion, the shaft of which has fastened to it one end of a hair-spring, the other end of the spring being fastened to the case. This hair-spring brings the pointer back to the zero mark on the dial when the pressure is removed. To this pinion-shaft is also attached a hand or pointer for indicating the pounds pressure per square inch on a circular dial, not shown in the engraving.

FIG. 440.

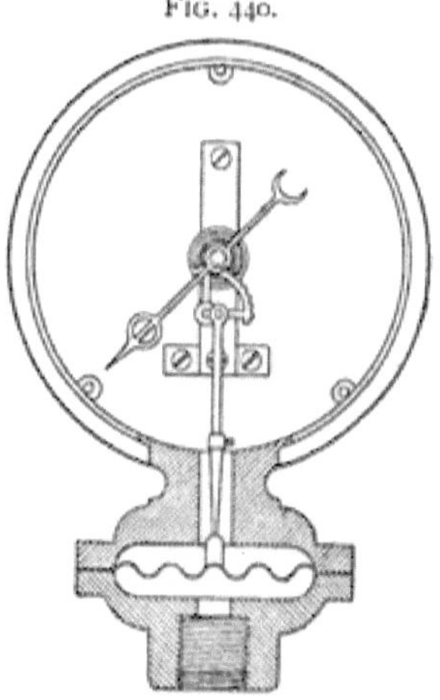

FIG. 441.

The commonest form of pressure-gauge now in use is a modification of the Bourdon gauge. This originally consisted of an oval tube bent as in Fig. 441. One end of this tube is fastened to a fixture at the bottom of the case, the other end is free to move according to the pressure within it. Steam-gauges thus constructed are objectionable in one respect,—the free end of the tube, nearly one-half of it, is below the line of drainage, which is liable to freeze and injure the tube; even if it does not burst it, the accuracy of the gauge will be seriously impaired, and, in consequence, it will be utterly worthless for the purpose intended.

Gauges are now made with two oval tubes, as in Fig. 442. As these tubes never exceed a semicircle, a perfect drainage is had at any time by simply opening the drainage-cock below the case. An important advantage is had in another respect. These tubes, being shorter, are less sensitive to vertical shocks, as in locomotive service. The lever and rack being attached to the free end of the tubes, and not pivoted to the case, the horizontal vibrations of the free ends of the tubes are not violently transmitted to the pointer, as they were when a single tube was employed and the quadrant pivoted to the case.

Each boiler should have its own steam-gauge. The connection should be with the steam-room of the boiler direct, and not with the steam-pipe leading to the engine. No steam-gauge should be used without a siphon between it and the boiler to protect the bent tubes from

FIG. 442. FIG. 443. FIG. 444.

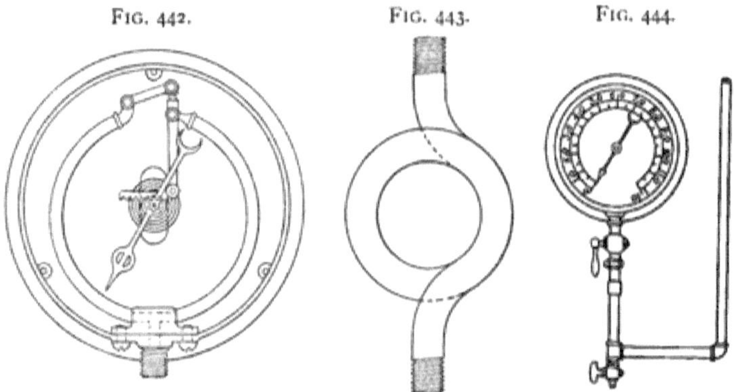

expansion by heat. The siphon interposes a body of water between the steam and the gauge which fully protects the latter from injury. The simplest form of siphon is a bent pipe, shown in Fig. 443. Such a pipe cannot drain dry. It must, therefore, be protected from the frost or fitted with a pet-cock to completely empty it when not in use in cold weather.

When a steam-gauge is piped, as shown in Fig. 444, the loop thus formed will fill with water of condensation and protect the gauge without a siphon attachment or the formation of other water-pocket. The insertion of a pet-cock in the lower opening of the bottom tee will afford complete drainage. There are a number of combined siphons and stopcocks in the market, of which Fig. 445 may be taken as representative of the class. The section shows a loose cap over the central pipe which extends into the chamber. This deflects the entering steam, which, condensing in the chamber, effectually prevents the live steam reaching the spring of the gauge. The cap over the pipe falls as the pressure is removed, making a siphon which empties the water from the chamber, thus preventing danger of bursting from the action of frost.

Shaw's mercury gauge is shown in sectional elevation in Fig. 446. The only piece of moving mechanism in the gauge is a double-headed piston inserted between two flexible diaphragms. This piston has ends of unequal areas. The larger piston is placed at the top, or measuring-chamber, the smallest one at the bottom, or entering-chamber, this latter communicating with the boiler or other vessel containing the fluid pressure to be measured. The steam pressure acts on the lower diaphragm, forcing it upward, carrying the double-headed piston with it. The larger end of this piston communicates to its diaphragm the same upward movement.

Fig. 446.

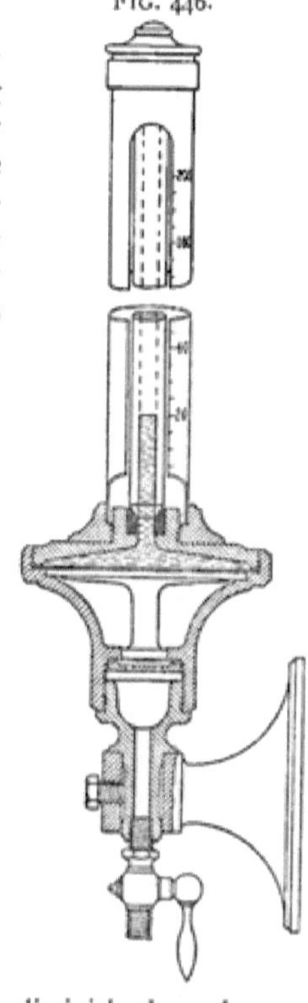

Fig. 445.

This upward movement has the effect to diminish the volume of the upper chamber, thereby forcing its contained mercury into the vertical glass tube, where it records measurements in pounds per square inch by means of a graduated scale, the divisions of which correspond to the applied pressures. It will be seen that the principle of action is that of differential areas, analogous to the short and long arms of a lever,—a difference of 10 to 1, for example, in area of the gauge-pistons, and their corresponding diaphragms give the same result as 10 to 1 in a lever. In either case the employment of one pound on the long end of the lever or on the larger area in the gauge will balance ten pounds on the short end of the lever or the small area of the gauge.

Dry Pipe.—Such a pipe is sometimes attached to the interior of stationary steam boilers not provided with a steam-dome or steam-drum. It consists of a pipe with closed ends, located inside of the boiler in the steam-space and close to the upper side of the shell, as in Fig. 447, which shows a dry pipe suspended from the shell, usually near the rear end of

FIG. 447.

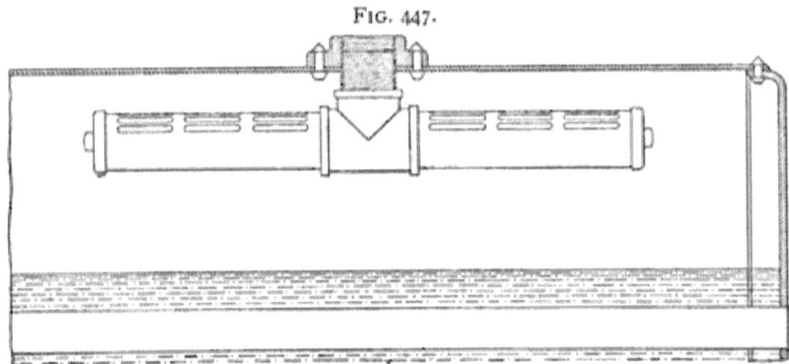

the boiler. Fig. 448 shows a dry pipe leading out of the boiler through one of the heads. The drawing does not show it, but the flanges at the head should be so bolted that the steam-pipe may be wholly detached without disturbing the flanged joint of the dry pipe next the boiler-head. There is no rule for either diameter or length, so that both dimensions are widely variable in practice, diameters varying from three to six inches and from one-eighth to seven-eighths that of the boiler

FIG. 448.

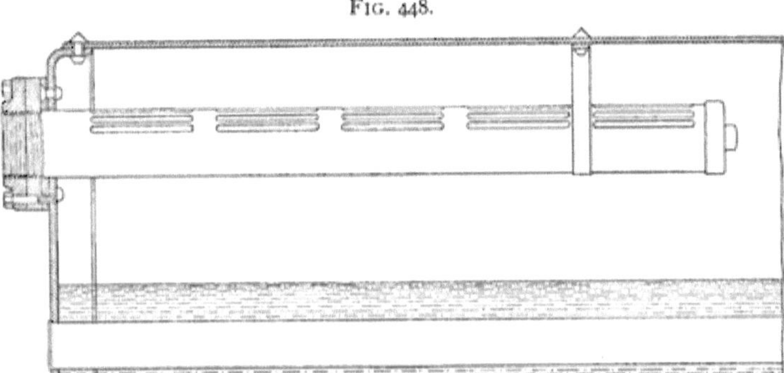

for length, but one-fourth the length of the boiler is not far from the average. The upper surface of the pipe is drilled or slotted with holes aggregating a little less than the area of the stop-valve ; one small hole should be drilled in the bottom at each end for drainage. By having a considerable number of holes distributed over a large pipe surface in a direction different from that taken by the rising water when priming, but little water will be carried over by the steam into this pipe, and dryer steam will be supplied than if this dry pipe was not attached.

Edgerton's separator for steam boilers is shown in Fig. 449. At any convenient place on the top of the boiler-shell is riveted a pipe-flange, in the lower half of which is a nipple extending into the boiler. On the bottom of this nipple is screwed the separator, which consists of three pans of the design shown in the engraving, the lower pan having its flange turned upward, the upper pan having its flange turned downward; intermediate between these two is the third pan. The distance apart at which these pans are to be set is fixed by wrought-iron ferrules on the outside of the bolts which hold the pans together. The course of the steam is around and downward over the flange of the lower pan; then, changing direction, it passes upward around and inside of the flange of the upper pan, and from thence to the steam-pipe and beyond. Small holes are drilled in the two lower pans, that any water of condensation may drop back into the boiler.

Fig. 449.

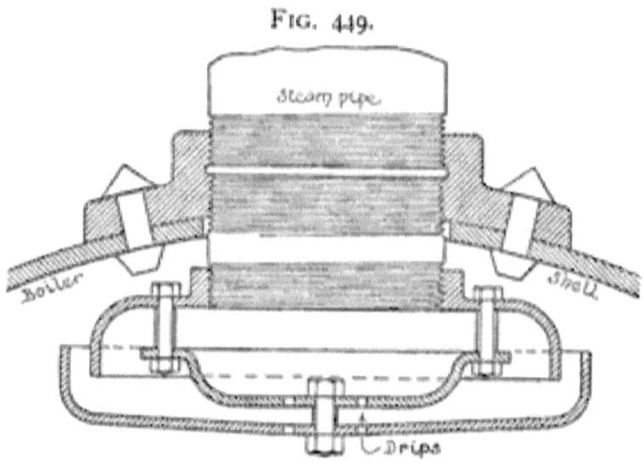

Steam Stop-Valves.—The diameter of steam outlets for horizontal tubular boilers is commonly 2 inches for a 36-inch boiler up to 6 inches for a 72-inch boiler. The least size of opening in a steam boiler should be such that the velocity of steam issuing through the steam-pipe shall not exceed 100 feet per second when the boilers are worked up to their limit.

A flange should be riveted to the top of the boiler-shell with which the steam-pipe is to be connected. This flange should be located preferably at least as far back as the centre or between the centre and the rear end of the boiler rather than immediately over the furnace. If a dry pipe is used, the steam outlet is, of course, wherever the dry-pipe outlet may be fixed by the designer of the boiler. The steam stop-valve should always be next to the boiler, so that if for any reason one boiler out of a battery of several is to be withdrawn from service, the valve will completely shut off the steam from such boiler.

A stop-valve must never be placed underneath a safety-valve if the closing of the former also closes communication with the latter; but if for any reason it is imperative that a stop-valve and safety-valve be connected with the same opening, then a combined valve, such as shown in Fig. 415, may be used, or a tee attached to the boiler with the safety-valve on the top and stop-valve on the horizontal branch.

Globe valves threaded for wrought-iron pipe are in very general use for steam-piping. These are usually of gun-metal for sizes less than four inches in diameter, and cast-iron bodies with gun-metal mountings for larger sizes. A globe valve is shown in section in Fig. 450, having a metal valve loosely attached to a screwed spindle, by which it may be opened or closed by turning the hand-wheel attached to the same spindle. The valve is loosely fitted to the spindle, that it shall be free to revolve around the stem and to secure a flexible joint favoring complete contact with the seat when screwed down. Leaky globe valves when fitted with hard-metal valves may be restored by regrinding. For grinding hard brass moderately fine emery may be used for the preliminary grinding, to be followed by and finished with powdered glass.

FIG. 450. FIG. 451.

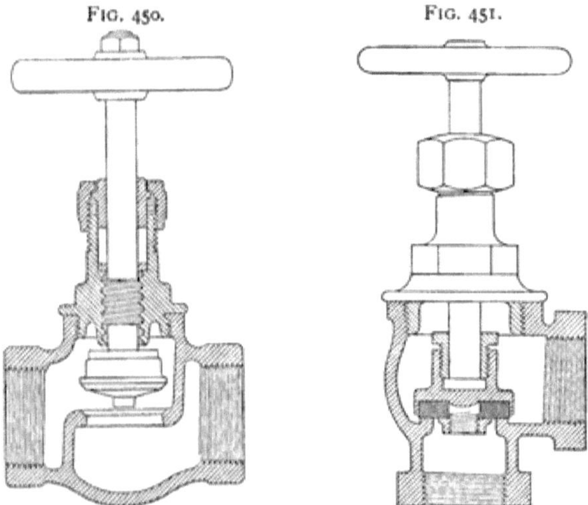

Soft metal disks, as well as those made of vulcanized rubber, are now much used as valve-faces for stop-valves. These have undergone the test of many years and have stood it well; such a disk is shown by double-hatching in the angle valve, Fig. 451. These disks, of hard rubber or soft metal, are renewable; a new disk can be easily applied by unscrewing the upper half of the valve, taking out the worn-out disk, and replacing it by a new one. The angle valve, just referred to, is a good form of valve to use on top of a boiler, because there are no pockets for the accumulation of water of condensation.

The Eastwood stop-valve, shown in Fig. 452, is not unlike an ordinary globe valve in its general features, but differs in the details of the valve-seat, which is bevelled outward. The valve-face is a copper disk fitting into a recess in the valve to give it a substantial backing. The inner face of this copper disk and the outer one of the valve-seat are ground to a tight joint. In the event of repairs incident to long wear or through any other cause a new valve-face is required, it can be easily furnished by unscrewing the old copper disk and inserting a new one to take its place.

Flanged stop-valves are not commonly employed for steam-pipes less than four inches in diameter. The body is ordinarily made of cast iron, into which is screwed a hard gun-metal seat. Such a valve is shown in Fig. 453. In this case the seat has radial arms, with a central

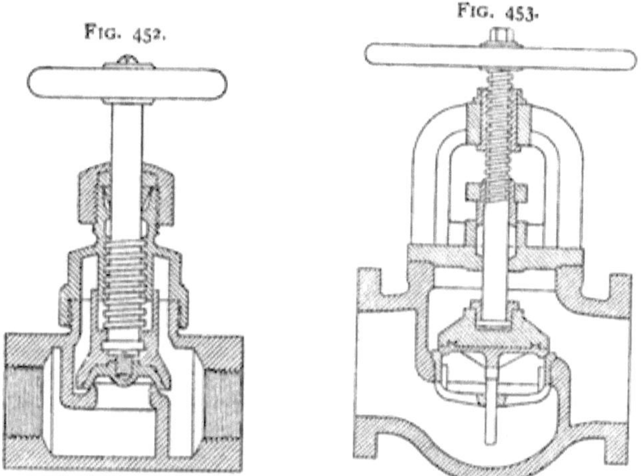

FIG. 452.

FIG. 453.

boss for guiding the spindle attached to the upper valve. The valve is made of hard gun-metal, reinforced by a cast-iron back, clearly shown in the illustration. The collar at the lower end of the spindle for raising and lowering the valve is convex, and, being loosely fitted in the back of the valve, with a gland above it, allows perfect freedom of movement between the stem, valve, and the valve-seat ; the valve can, therefore, be closed without any lateral strains. Large stop-valves are usually fitted with a yoke extending above the bonnet for the purpose of fixing a nut, through which a screwed spindle passes for operating the valve-disk within. This arrangement prevents any injurious action of the steam upon the screw-threads by removing them completely from the interior of the valve.

A combined stop- and check-valve for steam-pipe is shown in Fig. 454. This valve is so designed that the valve is closed when the pressure in the boiler is below that of the steam main, although it may be positively closed at all times if desired. This check-valve principle is, of course, a decided advantage in case of an accident to one of the boilers, the valve closing and preventing a loss of pressure in the mains and the liability of crippling the entire plant. Referring to the sectional view of the valve, it may be seen that the valve-disk is free to move within certain limits upon the stem. The upper part of the valve-disk is bored out and fitted with a cover which screws on. The valve-stem passes through a hole in the cover and through the lower part of the valve and into a spider-bearing or guide at the throat of the valve. A piston containing two rings is screwed upon the stem, the periphery of the rings and piston bearing against the inner surface of the valve-disk as shown to prevent it from chattering, the whole acting like a dash-pot.

FIG. 454.

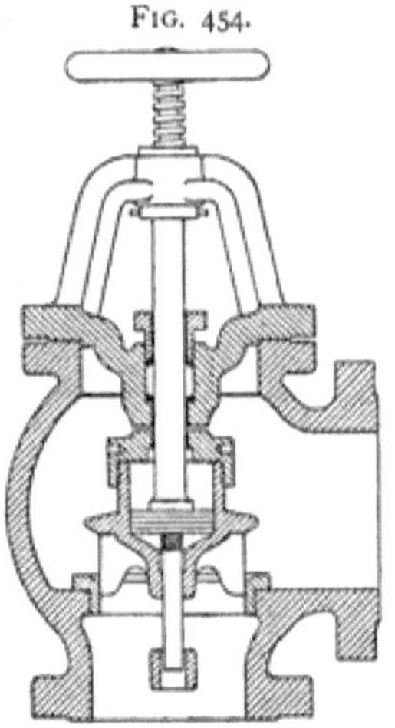

Expansion-Joints.—One effect of heat upon iron is to increase its volume; expansion, then, especially in the case of long steam-pipes, must not be overlooked. The expansion of iron pipe for each 100 feet in length from the freezing temperature will approximate as follows:

Temperature, Fahr.	32°	212°	259°	287°	307°	324°	335°
Steam pressure above atmosphere	0	0	20	40	60	80	100
Elongation in inches	0	1.44	1.82	2.04	2.20	2.34	2.42

It will be seen that the linear expansion of steam-pipes must be provided for or leaky joints, perhaps broken fittings, will result.

Steam-pipes fitted with expansion-joints should be securely anchored somewhere near the extreme ends of the pipe, leaving the centre free to expand and contract according to the temperature of the steam. The necessity for anchoring at each end is owing to the fact that the two ends of a common expansion-joint are not bolted together. The internal area of the pipe multiplied by the steam pressure gives that pressure exerted at each end of each pipe tending to force them apart.

FIG. 455.

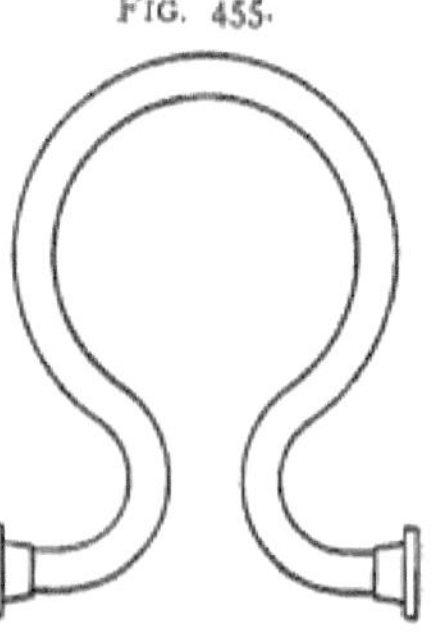

A copper pipe bent as shown in Fig. 455, with flanges joining the ends of wrought-iron pipe, is now used with satisfactory results. The distance from face to face of the flanges may be 3 feet for 2-inch pipe up to 7 feet for 6-inch pipe; the distance from the centre of the line of wrought-iron pipes to the centre of the bent copper pipe at its highest point in the bend may be 18 inches for a 2-inch pipe up to 48 inches for a 6-inch pipe. The radius for bending copper pipes should not be less than 12 inches for a 2-inch pipe nor less than 30 inches for a 6-inch pipe, and both of these should be greater if space will permit. Dimensions for pipes of intermediate diameters to the above can be had by simple interpolation.

The expansion-joint shown in Fig. 456 is probably in more general use than any other. The engraving shows flanged ends, and this detail

FIG. 456.

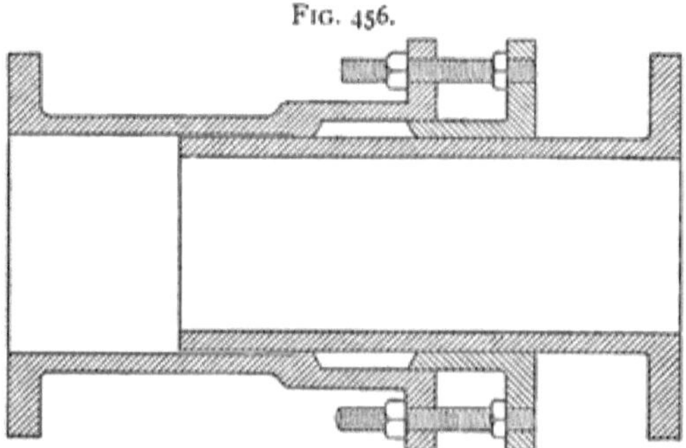

is recommended, though for sizes up to 2½ inches screwed ends are commonly used when the expansion-joint is made of brass. The ordinary traverse of the slip-joint for a 2-inch pipe is 2½ inches, and for a 6-inch pipe it is 5 inches. Special traverse greater than the above can be had when ordering an expansion-joint, but the above will cover all ordinary requirements.

Pearson's expansion-joint is shown in Fig. 457. The steam mains are anchored in two places, and at a point midway is placed the expansion-joint, which was designed to eliminate the strain on anchorages

FIG. 457.

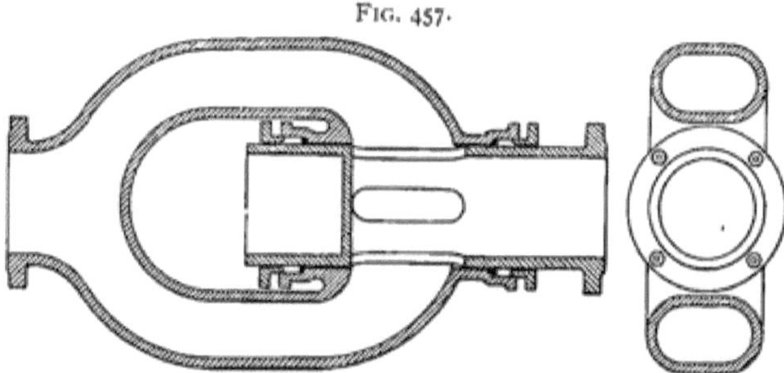

common to the usual form of cylindrical slip-joint, which the pressure on the ends of a pipe containing an ordinary expansion-joint tends to pull apart. The entering pipe in this joint slips through stuffing-boxes made in the ordinary manner. One end of this slip-pipe is open for connecting with the steam-pipe, the other end is fitted blank. Openings are made through the slip-pipe, as shown in the engraving, permitting a flow of steam into the branched pipe carrying the stuffing-boxes, the two branches being united at the further end into a common opening and flanged, to connect with the steam-pipe to any point beyond. As there are no free ends in this expansion-joint open to the atmosphere, it will be seen that this arrangement constitutes a balanced joint.

Smith's balanced expansion-joint, shown in Fig. 458, consists of three parts: First, a sliding pipe having two diameters, the larger diameter being of a short length, making it similar to a piston centrally located on a large hollow piston-rod. The sliding pipe corresponds to

FIG. 458.

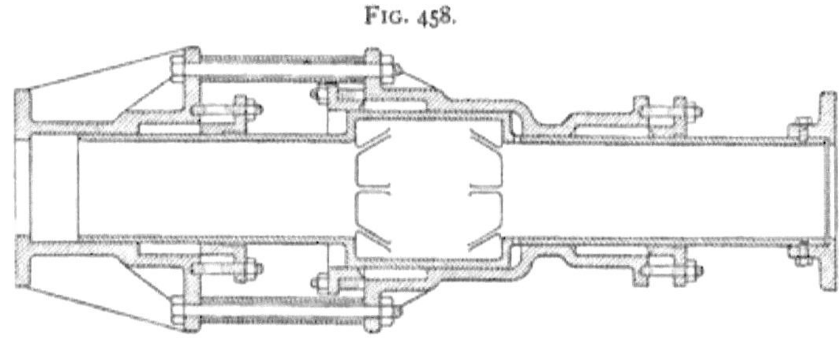

the normal diameter of the steam-pipe to which it is attached and of which it is a continuation. The larger or piston diameter is such that its area is just double that of the sliding pipe, measured by its outside diameter. Second, a sleeve bored to fit both the piston and the sliding pipe; this sleeve is provided with a stuffing-box at each end. There are openings in the sliding pipe by which steam has access behind the piston, the pressure tending to force the piston outward through the large stuffing-box. Third, immediately outside of the piston stuffing-box is another bored sleeve accurately fitting the sliding pipe. This sleeve is also provided with a stuffing-box, and is held by a suitable flange at a fixed distance from the opposite or piston stuffing-box by means of distance pieces made of iron pipe, through which pass the bolts for holding both sleeves rigidly together. The relative areas of these two openings into the atmosphere are as 2 to 1. The rear end of this sleeve is fitted with a flange for connection with steam-pipe.

In an ordinary expansion-joint the steam exerts a pressure equivalent to the outside area of the sliding pipe multiplied by the steam pressure. This must be resisted by an abutment somewhere in the pipe system, or the sliding pipe will be blown out of its stuffing-box. In the expansion-joint now under consideration the steam pressure is balanced, because the tendency of the piston to move in one direction is offset by that of the sliding pipe to move in the opposite direction, both areas being alike in extent and subjected to the same pressure, the two forces thus counterbalancing each other. This joint is thus free to move according to the expansion of the pipes to which it may be attached, but is in equilibrium so far as internal steam pressure is concerned. The movement, then, of the sliding pipe through the stuffing-boxes will only be such as is due to the expansion of the two pipes thus joined, which amount depends upon the temperature of the steam and the length of the pipe. The expansion-joint herein illustrated was designed for a working steam pressure of 225 pounds per square inch, the steam-pipe being 6 inches in diameter, the joint permitting a maximum travel of 3 inches.

Damper.—This may be located at any convenient place between the exit of the gases from the boiler setting and the chimney, or in some cases it is placed in the chimney itself. Two forms of dampers are shown. Fig. 459 represents the sliding form and Fig. 460 the butterfly

FIG. 459.

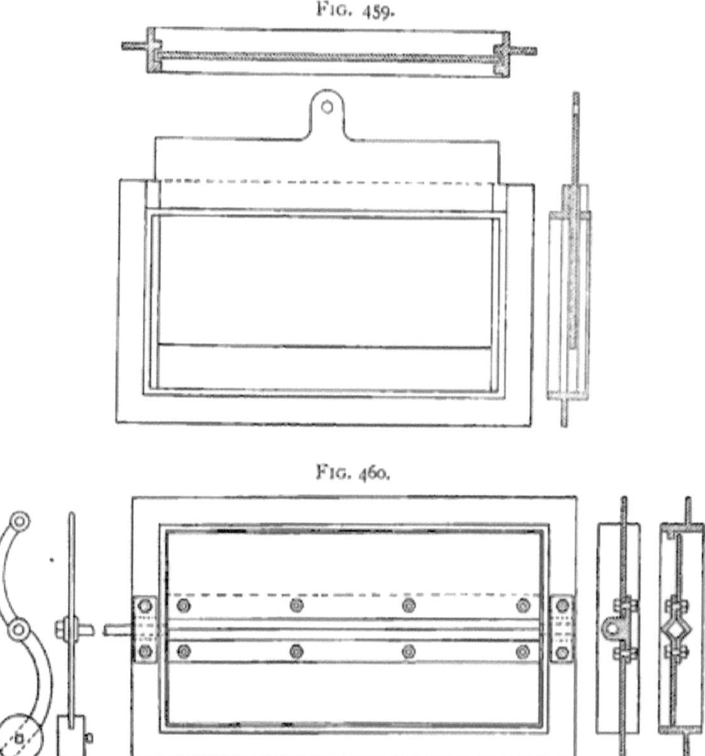

FIG. 460.

type. The latter is generally preferred because of its ease of operation, there being no sliding friction, consequently little liability for it to stick fast or become otherwise inoperative.

Regulating the furnace draught by a damper is better than doing it with the ash-pit doors, unless the latter are absolutely air-tight. The reason needs simply to be suggested. Combustion can only proceed so long as the fire receives fresh accessions of oxygen; by closing the damper the non-supporting products of combustion completely envelop the fuel and combustion stops; no air can reach the fire, because it cannot dislodge the gases there present. If, on the other hand, the damper be wide open, and the ash-pit doors as ordinarily constructed be closed, the chimney draught will carry off the products of combustion above the fire and form a partial vacuum in the ash-pit. The ash-pit doors, not being absolutely air-tight, permit a leakage, and combustion goes on the same as usual, except as to intensity, governed by the quantity of air thus admitted.

An Automatic Damper Regulator is advantageous if the automatic device is a good one; it will furnish a more uniform boiler pressure than is likely to be had by hand regulation. Automatic damper regulators are all constructed on one general principle, the boiler pressure acting upon a flexible diaphragm or a piston, which, by suitable connections, operates the damper, opening or closing it according to the steam pressure at which the damper regulator is intended to work. Among the earlier inventions for regulating the draught by the pressure of steam in the boiler was that of Clark (1854). This damper regulator was open to some objection and was improved upon some twenty years

FIG. 461.

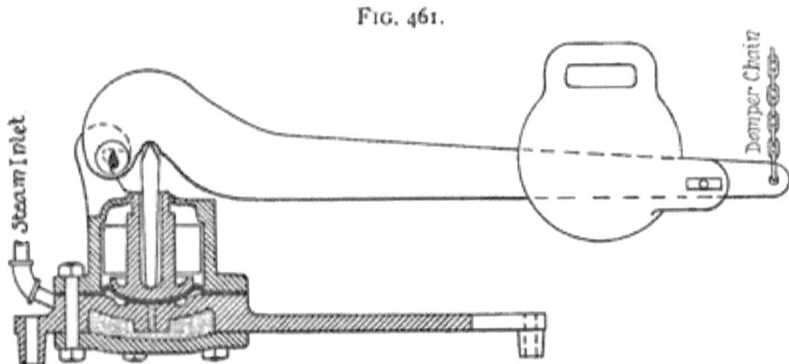

later by Le Van, whose regulator is shown in Fig. 461, the improvement consisting in a protection offered the flexible diaphragm from the injury which it sustained by the direct contact of the steam. There is combined with the cylinder or pressure-chamber of the diaphragm a water-chamber, a steam-supply pipe connected near its top, a water-pipe extending from the lower surface of the diaphragm to the interior of the water-chamber at a point below the opening of the steam-supply pipe.

In the operation of the device it is obvious that the water of condensation from the steam-supply pipe will be forced up into the projecting pipe and against the diaphragm by the steam pressure acting upon the surface of the water. This method of operation effectually prevents the direct contact of the steam with the diaphragm, because there is always a body of water interposed between the diaphragm and the steam. Should the pressure in the boiler rise above the desired limit, the weighted lever will rise, and by means of a chain or rope leading to the damper close the latter and keep it closed until the pressure of steam falls below its assigned limit, after which the lever will fall and, by reversing the direction of the chain or rope, will open the damper and permit a more rapid rate of combustion.

Hydraulic Damper-Regulator.—The engraving, Fig. 462, represents the Mason regulator for controlling a damper by the variation of boiler pressure, but the motive-power employed in opening or closing the damper is water pressure from the street main, from an overhead tank, or from the boiler itself. The advantage of using water pressure for damper regulation is the constant and non-variable movement obtained. In this regulator the steam from the boiler enters through a connection shown to the left of the base and passes into a chamber in the bottom of the base and under a heavy rubber diaphragm. The steam pressure forces the diaphragm upward; this upward pressure is counterbalanced by a heavy weight placed on the lower horizontal lever; this weight can be moved out or in to suit any boiler pressure. When the boiler pressure rises above the normal, at which the weighted lever is set, the lever is lifted upward, carrying with it a vertical valve-rod, shown attached to the weighted lever at the bottom and the com-

FIG. 462.

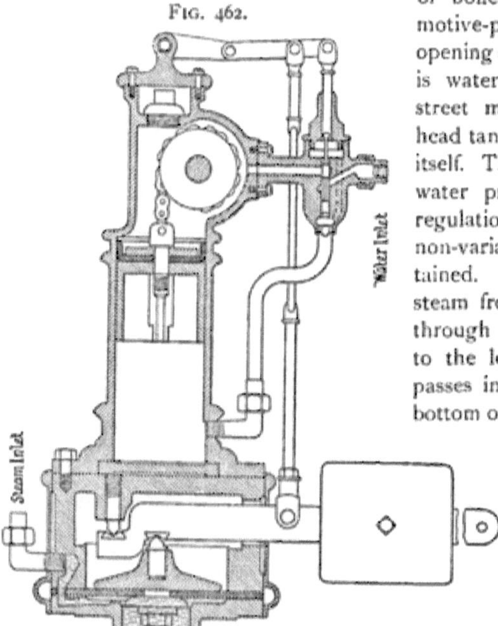

pensating lever at the top, this movement opening the water-valve and allowing the water pressure to pass through the ports of the valve-chamber to the top of the main piston in the cylinder forming the central part of this apparatus. This piston by its downward movement winds up a chain to which the damper is attached, thereby closing the damper. As the chain-wheel travels around, the compensating arm shown at the top of the apparatus is thrown outward by a cam attached to the chain-wheel. Raising one end of the lever fulcrumed on the small valve-rod at the top tends to close the port of the water-valve. This serves as a compensating arrangement, and does not allow the damper to be entirely closed on slight changes of pressure. The reverse action takes place when the boiler pressure drops and the water pressure is shut off. A weight is placed on the damper in the flue heavy enough to draw the wheel back again to its first position. The bottom of the water-valve chamber and the main cylinder are connected with a pipe which carries away whatever water remains after the piston has worked in either direction, also shown in the engravings.

FIG. 463.

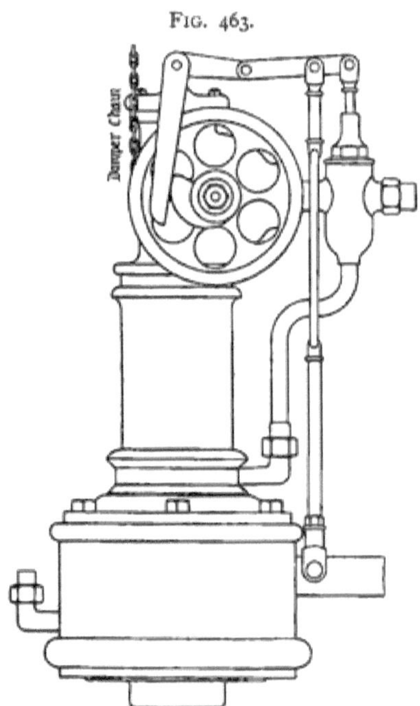

Whistle.—Nearly all lists of boiler furnishings include a steam whistle. Fig. 464 is a sectional drawing representative of steam whistles as a class. A whistle as shown consists of a sounding-bell attached to a spindle by the base, in which is a thin annular orifice for the escape of the steam into the atmosphere, and a valve, usually included in the base, for regulating or controlling the pressure of steam escaping from the base. The valve is opened by the action of the lever forcing it off its seat against the steam pressure. The spring on the back of the valve keeps the latter closed and counterbalances any tendency of the lever to keep the valve open when there is no pressure behind it. The central spindle rising vertically from the base has screw-threads at its top, on which is screwed the sounding-bell. The object of this screw is to secure a vertical adjustment of the bell suited to the steam pressure and tone desired.

The sound from a steam whistle is produced by vibrations set up in the bell incident to the action of the steam escaping into the atmosphere from a thin annular opening in the base of the whistle. A vibrating body, before it can act as a sounding body, must produce alternate compression and rarefaction in the air, and these must be well marked. Relatively deep, grave sounds are produced by slower vibrations. The larger the diameter and the greater the height of the bell, the deeper or lower will be the note sounded. Small diameters and short heights of bell produce a shrill note, and between these two may be had any tone desired.

Fig. 464.

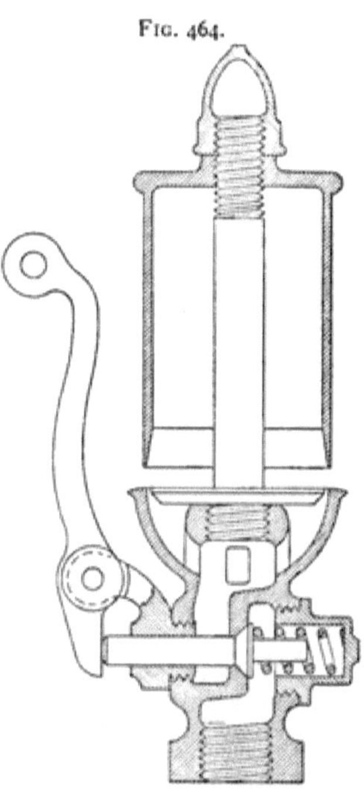

Malleable - Iron Unions are much used for coupling together such wrought-iron pipes as are likely to afterwards require disconnecting. The sectional elevation, Fig. 465, shows an ordinary union. It consists of a collared fitting which screws on one of the pipes to be joined. On this fitting is an internally threaded coupling free to revolve about the under side of the collar, against which it makes close contact. The upper fitting is internally and externally threaded, the inside thread for the pipe it is to connect, the outside thread fitting into that of the loose coupling below. A vulcanized rubber or soft metal washer makes a flexible seating for the two halves of the union, which are drawn together by the loose coupling, the exterior of the latter being provided with hexagon or octagon sides for the use of a wrench.

Fig. 465. Fig. 466. Fig. 467.

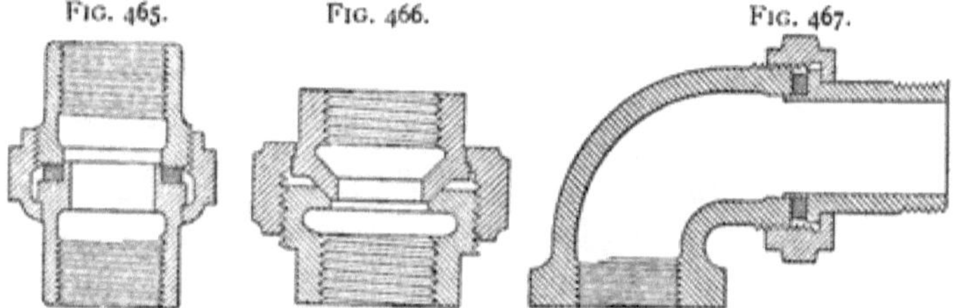

The Eastwood ground-joint union is shown in Fig. 466. This union has a conical joint which can be ground tight and dispense with packing.

An elbow union is shown in Fig. 467. It is much used in steam-heating practice, and deserves a wider adoption in the smaller sizes of steam- and water-piping, such as for pumps, injectors, etc.

Steam-Pipe.—Lap-welded wrought-iron pipes with cast-iron fittings are almost exclusively employed in the piping of steam-plants. Such pipes are made in all needed diameters and are uniform in all working dimensions. The following, Table L., gives the standard dimensions for nominal diameters from ⅛ inch to 12 inches :

TABLE L.

STANDARD DIMENSIONS OF WROUGHT-IRON PIPE.

Ordinary Rating.	Diameters. Internal.	Diameters. External.	Thickness.	Circumference. Internal.	Circumference. External.	Area. Internal.	Area. External.	Length for 1 Cubic Foot, Capacity.	Weight of Pipe per Foot.	Threads. Number per Inch.	Threads. Length of Perfect Screw.	Threads. Whole Length of Thread.
Ins.	Ins.	Ins.	Ins.	Ins.	Ins.	Sq. In.	Sq. In.	Feet.	Lbs.		Ins.	Ins.
⅛	.27	.41	.068	.85	1.27	.06	.13	2500	.24	27	.19	⅜
¼	.36	.54	.088	1.14	1.70	.10	.23	1385	.42	18	.29	½
⅜	.49	.68	.091	1.55	2.12	.19	.36	751.5	.56	18	.30	⅝
½	.62	.84	.109	1.96	2.65	.31	.55	472.4	.85	14	.39	¾
¾	.82	1.05	.113	2.59	3.30	.53	.87	270.0	1.13	14	.40	⅞
1	1.05	1.32	.134	3.29	4.13	.86	1.36	166.9	1.67	11½	.51	13/16
1¼	1.38	1.66	.140	4.34	5.22	1.50	2.16	96.25	2.26	11½	.54	1
1½	1.61	1.90	.145	5.06	5.97	2.04	2.84	70.65	2.69	11½	.55	1 1/16
2	2.07	2.38	.154	6.49	7.46	3.36	4.43	42.36	3.67	11½	.58	1⅛
2½	2.47	2.88	.204	7.75	9.03	4.78	6.49	30.11	5.77	8	.89	1 3/16
3	3.07	3.50	.217	9.64	11.00	7.39	9.62	19.49	7.55	8	.95	1⅜
3½	3.55	4.00	.226	11.15	12.57	9.89	12.57	14.56	9.06	8	1.00	1⅜
4	4.03	4.50	.237	12.65	14.14	12.73	15.90	11.31	10.73	8	1.05	1 7/16
4½	4.51	5.00	.246	14.15	15.71	15.94	19.64	9.03	12.49	8	1.10	1 7/16
5	5.05	5.56	.259	15.85	17.48	19.99	24.30	7.20	14.56	8	1.16	1 9/16
6	6.07	6.63	.280	19.05	20.81	28.89	34.47	4.98	18.77	8	1.26	1⅝
7	7.02	7.63	.301	22.06	23.95	38.74	45.66	3.72	23.41	8	1.36	1¾
8	7.98	8.63	.322	25.08	27.10	50.04	58.43	2.88	28.35	8	1.46	1⅞
9	8.93	9.63	.348	28.07	30.24	62.73	72.76	2.26	33.70	8	1.57	2
10	10.02	10.75	.366	31.48	33.77	78.84	90.76	1.80	40.64	8	1.68	2⅛
11	11.22	12.00	.388	35.26	37.70	98.94	113.10	1.46	47.73	8	1.80	2¼
12	12.18	13.00	.410	38.26	40.84	116.54	132.73	1.24	54.66	8	1.88	2⅜

Pipes 1 inch and below are butt-welded and proved to 300 pounds per square inch hydraulic pressure.

Pipes 1¼ inches and larger diameters are lap-welded and proved to 500 pounds per square inch hydraulic pressure.

Threads for wrought-iron pipes are cut tapering at ¾ inch per foot of total taper for all sizes up to and including 9 inches ; for pipes 10 inches in diameter and larger the taper is ⅜ inch per foot.

The table gives dimensions of 11-inch pipe ; this is regarded in the trade as an odd size, and fittings are rarely, if ever, made for it except to order. The common practice is to ignore it and take 12 inch pipe instead.

Flange Unions.—No uniformity of dimensions is followed by manufacturers of cast-iron flanges for wrought-iron pipe. The differences may not be great, but they are sufficient to prevent interchange of parts. Two sizes of flanges are commonly necessary in putting up a line of large steam-pipe,—one set of flanges to match the flanged valves and another set for ordinary connections, the valve-flanges being commonly of larger diameter than the size usually adopted for merely connecting one pipe with another. The illustration, Fig. 468, and the tabular dimensions accompanying relate to pipe-connections only. Valve-flanges and cast-iron connections differ so much in diameter that special flanges usually have to be made for them.

FIG. 468.

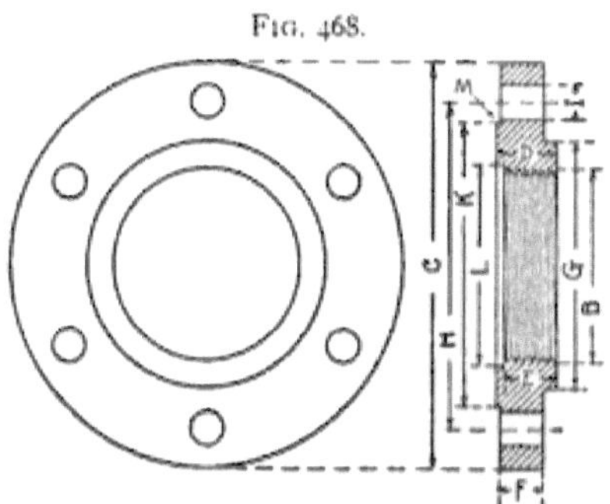

TABLE LI.

FLANGE UNIONS.

Reference letters correspond to those shown in Fig. 468. M, 1/16-inch for all sizes up to 6 inch, and 1/8-inch for larger pipes.

A.	B.	C.	D.	E.	F.	G.	H.	I.	J.	K.	L.
Nominal Diameter of Pipe.	Outside Diameter of Pipe.	Diameter of Flange	Depth of Flange.	Depth of Thread.	Thickness of Flange.	Diameter of Boss.	Bolt Circle.	Bolt Hole.	Number of Bolts.	Diameter of Joint, outside.	Diameter of Joint, inside.
Ins.	Ins.	Ins.	Ins.	Ins.	Ins.	Ins.	Ins.	Ins.		Ins.	Ins.
1	1.3	5	5/8	3/4	5/8	2 1/8	3 3/8	[illegible]	3	2 3/4	1 1/2
1 1/4	1.7	5 1/8	7/8	3/4	5/8	2 1/2	4	[illegible]	3	3 1/8	1 7/8
1 1/2	1.9	5 3/4	1	7/8	3/4	2 7/8	4 3/8	[illegible]	4	3 1/2	2 1/8
2	2.4	6 1/2	1 1/8	1	7/8	3 3/8	5	[illegible]	4	4 1/4	2 1/2
2 1/2	2.9	7 1/4	1 1/8	1	7/8	4	5 3/4	[illegible]	4	5	3 1/8
3	3.5	7 3/4	1 1/4	1 1/8	1	4 5/8	6 1/4	[illegible]	6	5 1/2	3 3/8
3 1/2	4.0	8 1/2	1 1/4	1 1/8	1	5 1/4	7	[illegible]	6	6 1/4	4 1/8
4	4.5	9 1/2	1 3/8	1 1/4	1	5 3/4	7 5/8	[illegible]	6	6 3/4	4 5/8
4 1/2	5.0	10 1/4	1 3/8	1 1/4	1	6 3/8	8 3/8	[illegible]	6	7 1/2	5 1/8
5	5.6	11 1/4	1 1/2	1 1/4	1 1/8	7	9	[illegible]	6	8	5 3/4
6	6.6	12 1/2	1 5/8	1 3/8	1 1/8	8 1/8	10 1/4	[illegible]	6	9 1/4	6 3/4
7	7.6	13 1/2	1 3/4	1 1/2	1 1/4	9 1/4	11 1/4	[illegible]	6	10 1/4	7 3/4
8	8.6	14 3/4	1 7/8	1 5/8	1 3/8	10 3/8	12 1/2	[illegible]	8	11 1/2	8 3/4
9	9.6	15 3/4	2	1 3/4	1 1/2	11 1/2	13 1/2	[illegible]	8	12 3/8	9 3/4
10	10.8	17 1/4	2 1/4	2	1 5/8	12 7/8	15	[illegible]	10	14	11
12	13.0	20	2 1/2	2 1/4	1 3/4	15 1/2	17 1/2	1 [illegible]	12	16 1/4	13 1/4

CHAPTER XI.

CHIMNEYS.

A CHIMNEY is employed in steam engineering as an attachment to a steam-boiler furnace for the purpose of creating and maintaining a draught through the body of burning fuel. Its practical value as compared with other methods is in the certainty and simplicity of its action, adapting itself automatically to the demands of a lesser or greater number of furnaces, with little or no loss of efficiency for either under- or overload as compared with its normal power-rating.

The fuels used for steaming purposes in this country are anthracite, semi-anthracite, and bituminous coals. The first of these is a very hard coal, and requires a powerful draught for its rapid combustion. Bituminous coals are found in every grade of quality from good to bad; they burn readily, and thus a less intensity of draught is needed. It is for this reason that a less height of chimney is permissible.

The rate of combustion per square foot of grate surface per hour will ordinarily vary from 9 to 13 pounds for anthracite coal, and from 12 to 20 pounds for bituminous coal,—the latter occasionally running much higher, nearly or quite doubling these figures. We shall not go far wrong if we assume an average rate of combustion of 12 pounds per hour for anthracite coal and 15 pounds per hour for bituminous coal in ordinary steam-boiler furnaces with good draught. The evaporative economy falls off rapidly under the higher rates of combustion of bituminous coal.

The temperature of air supplied boiler furnaces may vary in certain localities from 10° below zero in winter to 95° Fahr. in summer. Assuming 62° Fahr. to be an average temperature, the corresponding weight of air will be 0.076 pounds per cubic foot, or 13.14 cubic feet to the pound at atmospheric pressure (14.7 pounds); in relation to water its density is about 1 to 820. If the combustion were perfect, with the minimum supply of air (12 pounds) the weight of the products of combustion would be 13 pounds, but for chimney calculations it is customary to assume the maximum of air supply, or 25 pounds of gas.

Chimney Draught is an effect produced by the difference in specific gravity of the cold air entering the furnace under the grate and the heated products of combustion escaping from the chimney. The difference in weight is explained by the fact that gases expand by heat; the atmosphere of an active fire must, therefore, be of less density than the outer air.

The draught of a chimney depends upon its height, and the difference between the specific gravity of the gases with which it is filled and that of the outside air. Draught properly begins at the level where the air passes through the fire, and not at the level of the ground at the base of the chimney. Its action increases at first with the temperature, but afterwards gradually diminishes with the temperature of the gases. For example, from 32° Fahr. to 300° Fahr. the draught augments very rapidly; from 300° Fahr. to 750° Fahr. the draught varies little, but arrives at the maximum when the temperature is about 585° Fahr. Some investigators place the temperature higher than this, but for steam-boiler purposes this temperature is as high as the products of combustion should be allowed to attain, and from 550° to 600° Fahr. may be taken as the maximum temperature in any ordinary calculations. Some steam boilers do not part with their gases until the temperature is reduced to 450° Fahr., and only occasionally is the reported temperature less.

Intensity of draught denotes the velocity of flow of air through the furnace. This property is secured by height of chimney, or by high temperatures of escaping gases, or both. The larger of the small sizes of anthracite coal require an intensity of draught corresponding to not less than seven-eighths to one inch of water, and even this is scarcely enough at times to secure the desired rate of combustion; whereas with ordinary free-burning coals in lumps as large as a hickory-nut or larger a draught of three-eighths to five-eighths inch of water will suffice for ordinary boiler settings.

Chimney Draught from Absolute Zero.—The best practice when dealing with problems relating to heated gases is to reckon temperatures from absolute zero (which on the Fahrenheit scale is —460°), because the expansion of a gas due to a given rise of temperature is found to be exactly proportional to the absolute temperature.

The best chimney draught is had (reckoning temperature from absolute zero) when the temperature of gas in the chimney is to that of external air as 1 is to 2. The absolute temperature of air at 62° Fahr. is $62 + 460 = 522°$; therefore we get the best draught when $522 \times 2 = 1044°$ absolute, or $1044° - 460° = 584°$ Fahr., a temperature above the melting-point of bismuth (518° Fahr.) and below that of lead (630° Fahr.). In general, when the escaping gases leave the boiler at a temperature of melting lead it indicates that heat is going to waste.

Example: Suppose a chimney 120 feet high, the temperature of escaping gases = 584° Fahr. and that of the atmosphere 62° Fahr., the draught in inches of water may be found thus:

$$120 \times \frac{460 + 584}{460 + 62} = \frac{125,280}{522} = 240 \text{ feet.}$$

The height of a column of escaping gas at 584° Fahr. equals the weight of a column of air at 62° Fahr. outside a chimney 120 feet high; then $240 - 120 = 120$. We will call this remainder the "motive column."

The relation of weight as compared with water (820 times heavier) may be expressed thus:

$$\frac{820 \times 240}{120} = 1640.$$

If we divide the motive column by this amount we have

$$\frac{120}{1640} = .0732 \text{ foot},$$

or ⅞ inch nearly, as the height of a column of water lifted by the action of a chimney corresponding to the height and temperature here given.

Temperature.—If we assume that air entering the ash-pit is increased in temperature from 62° Fahr. to 2500° Fahr. in its passage through the bed of burning fuel, its volume will have been increased 5.663 times; that is, the volume occupied by 1 pound of air at 62° Fahr. is 13.14 cubic feet and would be expanded to 74.40 cubic feet at the increased temperature; or, the weight of 1 cubic foot would be less in the proportion of 0.0761 to 0.0134 pounds. This sudden expansion, occurring as it does within a distance of 6 inches or thereabouts, and probably in less than a half second of time, means the acceleration of its velocity, due to increase in volume alone, from 1 foot below the grates to more than 5½ feet per second at the surface of the fire.

This sudden rise in temperature does not immediately affect the draught, for it is confined to the furnace alone, and this is always partially, and sometimes wholly, surrounded by absorbing surfaces, so that after the passage of the heated gases along the absorbing surfaces the temperature of the escaping products of combustion is lowered from the original temperature to about ¼ or less at the moment of entering the chimney. It is this final temperature only which is to be taken into account in draught calculations. This continual loss of heat by coming in contact with absorbing surfaces has the effect to retard the flow of gases from the furnaces to the chimney, but another and more serious retardation of flow of air into the furnace occurs in the passage of air through the fire itself, and this is greatly intensified in the case of fine anthracite coals or of caking bituminous coals, the former often requiring a strong and positive blast to force the air through the fuel. With the latter fuel a frequent breaking up of the fire becomes almost a necessity to maintain a proper rate of combustion. For natural draught the temperature of the escaping gases should not generally exceed 550° Fahr.; all things considered, this is regarded as a fair and economical working temperature. Variations above and below this temperature, say 50° Fahr., have but little effect on the draught so far as the economical result of furnace working is concerned.

An approximate method of determining high temperatures consists in suspending strips of metal, the melting-points of which are known, in the flues leading to the chimney, or perhaps in the chimney itself.

Three different metals, such as tin, which melts at 455° Fahr., bismuth, which melts at 518° Fahr., and lead, which melts at 630° Fahr., to which may be added zinc, which melts at 793° Fahr., and antimony, which melts at 810° Fahr., are generally employed in making such tests. In the selection of metals one is chosen of which the melting-point is nearest the supposed temperature, then another piece the melting-point of which is below, and a third of which the melting-point is above it. For example, if the temperature be supposed to be about 600° Fahr., then lead will be chosen for the central strip, and one of bismuth for the lower, and one of zinc for the higher temperature. The temperature can readily be approximated when it is known that it is between two of the three metals exposed. If the metal having the lowest melting-point is melted and the other two are uninjured by the heat, the temperature must be assumed to be above the one and below the other; similarly, if two of the bars are melted, the temperature must then be somewhere between the highest and the intermediate points of fusion. This method is not entirely accurate, but sufficiently so for all ordinary purposes.

The Area of a Chimney was made the subject of experimental research by Mr. Isherwood, who found that the ordinary variable limits were from one-sixth to one-ninth that of the grate area. Upon these experiments is based the common recommendation that the area of a chimney be one-eighth that of the grate surface.

The grate area for a steam boiler bears an approximately fixed relation to that of the heating surface, averaging not far from one-thirty-fifth that of the latter.

The chimney area must bear some relation to the quantity of coal burnt. In practice it is found that for sizes up to 1000 horse-power the most satisfactory chimneys are those in which from 1½ to 2 square inches of chimney area are had for each pound of coal burnt per hour. According to this rule a chimney suitable for 1000 pounds of coal per hour would vary between $1000 \times 1.5 = 1500$ square inches, or 43¾ inches diameter, and $1000 \times 2 = 2000$ square inches, or 50½ inches diameter.

Example: Required a grate and chimney area suitable for 1000 pounds of coal per hour, the rate of combustion to be 12 pounds per square foot of grate per hour:

$$\frac{1000}{12} = 83.3 \text{ grate area in square feet.}$$

$$\frac{83.3}{8} = 10.4 \text{ square feet} = 1498 \text{ square inches} = 43\tfrac{3}{4} \text{ inches, diameter of chimney,}$$

which corresponds to 1½ square inches of chimney area for each pound of coal burnt per hour. It also corresponds to one-eighth that of the grate area.

Height of Chimneys.—In order to give the required draught to a steam-boiler furnace of say 50 horse-power, the chimney should not be less than 60 or 65 feet in height above the ground surface at its base, and need not exceed 150 feet for boiler furnaces of 1000 horse-power, unless there is higher land or buildings in the immediate neighborhood which would affect its draught. Anthracite coals require a higher chimney than free-burning bituminous coals. The area having been fixed, either by quantity of coal burned or in proportion to grate area, a common rule is to make the height of a small chimney twenty-five times its diameter, with a gradual decrease in the ratio for larger chimneys; thus, a 4-foot chimney may be 100 feet in height, a 5-foot chimney 120 feet in height, a 6-foot chimney 135 feet in height, an 8-foot chimney 160 feet, and a 10-foot chimney 175 feet.

The formulas usually given for the height of chimneys do not always furnish the proportions which certain localities seem to require, and as a result there is more of empiricism than calculation in this detail. Many large chimneys are higher than necessary when considered merely as a means of obtaining or maintaining a good draught. It is doubtful whether there are any advantages sufficient to pay the extra cost for making a steam-boiler chimney more than 150 feet high for internal diameters less than 100 inches.

Horse-Power of a Chimney.—By this is meant a chimney of suitable diameter and height for the grate surface required for the combustion of fuel necessary to develop the stated amount of horse-power in steam boilers of well-known types. The standard horse-power in steam engineering is the evaporation of 30 pounds of water per hour from a feed-water temperature of 100° Fahr. into steam at 70 pounds gauge pressure. This is equivalent to 34½ pounds of water evaporated per hour from and at 212° Fahr.

For the chimney calculations in Table LII., the tabular dimensions of horizontal tubular boilers given in Tables XL., XLII., and XLIV. are assumed to meet all the requirements of ordinary single-cylinder engines, non-condensing and non-compound. For single-cylinder condensing and for compound engines the boiler-power will be somewhat in excess of the requirements of the engines.

The diameter of chimney given for each power is ample for any fuel, whether anthracite or bituminous coal. No length of side is given for a square chimney of equal area because the same diameter should be used for both. The corners of a square chimney, especially for those below 60 inches diameter, count for very little, the general configuration of flow of gases upward is in a round column. It is doubtful if the corners serve any useful purpose, and may be wholly neglected in computing chimney area.

The height of chimneys for the several powers given will be found suitable for the stipulated grades of bituminous coal and for anthracite coal not finer than buckwheat size. If finer coals are used the fire may require assistance, which can be furnished by either a steam jet or fan blower.

TABLE LII.

TABLE OF CHIMNEY DIMENSIONS FOR A SINGLE BOILER AND FURNACE FROM 20 TO 100 HORSE-POWER, BASED UPON THE TABULAR DIMENSIONS GIVEN IN TABLES XL., XLII., AND XLIV., IN WHICH THE GRATE AREA IS ASSUMED TO BE NINE TIMES THAT OF THE TUBE AREA FOR THE SMALLEST BOILER, DIMINISHING TO SEVEN TIMES THE TUBE AREA FOR THE LARGEST BOILER. A COMMERCIAL HORSE-POWER RATING APPROXIMATING 15 SQUARE FEET OF HEATING SURFACE PER HORSE-POWER IS ASSUMED FOR ALL BOILERS IN THIS SERIES.

Horse-Power.	Reference Number.	Boiler and Furnace Details.								Chimney.	
		Boilers.		Total Heating Surface.	Grate.		Chimney Area.	Nearest Diameter, Round Chimney.	Height in Feet.	
		Diameter.	Length.		Area.	Ratio to Heating Surface.			Bituminous Coal, free burning.	Anthracite Coal, small sizes.
		Inches.	Feet.	Sq. Feet.	Sq. Feet.		Sq. Feet.	Inches.		
20	2	38	10	311	12.15	25.60	2.02	20	50	60
25	3	40	12	396	12.87	30.75	2.15	20	50	60
30	4	42	14	488	13.68	35.69	2.28	20	55	65
35	5	44	14	515	14.40	35.73	2.40	22	55	65
40	7	48	14	611	17.46	35.01	2.91	24	55	70
45	8	50	14	682	19.71	34.58	3.29	25	55	70
50	10	54	14	778	22.77	34.19	3.67	26	60	70
55	11	56	14	827	24.30	34.03	3.80	27	60	75
60	19	58	16	908	24.16	37.58	3.80	27	60	75
65	20	60	16	971	26.00	37.36	3.94	27	60	80
70	27	60	18	1055	28.72	36.85	4.35	28	65	80
75	29	64	18	1142	31.12	36.70	4.58	29	65	80
80	30	66	18	1242	34.16	36.36	4.88	30	65	85
85	30	66	18	1242	34.16	36.36	4.88	30	65	85
90	31	68	18	1361	37.84	35.76	5.00	30	70	90
95	33	72	16	1421	39.55	35.98	5.00	30	70	90
100	33	72	18	1598	39.55	40.48	5.00	30	70	90

The diameter for chimneys from 100 to 1000 horse-power, as given in Table LIII., is estimated from an entirely different stand-point from that of Table LII. It is here assumed that 4 pounds of coal per horse-power per hour may be required, and for large boiler plants this quantity is correct for anthracite coal, and averages tolerably close for bituminous coals, inasmuch as the latter varies widely in evaporative power. The rate of combustion, assumed to be 12 pounds per square foot of grate surface per hour, is, for steam plants for the last half of the table, somewhat less than the average for bituminous coal. There are steam-boiler furnaces burning double that weight, and occasionally more than three times the assumed rate ; but as capacity tests and economy tests are often widely at variance with each other, no opinion can be here expressed upon what might be considered excessive rates of combustion unless the ratio of grate to heating surface is known, and especially the temperature of the escaping gases. If the temperature of the latter be higher than 600° Fahr., heat is going to waste, and the rate of combustion ought to be lowered until the temperature is reduced to a more economical figure. This table, in common with the preceding one, makes no difference in the diameter of a round chimney and that of a square chimney to make the areas equal, for the reason given on page 394.

TABLE LIII.

TABLE OF CHIMNEY DIMENSIONS FOR TWO OR MORE BOILERS SET IN BATTERY AND WORKING TOGETHER; HORSE-POWER IS BASED ON 4 POUNDS OF COAL PER HOUR PER HORSE-POWER. THE RATE OF COMBUSTION IS ASSUMED TO BE 12 POUNDS PER SQUARE FOOT OF GRATE SURFACE PER HOUR; THE PROPORTION OF GRATE TO CHIMNEY AREA VARIES FROM ONE-SEVENTH FOR THE 100 HORSE-POWER BOILER TO ONE-TENTH FOR THE 1000 HORSE-POWER BOILER.

Horse-Power.	Coal per Hour.	Area of Grate at 12 Pounds per Square Foot per Hour.	Area of Chimney.	Nearest Diameter, Round Chimney.	Height in Feet.	
					Bituminous Coal, free burning.	Anthracite Coal, small sizes.
		Square Feet.	Square Feet.	Inches.		
100	400	33.33	4.76	30	70	90
125	500	41.67	5.18	31	75	90
150	600	50.00	6.82	36	75	95
175	700	58.33	7.78	38	80	100
200	800	66.67	8.69	40	80	100
250	1000	83.33	10.64	44	85	105
300	1200	100.00	12.50	48	85	105
350	1400	116.67	14.18	51	90	110
400	1600	133.33	16.00	55	90	115
450	1800	150.00	17.65	57	90	115
500	2000	166.67	19.25	60	95	120
550	2200	183.33	20.65	62	95	120
600	2400	200.00	22.22	64	100	125
650	2600	216.67	23.65	66	100	125
700	2800	233.33	25.01	68	105	130
750	3000	250.00	26.32	70	105	135
800	3200	266.67	27.61	72	110	135
850	3400	283.33	28.82	73	110	140
900	3600	300.00	30.00	74	115	145
950	3800	316.67	31.67	76	115	145
1000	4000	333.33	33.33	78	120	150

Percentage of Chimney Area.—The larger percentage of chimney area for the 100 horse-power boiler over the succeeding ones is on account of the friction of the gases against the sides of a small chimney. The usual recommendation is to add a constant of 4 inches to any diameter found to be necessary for a chimney; but this increase in diameter is not necessary for the larger chimneys, especially when based on one-eighth the grate area,—in fact, the necessity for any increase in diameter for friction is not apparent for chimneys over 48 inches in diameter, when based on the above proportion, and at 60 inches in diameter a gradual reduction may be made in the ratio of chimney area to grate area.

Chimney Design.—A chimney-shaft may be round, octagonal, square, or any other form to suit the taste of the designer or its adaptation to the conditions which call for its erection. A round chimney is to be preferred to any other interior cross-section, because that is the natural form of a column of ascending gases from a fire. Square chimneys are a practical or commercial necessity when of small diameters, say 48 inches and less, because of the rectangular shape of common bricks and the fact that bricks for circular walls, except fire-bricks, must be made to order. Some square chimneys have a round inner lining, and this detail has much to commend it; the lower part is commonly of fire-brick 6 inches thick, the upper of red brick 4¼ inches thick. This lining is not built into, but just clears the inner side of the square, leaving the corners open. It may extend to the top of the chimney with advantage. Such a chimney is easy of construction and of comparatively low cost.

The inside diameter of a chimney need not be otherwise than parallel from bottom to top, although chimneys are made both larger and smaller at the top than at the bottom, and without any perceptible loss of effect. When the lining of a chimney extends only a portion of the height, the area is considerably increased where the lower diameter debouches into the upper portion of the shaft; this is not known to have any injurious effect upon the draught.

The Stability of a Chimney is of the utmost importance. It seems almost superfluous to say that a chimney should be made of the best materials and constructed in the best manner. The wind pressure is the greatest resistance to be overcome by a chimney, and this is reckoned at 55 pounds per square foot as a maximum. Except in the case of whirlwinds or cyclones, no such wind pressure is experienced in this country. Its ability to withstand wind pressure varies with the shape of the chimney, a round chimney offering least resistance to the wind, an octagon a greater resistance, and a square chimney most of all. Without entering upon the mathematics of the stability of chimneys, it may be said that complete stability may be had in brick chimneys if the outside diameter at the base is one-tenth of its height for a square chimney, one eleventh of the height for an octagon chimney, and one-twelfth of its height for a round chimney, with a uniform taper of ¼ inch per foot throughout the whole height, excepting, of course, the base or pedestal upon which the shaft rests. The thickness of the brickwork at the top of a chimney should not in any case be less than one brick, say 8 to 9 inches, depending upon the commercial sizes of bricks in the locality where the chimney is to be erected ; and this thickness is ample for all heights up to 160 feet, above which the walls should be a brick and a half thick at the top. The thickness here given may continue downward from the top for say 25 feet, when a half-brick should be added to the thickness for another 25 feet, and so on down to the foundation or to the pedestal upon which the shaft rests.

Metal Chimneys.—The superiority of metal over brick for high chimneys is now being urged by many engineers. The arguments, in general, are these : the enormous quantity of brick and material necessary for high brick chimneys increases the cost and weight of such construction, requiring massive foundations and much space ; should these foundations sink but a trifle, there will result cracks in the chimney, thus impairing the draught and endangering the stability ; the sudden changes in temperature to which all chimneys are subjected are also destructive to brick chimneys ; being unprotected by metal shells, the sudden contraction of the side exposed to the cold blast of wind, rain, or sleet causes them to crack ; in a brick chimney the only resistance to wind pressure is that due to its weight, and as most brick chimneys are square or octagonal in cross-section the resistance is greater, since the surface exposed to the wind is flat.

Metal chimneys are built with or without fire-brick lining, depending upon the temperature of the escaping gases, the thickness of the lining varying for the different diameters and height. The weight of metal chimneys when lined is in most cases sufficient to withstand overturning by ordinary wind pressure, but the precaution of bolting securely to a good foundation should not be omitted ; this, it will be understood, is in addition to the customary fastening to the foundation plate, upon which the metal shell rests.

A Ladder should be provided for a chimney, which may be either on the outside or the inside, so as to gain access to the top for any examination or repairs that may be required in after service.

A Manhole with an iron door and frame should always be constructed in the base of a chimney for admission for examination, cleaning, and repairs. The top of this opening should be arched, so that the strength of the chimney shall not be impaired.

INDEX.

THE END.

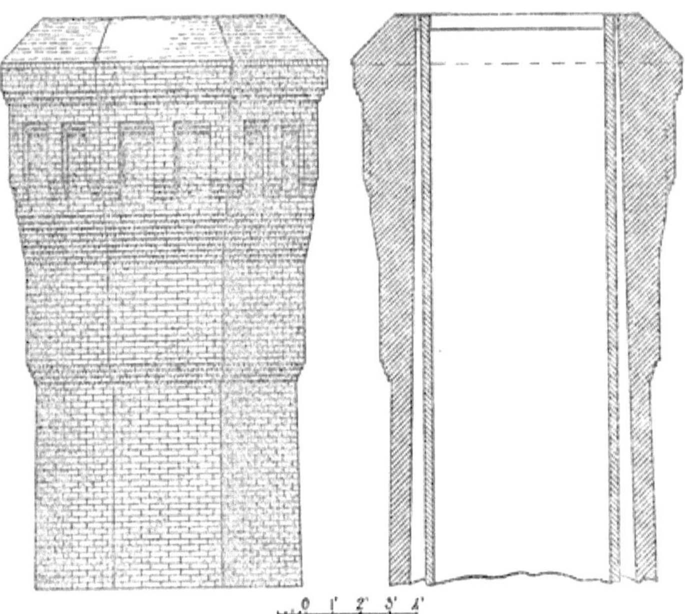
0 1' 2' 3' 4'

SPECIMEN PLATE.

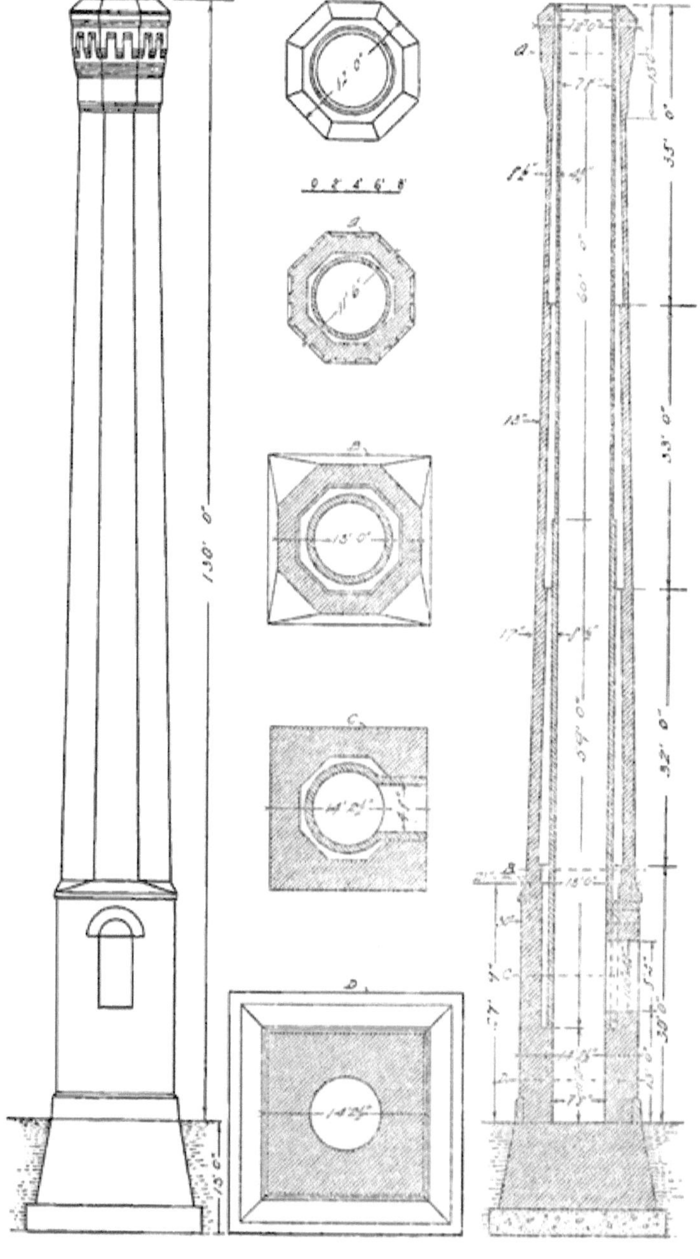

www.ingramcontent.com/pod-product-compliance
Lightning Source LLC
Chambersburg PA
CBHW020540300426
44111CB00008B/744

* 9 7 8 3 7 4 3 4 6 7 0 2 6 *